Maple® V Quick Reference

Maple® V Quick Reference

Nancy R. Blachman
Variable Symbols, Inc. and Stanford University

and

Michael J. Mossinghoff

Brooks/Cole Publishing Company
Pacific Grove, California

I(T)P ™
The trademark ITP is used under license.

Brooks/Cole Publishing Company
A Division of Wadsworth, Inc.

Printed in the United States of America.

10 9 8 7 6 5 4 3 2

Library of Congress Cataloging-in-Publication Data
Blachman, Nancy R.
Maple V Quick Reference / Nancy R. Blachman and Michael J. Mossinghoff.
p. cm.
Includes bibliographical references and index.
ISBN 0-534-20478-3
1. Maple (Computer file) 2. Mathematics–Data processing.
I. Mossinghoff, Michael J. [date]. II. Title.
QA76.95.B52 1994
510'.285'5369–dc20 94-10459
CIP

Sponsoring Editor: *Jeremy Hayhurst*
Editorial Assistant: *Elizabeth Barelli Rammel*
Production Editors: *Marlene Thom and Kirk Bomont*
Manuscript Editor: *Barbara Kimmel*
Cover Design: *Susan Haberkorn*
Printing and Binding: *Malloy Lithographing, Inc.*

Preface

The *Maple V Quick Reference* provides a concise, comprehensive reference to the Maple computer algebra system. It will help you find the commands you need and use them effectively. It is suitable for users at every level of experience with Maple.

The *Maple V Quick Reference* includes:

- An overview and introduction to Maple
- Information on Maple's user interface
- Lists of command names and options organized by subject
- Descriptions of all Maple commands (arranged alphabetically)
- Details on freely available Maple programs
- Answers to frequently asked questions
- Maple analogues for common Mathematica commands
- Pointers to other resources
- A glossary of terms

HOW TO USE THIS GUIDE

Chapter 1, **Introduction to Maple**, gives an overview and introduction to Maple. It describes Maple's syntax and provides many examples of numeric and symbolic calculations, graphics, and programming in Maple. If you are not familiar with Maple or would like a quick review, start with this section.

Chapter 2, **User Interface**, describes the graphical worksheet interface and the text-based command-line interface for Maple.

Chapter 3, **Subject Listing**, groups all the commands in Maple into categories according to their function. You can use this chapter to browse through all of the commands in Maple or to find the name of a particular function or option.

Chapter 4, **Descriptions of Commands**, lists all Maple commands alphabetically and gives a synopsis and template for each. Most listings include at least one example.

Chapter 5, **Electronic Resources**, describes the share library, electronic mailing lists, news groups, and other resources available electronically.

Chapter 6, **Frequently Asked Questions**, answers some questions users commonly ask about Maple.

Chapter 7, **Mathematica to Maple**, may be useful if you are already familiar with Mathematica. It compares the two systems and lists Maple analogues for many common Mathematica commands.

Chapter 8, **Learning More**, lists other sources of information on Maple, including several publications and user groups.

Chapter 9, **Glossary**, defines some of the terms used in this guide.

Chapter 10, **Tables**, summarizes Maple's operators and special punctuation marks.

HOW THIS GUIDE WAS PRODUCED

Maple commands were tested on several platforms for this guide, including Sun and NeXT workstations and Apple Macintosh and IBM-compatible personal computers, using Maple V, Maple V Release 2, and Maple V Release 3. This book contains information from the Maple help files and the *Maple V Library Reference Manual* developed by Waterloo Maple Publishing. The book was designed by Nancy Blachman, Michael Mossinghoff, and Peter Altenberg and was typeset with the LaTeX document preparation system. PostScript illustrations were produced by Maple, slightly massaged, and included in the book with Radical Eye Software's epsf macros. Unix utilities were used to produce LaTeX input for Chapters 3, 4, and 7 from databases created by the authors. Larry Wall's Perl utility was especially useful in the production of these chapters. The dvips program by Radical Eye Software rendered a PostScript file from the TeX output.

ACKNOWLEDGMENTS

Nancy and Michael — We are grateful to all who gave us suggestions for this guide. We would especially like to thank Kate Atherley of Waterloo Maple Software, Bill Bauldry of Appalachian State University, Nelson Blachman, who recently retired from GTE, Jonathan Brown of Wadsworth, Inc., Robert Campbell of Variable Symbols, Stan Devitt of the University of Saskatchewan and Waterloo Maple Software, Michael Fennel, Dave Hart of Indiana University, Jeremy Hayhurst of Brooks/Cole, Benton L. Leong of Waterloo Maple Software, Michael B. Monagan of ETH, Darren Redfern of Practical Approach, Peter Reynolds of the Office of Naval Research, Carol Scheftic of Carnegie Mellon University and Brooks/Cole, Glenn A. Sowell of the University of Nebraska at Omaha, Paul Swets of the University of Texas at Austin, Jenny Watson of Clecom Ltd., and Sue Worden of the University of Texas Computation Center.

We would like to thank Daniel Drucker of Wayne State University, Abi Fattahi of Whittier College, Betty Field of Maricopa Community Colleges, John Ramsay of the College of Wooster, and Daniel Schwalbe of Macalester College for reviewing an early draft of the manuscript and providing many ideas, comments, and suggestions that we incorporated into this guide.

We appreciate Ed W. Sznyter of Bābel Press and Cameron Smith's assistance with TeX. We thank Peter Altenberg for helping with the graphic design of this book.

We would also like to thank the developers of Maple, the authors of the Maple Library Reference Manual (Bruce W. Char, Keith O. Geddes, Gaston H. Gonnet, Benton L. Leong, Michael B. Monagan, and Stephen M. Watt), and Ron Neumann of Waterloo Maple Software for their assistance.

We also thank Nancy Champlin Conti and Elizabeth Barelli Rammel, editorial assistants at Brooks/Cole, and Kirk Bomont, the production editor, for going beyond the call of duty.

Nancy — I would also like to thank Mark Yoder and Robert López of Rose-Hulman Institute of Technology for inviting me to their NSF-sponsored workshop on revitalizing science, mathematics, and engineering education by using computer-algebra systems. At the workshop I received insightful suggestions from Herb Brown of the State University of New York at Albany, Duane Broline of Eastern Illinois University, Brian Evans, who is currently doing post-doctoral research at the University of California at Berkeley, Estela Llinas of the University of Pittsburgh at Greensburg, and Joel Trussell of North Carolina State University.

I thank Michael Morgan of Morgan/Kauffman Publishing Company and Bob Evans of Brooks/Cole for recommending my Maple books to Jeremy Hayhurst, and Wayne Oler of Wadsworth, Inc. for helping me make up my mind to go with Brooks/Cole. I am pleased that Jeremy Hayhurst took an interest in my manuscript, for it has been a delight working with him.

I would like to thank my parents, Nelson and Anne Blachman, for their support and encouragement.

Last, but definitely not least, I would like to thank Anupama Murty for contributing to Chapter 4 of this guide, and Helen Chernicoff for handling Variable Symbols business while I was working on this guide.

Michael — In addition, I would like to thank the Department of Mathematics at the University of Texas at Austin, especially Rafael de la Llave and Jeffrey Vaaler. I would also like to thank Heather Lacey of Wadsworth, Inc. for her assistance. Most of all, I would like to thank my wife Kristine for her patience, support, and encouragement.

A FINAL WORD

Because of the growing popularity of Maple, we decided to develop a guide to help Maple users. We hope you find the *Maple V Quick Reference* useful. If you come across any errors or omissions or if you have suggestions for how this guide can be improved, please tell us.

Nancy Blachman
Variable Symbols, Inc.
6537 Chabot Road
Oakland, CA 94618-1618
Email: `nb@cs.stanford.edu`
Fax: 510-652-8461
Telephone: 510-652-8462

Michael Mossinghoff
Department of Mathematics
University of Texas at Austin
Austin, TX 78712
Email: `mossingh@math.utexas.edu`

Contents

Maple® V Quick Reference

Chapter 1
Introduction to Maple

Maple is a computer algebra system: an advanced software program for exact symbolic manipulation, precise numerical computation, graphics, and mathematical programming. It was developed by researchers and students at several institutions, including the University of Waterloo, Drexel University, the Swiss Federal Institute of Technology (ETH) in Zürich, and Waterloo Maple Software. Maple is used throughout the world in high school and college mathematics classes, corporations and universities for education, engineering applications, and scientific research. Maple features:

- Advanced algorithms for symbolic manipulation, including all the standard operations in algebra, calculus, and linear algebra.
- A large collection of numerical functions, operating on numbers of arbitrary magnitude and precision.
- Two- and three-dimensional graphics for plotting functions and displaying data.
- Packages of mathematical functions encoding operations and algorithms in such areas as algebra, calculus, geometry, linear algebra, discrete mathematics, number theory, statistics, and mathematical physics.
- A powerful programming language for writing your own applications.
- The ability to inspect and even modify the source code of most its mathematical functions.
- Special routines for students for solving problems in algebra and calculus one step at a time.
- An attractively priced student edition.

Maple runs on many kinds of computers, including products from Apple, Convex, Cray, DEC, Hewlett-Packard, IBM, MIPS, NeXT, Sequent, Silicon Graphics, and Sun. It runs on DOS personal computers with or without Microsoft Windows. Maple commands and programs are portable: you can run the same Maple program on a personal computer, workstation, mainframe, and supercomputer.

This chapter offers an introduction to the Maple computer algebra system. It begins by describing how to use Maple interactively and how to get help on-line. Then it presents some of Maple's numeric and symbolic capabilities and its facilities for making plots and graphs. Maple's data types, control statements, and data structures are discussed next, and the Maple library is described in some detail. The chapter also includes an introduction to writing your own functions and procedures in Maple, followed by a summary of input and output. A set of exercises provides an opportunity for additional practice. The last section lists several companies that sell Maple software.

This guide was developed using Maple V Release 3, Maple V Release 2, and Maple V. All new features in Maple V Release 3 are clearly labeled. Maple V Release 3 contains many improvements in mathematical functions, graphics, programming, and on-line help over previous versions of Maple.

1 THE STRUCTURE OF MAPLE

There are three main components to the Maple computer algebra system. First, the kernel is Maple's computational engine. It is loaded whenever you start a Maple session. It performs the most basic tasks, like evaluating expressions, doing low-level algebraic manipulation, and handling input and output. The kernel is written in the C programming language and is rather small, generally less than a megabyte when compiled.

Second, the Maple library contains the vast majority of Maple's mathematical knowledge. The functions in the library are written in Maple's own programming language, and you can readily inspect this code. Some library routines are read into your session by the kernel as they are needed. Others you load explicitly when you need them. Some are collected together in *packages*. You can load an entire package at once to gain access to dozens of routines in a subject like graphics or linear algebra. More details on the structure of the Maple library appears in Section 11 on page 62.

The third component of the Maple system is the user interface. Unlike the kernel or the library, the user interface can be rather different on different computers. Under Maple V Release 2 and Release 3, all computers with sufficient graphics capability use a *worksheet* interface. Worksheets let you mix Maple commands and results, graphics, and explanatory text in a single document. Chapter 2, User Interface, describes the two main kinds of user interface to the Maple system.

2 NOTATION

We often give a template for a Maple function to describe the number and types of parameters that it requires. In a template, words and symbols in `typewriter` font must be typed exactly as shown, but words in *italics* should be relaced by particular Maple expressions. For example, the Maple function for computing limits is called `limit`. The basic template for limit shows that it requires two arguments: first, an expression; second, an equation.

```
limit(expr, var = val);
```

This chapter contains many examples of Maple commands. Maple commands are written in typewriter font and are always preceded by a Maple prompt. On most systems, this prompt is a >, but some versions of Maple may use •, a solid triangle, or an enlarged > as the prompt. We use > as the Maple prompt throughout this guide.

Results from Maple are printed in mathematical format centered on the page. Maple V Release 2 and Release 3 print results this way on computers that support the worksheet interface. For example, we find the limiting value of $1/(1+\frac{r}{m})^m$ as m goes to infinity with the command:

```
> limit(1/(1+r/m)^m, m = infinity);
```

$$\frac{1}{e^r}$$

On systems without the worksheet interface, such as a simple terminal, and with earlier versions of Maple, this result would print as:

```
   1
------
exp(r)
```

Occasionally, we refer to pressing a certain key on your keyboard by enclosing the key name in a box. For example, pressing the enter key is denoted by [ENTER].

3 MAPLE SESSIONS

This section presents the basic commands and conventions you need to know to start using Maple interactively. It describes how to begin and end a Maple session, how to enter Maple commands, and how to interrupt a lengthy calculation.

3.1 Starting Maple

How you start Maple depends on the type of computer you are using. If you are using Maple under the X Window System (or one of its specializations, like DECwindows, OpenWindows or OSF/Motif), type `xmaple &` in a command window. From DOS or a Unix or VMS terminal, start Maple by typing `maple`. Other systems let you start Maple by clicking on a maple leaf icon. The following table summarizes how to start Maple on several computers.

Type of Computer	Command to Start Maple
DOS	Type `maple`
MS-Windows, Macintosh, NeXT	Double click on the Maple icon
X Window Systems	Type `xmaple &`
Unix terminal	Type `maple`
VMS terminal	Type `maple`

If Maple does not start, make sure that the Maple program is in a directory in your search path. On Unix systems, check your search path by entering `echo $PATH` or `echo $path`. Consult with your system administrator or read your Maple installation guide for assistance with updating your search path and starting Maple.

When you start Maple in a graphics environment, a new window appears for your Maple session. (A picture of this window as it appears under the X Window System appears on page 84 in Figure 2.1.) The Maple logo, with a picture of Sir Isaac Newton, may appear briefly on your screen as the program initializes (Figure 1.1).

Figure 1.1: The Maple V logo

When you start Maple on a terminal, a prolog similar to this one appears:

```
    |\^/|     Maple V Release 2 (University of Texas at Austin)
._|\|   |/|_. Copyright (c) 1981-1993 by the University of Waterloo.
 \  MAPLE  /  All rights reserved. Maple and Maple V are registered
 <____ ____>  trademarks of Waterloo Maple Software.
      |       Type ? for help.
>
```

After the Maple kernel has been loaded, Maple prompts the user to indicate its readiness to accept commands.

Some versions of Maple come with two user interfaces. A graphical user interface is good for interactive Maple sessions, whereas a command-line version is better for batch jobs. On Unix systems the command-line version is called `maple` and should be in the directory where Maple is installed on your system.

3.2 Entering Maple Commands

Here we ask Maple for 5^{10}.

```
> 5^10;
```

$$9765625$$

With few exceptions, every Maple command you enter must end with a semicolon (`;`) or a colon (`:`). If you end a command with a colon, the calculation is performed but the result is not printed on the screen. A colon is used whenever it is not important to see the result of a command, such as assigning a value to a variable or reading a function from the library. Because the end of a command is marked with special punctuation, you can type several commands on one line or use several lines to enter a single command. If you should forget a semicolon, enter it on the next line.

```
> 2^10; 2^20; 2^30;
```

$$1024$$

$$1048576$$

$$1073741824$$

```
> 13 + 29
> ;
```

$$42$$

To request that a calculation be performed, press the RETURN or ENTER key. (The key you use depends on your computer.) After Maple has finished calculating, it prints another prompt to show it is ready to read another command.

There are many functions available in Maple. Some command names are almost self-explanatory.

Function	**Description**
`factor`	Factor a polynomial
`limit`	Calculate the limiting value of an expression
`plot`	Draw a two-dimensional plot
`round`	Round a number to its closest integer
`simplify`	Simplify an expression
`solve`	Solve equations

Many names are abbreviations. Some are very common in mathematics.

Function	**Description**
`abs`	Absolute value
`diff`	Differentiate
`dsolve`	Solve differential equations
`ifactor`	Factor an integer
`igcd`	Greatest common divisor of integers
`int`	Integrate
`tan`	Tangent

Invoke a Maple function by placing its arguments in parentheses after the function name. For example, we calculate a square root with the function `sqrt` and compute the derivative of the sine function with the function `diff`.

```
> sqrt(676) - 1;
```

$$25$$

```
> diff(sin(x), x);
```

$$\cos(x)$$

Maple is case sensitive: the command `diff(sin(x), x)` is quite different from `Diff(sin(x), x)`. The names of most Maple functions begin with a lowercase letter. Many functions whose names begin with uppercase letters, like `Diff`, `Int`, and `Limit`, are *inert* – they are formal operations that do not perform any calculations. For example, we can specify the derivative of the sine function without computing the answer with the inert function `Diff`.

```
> Diff(sin(x), x);
```

$$\frac{\partial}{\partial x}\sin(x)$$

Notice the second input line in the worksheet on page 84 in Figure 2.1. The inert function `Int` displays the integral itself in standard mathematical notation.

Spaces in Maple commands are usually ignored. You may want to put some spaces in your commands to make them easier to read.

The best way to learn about Maple is by practicing and experimenting with its commands. The rest of this chapter contains many examples showing some of the capabilities of Maple. Try entering these commands yourself. Experiment with them. When typing in examples from this guide, be careful to use the same:

- spelling and capitalization,
- types of parentheses, brackets and braces,
- punctuation (quotes, commas, semicolons and colons).

3.3 Using Previous Results

You may want to use a result you obtained in Maple as input to another command. The double-quote symbol " refers to the previously calculated result.

```
> int(cos(x) * sin(x)^5, x);
```

$$\frac{1}{6}\sin(x)^6$$

```
> diff(", x);
```

$$\cos(x)\sin(x)^5$$

Similarly, you can refer to the second to last result with "" and the third to last result with """.

If you want to have access to all results computed in a Maple session, load the library function `history` with the command `readlib(history)`, then enter `history()` to initiate the history mechanism. Just above your prompt, or in place of it, you will see "O*n* :=" (the letter O, for output, followed by an integer, *n*). Thereafter, the *n*th result you compute is stored in the variable O*n*, and you can refer to it with this name in future commands. Once the history mechanism has been started, "" and """ are not available. Type `off` to turn off this feature.

```
> readlib(history):
> history();

O1 := 3^10;
                              59049

O2 := O1 + 1;
                              59050

O3 := off;
>
```

3.4 Quitting Maple

To end a Maple session, type `quit`, `stop`, or `done`. You do not have to end these commands with a semicolon or colon. The following table lists some other ways to quit Maple on different computers.

Type of Computer	Quit Command
All	`quit`, `stop`, or `done`
DOS	Press the [F3] key
MS-Windows	Select the **Exit** command in the **File** menu, or press [ALT]-[F4]
Macintosh	Select the **Quit** command in the **File** menu, or press [COMMAND]-q
X Window Systems	Select the **Exit** command in the **File** menu, or press [ALT]-x
Unix	Press [CONTROL]-/
NeXT	Select the **Quit** menu command, or press [COMMAND]-q

3.5 Interrupting a Calculation

Interrupt a lengthy Maple calculation by pressing your computer's interrupt key. On a few computers, pressing the interrupt key aborts your Maple session completely. On most systems, though, you receive another prompt and remain in Maple after aborting a command. Here are the Maple interrupt commands for a few computers.

Type of Computer	Interrupt Command
DOS	[CONTROL]-[BREAK] or [CONTROL]-c
MS-Windows	[CONTROL]-[BREAK]
Macintosh	[COMMAND]-. (command key and period)
X Window Systems	**Interrupt** button
Unix	[CONTROL]-c
NeXT	**Interrupt** command in the **Kernel** menu, or [COMMAND]-.
VMS	[CONTROL]-z

In Maple V on Unix systems, typing the interrupt command twice quits your session entirely. This has been removed in Maple V Release 2 for Unix.

4 ON-LINE HELP

There is extensive help available to you in a Maple session. You can request documentation on any Maple function with the ? command, look up names of Maple commands with a topic browser, and gain an overview of Maple's features with an on-line tutorial.

4.1 Help Command

Typing a question mark (?) followed by a command name requests help on that command in Maple. Help text on a command usually includes a detailed description of the command, a template with the number and types of parameters it requires, several examples of its use, and any related commands or help topics. A help request is one of the exceptional Maple commands that does not have to end with a semicolon.

Below we obtain help on the greatest common divisor function for integers, `igcd`. Depending on your computer, this help text may appear in a separate window.

```
> ?igcd

FUNCTION: igcd - greatest common divisor of integers
FUNCTION: ilcm - least common multiple of integers

CALLING SEQUENCE:
   igcd(x_1,x_2,...);
   ilcm(x_1,x_2,...);

PARAMETERS:
   x_1,x_2,... - any integers

SYNOPSIS:
- The function igcd computes the greatest common divisor of an
  arbitrary number of integers.  The function ilcm computes the
  least common multiple of an arbitrary number of integers.

EXAMPLES:
> igcd();
                                 0

> ilcm();
                                 1

> igcd( -10, 6, -8 );
                                 2
```

```
> ilcm( -10, 6, -8 );
                                    120

> igcd( a, b );
                                 igcd(a, b)

SEE ALSO:  gcd, lcm
```

You can also use the ? command to look up names of commands. If your help request doesn't match any topic, Maple lists the names of all topics that start with your request. For example, below we list all the help topics that begin with the letter "j":

```
> ?j
  Try one of the following topics:
```

$$\{join, jordan, jacobi, J, jacobian, JordanBlock\}$$

If there aren't any commands that start with your request, Maple may suggest some alternate spellings:

```
> ?jorden
  Try one of the following topics:
```

$$\{jordan, JordanBlock\}$$

There are also help topics in Maple that discuss some general aspects of Maple, like data types and data structures, graphics, procedures, and the Maple library. This table summarizes some of the ways to use the help command.

Command	**Help Topic**
?*name*	*name*, or list topics that begin with *name*
?*name*[*subtopic*]	*subtopic* in the context of *name*
?*name*, *subtopic*	*subtopic* in the context of *name*
?*pkg*[*name*]	*name* in the package *pkg*
?	The help command itself
?index	Index of help topics
?library	Standard library functions
?libmisc	Miscellaneous library functions
?packages	Brief description of all packages
?datatypes	Basic data types
?expressions	Maple expressions
?statements	Maple statements
?index,procedures	Maple procedures
?index,tables	Tables and arrays
?misc	List of miscellaneous topics
?updates	Summary of new features in each update of Maple

4.2 Help Topic Browser

The help topic browser helps you find names and descriptions of Maple commands. All Maple commands are organized into categories in the browser, with four basic divisions: Graphics, Mathematics, Programming and System.

Picking one of these topics presents you with a list of subtopics, which may themselves have subtopics, and so on. Eventually you find lists of Maple commands. When you highlight a command, a brief synopsis of its function appears at the bottom of the browser. You can then call up help text on the function. Figure 2.2 on page 90 is a picture of the topic browser as it appears under the X Window System.

The precise method of starting the topic browser, navigating through topics, and selecting commands varies by computer. Under DOS, the F1 key starts the browser, the arrow keys let you navigate through the topics, and the ENTER key selects a topic. Other platforms use the mouse to find and select topics.

The topic browser in Maple V Release 3 includes a search function for finding names of commands. It searches the synopsis of each command and a keyword section of the help index for a given word or string and presents a list of matching commands. Some Maple V Release 2 platforms, like DOS, also include this feature.

A topic browser is included in all platforms with Maple V Release 2 and Release 3.

4.3 Tutorial

Maple V Release 2 and Release 3 include an on-line tutorial for Maple beginners. Picking topics from a menu lets you focus on a particular subject or work through all the topics in the tutorial in sequence. The tutorial covers the essential aspects of Maple and provides numerous examples of calculations, symbolic manipulation, and graphics. Special sections discuss commands for doing calculus and linear algebra in Maple. Quizzes test your understanding of basic Maple syntax and commands.

Start the tutorial with the command `tutorial()`.

4.4 More Help Commands

Maple V Release 3 includes four new help commands: `info`, `usage`, `example`, and `related`. The `info` command displays the short description of a function found at the top of its help page.

```
> info(abs);
  FUNCTION: abs - The absolute value function
```

The `usage` command displays the calling sequence for a function and a description of its parameters. This function may be invoked with a double question mark.

```
> ??abs
  CALLING SEQUENCE:
     abs(x)
```

```
PARAMETERS:
   x - an expression
```

The `example` command displays the examples section for a function from the on-line help text. It may be invoked with a triple question mark: ???*function*.

Finally, the `related` command shows related functions and help topics.

```
> related(abs);
  SEE ALSO: evalc, signum, inifcns
```

5 NUMERICAL CALCULATIONS

You can use Maple as a calculator to evaluate mathematical expressions. However, the numerical capabilities of Maple are much more extensive than that of an ordinary calculator. Maple has an extensive library of numerical functions. It can compute exact results, or approximations with arbitrarily high precision. This section describes some of these numerical capabilities.

5.1 Basic Calculations

Most people are used to writing common mathematical expressions with special symbols like +, -, and <. Maple uses this standard notation for doing arithmetic and making comparisons. These symbols are examples of *programming language operators* in Maple. A complete list of these operators and other special characters appears on the inside front cover of this guide. The following table lists the common operators for doing arithmetic, making comparisons, and assigning values to variables.

Operator	Description	Example
!	Factorial	`4!`
^	Exponentiation	`2^5`
+	Addition	`13 + 29`
-	Subtraction	`69 - 27`
-	Negation	`-42`
*	Multiplication	`6 * 7`
/	Division	`168 / 4`
<	Less than	`a < b`
<=	Less than or equal	`a <= b`
>	Greater than	`a > b`
>=	Greater than or equal	`a >= b`
=	Equals	`x^2 + 5*x = 6`
<>	Does not equal	`a <> b`
:=	Assignment	`a := 3`

You can calculate many standard mathematical functions with Maple commands. Some of these are listed in the following table.

Mathematics	Maple	Name
$\|x\|$	`abs(x)`	Absolute value
$\sqrt{x}$	`sqrt(x)`	Square root
$\binom{n}{m}$	`binomial(n, m)`	Binomial coefficient
e^x	`exp(x), E^x`	Exponential function
$\log x, \ln x$	`log(x), ln(x)`	Natural logarithm
$\sin x, \cos x, \tan x$	`sin(x), cos(x), tan(x)`	Trigonometric functions
$\Gamma(z)$	`GAMMA(z)`	Gamma function
$J_v(x)$	`BesselJ(v, x)`	Bessel function
$2/\sqrt{\pi}\int_0^x e^{-t^2}\,dt$	`erf(x)`	Error function
$\zeta(s)$	`Zeta(s)`	Riemann Zeta function

5.2 Precedence and Associativity

Maple generally follows standard precedence and associativity rules when doing arithmetic. For example, exponentiations are performed before multiplications and divisions, which are evaluated before additions and subtractions. Parentheses can be used to alter or emphasize the order of operations in an expression.

```
> 1 + 2 * 3;
```

$$7$$

```
> (1 + 2) * 3;
```

$$9$$

Sometimes parentheses are required. Negation has the same precedence as addition, so we must put parentheses around the `-3` in the following expression:

```
> 25 + -3;
  syntax error:
  25 + -3;
       ^
> 25 + (-3);
```

$$22$$

The carat (`^`) points to the location of the syntax error.

Parentheses are often required with the exponentiation operator. If you give Maple a chain of exponents to evaluate, Maple won't assume that the chain should be evaluated left-to-right or right-to-left; instead, it returns an error. (The `^` operator is said to be *nonassociative* in Maple.)

```
> 2^3^4;
  syntax error:
  2^3^4;
     ^
> (2^3)^4;
```

$$4096$$

You must also use parentheses when an exponent is negative.

```
> 2^-5;
  syntax error:
  2^-5;
    ^
> 2^(-5);
```

$$\frac{1}{32}$$

All Maple operators are listed in order of precedence in a table in the back of this guide (page 480). The same table lists associativity rules for each operator.

5.3 Exact and Approximate Numbers

Maple distinguishes between *exact* and *approximate* values. Integers are exact numbers. In Maple, an integer is a whole number with no decimal point. Quotients of integers (fractions) and complex numbers with integer parts (Gaussian integers) are also exact. Approximate numbers are characterized by the presence of a decimal point. Maple calls these numbers floating-point numbers, or floats.

The following table lists some of the types of numbers known to Maple.

Type	Description	Exact/Approximate	Examples
`integer`	Integer	exact	`3`, `-47`
`float`	Floating-point number	approximate	`3.`, `.2*10^5`
`fraction`	Fraction of integers	exact	`3/4`, `-7/3`
`complex`	Complex number	either	`3 + 4*I`, `1.7*I`

If you enter a mathematical expression with only exact values, Maple returns an exact result.

```
> 4 + 2/4 + 24/144;
```

$$\frac{14}{3}$$

```
> sqrt(68);
```

$$2\sqrt{17}$$

```
> arctan(1);
```

$$\frac{1}{4}\pi$$

Maple prints an exact result even if it doesn't fit on one output line. A backslash, \, indicates that the output continues on the next line.

```
> 100!;
```

933262154439441526816992388562667004907159682643816214685929638952175\
5999932299156089414639761565182862536979208272237582511852109168640\
000000000000000000000000

When an arithmetic expression contains an approximate number, Maple returns an approximate result.

```
> 4. + 2/4 + 24/144;
```

$$4.666666667$$

```
> sqrt(68.);
```

$$8.246211251$$

```
> arctan(1.);
```

$$.7853981634$$

To compute a numerical approximation of an exact value, use the function `evalf`. The name `evalf` is short for "evaluate using floating-point arithmetic."

```
> evalf(sqrt(68));
```

$$8.246211251$$

An optional second argument to `evalf` specifies the desired precision. Of course, the more precision you request, the longer it takes Maple to compute the result. Here we ask for $\sqrt{68}$ to 100 decimal digits of precision.

```
> evalf(sqrt(68), 100);
```

8.246211251235321099642819711948154050294398450747240868797267146189\
9086926752431871757273016213685594

If you want several approximate values in your session computed to a certain number of significant digits, set the global variable `Digits` to this number. Its default value is 10.

```
> Digits := 30;
```

$$Digits := 30$$

```
> evalf(sqrt(68));
```

$$8.24621125123532109964281971194$$

Remember to reset `Digits` if you no longer want results computed to 30 places of precision.

```
> Digits := 10:
```

To convert an approximate real number to an exact fraction, use the `convert` function. This versatile function converts numbers or expressions from one format to another.

```
> convert(3.1415926, fraction);
```

$$\frac{86598}{27565}$$

5.4 Symbolic Constants

Maple has names for some common mathematical constants.

Constant	**Maple Name**	**Description**	**Value**
π	`Pi`	Ratio of circumference to diameter in a circle	≈ 3.141592654
e	`E` *or* `exp(1)`	Exponential constant	≈ 2.718281828
γ	`gamma`	Euler's constant	$\approx .5772156649$
$\sum_{k=0}^{\infty} \frac{(-1)^k}{(2k+1)^2}$	`Catalan`	Catalan's constant	$\approx .9159655942$
i or j	`I`	Imaginary unit	$\sqrt{-1}$
∞	`infinity`	Real or complex infinity	

There are also three logical constants: `true`, `false`, and `FAIL`, which are discussed in "Booleans" on page 45.

Maple considers the symbols for real constants, like `Pi` and `E`, to be exact values. You can compute approximations to these values to whatever precision you need with `evalf`.

```
> evalf(E, 40);
```

$$2.718281828459045235360287471352662497757$$

5.5 Exact Numerical Functions

Maple has many routines for operating on exact numbers. Below we test an integer for primality, factor an integer, and find the integer part of the square root of an integer.

```
> isprime(27! + 1);
```

$$true$$

```
> ifactor(2^(2^6) + 1);
```

$$(67280421310721)(274177)$$

```
> isqrt(binomial(100, 50));
```

$$317633978890112$$

5.6 Approximating Numerical Functions

Maple provides many routines that find approximate answers, including functions for finding roots of functions, solving systems of equations approximately, approximating integrals, and solving differential equations numerically. This section gives a few examples of some of these functions.

The function `fsolve` finds roots of functions and solves systems of equations using approximating methods. Given a polynomial equation, `fsolve` finds all real roots of the polynomial. (Supply the option `complex` as a third argument to look for complex roots, too.)

```
> fsolve(x^3 - 2*x^2 + 1 = 0, x);
```

$$-.6180339887, 1., 1.618033989$$

If the equation is not a polynomial, `fsolve` searches for a single real root. If you specify an interval with a third argument, it searchs for a solution only in this (open) interval. If no roots are found, Maple returns the `fsolve` command unevaluated.

```
> fsolve(BesselJ(0, x), x, 7..8);
```

$$\mathrm{fsolve}(\mathrm{BesselJ}(0, x), x, 7..8)$$

```
> fsolve(BesselJ(0, x), x, 5..9);
```

$$8.653727913$$

Maple always tries to evaluate definite integrals by using symbolic techniques. If a function has no singularities in the given interval, the function `int` tries to compute an antiderivative, then evaluate it at the endpoints of integration. Of course, a function may not have an antiderivative with a simple, closed form. In this case, `int` may express the answer in terms of a special function, such as the error function "erf," or it may simply return the integral unevaluated.

```
> int(exp(-x^2), x=0..1);
```

$$\frac{1}{2}\sqrt{\pi}\,\mathrm{erf}(1)$$

```
> int(1/(x + ln(x)), x=1..exp(1));
```

$$\int_{1}^{\mathrm{e}} \frac{1}{x+\ln(x)}\,dx$$

If an integral is returned unevaluated, it is quite possible that it has no closed-form antiderivative in terms of elementary functions. Maple uses an advanced integration algorithm (the Risch algorithm) that can prove that there is no closed-form solution for a large class of integrals involving exponential, logarithmic, trigonometric and algebraic expressions.

You can still find an approximate value for the integral using numerical techniques by applying `evalf`.

```
> evalf(", 15);
```

$$.796960924149206$$

If you want to integrate a function numerically without first attempting symbolic integration, use `evalf` with the inert integration function, `Int`.

```
> evalf(Int(ln(ln(x)), x=exp(1)..100));
```

$$124.4869388$$

You can solve differential equations numerically with the `numeric` option on the `dsolve` command. It returns a procedure that can be used to find approximate values on the solution curve, or to plot this curve.

```
> de := (D@@2)(y)(x) + D(y)(x) - x*y(x) = 0;
```

$$de := \mathrm{D}^{(2)}(y)(x) + \mathrm{D}(y)(x) - x\,y(x) = 0$$

```
> f := dsolve({de, D(y)(0)=1, y(0)=0}, y(x), numeric);
  f := proc(rkf45_x) ... end
```

```
> f(0);
```

$$\left[x = 0, y(x) = 0, \frac{\partial}{\partial x} y(x) = 1.\right]$$

```
> f(3);
```

$$\left[x = 3, y(x) = 4.351539384924291, \frac{\partial}{\partial x} y(x) = 5.289102578946162\right]$$

These calculations were performed using Maple V Release 3. Under Release 2, the format of the returned expressions is slightly different.

```
> plots[odeplot](f, [x, y(x)], -3..3);
```

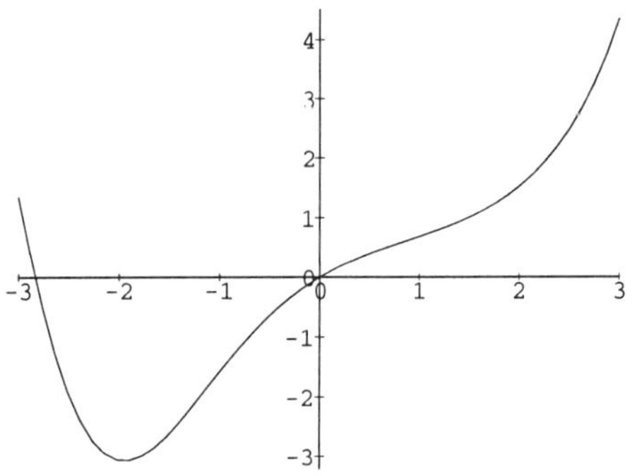

Figure 1.2: The numerical solution of a differential equation

The ordinary differential equation plotting function, `odeplot`, in the `plots` package graphs the solution curve (Figure 1.2). For more information on the `plots` package, see Section 7, Graphics, on page 22.

Maple performs many more numerical calculations besides those listed here. Maple's accuracy is not limited by the architecture of a computer, nor is the size of the output limited by the width of a screen. Maple returns exact results when given exact data and computes approximate results to very high accuracy on request.

6 SYMBOLIC MANIPULATION

Maple has a large number of routines for manipulating symbolic expressions. We examine a few of them in this section for operating on polynomials and rational functions, solving systems of equations exactly, and doing common operations in calculus.

6.1 Polynomials and Rational Functions

Maple has many routines for manipulating polynomials and rational functions. Here we expand and factor a polynomial in one variable.

```
> expand((x+1)^10 + (x-1)^10);
```

$$2x^{10} + 90x^8 + 420x^6 + 420x^4 + 90x^2 + 2$$

```
> factor(");
```

$$2\left(x^2 + 1\right)\left(x^8 + 44x^6 + 166x^4 + 44x^2 + 1\right)$$

Next, we factor a polynomial in two variables, then compute the partial-fraction decomposition of a rational function.

```
> factor(x^2 + 2*x^2*y - 2*x*y^2 - y^2);
```

$$(x + 2xy + y)(-y + x)$$

```
> convert(1/(x^3 + 5*x^2 + x - 3), parfrac, x);
```

$$-\frac{1}{6}\frac{1}{x+1} + \frac{1}{6}\frac{3+x}{x^2+4x-3}$$

6.2 Solving Equations

Maple finds exact solutions to polynomial equations of degree four or less (as well as some polynomial equations of higher degree).

```
> solve(x^3 - 3*x^2 - 17*x + 51 = 0, x);
```

$$3, \sqrt{17}, -\sqrt{17}$$

Maple solves simultaneous equations.

```
> solve({5*x + 6*y = 7,  5*x + 7*y = 8}, {x, y});
```

$$\left\{y = 1, x = \frac{1}{5}\right\}$$

If the result of a symbolic computation is rather lengthy, Maple may abbreviate the result by letting special symbols named %1, %2, and so on stand for expressions that occur several times.

```
> solve(x^3 - x^2 - x - 1 = 0, x);
```

$$\%1^{1/3} + \frac{4}{9}\frac{1}{\%1^{1/3}} + \frac{1}{3},$$

$$-\frac{1}{2}\%1^{1/3} - \frac{2}{9}\frac{1}{\%1^{1/3}} + \frac{1}{3} + \frac{1}{2}I\sqrt{3}\left(\%1^{1/3} - \frac{4}{9}\frac{1}{\%1^{1/3}}\right),$$

$$-\frac{1}{2}\%1^{1/3} - \frac{2}{9}\frac{1}{\%1^{1/3}} + \frac{1}{3} - \frac{1}{2}I\sqrt{3}\left(\%1^{1/3} - \frac{4}{9}\frac{1}{\%1^{1/3}}\right),$$

$$\%1 := \frac{19}{27} + \frac{1}{9}\sqrt{33}$$

6.3 Calculus

Maple computes many definite integrals exactly. An answer may be expressed in terms of a constant known to Maple, like π, the exponential constant e, or Catalan's constant.

```
> int(sin(x)^4, x=0..Pi);
```

$$\frac{3}{8}\pi$$

```
> int(x^4*exp(x), x=0..1);
```

$$9\,\mathrm{e} - 24$$

```
> int(ln(x)/(1+x^2), x=0..1);
```

$$-\mathit{Catalan}$$

Maple differentiates functions and computes indefinite integrals, too. There are two functions for finding derivatives. The first two examples that follow use `diff`, which differentiates expressions with respect to the named variable or variables. The first calculates an ordinary first derivative; the second computes a mixed partial derivative of a function of two variables. The last example uses the differential operator `D`, which operates on functions.

```
> diff(x*ln(x) - x, x);
```

$$\ln(x)$$

```
> diff(sin(x*y), x, y);
```

$$-\sin(xy)xy + \cos(xy)$$

```
> D(sin);
```

$$\cos$$

The int function computes an indefinite integral if you don't specify an interval of integration. In the following example, we compute an indefinite integral of a rational function, then check the result by differentiating. The normal function with the expanded option simplifies the expression so we can check that the derivative of the integral is really the original function.

```
> int(1/(x^3 + 5*x^2 + x - 3), x);
```

$$-\frac{1}{6}\ln(x+1)+\frac{1}{12}\ln\left(x^2+4x-3\right)-\frac{1}{42}\sqrt{7}\operatorname{arctanh}\left(\frac{1}{14}(2x+4)\sqrt{7}\right)$$

```
> diff(", x);
```

$$-\frac{1}{6}\frac{1}{(x+1)}+\frac{1}{12}\frac{2x+4}{x^2+4x-3}-\frac{1}{42}\frac{1}{1-\frac{1}{28}(2x+4)^2}$$

```
> normal(", expanded);
```

$$\frac{1}{x^3+5x^2+x-3}$$

In the following integral, α and β are treated as constants.

```
> int(exp(alpha*x)*sin(beta*x), x);
```

$$-\frac{\beta e^{\alpha x}\cos(\beta x)}{\alpha^2+\beta^2}+\frac{\alpha e^{\alpha x}\sin(\beta x)}{\alpha^2+\beta^2}$$

Sometimes the value of a definite integral depends on a property of an undefined symbol. For example, the value of the following integral depends on the sign of r.

```
> int(exp(-r*x), x=0..infinity);
```

$$\lim_{x\to\infty-}-\frac{e^{(-rx)}}{r}+\frac{1}{r}$$

In Maple V Release 2 and Release 3, the assume command lets you give an attribute to a variable without assigning it a value. If we assume that r is positive, we can compute the integral.

```
> assume(r > 0);
> r;
```

$$r\sim$$

The tilde (~) reminds us that r has been given an attribute with assume.

```
> int(exp(-r*x), x=0..infinity);
```

$$\frac{1}{r\sim}$$

In addition to integration and differentiation, Maple finds limits.

```
> limit(sin(x)/x, x=0);
```

$$1$$

It computes Taylor series expansions. This command finds the Taylor series for the natural logarithm function about the point $x = 1$ up to order 7.

```
> taylor(ln(x), x=1, 7);
```

$$x - 1 - \frac{1}{2}(x-1)^2 + \frac{1}{3}(x-1)^3 - \frac{1}{4}(x-1)^4 + \frac{1}{5}(x-1)^5 - \frac{1}{6}(x-1)^6 + \mathrm{O}\left((x-1)^7\right)$$

Maple solves many ordinary differential equations exactly. Solutions for differential equations given without boundary conditions are given in terms of a suitable number of arbitrary constants named *_C1*, *_C2*, and so on. These names start with an underscore so that they are less likely to conflict with any of the names you define. Here we solve the differential equation $y'(x) + y(x)\tan x = \sec x$.

```
> dsolve(D(y)(x) + y(x)*tan(x) = sec(x), y(x));
```

$$y(x) = \sin(x) + \cos(x)\,\_C1$$

If Maple cannot solve a differential equation with boundary conditions exactly, numerical methods may be invoked, as shown on page 17.

7 GRAPHICS

Maple has a powerful facility for drawing curves, surfaces, and data plots. This section describes some common commands for drawing graphs and plotting data in two and three dimensions. It also discusses how to save plots for future display.

On most Maple platforms, you can adjust the appearance of a plot interactively. You can change the axes, plotting style, and scaling type of any graph. With three-dimensional plots, you can also adjust the shading and lighting schemes, projection, and orientation of the surface. Under the worksheet interface, you can paste a plot directly into your session once you are satisfied with it.

If you are using a simple text terminal or a computer without graphics, you may still be able to plot curves and surfaces. Check whether your terminal can emulate a graphics terminal that Maple supports, like Tektronix or ReGIS. If you have access to a graphics printer like an HP LaserJet or a PostScript printer, see the last part of this section on how to redirect your graphics to a file in the proper format.

7.1 Two-Dimensional Graphics

The `plot` command produces two-dimensional plots. This section lists some of the ways you can call `plot` and gives several examples of its use.

There are two ways to call `plot`. To plot an expression in a single variable over the interval $[a, b]$, use a command of the form:

`plot(`*expr*`, `*var*` = `*a*`..`*b*`);`

To plot a function requiring a single argument over an interval, use a command of the form:

`plot(`*f*`, `*a*`..`*b*`);`

For example, the following commands produce identical plots: a single period of the sine curve (Figure 1.3).

```
> plot(sin(x), x = 0..2*Pi);
> plot(sin, 0..2*Pi);
```

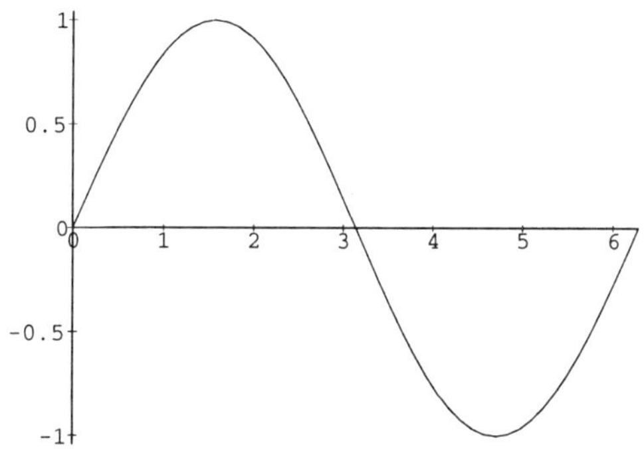

Figure 1.3: The sine curve

You can plot an expression over a half line or the entire real line by making one or both of the endpoints of your domain infinite (Figure 1.4).

```
> plot(x^3 - x, x=-infinity..infinity);
```

A second interval selects the range of values displayed along the y-axis (Figure 1.5).

```
> plot(tan(x), x = 0..10, -5..5);
```

Draw a parametric plot with a command like:

`plot([`$expr_x(t)$`, `$expr_y(t)$`, `*t*` = `*a*`..`*b*`]);`

For example (Figure 1.6),

```
> plot([cos(t), 2*sin(t), t=-Pi..Pi], scaling=CONSTRAINED);
```

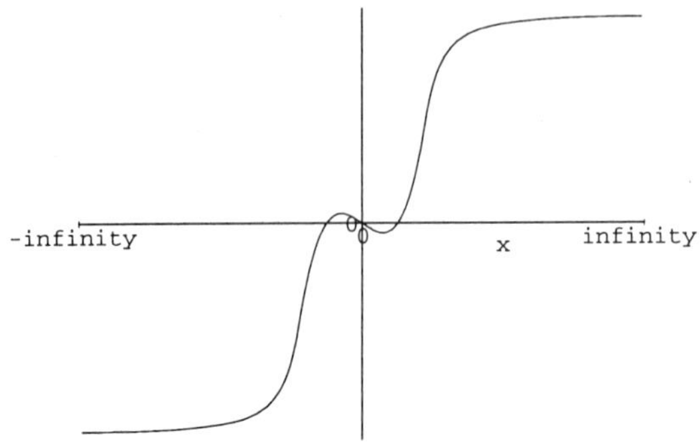

Figure 1.4: A plot extending to infinity

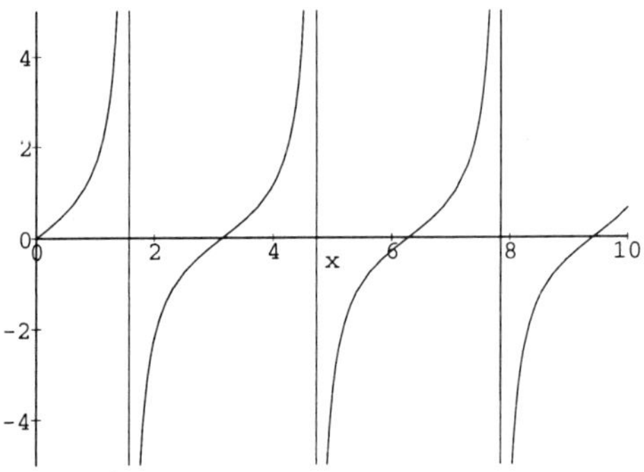

Figure 1.5: The tangent function

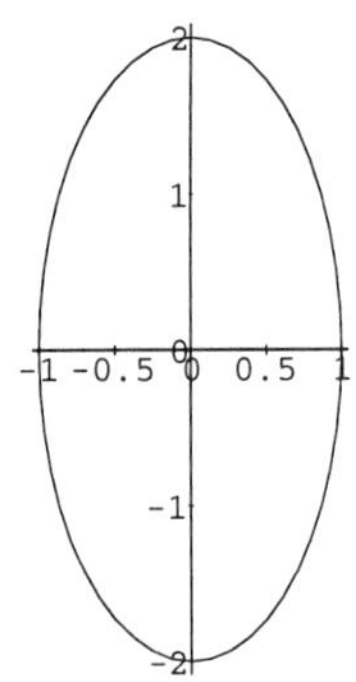

Figure 1.6: A parametric plot

Setting the `scaling` option to `CONSTRAINED` prevents distortion of the graph, ensuring that a unit along the x-axis looks the same as a unit along the y-axis. You can also draw a parametric plot in polar coordinates:

plot([$expr_r(t)$, $expr_\theta(t)$, t = a..b], coords=polar);

For example (Figure 1.7),

```
> plot([1 + sin(5*t)/5, t, t=-Pi..Pi], coords=polar,
>      scaling=CONSTRAINED);
```

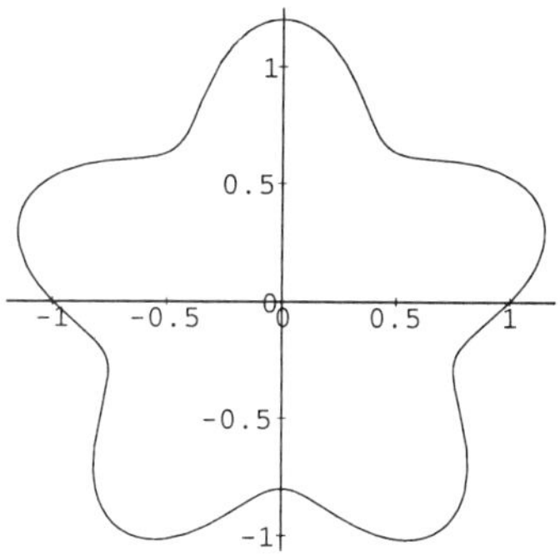

Figure 1.7: A parametric plot in polar coordinates

You can plot a list of points by naming them explicitly. By default, successive points are joined by lines. If you specify the option `style=POINT`, Maple instead draws a small cross at each data point. In Maple V Release 3, the `symbol` option lets you name the symbol drawn at each data point.

plot([x_1, y_1, x_2, y_2, x_3, y_3, ...], style=POINT);

For example, we plot a few points on a parabola (Figure 1.8).

```
> plot([0, 0, 1/4, 1/16, 1/2, 1/4, 3/4, 9/16, 1, 1,
>        5/4, 25/16, 3/2, 9/4, 7/4, 49/16, 2, 4], style=POINT);
```

Finally, you can plot several functions on the same graph, even an expression together with a parametric plot.

plot({$expr_1$, $expr_2$, ...}, var = a..b);

plot({$expr_1(s)$, [$expr_x(t)$, $expr_y(t)$, t = a..b]}, s = c..d);

Here we plot two trigonometric functions on the same graph (Figure 1.9), then a line plus an arc of a circle, where the arc is given parametrically (Figure 1.10). The `numpoints`

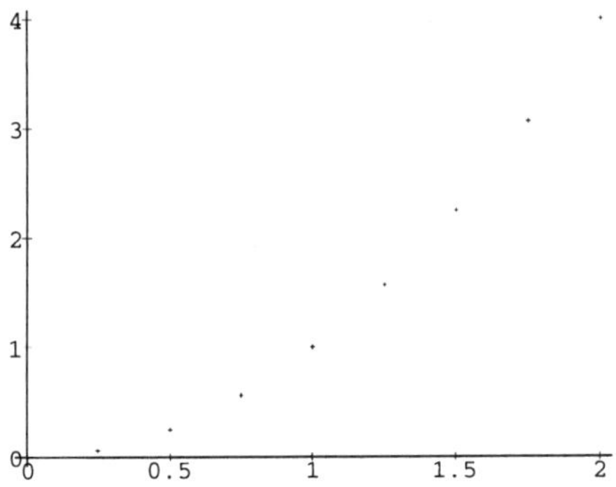

Figure 1.8: Plotting points

option in the first of these improves the resolution of the curve by guaranteeing that at least 100 points are sampled instead of the usual minimum of 49.

```
> plot({4*cos(x)*sin(x)^2, -sec(x)}, x=-2*Pi..2*Pi, -4..4,
>      numpoints=100);
```

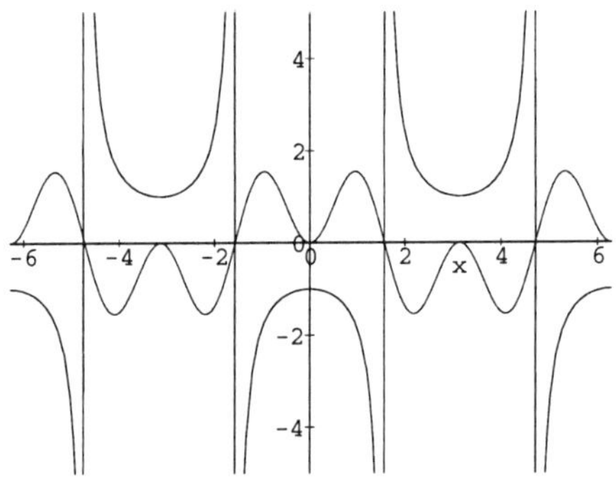

Figure 1.9: Graphing two functions

```
> plot({s, [1+cos(t), sin(t), t=Pi/2..Pi]}, s=0..1,
>      scaling=CONSTRAINED);
```

Plot Options

Options control the way a plot is drawn, as well as the placement of cosmetic features. Below we list all the options for the `plot` command. Some of these options, namely **`axes`**, **`linestyle`**, **`scaling`**, **`style`**, **`symbol`**, and **`thickness`**, can be adjusted interactively on most Maple platforms.

a..*b*
: Range of values displayed along the y-axis (default: show the entire graph).

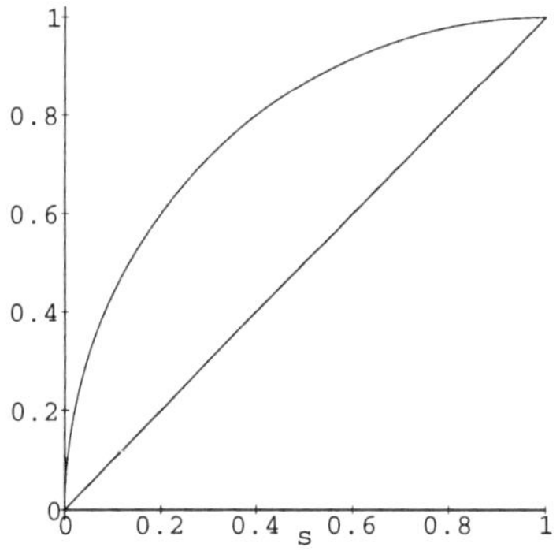

Figure 1.10: Graphing a function and a parametric curve

`axes=BOXED`
Enclose the entire graph in a bounding box.

`axes=FRAME`
Draw axes along the outer edges of the graph.

`axes=NONE`
Don't draw axes.

`axes=NORMAL`
Draw axes through the graph, usually intersecting at the origin (the default).

`axesfont=[`*family*`,`*style*`,`*size*`]`
Font for the labels of the tick marks on the axes. See `font` for more details. This option is new in Maple V Release 3.

`color=`*c*
Color of plot: `COLOR(RGB,` *r*`,` *g*`,` *b*`)`, `COLOR(HUE,` *h*`)`, `aquamarine`, `black`, `blue`, `brown`, `coral`, `cyan`, `gold`, `gray`, `green`, `grey`, `khaki`, `magenta`, `maroon`, `navy`, `orange`, `pink`, `plum`, `red`, `sienna`, `tan`, `turquoise`, `violet`, `wheat`, `white`, or `yellow` (default: `black`).

`colour=`*c*
Synonym for `color`.

`coords=polar`
Draw a polar plot given parametrically.

`discont=true`
Plots the expression over subintervals where it is continuous. This option works only with expressions, not functions. It is new in Maple V Release 3 (default: false).

`font=[`*family*`,`*style*`,`*size*`]`
Font for text objects. *family* may be `TIMES`, `COURIER`, `HELVETICA`, or `SYMBOL`. For `TIMES`, *style* may be `ROMAN`, `BOLD`, `ITALIC`, or `BOLDITALIC`. For `COURIER` or

HELVETICA, *style* may be BOLD, OBLIQUE, BOLDOBLIQUE, or may be omitted. For SYMBOL, *style* should be omitted. *size* specifies the size in points of the font used. This option is new in Maple V Release 3.

labelfont=[*family*,*style*,*size*]
Font for the axes labels. See font for more details. This option is new in Maple V Release 3.

labels=[str_x, str_y]
Labels for the x- and y-axes (default: the name of the independent variable on the x-axis, and no label on the y-axis).

linestyle=*n*
The dash pattern used to render lines in the plot. The default pattern (normally a solid line) is drawn if *n* is 0; varying dash patterns are used for greater values of *n*. This option is new in Maple V Release 3.

numpoints=*n*
Minimum number of points at which the function is sampled (default: 49).

resolution=*n*
Number of pixels across the display device (default: 200). Determines the maximum number of times an interval is subdivided when plotting a curve.

scaling=CONSTRAINED
Constrains the scaling to maintain a 1:1 aspect ratio.

scaling=UNCONSTRAINED
Maximizes the drawing of the graph at the expense of maintaining proper scaling (the default).

style=LINE
Connect the points sampled or listed with lines (the default).

style=PATCH
Patch style for drawing polygons.

style=POINT
Show only the points sampled or listed. By default, the points are drawn as small crosses (see symbol).

symbol=BOX
Draw a box at each point when style=POINT. This option is new in Maple V Release 3.

symbol=CIRCLE
Draw a circle at each point when style=POINT. This option is new in Maple V Release 3.

`symbol=CROSS`
Draw a cross at each point when `style=POINT`. This option is new in Maple V Release 3.

`symbol=DIAMOND`
Draw a diamond at each point when `style=POINT`. This option is new in Maple V Release 3.

`symbol=POINT`
Draw a small dot at each point when `style=POINT`. This option is new in Maple V Release 3.

`thickness=`*n*
Thickness of lines in the plot. *n* may be 0 (the default), 1, 2, or 3. This option is new in Maple V Release 3.

`tickmarks=`[n_x, n_y]
Minimum number of marks along the horizontal and vertical axes. Each component should be a nonnegative integer, or the name `default` for a default value (default: selected by system).

`title=`*string*
Title for the graph (default: no title).

`titlefont=`[*family*, *style*, *size*]
Font for the title of the plot. See `font` for more details. This option is new in Maple V Release 3.

`view=`[x_1..x_2, y_1..y_2]
Limit the region of the curve to be viewed (default: entire curve). This option is new in Maple V Release 3.

`xtickmarks=`*n*
Minimum number of marks along the horizontal axis (default: selected by system).

`ytickmarks=`*n*
Minimum number of marks along the vertical axis (default: selected by system).

The Maple library contains many routines for drawing curves and surfaces in the `plots` package. Load this package with the command `with(plots)`. Routines from this package for making two-dimensional graphics are listed in the table on page 30. An (R2) indicates that the function was introduced in Maple V Release 2.

Here are some examples of a few of these functions. Section 7.3, Saving and Printing Plots (page 42), includes an example of `display`. The `polarplot` command below graphs the "nephroid of Freeth" (Figure 1.13).

```
> with(plots):
> animate(x^n, x=0..1, n=1..6, frames=30);

> conformal(z^4, z=0..1+I);
```

Function		Description
`animate`	(R2)	Create an animation of two-dimensional plots of functions
`conformal`		Conformal plot of a complex function (Figure 1.11)
`display`		Redisplay a plot, or display several plots on one graph
`fieldplot`	(R2)	Plot a two-dimensional vector field
`gradplot`	(R2)	Plot a two-dimensional gradient field (Figure 1.12)
`implicitplot`	(R2)	Plot a curve given implicitly by an equation in two variables
`loglogplot`	(R2)	Plot both axes with a logarithmic scale
`logplot`	(R2)	Plot the vertical axis with a logarithmic scale
`odeplot`	(R2)	Plot the output from `dsolve` with the `numeric` option (see page 17)
`polarplot`		Plot a curve in polar coordinates (Figure 1.13)
`polygonplot`	(R2)	Plot one or more polygons
`sparsematrixplot`		Displays a point in a two-dimensional lattice at the position of each nonzero element of a matrix, showing its sparsity pattern.
`textplot`	(R2)	Position text in two-dimensional space

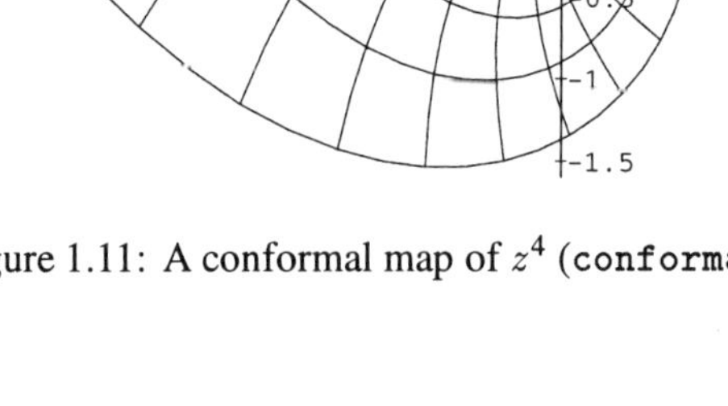

Figure 1.11: A conformal map of z^4 (`conformal`)

```
> gradplot(sin(x)*sin(y), x=-Pi..Pi, y=-Pi..Pi, arrows=SLIM,
>     axes=BOXED);
```

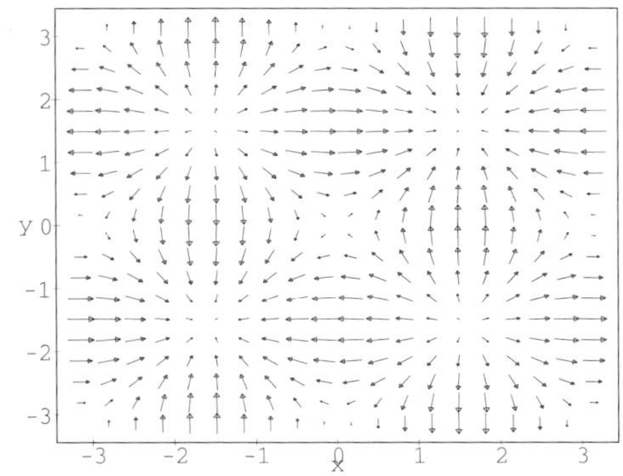

Figure 1.12: A gradient field (`gradplot`)

```
> polarplot(1 + 2*sin(t/2), t=0..4*Pi, scaling=CONSTRAINED,
>     axes=FRAME);
```

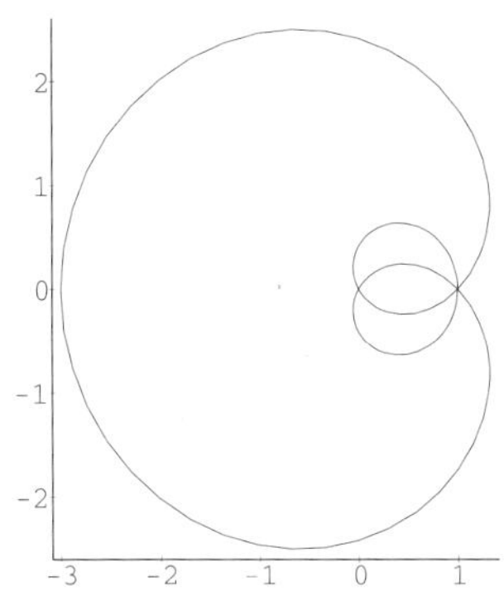

Figure 1.13: The nephroid of Freeth (`polarplot`)

Maple V Release 2 and Release 3 also include a `DEtools` package for graphing solutions to ordinary and partial differential equations in two and three dimensions.

7.2 Three-Dimensional Graphics

This section discusses some functions for producing three-dimensional plots. The function `plot3d` graphs a surface in three dimensions.

Just as with `plot`, there are two ways to call `plot3d`. To plot an expression in two variables over a rectangular region in the plane, use a command of the form:

`plot3d(`*expr*`,` var_1 `=` *a*`..`*b*`,` var_2 `=` *c*`..`*d*`);`

For example (Figure 1.14),

```
> plot3d(sin(x*y)/(x*y), x=-Pi..Pi, y=-Pi..Pi);
```

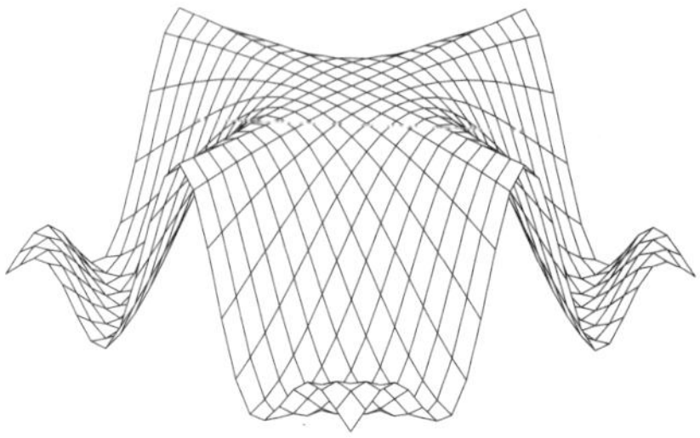

Figure 1.14: $\sin(xy)/xy$ on $[-\pi, \pi] \times [-\pi, \pi]$

To plot a function requiring two arguments, use a command of the form:

plot3d(*f*, *a*..*b*, *c*..*d*);

For example, we plot Maple's `Beta` function, defined as $\text{Beta}(x, y) = \Gamma(x)\Gamma(y)/\Gamma(x + y)$ (Figure 1.15).

```
> plot3d(Beta, 1..2, 1..2, axes=BOXED);
```

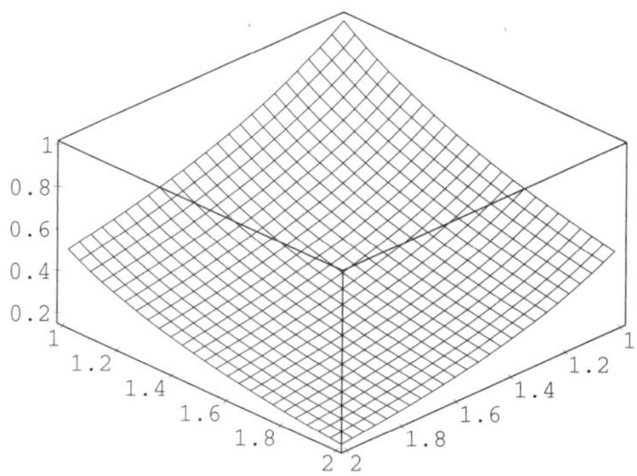

Figure 1.15: The Beta function

You can draw a parametric surface with a command like:

plot3d([*expr*$_x$(*s*, *t*), *expr*$_y$(*s*, *t*), *expr*$_z$(*s*, *t*)], *s* = *a*..*b*, *t* = *c*..*d*);

The notation for a two-dimensional parametric plot is not consistent with the notation for a three-dimensional parametric plot. In the former case, the range needs to be inside the square brackets; in the latter case the ranges go outside. For example, we graph a helical surface parametrically (Figure 1.16):

```
> plot3d([s*cos(t), s*sin(t), t], s=1..3, t=0..4*Pi, axes=FRAME);
```

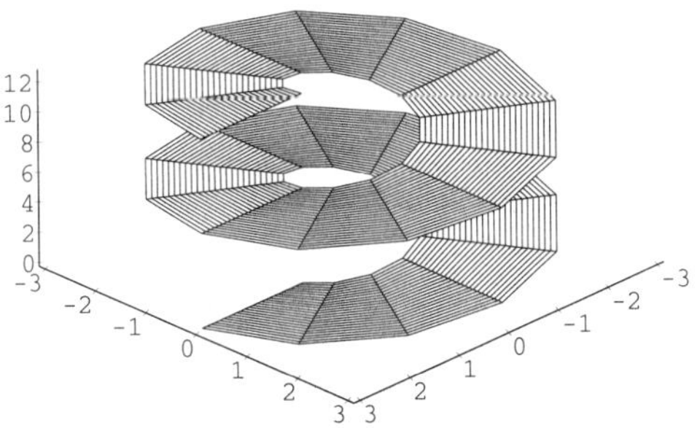

Figure 1.16: A parametric surface

You can create a contour plot of a surface by using the style=CONTOUR option (Figure 1.17).

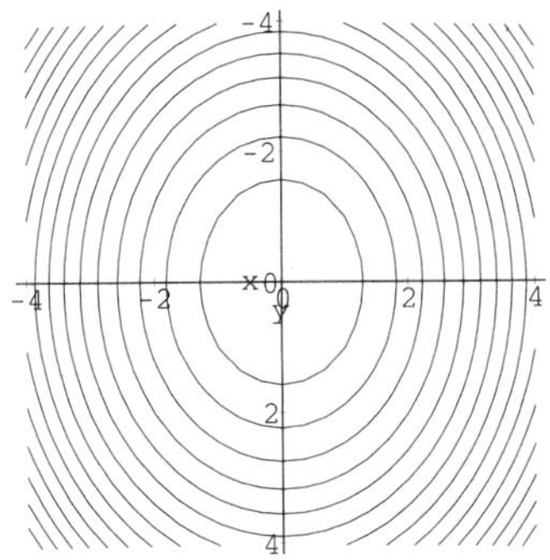

Figure 1.17: A contour plot

```
> plot3d(2*x^2 + 3*y^2, x=-4..4, y=-4..4, style=CONTOUR,
>     scaling=CONSTRAINED, orientation=[0,0], axes=NORMAL);
```

Here, the scaling option prevents distortion of the contours, and the orientation option selects the usual view for a contour graph: looking directly down onto the surface.

You can plot several surfaces on the same graph. The last three options in the next example (Figure 1.18) shade the two surfaces plotted. The default shading scheme is turned off with shading=NONE, then one weak ambient white light and one strong directed white light highlight the aquamarine surfaces. If you don't want to type in this entire command, omit the first five option settings, then adjust them yourself interactively after the graph is displayed (see page 92).

```
> plot3d({exp(-y^2)*sin(x) - 1, exp(-y^2)*cos(x) + 1}, x=-Pi..Pi,
>     y=-2..2, scaling=CONSTRAINED, axes=FRAME, orientation=[60,75],
>     shading=NONE, style=PATCHNOGRID, color=aquamarine,
>     ambientlight=[0.5,0.5,0.5], light=[45,90,1,1,1]);
```

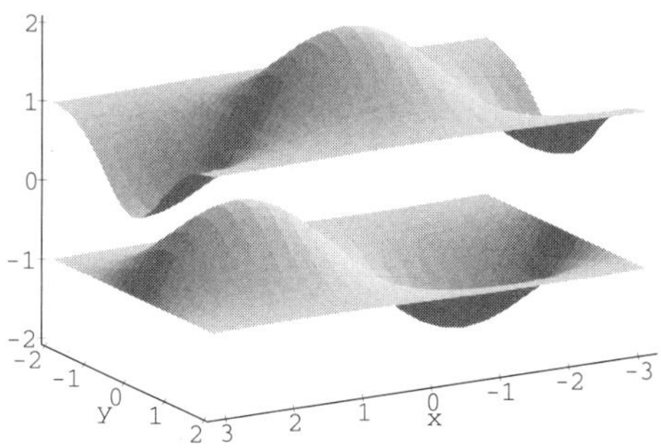

Figure 1.18: Two surfaces and user-defined lighting

Plot3d Options

All of the options for the `plot3d` command are listed below. Several of them, namely `axes`, `light`, `linestyle`, `orientation`, `projection`, `scaling`, `shading`, `style`, `symbol`, and `thickness`, can be adjusted interactively on most Maple platforms (see page 92).

`ambientlight=[`*r*, *g*, *b*`]`
: Use an undirected light with the given red, green, and blue intensities to light the surface. The parameters *r*, *g*, and *b* are numbers between 0 and 1. (Default: none)

`axes=BOXED`
: Enclose the entire graph in a bounding box.

`axes=FRAME`
: Draw axes along the outer edges of the graph.

`axes=NONE`
: Don't draw axes (the default).

`axes=NORMAL`
: Draw axes through the graph (usually intersecting at the origin).

`axesfont=[`*family*`,`*style*`,`*size*`]`
: Font for the labels of the tick marks on the axes. See `font` for more details. This option is new in Maple V Release 3.

`color=`*c*
Color of plot: a procedure or expression in the plot variables that returns a color, or a constant color: `COLOR(RGB,` *r*`,` *g*`,` *b*`)`, `COLOR(HUE,` *h*`)`, `aquamarine`, `black`, `blue`, `navy`, `coral`, `cyan`, `brown`, `gold`, `green`, `gray`, `grey`, `khaki`, `magenta`, `maroon`, `orange`, `pink`, `plum`, `red`, `sienna`, `tan`, `turquoise`, `violet`, `wheat`, `white`, or `yellow` (default: `black`). See page 34 for an example of the use of the `color` option.

`colour=`*c*
Synonym for `color`.

`contours=`*c*
Specify contours for a contour plot. *c* is the number of contours (default: 10) or a list of z-values to use for contours. This option is new in Maple V Release 3.

`coords=cartesian`
Use a cartesian (rectangular) coordinate system (the default).

`coords=cylindrical`
Use a cylindrical coordinate system.

`coords=spherical`
Use a spherical coordinate system.

`font=[`*family*`,`*style*`,`*size*`]`
Font for text objects. *family* may be `TIMES`, `COURIER`, `HELVETICA`, or `SYMBOL`. For `TIMES`, *style* may be `ROMAN`, `BOLD`, `ITALIC`, or `BOLDITALIC`. For `COURIER` or `HELVETICA`, *style* may be `BOLD`, `OBLIQUE`, `BOLDOBLIQUE`, or may be omitted. For `SYMBOL`, *style* should be omitted. *size* specifies the size in points of the font used. This option is new in Maple V Release 3.

`grid=[`n_x`,` n_y`]`
Use an n_x by n_y mesh of points to generate the surface (default: 25 by 25).

`labelfont=[`*family*`,`*style*`,`*size*`]`
Font for the axes labels. See `font` for more details. This option is new in Maple V Release 3.

`labels=[`str_x`,` str_y`,` str_z`]`
Labels for the x-, y- and z-axes (default: the name of the independent variables on the x- and y-axes, and no label on the z-axis).

`light=[`ϕ`,` θ`,` *r*`,` *g*`,` *b*`]`
Use a directed light source with the given red, green, and blue intensities at the indicated position (in spherical coordinates) to light the surface. The parameters *r*, *g*, and *b* are numbers between 0 and 1. (Default: none)

`linestyle=`*n*
The dash pattern used to render lines in the plot. The default pattern (normally a solid line) is drawn if *n* is 0; varying dash patterns are used for greater values of *n*. This option is new in Maple V Release 3.

`numpoints=`n
Use a square grid with at least n total points to generate the surface (default: 625).

`orientation=[`θ`,` ϕ`]`
Viewpoint orientation, where θ and ϕ are angles, specified in degrees in a spherical coordinate system. (Default: [45, 45])

`projection=`p
Specify the type of projection. The parameter p is a number between 0 and 1. A value of 1 denotes orthogonal projection (the default) and 0 denotes a wide-angle perspective.

`projection=FISHEYE`
Use a wide-angle perspective, equivalent to `projection=0`.

`projection=NORMAL`
Use normal perspective, equivalent to `projection=0.5`.

`projection=ORTHOGONAL`
Use an orthogonal projection, equivalent to `projection=1` (the default).

`scaling=CONSTRAINED`
Constrains the scaling to maintain a 1:1:1 aspect ratio.

`scaling=UNCONSTRAINED`
Maximizes the drawing of the graph at the expense of maintaining proper scaling (the default).

`shading=XYZ`
Each of the axes has its own range of colors. The default is device dependent.

`shading=XY`
Shade the surface in one color range along the x-axis and another color range along the y-axis.

`shading=Z`
Shade the surface according to the height or z-value.

`shading=ZHUE`
Shade the surface according to the height or z-value by using a rainbow of colors along the z-axis.

`shading=ZGRAYSCALE`
Shade the surface according to the height or z-value by using a gray scale along the z-axis.

`shading=NONE`
No shading scheme.

`style=CONTOUR`
Contour plot of surface.

`style=HIDDEN`
: Draw a wireframe, but remove all line segments that are hidden behind other parts of the surface (the default).

`style=LINE`
: Same as `WIREFRAME`.

`style=PATCH`
: Draw the surface with shaded rectangular patches on a wireframe grid.

`style=PATCHCONTOUR`
: Combination of `PATCH` and `CONTOUR` styles.

`style=PATCHNOGRID`
: Draw the surface with shaded rectangular patches with no wireframe grid.

`style=POINT`
: Plot only the points sampled. By default, the points are drawn as small crosses (see `symbol`).

`style=WIREFRAME`
: Draw a wireframe surface, connecting adjacent sampled points with line segments.

`symbol=BOX`
: Draw a box at each point when `style=POINT`. This option is new in Maple V Release 3.

`symbol=CIRCLE`
: Draw a circle at each point when `style=POINT`. This option is new in Maple V Release 3.

`symbol=CROSS`
: Draw a cross at each point when `style=POINT`. This option is new in Maple V Release 3.

`symbol=DIAMOND`
: Draw a diamond at each point when `style=POINT`. This option is new in Maple V Release 3.

`symbol=POINT`
: Draw a small dot at each point when `style=POINT`. This option is new in Maple V Release 3.

`thickness=`n
: Thickness of lines in the plot. n may be 0 (the default), 1, 2, or 3. This option is new in Maple V Release 3.

`tickmarks=[`n_x`,` n_y`,` n_z`]`
: Minimum number of marks along x-, y-, and z-axes. Each component should be a nonnegative integer, or the name `default` for a default value (default: selected by system).

`title=`*string*
Title for the graph (default: none).

`titlefont=[`*family*`,`*style*`,`*size*`]`
Font for the title of the plot. See `font` for more details. This option is new in Maple V Release 3.

`view=`z_1`..`z_2
Limit the region of the surface to be viewed to the range of z-values $z_1 \,..\, z_2$ (default: entire surface).

`view=[`x_1`..`x_2`, `y_1`..`y_2`, `z_1`..`z_2`]`
Limit the region of the surface to be viewed (default: entire surface).

The following three examples demonstrate a few more of these options: a sphere given in spherical coordinates (Figure 1.19), a saddle surface in cartesian coordinates displayed with a combination patch and contour style (Figure 1.20), and a slanted double cone given in cylindrical coordinates (Figure 1.21). (If you have a black and white monitor, try using `shading=ZGRAYSCALE` instead of `Z` or `ZHUE`.)

```
> plot3d(1, theta=0..2*Pi, phi=0..Pi, coords=spherical,
>        scaling=CONSTRAINED);
```

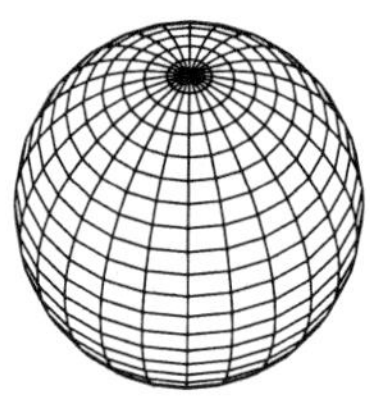

Figure 1.19: Spherical coordinates

```
> plot3d(x^2 - y^2, x=-7..7, y=-5..5, style=PATCHCONTOUR,
>        orientation=[80,45], shading=ZHUE, axes=BOXED);
```

```
> plot3d(z*sin(theta), theta=0..Pi, z=-4..4, style=HIDDEN,
>        coords=cylindrical, scaling=CONSTRAINED, shading=Z,
>        orientation=[45,75], axes=BOXED);
```

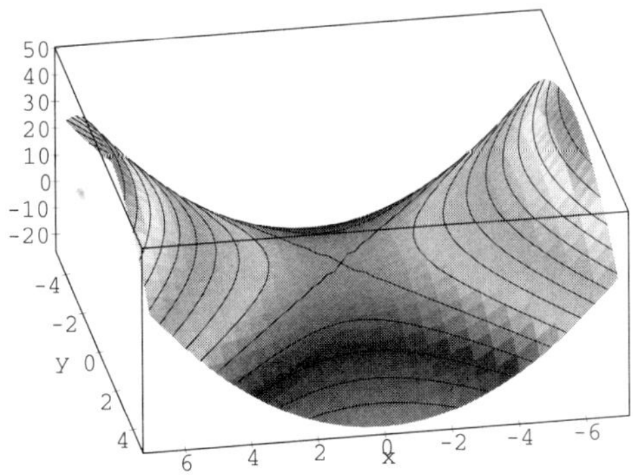

Figure 1.20: A shaded saddle surface with contours

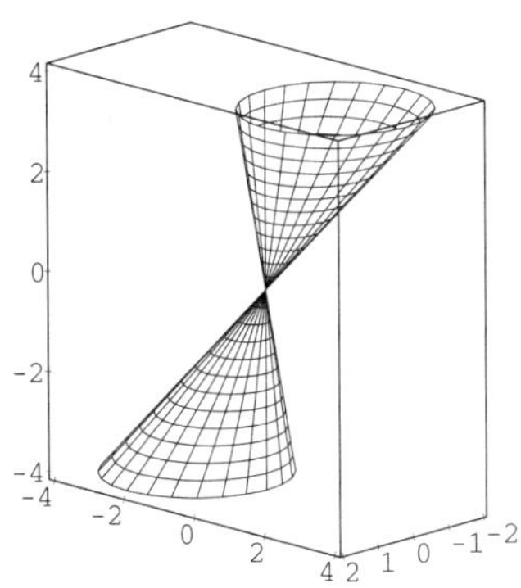

Figure 1.21: Cylindrical coordinates

The `plots` package also contains many routines for drawing three-dimensional graphs. These are listed in the table below. An (R2) indicates that the function was introduced in Maple V Release 2. Remember to load this package first by entering `with(plots)`.

Function		Description
`animate3d`	(R2)	Create an animation of three-dimensional plots of functions
`contourplot`	(R2)	Contour plot of surface
`cylinderplot`		Plot a surface in cylindrical coordinates
`densityplot`	(R2)	Density plot of surface
`display3d`		Redisplay a plot, or display several plots on one graph
`fieldplot3d`	(R2)	Plot a three-dimensional vector field
`gradplot3d`	(R2)	Plot a three-dimensional gradient field
`implicitplot3d`	(R2)	Plot a surface given implicitly by an equation in three variables (Figure 1.22)
`matrixplot`		Plot a surface with heights given by the entries in a matrix
`odeplot`	(R2)	Plot output from `dsolve` with the `numeric` option
`pointplot`		Plot points in three dimensions
`polygonplot3d`	(R2)	Plot one or more polygons in three-dimensional space
`polyhedraplot`	(R2)	Plot one or more polyhedra (Figure 1.23)
`spacecurve`		Plot a curve given parametrically in three-dimensional space (Figure 1.24)
`sphereplot`		Plot a surface in spherical coordinates
`surfdata`	(R2)	Create a surface from a list of data
`textplot3d`	(R2)	Position text in three-dimensional space
`tubeplot`		Draw tubes around curves (Figure 1.25)

Here are some examples of a few of these functions.

```
> with(plots):
> implicitplot3d(x^2 + y^2 + z^2/2 = 1,
>     x=-1..1, y=-1..1, z=-1.5..1.5, grid=[8,8,8], axes=FRAME,
>     scaling=CONSTRAINED, style=PATCH, shading=ZHUE);

> polyhedraplot([0,0,0], polytype=dodecahedron, style=WIREFRAME,
>     orientation=[40,80], scaling=CONSTRAINED);

> spacecurve([(sin(20*t)+3)*cos(t),(sin(20*t)+3)*sin(t),cos(20*t)],
>     t=0..2*Pi, numpoints=300, scaling=CONSTRAINED, shading=XY,
>     axes=FRAME);

> tubeplot([cos(t), sin(t), t], t=0..4*Pi, radius=.3, tubepoints=15,
>     axes=FRAME);
```

See the resources listed in Chapter 8 beginning on page 465 for more examples of curves and surfaces you can make with Maple.

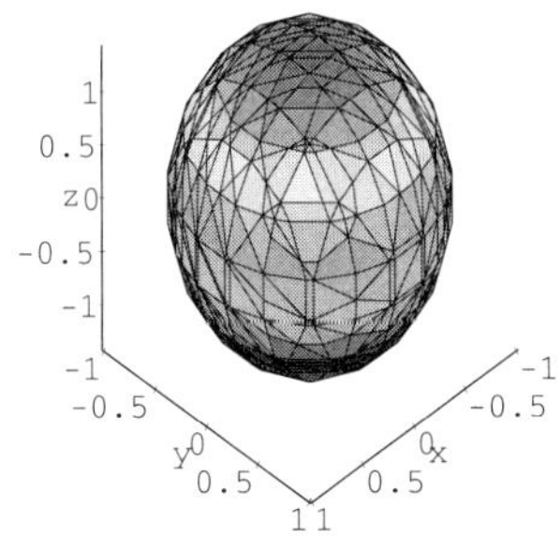

Figure 1.22: An implicit plot in three dimensions (`implicitplot3d`)

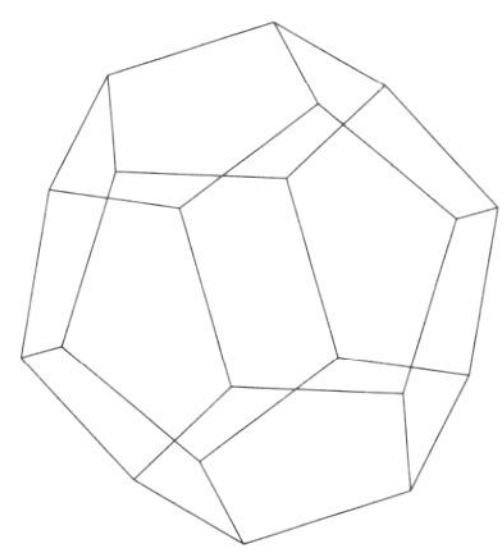

Figure 1.23: A dodecahedron (`polyhedraplot`)

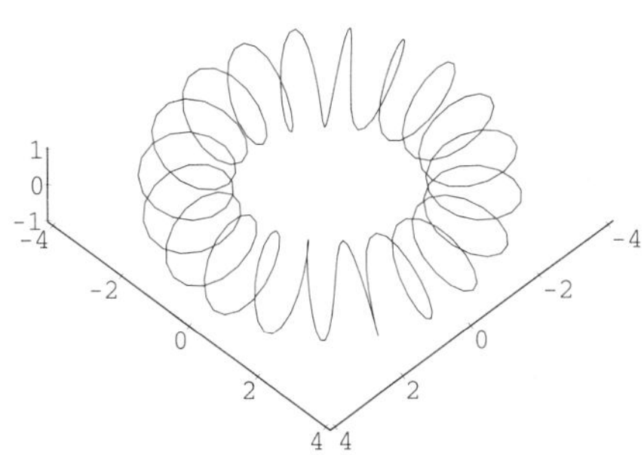

Figure 1.24: A parametric curve in three dimensions (`spacecurve`)

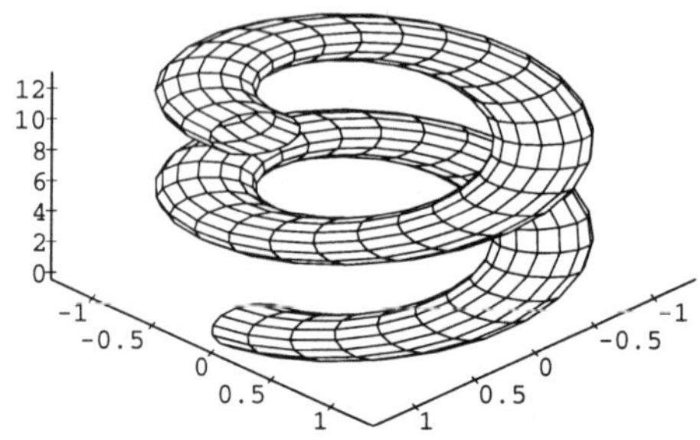

Figure 1.25: A tube plot

7.3 Saving and Printing Plots

Maple creates a special data structure each time you plot a curve or a surface. Maple uses this structure to redraw a plot quickly if you change some of its cosmetic features, like the color scheme or axis style. You can save this data structure in a variable and use it to redisplay the plot yourself later. Here we create three plots and store their associated data structures in variables. The `plot` commands end with colons to prevent each curve from being displayed.

```
> midcurve := plot(exp(-x^2)*cos(12*x), x=-2..2):
> topcurve := plot(exp(-x^2), x=-2..2):
> botcurve := plot(-exp(-x^2), x=-2..2):
```

Evaluating one of these variables displays the plot (Figure 1.26):

```
> midcurve;
```

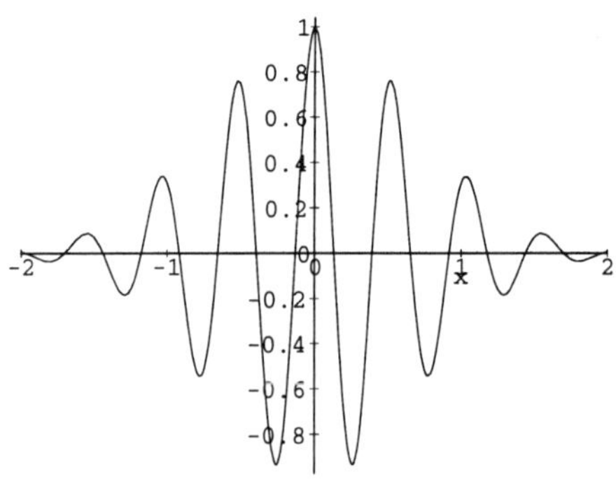

Figure 1.26: `midcurve`

The `display` function in the `plots` package displays several curves on the same graph (Figure 1.27). Specifying a cosmetic option like `axes` or `scaling` to `display` overrides options or defaults set in the individual plots.

```
> with(plots):
> display([topcurve, botcurve, midcurve], axes=NONE,
>      scaling=CONSTRAINED);
```

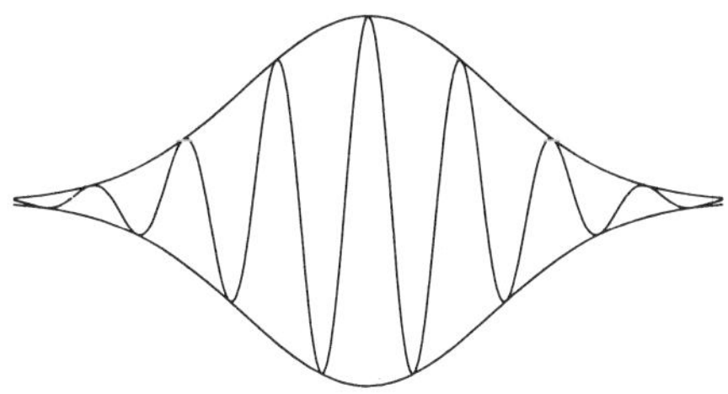

Figure 1.27: Displaying multiple plots

If you want to redisplay the curves in another Maple session, save these variables in a file.

```
> save topcurve, midcurve, botcurve, `myfile.m`;
```

The `read` command reads this file into another Maple session. Then the same variables are available for displaying the curves.

```
> read `myfile.m`;
> with(plots):
> display([topcurve, botcurve, midcurve], axes=NONE,
>      scaling=CONSTRAINED);
```

Note that `save` and `read` do not require parentheses around their arguments.

Section 13, Input and Output, on page 72 includes more details on saving results in files and restoring them in other sessions.

The plot data structure is only Maple's internal representation of a plot. It is not something you can send to a printer to produce a picture. To print a plot, you must first inform Maple of the type of your printer. On many systems, this can be chosen interactively after the graph is displayed on the screen. Under the worksheet interface, look for the File menu in the graphics window. The Print or Printer Setup command on it lets you enter a graphics format and file name for saving the plot, and sometimes a printer name for printing the plot directly.

On other systems, such as the command-line interface on Unix systems, you must use the `plotsetup` or `interface` command to direct graphics to a file in a particular format for printing. The option `plotdevice` lets you name the graphics format or printer type, and the option `plotoutput` lets you specify the name of the file. The help command `?plot[device]` lists the names of all the printing formats supported by Maple.

For example, the following commands instruct Maple to send graphics to the file `myplot.ps` in PostScript format.

```
> interface(plotdevice=postscript);
> interface(plotoutput='myplot.ps');
```

The next plot is stored in the file `myplot.ps` and does not appear on the screen.

```
> plot(sin(x)*cos(2*x), x=-2*Pi..2*Pi);
```

Subsequent plots are also stored in this file, but the file is erased before each save so that only the most recent graph is stored there. Therefore you should reset `plotoutput` after each picture you want to keep.

After quitting Maple, you can print the file `myplot.ps` on your PostScript printer.

8 DATA TYPES

Unlike many programming languages, you do not declare the type of each variable and function that you use in Maple. Still, Maple determines the type of every expression you enter. You can test this type and convert between types. The following table lists some of the basic data types in Maple.

Type	Description	Examples
`numeric`	Any numeric type	`3.7`, `4`, `8*I`
`integer`	Integer	`0`, `14`
`posint`	Positive integer	`14`
`negint`	Negative integer	`-4`
`fraction`	Fraction of integers	`-4/7`, `4/3`
`rational`	Integer or fraction	`-15`, `3/11`
`float`	Floating-point number	`31.7`, `-0.05`
`complex`	Any complex number	`4.1 + I`
`complex(integer)`	Gaussian integer	`5 - 3*I`
`boolean`	Logical value	`true`, `false`, `FAIL`
`string`	Literal string	`'This is a string.'`
`polynom`	Polynomial	`x^8 + 3.9*x - 17`
`series`	Power series	`1 + x + x^2/2 + O(x^3)`
`ratpoly`	Rational expression	`(1 + x^2)/(x - 4)`

You can test the type of any expression with the `type` command.

```
> type(3., integer);
```

false

```
> type(3., float);
```

true

```
> type(3, posint);
```

true

```
> type(x+5, polynom);
```

$$true$$

The convert command lets you convert expressions from one type to another.

```
> convert(355/113, float);
```

$$3.141592920$$

```
> series(exp(x), x, 5);
```

$$1 + x + \frac{1}{2}x^2 + \frac{1}{6}x^3 + \frac{1}{24}x^4 + \mathrm{O}\left(x^5\right)$$

```
> convert(", polynom);
```

$$1 + x + \frac{1}{2}x^2 + \frac{1}{6}x^3 + \frac{1}{24}x^4$$

We have already discussed some of the numeric types in Section 5, Numerical Calculations, on page 11. The following sections discuss two more important basic data types, boolean and string.

8.1 Booleans

A *boolean* is a logical value. There are three boolean constants in Maple: **true**, **false**, and FAIL. Well-posed tests return **true** or **false**; ill-posed tests or undetermined results return FAIL.

A test can be simple or compound. A simple test uses a relational operator (=,<>, <, >, <=, >=) to test for an equality or an inequality or calls a function that returns a boolean value. Two such functions in Maple are **type** for testing the type of a function, and **isprime**, for checking whether an integer is prime. A compound test is one or more simple tests linked by the logical operators **and**, **or**, and **not**. The **not** operator has the highest precedence, followed by **and**, then **or**. For example, the following compound test returns *true*:

```
> not (1=2 or isprime(6)) and 3<>4;
```

$$true$$

Maple uses *short-circuit* evaluation of compound boolean expressions: evaluation of a boolean expression stops as soon as its value is known. For example, consider an expression of the form

condition-1 or *condition-2* or *condition-3*

If *condition-1* evaluates to *true*, the entire expression is *true*, so the other conditions are not evaluated.

If you type a simple relation at the command prompt, it is left unevaluated. Call **evalb**, "evaluate boolean," to determine its truth or falsehood.

```
> 1 < 2;
```

$$1 < 2$$

```
> evalb(");
```

$$true$$

Maple leaves a test like this unevaluated because many commands, like the solve functions, require a symbolic relation as an argument.

8.2 Strings

There are two kinds of strings in Maple. A *simple string* is a combination of letters, digits, and underscore characters (_) that begins with a letter or an underscore. A *quoted string* is any combination of characters surrounded by back quotes (` `). A quoted string may contain blank spaces, special characters like + and [, and even newline characters. In Maple V Release 2, the maximum length of a string is 499 characters. There is no restriction on the length of a string in Maple V Release 3. The empty string is denoted by two adjacent back quotes ``.

Any string is a valid name for a function or a variable. Simple strings are most commonly used for names, but a quoted string is equally valid. Avoid using simple strings that begin with the underscore character, since Maple uses names like these for global variables.

In the following example, we assign values to one simple string and one quoted string. Notice that we include a back quote character in a quoted string by typing two successive back quotes inside the string.

```
> simple := 10;
```

$$simple := 10$$

```
> `strange +/`` quoted-string.` := 20;
```

$$strange\ +/`\ quoted\text{-}string. := 20$$

```
> simple + `strange +/`` quoted-string.`;
```

$$30$$

Use the string concatenation operator (.) to join strings. Its left operand must be a string, but its right operand can be any Maple expression. When the right operand of . is an integer, it is converted to a string and concatenated to the left operand. This operator is used for constructing messages printed from programs and for creating new symbols.

```
> student . 53;
```

student53

```
> z := 1000:
> `The value of z is ` . z;
```

The value of z is 1000

```
> `` . z . ` is the value of z.`;
```

1000 is the value of z.

Notice in the last example that we start with the empty string. This is necessary because the first argument of a concatenation must be a name. In fact, the left argument of a concatenation is not evaluated, so without the empty string we would get:

```
> z . ` is the value of z.`;
```

z is the value of z.

The concatenation operator also operates on expression sequences (page 52) and ranges (page 53).

9 STATEMENTS

In this section, we describe the Maple statements for assigning and freeing variables, testing values of expressions, and executing a series of statements in a loop. The syntax of these statements is similar to that of the ALGOL 68 programming language, and is somewhat reminiscent of Pascal, too. These statements are useful in interactive sessions and are especially important in programming.

9.1 Assigning and Freeing Variables

You can assign a Maple expression to a variable with the assignment operator, `:=`. Assignment statements have the form

name := *expression*;

Any string is a valid variable name in Maple. Most variable names are simple strings that start with a letter, but quoted strings may be used as well, as you saw in the section on Strings on page 46.

Some sample assignments follow. In the second example, we use the inert form of the integrate command so that the integral is not evaluated.

```
> a := 3.17 + 5;
```

$$a := 8.17$$

```
> easyIntegral := Int(x^3, x);
```

$$\mathit{easyIntegral} := \int x^3\, dx$$

If an assigned variable appears in a subsequent expression, its value is substituted when the expression is evaluated.

```
> a^2;
```

$$66.7489$$

Single right quotes surrounding any expression prevent the expression from being evaluated (though some trivial simplifications may occur). The following example uses right quotes to prevent a substitution from being made on the variable a.

```
> 'a^2 - 2 - 4';
```

$$a^2 - 6$$

Now the expression has no single quotes surrounding it, so requesting the last result with the ditto operator evaluates the expression fully.

```
> ";
```

$$60.7489$$

You can free a variable—that is, disassociate it from its assigned value—by assigning the variable its own name. You must enclose the right side in single quotes to keep the variable a from being evaluated.

```
> a := 'a';
```

$$a := a$$

```
> a;
```

$$a$$

The `unassign` function frees many variables at once. You have to load this function from the library before using it (see Section 11, on page 62 for more about the structure of the Maple library). Again, you must surround each variable name with right quotes to prevent it from being evaluated when unassigning. Here we unassign `easyIntegral`, as well as `simple` and `z` from the last section.

```
> readlib(unassign):
> unassign('easyIntegral', 'simple', 'z');
```

The `restart` command frees every variable at once. It effectively restarts your session, clearing every assignment you have made and forgetting every definition you have read in. This command was introduced in Maple V Release 2.

You can look at all of the symbols that are currently defined in your session with the command `anames()`.

9.2 Statement Separators and the Empty Statement

The semicolon and colon are *statement separators* in Maple. They are required to separate successive statements, similar to the role of the semicolon in the Pascal programming language. No punctuation is needed after the last statement in a sequence of statements. This explains why no semicolon is required after a `quit` command. In the sections that follow, we'll see many examples of sequences of statements inside `if` blocks, `do` loops, and `proc` definitions where the last statement has no trailing punctuation.

There is no penalty for placing a semicolon or colon after a statement where it is not required. An extra punctuation mark like this is interpreted as the "empty statement," which does nothing.

9.3 Conditions

The `if` statement lets you test the value of an expression in Maple. It is an essential part of the Maple programming language.

```
if condition1 then
     statement-sequence1
elif condition2 then
     statement-sequence2
elif condition3 then
     statement-sequence3
 :
else
     default-statement
fi;
```

Here, $condition_1$ is tested first. If it evaluates to *true*, $statement\text{-}sequence_1$ is executed. Otherwise, $condition_2$ is tested, and so forth. If none of the conditions are *true*, the *default-statement* is executed. Note Maple uses a three-valued logic when evaluating boolean expressions. A well-posed condition returns *true* or *false*; an ill-posed condition may return *FAIL*.

The `elif` parts and `else` part are all optional. Every `if` statement must end with the keyword `fi` (`if` written backwards).

A condition can be any boolean expression, such as a test for equality. The condition in an `if` statement is automatically evaluated, so you don't need to use `evalb`. Because of Maple's short-circuit evaluation of compound tests, we can write a statement like the following without fear of dividing by zero.

```
> if a <> 0 and b/a > c then c := 2*c fi;
```

9.4 Iteration

Repeat a sequence of Maple statements by putting them in a `do` loop. There are several kinds of `do` loops in Maple. All of them require the keyword `do` just before the block of statements to be repeated, and all must end with the keyword `od` (`do` backwards).

The first kind of `do` statement is the `for` loop. There are two varieties of `for` loops. One keeps track of an indexing variable while the loop is being executed. This variable takes on successive values of a sequence determined by values given in the `from`, `by`, and `to` clauses of the statement. For example, the following loop prints the first 4 prime numbers in reverse order. The indexing variable `i` takes on the successive values 4, 3, 2, 1, and 0. The loop halts as soon as `i` becomes 0.

```
> for i from 4 by -1 to 1 do
>     ithprime(i)
> od;
```

$$7$$

$$5$$

$$3$$

$$2$$

The `from`, `by`, and `to` clauses are all optional. Their default values are 1, 1, and infinity, respectively. These clauses may appear in any order under Maple V Release 2 and Release 3.

The other version of the `for` loop operates on successive parts of a Maple expression. It is commonly used to operate on each element of a list or set in turn.

```
for var in expr do
    statement-sequence
od;
```

See page 55 for an example of this type of `for` loop.

The second type of `do` statement is the `while` loop. It repeats a sequence of statements as long as a given condition holds.

```
while condition do
    statement-sequence
od;
```

For example, we can write the original `for` loop example as a `while` loop:

```
> i := 4:
> while i > 0 do
>     ithprime(i);
>     i := i-1
> od;
```

$$7$$

$$i := 3$$

$$5$$

$$i := 2$$

$$3$$

$$i := 1$$

$$2$$

$$i := 0$$

This time we get more lines of output because the body of the loop has two statements. By default, we see the result of every statement that is either entered directly to Maple or nested one level deep inside a compound statement, like a `do` or an `if`. We would not see results of statements nested two or more levels deep, for example, statements residing in an `if` statement that is inside a `do` loop. The `interface` variable `printlevel` determines the depth of results printed: see Section 12.3, Debugging (page 70).

The `for` and `while` clauses can appear together in a single loop. For example, the following loop finds the smallest prime number that ends with the digits 21 and stores it in the variable `n`.

```
> for n from 21 by 100 while not isprime(n) do od;
> n;
```

$$421$$

We free `i` and `n` for future use.

```
> i := 'i':  n := 'n':
```

10 DATA STRUCTURES

In this section, we discuss the basic data structures available in Maple: expression sequence, range, list, set, array, matrix, vector, and table. We discuss how to create and use these structures and indicate special functions designed to operate on them.

10.1 Expression Sequence

The *expression sequence* is a fundamental data structure in Maple. A sequence is simply a collection of Maple expressions separated by commas. The special sequence with no elements at all is denoted by NULL.

Some functions return an expression sequence.

```
> solve(x^2 - x + 1 = 0, x);
```

$$\frac{1}{2}+\frac{1}{2}I\sqrt{3},\frac{1}{2}-\frac{1}{2}I\sqrt{3}$$

The arguments in any function call form an expression sequence. Below we assign a sequence of indeterminates to a variable, then compute a mixed third partial derivative of an expression with respect to these indeterminates.

```
> vars := x, y, z:
> diff(x^2 * (y+z)^3, vars);
```

$$12x(y+z)$$

The selection operator, [ ], lets you access part of a sequence. Here we request the second element of the sequence vars.

```
> vars[2];
```

$$y$$

The sequence function seq in Maple generates a sequence according to some rule. It is similar to the for loop, but it is more efficient at producing sequences. By using seq, we create a sequence of monomials in x with exponents that range from the 20th prime through the 27th prime.

```
> seq(x^ithprime(n), n=20..27);
```

$$x^{71},x^{73},x^{79},x^{83},x^{89},x^{97},x^{101},x^{103}$$

Make sure that the iteration variable in a seq statement (n in the example above) does not already have a value. If it does, the command will not work unless you surround each occurrence of n with single right quotes to prevent its evaluation.

The sequence operator, $, provides a convenient shortcut to the seq command. We generate the same sequence of monomials with this operator. First we have to clear the definition of n, since the seq command leaves the indexing variable assigned.

```
> n := 'n':
> x^ithprime(n) $ n=20..27;
```

$$x^{71},x^{73},x^{79},x^{83},x^{89},x^{97},x^{101},x^{103}$$

The $ operator is commonly used in the diff function to request a derivative of high order. The simple expression x$8 returns a sequence of eight x's, so the following command computes an eighth derivative.

```
> diff(x^10, x$8);
```

$$1814400x^2$$

In Maple V Release 2 and Release 3, you can use the string concatenation operator on sequences to create new symbols.

```
> t . (seq(binomial(k, 2), k=2..10));
```

$$t1, t3, t6, t10, t15, t21, t28, t36, t45$$

We free `k` for future use.

```
> k := 'k':
```

10.2 Range

A *range* in Maple is an expression of the form $a \,..\, b$, where a and b are any Maple expressions. It represents some set of numbers between a and b, with the precise meaning depending on context. We have already seen several examples of the range expression.

In `seq` and `sum`, $a \,..\, b$ represents the numbers a, $a + 1$, $a + 2$, $\ldots$, $a + k$, where $a + k$ equals b, or more generally $a + k$ is as close as we can get to b without exceeding b. Below we sum the first ten squares with `sum`. The summand and the variable of summation are both enclosed in single right quotes in case `i` has already been assigned a value.

```
> sum('i^2', 'i'=1..10);
```

$$385$$

In `plot` and `int`, the range $a \,..\, b$ represents the real interval $[a, b]$.

```
> int(4*x*ln(x), x=1..E);
```

$$(e)^2 + 1$$

Maple sometimes puts unecessary parentheses in a result, as in the case above.

In `iscont`, a command that tests functions for continuity, the range $a \,..\, b$ represents the open real interval (a, b) by default.

```
> readlib(iscont):
> iscont(tan(x), x=-Pi/2..Pi/2);
```

true

We have to read in `iscont` from the library first with `readlib`.

A range can even have complex values. The `conformal` function in the `plots` package requires a range where a and b are complex numbers. They specify a rectangle in the complex plane over which a complex-valued function is conformally mapped.

```
> with(plots):
> conformal(z^4, z=0..1+I);
```

For more information on packages and library functions, see Section 11 on page 62.

A few Maple functions may return a range.

```
> limit(sin(1/x), x=0);
```

$$-1..1$$

The sequence operator, $, turns a range into a sequence.

```
> $1..5;
```

$$1, 2, 3, 4, 5$$

```
> $3.7 .. 9.2;
```

$$3.7, 4.7, 5.7, 6.7, 7.7, 8.7$$

The string concatenation operator distributes over ranges, returning a sequence of symbols.

```
> t . (1..5);
```

$$t1, t2, t3, t4, t5$$

10.3 List

A *list* in Maple is an expression sequence enclosed in square brackets, `[ ]`. If your keyboard does not have square brackets, use the alternate delimiters: parentheses and vertical bars, `(| |)`. A list may contain any combination of Maple expressions, including other lists. The same expression may appear many times in a list. Items in a list are ordered: Maple never rearranges the elements of a list unless you explicitly request it to do so. The empty list is denoted by `[]`.

The selection operator picks elements out of lists, just as with expression sequences. For example, the ith element of a list L is `L[i]`.

The `op` command can also be used to select parts of lists. In fact, the `op` function parses any Maple expression except expression sequences. The command:

```
op(expr);
```

returns a sequence of all the components, or operands, in the given expression. If *expr* is a list, this command returns the sequence of all the elements in the list. The `op` command with two arguments:

```
op(index, expr);
```

or

op(*range*, *expr*);

selects a particular element or range of elements from an expression. So the *i*th element of a list *L* can also be obtained with op(*i*, *L*).

A related command is nops. It returns the number of operands in an expression.

nops(*expr*);

When *expr* is a list, this is the length of the list.

The following examples exhibit the use of the selection operator and the op and nops functions on lists.

```
> list1 := [Pi, exp(1), Catalan]:
> list2 := [Pi, mu, epsilon]:
> op(list1);
```

$$\pi, \mathrm{e}, Catalan$$

```
> op(1, list1);
```

$$\pi$$

```
> biglist := [op(list1), op(list2)];
```

$$biglist := [\pi, \mathrm{e}, Catalan, \pi, \mu, \epsilon]$$

```
> nops(biglist);
```

$$6$$

```
> biglist[3..5];
```

$$Catalan, \pi, \mu$$

```
> op(3..5, biglist);
```

$$Catalan, \pi, \mu$$

Notice that the selection operator accepts a range of indices, too.

One version of the do loop is often used with lists. The for statement with the in keyword lets you operate on each element of a list in turn. For example, the following loop calculates the first, fourth, 26th, and 440th primes.

```
> for i in [1, 4, 26, 440] do ithprime(i) od;
```

$$2$$

$$7$$

$$101$$

$$3079$$

The `select` command picks elements out of a list that satisfy a certain criterion and returns a new list consisting of these elements. The criterion should be a procedure that returns a boolean value. For example, we extract all the prime numbers from a list of integers between 3000 and 3100 by using the criterion `isprime`, and all the perfect squares in the same list by using a `type` test.

```
> select(isprime, [$3000..3100]);
```

$$[3001, 3011, 3019, 3023, 3037, 3041, 3049, 3061, 3067, 3079, 3083, 3089]$$

```
> select(type, [$3000..3100], square);
```

$$[3025]$$

The `member` command tests whether a particular expression is present in a list. It returns *true* or *false*.

```
> member(a, [a, b, c]);
```

true

The `map` command applies a given function to each element of a list. It returns a new list whose elements are the images of the original list elements under the given function.

```
> map(sin, [0, Pi/6, Pi/3, Pi/2, 2*Pi/3, 5*Pi/6, Pi]);
```

$$\left[0, \frac{1}{2}, \frac{1}{2}\sqrt{3}, 1, \frac{1}{2}\sqrt{3}, \frac{1}{2}, 0\right]$$

The `convert` function has subcommands for operating on lists. You can add up all of the elements in a list with `convert(`*list*`, '+')`. For example, we compute the mean of a list of numbers.

```
> grades := [73, 85, 69, 99, 58, 78, 84]:
> convert(grades, '+') / nops(grades);
```

$$78$$

Similarly, `convert(`*list*`, '*')` computes the product of the elements in a list.

10.4 Set

A *set* in Maple is a sequence of distinct expressions enclosed in curly braces, `{ }`. If your keyboard does not have curly braces, use the alternate delimiters: parentheses and asterisks, `(* *)`. In contrast to a list, the items in a set are unordered: Maple may rearrange the elements in a set for its own purposes. Duplicates in a set are automatically eliminated. The empty set is denoted by `{ }`.

The selection operator, the `for/in` loop, and the commands `op`, `nops`, `select`, `member`, `map`, `convert(`*set*`, '+')` and `convert(`*set*`, '*')` introduced in the last section work analogously on sets. Maple also provides the special set operators `intersect`, `union`, and `minus`.

```
> set1 := {0, Pi, 2*Pi, 3*Pi}:
> set2 := {Pi, exp(1), Catalan}:
> set1 union set2;
```

$$\{0, Catalan, \pi, \mathrm{e}, 2\pi, 3\pi\}$$

```
> set1 intersect set2;
```

$$\{\pi\}$$

```
> set1 minus set2;
```

$$\{0, 2\pi, 3\pi\}$$

```
> map(sin, set1);
```

$$\{0\}$$

```
> member(gamma, set2);
```

false

```
> evalf(convert(set2, '*'));
```

$$7.822102731$$

The `symmdiff` function computes the symmetric difference of two sets, so `symmdiff(`A`, `B`)` is equivalent to `(`A` minus `B`) union (`B` minus `A`)`. Unlike the other set operators, `symmdiff` is not an infix operator. This function was introduced in Maple V Release 2. You must read it from the library before using it. See Section 11 (page 62) for more about the structure of the Maple library.

```
> readlib(symmdiff):
> symmdiff(set1, set2);
```

$$\{0, Catalan, \mathrm{e}, 2\pi, 3\pi\}$$

A set can be converted into a list by `[op(`*set*`)]` or `convert(`*set*`, list)`; a sequence into a set by `{`*sequence*`}`; and a list into a set by `{op(`*list*`)}` or `convert(`*list*`, set)`.

10.5 Array, Matrix, and Vector

An *array* is a multidimensional data structure indexed by ranges of integers. For example, we define a two-dimensional array with three rows and two columns, with rows indexed by integers 0, 1, 2 and columns by −3, −2 with the command:

```
> a := array(0..2, -3..-2):
```

Then we assign some values to the array by using the selection operator:

```
> a[0, -3] := 10:
> a[0, -2] := 11:
```

If you want to look at all of the values of an array, use `print`.

```
> print(a);
```

array(0 .. 2, -3 .. -2, [
(0, -3) = 10
(0, -2) = 11
(1, -3) = $a_{1,-3}$
(1, -2) = $a_{1,-2}$
(2, -3) = $a_{2,-3}$
(2, -2) = $a_{2,-2}$
])

The `array` command accepts additional parameters: an indexing function and a list of initial values. An indexing function constrains the values of an array in some way, and allows certain kinds of arrays to be stored efficiently. There are five predefined indexing functions that may be used to define matrices. A *matrix* in Maple is a two-dimensional array where indexing begins at 1 in both dimensions.

The following table lists the five indexing functions defined in Maple and summarizes their constraints on matrices. They have analogous effects on higher-dimensional arrays.

Function	Constraint
`symmetric`	$A[i,j] = A[j,i]$
`antisymmetric`	$A[i,j] = -A[j,i]$
`diagonal`	Off-diagonal elements are 0
`identity`	Diagonal elements are 1, all others are 0
`sparse`	Entries not explicitly assigned are assumed to be 0

We define a 2 × 2 matrix with initial values.

```
> m := array(1..2, 1..2, [[x, 10], [10, y]]);
```

$$m := \begin{bmatrix} x & 10 \\ 10 & y \end{bmatrix}$$

As this matrix is symmetric, we could request the `symmetric` indexing function to store the matrix more efficiently. In this case, it is better to use an alternate syntax for specifying the initial values. It avoids needlessly writing the off-diagonal elements twice.

```
> m := array(symmetric, 1..2, 1..2, [(1,1)=x, (2,2)=y, (1,2)=10]);
```

$$m := \begin{bmatrix} x & 10 \\ 10 & y \end{bmatrix}$$

Here is a 3×3 identity matrix.

```
> id3 := array(identity, 1..3, 1..3):
> print(id3);
```

$$\begin{bmatrix} 1 & 0 & 0 \\ 0 & 1 & 0 \\ 0 & 0 & 1 \end{bmatrix}$$

You can also use the "evaluate over matrices" command `evalm` to print a matrix. This command also lets you perform some elementary operations on matrices, like addition, multiplication, and inversion—see the table at the end of this section on page 60.

The `matrix` function in the linear algebra package `linalg` provides a simpler way to define matrices. Similarly, the `vector` function in this package makes it easier to create *vectors*, one-dimensional arrays where indexing begins at 1. To use these functions and others in this package as we type them in the examples here, first load this package with the command:

```
> with(linalg):
```

Now we define some vectors and a matrix.

```
> v1 := vector([1, 1, 1]):
> v2 := vector([1, 2, 4]):
> v3 := vector([1, 3, 9]):
> m  := matrix([v1, v2, v3]);
```

$$m := \begin{bmatrix} 1 & 1 & 1 \\ 1 & 2 & 4 \\ 1 & 3 & 9 \end{bmatrix}$$

Since we have already loaded the linear algebra package, there are several commands now available to us for operating on matrices and vectors.

```
> det(m);
```

$$2$$

```
> dotprod(v2, v3);
```

$$43$$

```
> transpose(m);
```

$$\begin{bmatrix} 1 & 1 & 1 \\ 1 & 2 & 3 \\ 1 & 4 & 9 \end{bmatrix}$$

Many functions interpret a vector as a column vector, even though it is printed horizontally.

```
> multiply(m, v1);
```

$$[3 \;\; 7 \;\; 13]$$

The following table lists some of the matrix and vector operations available in Maple. The `evalm` function and the matrix multiplication operator `&*` are always available in a Maple session; the other functions in the table are defined in the `linalg` package.

Mathematics	Maple
$A+B$	`add(A, B)` *or* `evalm(A + B)`
$\vec{u}+\vec{v}$	`add(u, v)` *or* `evalm(u + v)`
AB	`multiply(A, B)` *or* `evalm(A &* B)`
$A\vec{v}$	`multiply(A, v)` *or* `evalm(A &* v)`
cA	`scalarmul(A, c)` *or* `evalm(c * A)`
$c\vec{v}$	`scalarmul(v, c)` *or* `evalm(c * v)`
A^{-1}	`inverse(A)` *or* `evalm(A^(-1))`
A^t	`transpose(A)`
A^*	`htranspose(A)`
adj$A$	`adjoint(A)`
Tr(A)	`trace(A)`
$\lvert A \rvert$	`det(A)`
per(A)	`permanent(A)`
$A\vec{x}=\vec{b}$	`linsolve(A, b)`
$\vec{u}^t A\vec{v}$	`innerprod(u, A, v)`
$\vec{u}\cdot\vec{v}$	`dotprod(u, v)`
$\vec{u}\times\vec{v}$	`crossprod(u, v)`
Eigenvalues	`eigenvals(A)`
Eigenvectors	`eigenvects(A)`
Null space	`kernel(A)` *or* `nullspace(A)`
Rank	`rank(A)`

10.6 Table

A *table* is a data structure whose entries are indexed by arbitrary Maple expressions. In contrast to arrays, no special initialization is required for tables. The selection operator creates new entries in the table as you need them.

```
> days[Jan] := 31:
> days[Feb] := 28:
> days[Mar] := 31:
> days[FirstQuarter] := days[Jan] + days[Feb] + days[Mar];
```

$$days_{FirstQuarter} := 90$$

The command `indices` lists all of the expressions under which values are stored in a table. The `entries` command lists all of the values stored in a table. You can print an entire table with `print`.

```
> indices(days);
```

$$[FirstQuarter], [Jan], [Mar], [Feb]$$

```
> entries(days);
```

$$[90], [31], [31], [28]$$

The indices and entries are always printed in the same order, although this ordering is unpredictable.

```
> print(days);
```

table([
FirstQuarter = 90
Jan = 31
Mar = 31
Feb = 28
])

You can define higher-dimensional tables just as easily. Below we redefine `days[Feb]` to be another table, making `days` a two-dimensional table. (This erases the previous value of `days[Feb]`.)

```
> days[Feb][Leap] := 29:
> days[Feb][NonLeap] := 28:
> print(days);
```

table([
Mar = 31
FirstQuarter = 90
Jan = 31
Feb = table([
Leap = 29
NonLeap = 28
])
])

11 MAPLE FUNCTIONS

Not all of the commands and functions in Maple are immediately available when you first start a session. Some are loaded automatically as soon as you need them; others you have to load yourself whenever you want to use them. This section describes the four basic types of Maple functions. For each category, we note whether functions of that type are available when you start a session, how to read them into your session if you need to, and whether you can look at the source code for these functions.

11.1 Built-in Functions

A small subset of Maple commands are compiled into the kernel. These functions are always available. Since they are compiled into the main program itself, you cannot look at the source code for these functions. The most basic Maple functions, like `op`, `type`, and `evalf`, are built-in functions, along with a few common mathematical commands, like `max` and `diff`.

You can ask to inspect the source code of any Maple function by setting the `interface` variable `verboseproc` to 2, then calling `print`. If you ask to print a built-in function, you get a response like this:

```
> interface(verboseproc=2);
> print(op);
  proc() options builtin; 117 end
```

We see only a short header. The integer 117 is an internal identifier for the `op` function. The phrase `options builtin` identifies the `op` function as a built-in command.

Remember to reset `verboseproc` back to its original value (1) when you are no longer interested in seeing how functions are implemented.

```
> interface(verboseproc=1);
```

11.2 Standard Library Functions

Standard library functions are loaded automatically whenever you first use them in a session. Most of the common mathematical commands in Maple, including `int`, `solve`, `plot`, and `exp`, are standard library commands. Maple's practice of loading definitions for functions only when they are needed helps reduce the amount of memory required by the Maple program. To the casual Maple user, a standard library function is no different from a built-in function, since no special work is required to bring its definition into your session. But there is one important difference: you can look at the source code of a standard library function. Here we display the source code for the function `ithprime`. The large integer that occurs in the program is the product of the first nineteen primes.

```
> interface(verboseproc=2);
> print(ithprime);
  proc(i)
  local j,k;
  options remember,`Copyright 1993 by Waterloo Maple Software`;
      if type(i,integer) and 0 < i then
          if i < 20 then
              op(i,[2,3,5,7,11,13,17,19,23,29,
                     31,37,41,43,47,53,59,61,67])
          elif i < 30000 then
              for j from ithprime(i-1)+2 by 2 do
                  if igcd(j,7858321551080267055879090) = 1 and
                     isprime(j) then RETURN(j) fi
              od
          else
              k := 29999;
              for j from 350353 by 2 do
                  if igcd(j,7858321551080267055879090) = 1 and
                     isprime(j) then
                      k := k+1; if k = i then RETURN(j) fi
                  fi
              od
          fi
      elif type(i,numeric) then
          ERROR(`argument must be a positive integer`)
      else 'ithprime(i)'
      fi
  end

> interface(verboseproc=1);
```

This function, like all others in the Maple library, is written in Maple's own programming language. You may want to refer back to the examples in this section when you read Section 12, Writing Your Own Functions.

11.3 Packages

A *package* in Maple is a collection of related functions stored together in the Maple library. Maple V Release 2 and Release 3 include more than two dozen packages containing hundreds of operations and algorithms in linear algebra, combinatorics, number theory, statistics, graphics, and more. A package is really a table of functions, and the table-lookup syntax invokes a procedure in a package directly. For example, this command invokes the `fibonacci` function in the combinatorics package `combinat` to find the 73rd fibonacci number:

```
> combinat[fibonacci](73);
```

$$806515533049393$$

The `with` command loads functions from packages into your session so they behave as standard functions. The next example loads a single function from the `combinat` package, then calls it without the table-lookup syntax.

```
> with(combinat, fibonacci):
> fibonacci(73);
```

$$806515533049393$$

If there is no second argument, the `with` command loads all the functions in the package at once. This is convenient if you expect to use many of them in your session.

```
> with(combinat):
> fibonacci(73);
```

$$806515533049393$$

```
> stirling2(20, 10);
```

$$5917584964655$$

The `stirling2` function is also in the `combinat` package. It computes the so-called Stirling numbers of the second kind. Given two integers n and m, the Stirling number $\left\{ {n \atop m} \right\}$ is the number of ways to partition a set of n elements into m nonempty subsets. It is computed in terms of binomial coefficients: $\left\{ {n \atop m} \right\} = \frac{1}{m!} \sum_{k=0}^{m} (-1)^{m-k} \binom{m}{k} k^n$.

Below we print the source code for `stirling2`. You can display the source code of any function in a package.

```
> interface(verboseproc=2);
> print(combinat[stirling2]);
  proc(n,m)
  local k;
  options remember,`Copyright 1993 by Waterloo Maple Software`;
      if type(n,integer) and type(m,integer) then
          if m < 1 or n < m then 0
          else iquo(convert([seq((-1)^(m-k)*binomial(m,k)*k^n,
```

```
                                  k = 0 .. m)],`+`),m!)
            fi
        elif m = 0 then 0
        elif n = m or m = 1 then 1
        elif n = m+1 then binomial(n,2)
        elif type(n,constant) and type(m,constant) then
            ERROR(`invalid arguments`)
        else 'procname(n,m)'
        fi
    end
```

```
> interface(verboseproc=1);
```

Subpackages

A package defined inside another package is a *subpackage*. Subpackages are new in Maple V Release 3. The `stats` package in Release 3 groups nearly all its functions into six subpackages. One of these is `describe`, which contains functions that compute various descriptive statistics.

As with packages, you can invoke a function in a subpackage with the long table-lookup syntax:

```
> stats[describe, mean]([2,4,9]);
```

$$5$$

After loading the package, you must still use the table-lookup syntax to access a function in a subpackage.

```
> with(stats):
> describe[mean]([2,4,9]);
```

$$5$$

Once the package is loaded, you can load selected functions from a subpackage or the entire subpackage at once using `with`, just as with packages.

```
> with(describe):
> mean([2,4,9]), median([2,4,9]);
```

$$5, 4$$

11.4 Miscellaneous Library Functions

A few Maple library functions must be read in explicitly with `readlib` before you can use them. These are the miscellaneous library functions. One such function is `C`, which translates expressions to a format usable in a C language program. If this function were part of the standard library, you would not be able to use `C` as a symbol in your mathematical expressions. An example of C translation appears in Section 13.4, Translating Expressions.

Another miscellaneous library function is `iscont`, which tests whether a given expression is continuous on a given interval. We call this function by issuing the commands:

```
> readlib(iscont):
> iscont(tan(x), x=0..Pi);
```

false

You can inspect the source code of any miscellaneous library function. The `readlib` function prints the source code if `verboseproc` is set to 2.

11.5 Maple Initialization Files

If you expect to use a particular package or miscellaneous library function regularly in the future, you may wish to add the appropriate `with` or `readlib` command to your personal Maple initialization file. In fact, you can put any sequence of Maple statements into this special file. For example, many people like to invoke the miscellaneous library function `history` to number their output lines in every session.

The name and location of the personal Maple initialization file on several computer systems are listed in the following table.

Computer	Personal Initialization File	Location
DOS	MAPLE.INI	Current directory
Macintosh	MapleInit	System folder
Unix	.mapleinit	Home directory
VMS	MAPLE.INI	Current directory

Most computers also have a system initialization file containing Maple commands that are executed each time anyone on the system starts Maple. It is usually located inside the Maple installation directory, often in the `lib` subdirectory.

Computer	System Initialization File	Location
DOS	MAPLE.INI	Maple LIB subdirectory
Macintosh	MapleInit	Maple folder
Unix	init	Maple lib/src subdirectory
VMS	MAPLE.INI	Maple LIB subdirectory

12 WRITING YOUR OWN FUNCTIONS

Writing a function to compute a simple formula is quite easy in Maple. It is also possible to write complicated programs with Maple's powerful programming language. Maple's own library is written in this language. We discuss two main methods for writing your own functions. We also discuss some techniques for finding and fixing errors in your code, and for measuring and improving the performance of your Maple programs. The last section gives some pointers to some more advanced topics discussed elsewhere in this guide.

12.1 Arrow and Angle Bracket Functions

The simplest way to define a new function is with the arrow operator, `->`. It is easy to define a mathematical function with this operator.

```
> f := x -> x^3 + x + 1;
```

$$f := x \to x^3 + x + 1$$

This defines a function `f` that takes a single argument and computes its image under the given polynomial. You can use `f` as you do any Maple function.

```
> f(3);
```

$$31$$

```
> fsolve(f(z)=0, z, complex);
```

$$-.6823278038, .3411639019 - 1.161541400I, .3411639019 + 1.161541400I$$

```
> int(f(x), x=0..1);
```

$$\frac{7}{4}$$

```
> D(f);
```

$$x \to 3x^2 + 1$$

Notice that the differential operator `D` applied to `f` returns a new function that computes the derivative of `f`.

You can also define a function of two or more variables with the arrow operator.

```
> g := (s, t) -> sqrt(s^2 + t^2);
```

$$g := (s, t) \to \text{sqrt}\left(s^2 + t^2\right)$$

```
> g(105, 208);
```

$$233$$

The `unapply` function applied to a mathematical expression returns an arrow function that computes the expression. For example,

```
> unapply(sin(2*Pi*x) + 5*x^3 - 3, x);
```

$$x \to \sin(2\pi x) + 5x^3 - 3$$

```
> unapply(sqrt(r^2 + s^2 + t^2), r, s, t);
```

$$(r, s, t) \to \sqrt{r^2 + s^2 + t^2}$$

You can also use an angle-bracket notation to define functions like this. Our functions `f` and `g` in this section could be defined as:

```
>  f := < x^3 + x + 1 | x >:
>  g := < sqrt(s^2 + t^2) | s, t >:
```

The differential operator applied to an angle-bracket function returns a function in the same notation.

```
> D(f);
```

$$< 3x^2 + 1 \mid x >$$

12.2 Procedures

You can write more complicated procedures by using the `proc` ... `end` syntax. The general form of a procedure definition in Maple is:

```
proc(formal-parameters)
    local local-variable-names;
    global global-variable-names;
    options options;
    statement-sequence
end;
```

The *statement-sequence* forms the body of the procedure. The `local`, `global`, and `options` parts are all optional. The `local` and `global` lines declare local and global variables. A local variable is known only within the body of the procedure. Its value is independent of any variable with the same name defined outside the procedure. A global variable is known throughout your Maple session. The `global` declaration is new in Maple V Release 3. In Release 2, all undeclared variables are assumed to be global.

In the rest of this section, we give a brief introduction to procedural programming in Maple through several examples. For additional examples, you may want to refer to the Maple source code printed in Section 11 on page 62, or to the examples in Chapter 6 (page 434).

First, we write a procedure to compute the same polynomial defined in the previous section.

```
> f := proc(x)
>   x^3 + x + 1
> end:
```

This defines `f` to be a function of a single variable. The `x` is a formal parameter for the procedure `f`: it is a placeholder for whatever actual argument `f` is called with. The value of the polynomial at this argument is the value returned by `f`. No semicolon is required after the polynomial, since it is the last (and indeed the only) statement in the body of the procedure. You can use `f` now as you do any Maple function.

Next, we write a function that appends an element to a list. It takes two arguments: the original list, and the element to be added.

```
> append := proc(l:list, e)
>     [op(l), e]
> end:
```

Here, the first argument, `l`, must be a list (this type checking feature was introduced in Maple V Release 2). We return a new list whose elements are all the operands of `l`, followed by the new element `e`. Because of the way Maple handles parameter passing, we cannot change the value of `l` inside the procedure. When a procedure is invoked, Maple evaluates the actual arguments supplied and replaces each occurrence of a formal parameter in the body of the procedure with the value of the corresponding actual argument.

```
> append([1, Pi, x], y^3);
```

$$[1, \pi, x, y^3]$$

```
> append(13, 4);
Error, append expects the 1st argument, l, to be of type list,
but received 13
```

Finally, we include a more complicated procedure that generalizes the final example given in the section on iteration (page 51). Given two positive integers n and m that have no common factors, we search for the smallest prime p that is congruent to n modulo m. That is, dividing p by m leaves a remainder n. Rephrasing the earlier example in this language, we found that 421 is the smallest prime congruent to 21 modulo 100. A prime satisfying the congruence always exists, by a theorem of Dirichlet.

```
> # Search for the smallest prime congruent to n mod m.
> dirichlet := proc(n:posint, m:posint)
>     local p;
>     if gcd(n, m) > 1 then
>         ERROR(`` . n . ` and ` . m . ` have a common factor.`)
>     else
>         for p from (n mod m) by m while not isprime(p) do od;
>         RETURN(p)
```

```
>     fi
> end:
```

The first line of the program is a comment. Maple ignores everything that comes after a # character on a line. Adding comments to your programs makes them easier to read and understand. We use a variable p inside the body of the procedure. It is declared `local` so that it cannot alter a symbol p that we may have already defined outside the procedure. We also use the ERROR statement to abort execution when an error arises. The displayed error message is built by using Maple's string concatenation operator. A RETURN statement makes the return value of the procedure explicit.

Try this procedure on a few sample inputs. The first one below re-creates the earlier example in the do loop section.

```
> dirichlet(21, 100);
```

$$421$$

```
> dirichlet(215, 1000);
Error, (in dirichlet) 215 and 1000 have a common factor.
> dirichlet(1024, 99999);
```

$$501019$$

For more examples of Maple programs, you can look at the package described in Chapter 6 beginning on page 434, inspect more code in the Maple library, or pick up one of the Maple books listed in Chapter 8 (page 465).

12.3 Debugging

When your Maple code does not give the result you expect, asking the following questions may help you isolate an error.

- Did you assign a value to one of the variables earlier in your session?
- Did you inadvertently assign a value to a Maple function or global variable?
- Have you loaded all required packages and miscellaneous library functions?
- Do all function calls have the correct number and type of arguments?
- Are your indexing variables distinct?
- Have you forgotten any punctuation?
- Does every parenthesis, bracket, and brace have a mate?
- Are all commands spelled correctly, with the correct capitalization?
- Does each statement work as intended?

The Mint program, distributed with most versions of Maple, is intended to help you find errors in your Maple programs. It is similar to the Unix program `lint` for detecting syntax errors in C programs. When you run the Mint program on a file containing your Maple program, it reports syntax errors and announces warnings of varying degrees of severity, including:

- Nested loops using identical indexing variables
- Local variable declared but never used

- Variable used but not declared local
- Wrong number of arguments in a procedure call
- System-defined name obscured by user-defined name
- Equals sign (=) used where assignment (:=) is probably intended

Mint can also generate a report summarizing how all the variables are used in your program.

The `debug` command traces the execution of specific procedures. After the command `debug(`*procedure-name*`)` has been entered, all the intermediate results computed by *procedure-name* are displayed whenever that procedure is invoked. Turn off tracing with `undebug(`*procedure-name*`)`. In Maple V, these commands are named `trace` and `untrace`. The names were changed because the linear algebra function `trace` obscured the debugging command `trace` whenever the `linalg` package was loaded.

The global variable `printlevel` determines the depth of intermediate results printed to the screen. Statements entered directly to a Maple session have depth 0. A procedure adds 5 to the depth level; a `do` or `if` block adds 1. The default value of `printlevel` is 1, so only results of top-level statements and statements inside a single `do` or `if` block are printed. For debugging, set this variable to a large integer to see the intermediate results in every procedure you call to a great depth. Setting `printlevel` to 1000 is not uncommon. If you set `printlevel` to 0, on the other hand, only the results of top-level statements are printed. A `printlevel` of -1 suppresses printing of all results.

The global variable `infolevel` lets you print descriptive messages that are encoded in Maple procedures with `userinfo` statements. These messages provide varying degrees of detail about functions being called and algorithms being applied. To see the messages for the function *fcn*, set `infolevel[`*fcn*`]` to a number between 1 and 5. The larger this number, the more detail you get. To see messages for all functions you invoke, set `infolevel[all]`. Here is an example of this feature.

```
> infolevel[factor] := 2:
> factor(x^2 + 2*x + 1);
  factor/polynom:   polynomial factorization: number of terms   3
  factor/quadfact:   applying the quadratic formula in   x
```

$$(x + 1)^2$$

You can put `userinfo` statements in your own programs, too. See page 434 in Chapter 6 for an example.

Explicit `print` commands in your program may also help you track down an error. For details on print commands, see Section 13.1, Printing Expressions, on page 73.

12.4 Timing and Optimizing

Use the `showtime` function to see the time and memory required by each Maple command you enter. Like `history`, `showtime` also saves your results in consecutively numbered variables. It is a miscellaneous library function, so you must load it first.

```
> readlib(showtime):
> showtime();
O1 := ifactor(20! + 1);
```

$$(117876683047)(20639383)$$

```
time   7.00   words   356635
O2 :=
```

Use the command `off` to turn this feature off.

You can *profile* your Maple procedures to determine which ones are using the most resources. With this information, you can optimize the ones that require the most time and space. Use the `profile` command to start collecting data on selected procedures. After some sample calls to these procedures, look at the collected data with the command `showprofile`. The `unprofile` command turns off profiling. Before using these functions you must load them with the command `readlib(profile)`.

Adding the `remember` option to a procedure may improve its performance considerably. With this option, the procedure "remembers" its value on each input you give it. Subsequent calls to the procedure with the same input need not be recomputed. They need only be recalled from a table of past results. For example, if a procedure `myprog` contains the declaration `options remember` and we have computed `myprog(10)` already, then re-executing `myprog(10)` simply returns the answer that was already computed. The savings is especially significant if the procedure is recursive: for example, if computing `myprog(11)` depends on the value of `myprog(10)`. Of course, this savings comes only at the expense of a larger memory requirement. Many Maple functions, including the two whose source we displayed earlier (`ithprime`, page 63, and `combinat[stirling2]`, page 64), use the `remember` option.

12.5 Advanced Topics

Some more advanced topics in Maple programming are discussed in Chapter 6, Frequently Asked Questions. For a description of Maple's scoping rules, see page 433. For information on creating help text for your programs, see page 439. For help on creating your own packages of functions and libraries of Maple routines, see page 434.

13 INPUT AND OUTPUT

In this section, we describe the different ways to format Maple expressions, some methods for reading commands from a file and saving results to a file, how to read data into a Maple program, and how to translate Maple expressions to formats suitable for use in other applications.

13.1 Printing Expressions

There are three basic forms of Maple output. The kind of output you see by default is given by the value of the `interface` variable `prettyprint`. When `prettyprint` is 0, Maple output is printed in a linear format. You can save results in this format and use them as input to Maple later.

```
> interface(prettyprint = 0):
> Int(x^7 / 7 + x^(a^3), x=sqrt(2)..Pi);
  Int(1/7*x^7+x^(a^3),x = 2^(1/2) .. Pi)
```

When `prettyprint` is 1, Maple output is printed in a multiline format. Exponents appear on separate lines, and symbols like integral signs and matrix brackets are approximated with text characters.

```
> interface(prettyprint = 1):
> Int(x^7 / 7 + x^(a^3), x=sqrt(2)..Pi);
                          Pi
                          /
                         |                    3
                         |          7       (a )
                         |   1/7 x   + x          dx
                        /
                        1/2
                       2
```

When `prettyprint` is 2, its default value, Maple output is printed in standard mathematical notation, with integral and sum signs, genuine matrix brackets, Greek letters, and square root signs.

```
> interface(prettyprint = 2):
> Int(x^7 / 7 + x^(a^3), x=sqrt(2)..Pi);
```

$$\int_{\sqrt{2}}^{\pi} \frac{1}{7}x^7 + x^{\left(a^3\right)}\,dx$$

There are three functions for printing expressions explicitly. The linear print function `lprint` prints an expression in the linear format that is suitable for Maple input. The `print` function prints an expression in the format indicated by the value of `prettyprint`. The `printf` function prints formatted output and is similar to the `printf` statement in C programs. It was introduced in Maple V Release 2. These functions are often called from Maple programs to print information about the state of a computation.

13.2 Saving to Files

The `save` command saves variable names in a file in the form of assignment statements. Values are saved in a format suitable for input to Maple. You can name particular variables that you want to save, or you can save all the variables that have been assigned a value in your session. The syntax of the **save** command and all the commands discussed in this section is given in the table below. Section 7.3 on page 42 has an example of a **save** command used to save the description of a graph in a file.

Always enclose file names in back quotes so that Maple does not interpret them as expressions. This is particularly important if your file name contains a special character like slash (/), period (.), or dash (-). If your file name ends with ".m", Maple saves the data in a special binary format. This format is not readable by people, but Maple reads files in this format faster than it reads a plain text file.

There are also Maple commands for saving individual expressions. The open command opens a file for output. The write and writeln commands write expressions to the file. Results are saved in Maple input format. The close command closes the output file. These functions are miscellaneous library functions and must be loaded with the command readlib(write) before they can be used.

The functions writeto and appendto echo all subsequent input commands and print all subsequent output to a file. You can make a transcript of a Maple session with these commands. Use the special file name terminal to restore printing to the screen.

The worksheet interface includes a **Save** command under the **File** menu that saves your entire Maple worksheet. See Chapter 2, page 85.

These commands are summarized in the following table.

Function	Description
save `*file*`	Save all assigned variables in *file*
save `*file*.m`	Save all assigned variables to the binary file *file*.m
save *x*, `*file*`	Save the assignment of *x* in *file*
save *varseq*, `*file*`	Save the assignments of the sequence of variables named in *varseq* in *file*
open(`*file*`)	Open *file* for output
write(*expr*)	Write *expr* to the output file
writeln(*expr*)	Write *expr* followed by a newline to the output file
close()	Close the output file
appendto(`*file*`)	Append all subsequent results to *file*
writeto(`*file*`)	Write all subsequent results to *file*
writeto(terminal)	Restore printing of results to the screen
Save (on **File** menu)	Save entire worksheet

13.3 Reading from Files

The read command reads Maple definitions and statements from a file. You can save a Maple program in a file and load it later with this statement. If the file name ends with ".m", Maple assumes it has the special binary format.

There are several commands available for reading data into a Maple session or program. The readstat command prints a prompt, then evaluates the Maple statement entered by the user. It returns the value of this statement.

The readline command reads a line of text from a file or from the terminal. This text does not have to be a Maple statement. Use the sscanf function to pick out individual fields in the string returned by `readline`, or the `parse` statement to evaluate the string as a Maple expression. These three commands were introduced in Maple V Release 2.

The `readdata` command reads columns of numbers from a file and returns a list of lists containing these data. You can specify whether these numbers are integers or floating-point numbers. For example, suppose the file `datafile` contains four columns of integers:

```
1  1   1   1
1  2   3   4
1  3   6  10
1  4  10  20
1  5  15  35
```

The following command reads these numbers as integers into a list of lists:

```
> readlib(readdata):
> readdata(`datafile`, integer, 4);
```

$$[[1, 1, 1, 1], [1, 2, 3, 4], [1, 3, 6, 10], [1, 4, 10, 20], [1, 5, 15, 35]]$$

Without the `integer` argument, or with `float` in its place, the numbers are read as floating-point numbers:

```
> readdata(`datafile`, 4);
```

$$[[1., 1., 1., 1.], [1., 2., 3., 4.], [1., 3., 6., 10.], [1., 4., 10., 20.], [1., 5., 15., 35.]]$$

The `readdata` function was introduced in Maple V Release 2. It is a miscellaneous library function.

Finally, under the worksheet interface, the **Open** command in the **File** menu reads a saved Maple worksheet. See Chapter 2, page 85.

These commands are summarized in the following table.

Function	Description
`read` `*file*`	Read the contents of *file*
`read` `*file*.m`	Read the contents of the binary file *file*.m
readstat(*string*)	Prompt the user with *string* and return the result of the Maple statement entered by the user
readline(`*file*`)	Read the next line in *file* and return this string
readline(terminal)	Read the next line from the standard input and return this string
sscanf(*string*)	Pick out fields in *string*
parse(*string*)	Convert *string* to a Maple expression
parse(*string*, statement)	Convert *string* to a Maple expression and evaluate it
readdata(`*file*`, *type*, *c*)	Read *c* columns of data of type *type* from *file*, returning a list of lists
Open (on **File** menu)	Read worksheet

13.4 Translating Expressions

Maple can translate expressions to several other formats. You can translate a Maple expression to a form suitable for use in a C or FORTRAN program, or in a LaTeX or troff document.

Command	**Translation**
`C(`*expr*`)`	C programming language
`fortran(`*expr*`)`	FORTRAN programming language
`latex(`*expr*`)`	LaTeX typesetting input
`eqn(`*expr*`)`	eqn notation for troff documents

Below we translate a Maple expression into C, FORTRAN, LaTeX, and eqn. This expression provides an excellent approximation to the harmonic number $1/1 + 1/2 + \cdots + 1/n$.

```
> h := n -> ln(n) + gamma + 1/(2*n) - 1/(12*n^2) + 1/(120*n^4);
```

$$h := n \rightarrow \ln(n) + \gamma + \frac{1}{2n} - \frac{1}{12n^2} + \frac{1}{120n^4}$$

The function `C` is a miscellaneous library function, so we must load it first.

```
> readlib(C):
> C(h(n));
  t0 = log(n)+0.5772156649015329+1/n/2-1/(n*n)/12+1/pow(n,4.0)/120;
```

By default, the `fortran` routine translates expressions to single-precision FORTRAN statements. If you set the global variable `precision` to `double`, double-precision code is generated.

```
> precision := double:
> fortran(h(n));
  t0 = dlog(n)+0.5772156649015329D0+1/n/2-1/n**2/12+1/n**4/120
```

```
> latex(h(n));
  \ln (n)+\gamma+{\frac {1}{2n}}-{\frac {1}{12n^{2}}}+
  {\frac {1}{120n^{4}}}
```

```
> eqn(h(n));
  {{ln (  "n" )}~+~gamma~+~{ 1 over {2 ~ "n" }}~-~{ 1 over
  {12 ~{  "n" sup 2 }}}~+~{ 1 over {120 ~{  "n"  sup 4 }}}}
```

14 EXERCISES

We hope that that you feel comfortable using Maple to perform numerical and symbolic computations, manipulate expressions, plot functions and data, and write simple functions. We recommend working the following exercises for practice with Maple.

1. Use on-line help to find the following.

   (a) List all the commands that begin with the letter `a`.

   (b) Find the names of all the packages that are distributed with Maple.

   (c) List the names of the functions defined in the `linalg` package.

2. Compute $262537412640768744 - e^{\pi\sqrt{163}}$ to 10, 20, 30, 35, and 40 places of precision.

3. Which is larger, π^e or e^π?

4. Verify the following identities:

   (a) $27^5 + 84^5 + 110^5 + 133^5 = 144^5$

   (b) $2682440^4 + 15365639^4 + 18796760^4 = 20615673^4$

   (c) $95800^4 + 217519^4 + 414560^4 = 422481^4$

   These identities disprove a conjecture that Leonhard Euler made in 1769. The first was discovered by L. J. Lander and T. R. Parkin in 1966, the second by N. D. Elkies in 1988, and the third by R. E. Frye in 1988.

5. Find the real root between 1 and 2 of the polynomial

   $$f(x) = x^{10} + x^9 - x^7 - x^6 - x^5 - x^4 - x^3 + x + 1$$

   Find all the real roots. Find all ten roots in the complex plane. Put these roots in a list and `map` the absolute value function `abs` onto each element. What do all of the complex roots have in common?

6. Consider the function

   $$f(x) = \frac{2x}{e^x - e^{-x}}$$

   (a) What is $\lim_{x\to 0} f(x)$?

   (b) Try to compute the definite integral $\int_0^1 f(x)\,dx$. First set `infolevel[int]` to 1 in order to see what the integrate function is doing.

   (c) Compute a Taylor series for $f(x)$ about $x = 0$ to order 10.

   (d) Integrate this series with respect to x.

   (e) Convert this expression to a polynomial.

   (f) Substitute 1 for x in the polynomial. We've just computed an approximation to the integral using the Taylor series of $f(x)$.

   (g) Use `evalf` and the inert function `Int` to force numerical evaluation of $\int_0^1 f(x)\,dx$. How does this compare to your result in part 6f?

(h) Use `int` to compute the integral

$$\int_0^1 \frac{x}{\sinh x} dx$$

Notice that the integrand is exactly the same as $f(x)$. Do you get a different answer? Does `int` call any routines that it didn't call the first time?

7. (a) Solve the differential equation

$$y'' + y' - 6y = 20e^x$$

subject to the initial conditions

$$y(0) = 5, y'(0) = 10$$

(b) Graph the solution on the interval $[-1, 1]$. Approximately where does this function attain a minimum?

(c) Plot the curve on a smaller interval, like $[-0.5, 0]$, to get a better idea of the location of this minimum.

(d) Find the approximate x-value of the minimum by setting the derivative of the function to zero and solving for x. Use `fsolve`, supplying an interval where you think the minimum occurs.

(e) Substitute this value back into the original function to find the minimum value of $y(x)$.

8. Load the linear algebra package `linalg` into your session.

(a) Define M to be a 5×5 matrix with the binomial coefficient $\binom{i-1}{j-1}$ in the ith row and jth column. Use the command:

```
> M := matrix(5, 5, (i,j) -> binomial(i-1,j-1));
```

(b) Display the entries in the matrix M graphically with the command:

```
> plots[matrixplot](M, heights=histogram, gap=0.3,
>     orientation=[-150,60]);
```

(c) Define another matrix S by multiplying M and its transpose, $S = MM^t$. (S is a *Pascal matrix.*)

(d) What is the determinant of S? Compute the inverse of S. Test whether S is positive definite with the `definite` function.

(e) Calculate the eigenvalues and eigenvectors of M. Use the functions `eigenvals` and `eigenvects` in the `linalg` package.

(f) Compute the characteristic polynomial of S with the `charpoly` function. Solve for the roots of this polynomial to find the eigenvalues of S.

(g) Define a vector $\vec{v}$ of five integers. Use `linsolve` to find the vector $\vec{x}$ that satisfies the matrix equation $S\vec{x} = \vec{v}$.

(h) Calculate the permanent of S with the `permanent` function. The *permanent* of a matrix is defined in a way similar to the determinant, but there is no alternating sign in the minor expansions.

9. (a) Plot the function $\sin(2\pi x) + x$ for x in the range $[0, 47]$.

   (b) Make the same graph with the option `numpoints` set to 60, and again with `numpoints` set to 29.

10. Enter the assignments
```
> x := 1:
> y := 2:
```
Now enter
```
> 'x + y';
```
What value is returned? Why? Now enter
```
> ";
```
What happens? Why? Then enter
```
> `x + y`;
```
followed by
```
> ";
```
Now what happens? Why?

11. In a town there are exactly three toy stores selling 5 popular toys. Store A sells the toys for \$15, \$17, \$18, \$32, and \$29. Store B sells the same 5 toys for \$14, \$18, \$22, \$29, and \$26. Store C always matches the best price in town. Write a Maple expression to determine store C's selling prices for the 5 toys. (Try writing a `for` loop first, then look up the `zip` command.)

12. Load the miscellaneous library function `finance`.

    (a) Compute the monthly payment required on a three-year, \$12,000 loan at an annual interest rate of 8%, compounded monthly:
```
> finance(amount=12000, interest=.08/12, periods=36);
```

    (b) Compute the amortization schedule for this loan. Use the function `amortization`. It is available whenever `finance` is loaded.
```
> amortization(12000, .08/12, payment);
```
Here, *payment* is the required payment calculated in part 12a.

13. Why can't `op` parse an expression sequence?

14. Write a function called `first` that returns the first element of a given list.

15. There are five ways to partition the number 4 as a sum of positive integers, disregarding the order of the terms: $1+1+1+1$, $1+1+2$, $1+3$, $2+2$, and 4. The `partition` function in the `combinat` package returns a list of all the possible partitions of an integer.

    (a) Load the `combinat` package, then test this function by entering:
```
> partition(4);
```

(b) Find all the partitions of 8. How many are there? If you don't mind several dozen lines of output, try 16.

(c) The `numbpart` function in the same package computes the number of partitions of an integer. Use this function to calculate the number of partitions of 16, 32, and 64.

(d) It is known that for large n, the number of partitions of n is approximately:

$$\frac{e^{\pi\sqrt{2n/3}}}{4n\sqrt{3}}$$

Define a function $p(n)$ that computes this value.

(e) How close is the approximation at $n = 64$? At $n = 128$?

(f) Use `plots[logplot]` to graph $p(n)$ on the interval $[1, 128]$ with a logarithmic scale on the vertical axis.

15 STUDENT AND ACADEMIC EDITIONS

There are two editions of Maple available: the student edition and the academic edition. Both come with the entire Maple library in Maple V Release 2 and Release 3, including all the packages of specialized mathematical functions. In the student edition, the library is stored in a compressed format to conserve disk space. The student edition runs just as fast as the academic edition.

The academic version is available on a wide variety of computers, including personal computers, workstations, mainframes, and supercomputers. The student version is currently available only on DOS personal computers (with or without Microsoft Windows) and Apple Macintosh personal computers.

The student edition is attractively priced, but imposes certain limits that are appropriate to student problems. These limits are summarized in the table below.

Limits of Maple V Release 2 Student Edition	
Memory usage	4 megabytes
Floating-point precision	100 digits
Array dimensions	3
Total array/matrix size	5120 items
Size of sequence	16382 terms
Size of sum or product	8191 terms

The student edition also comes with fewer contributed programs from the Maple share library. However, these programs are freely available from electronic sources (see Chapter 5, page 407).

The academic edition has much weaker restrictions.

Limits of Maple V Release 2 Academic Edition	
Memory usage	No limit
Floating-point precision	500000 digits
Array dimensions	No limit
Total array/matrix size	No limit
Size of sequence	131070 terms
Size of sum or product	65535 terms

16 HOW TO ORDER MAPLE

For more information on Maple, or to order the software, visit your local software dealer or contact a distributor. Waterloo Maple Software is the primary distributor of Maple, and Brooks/Cole is the exclusive distributor of the student edition of Maple within the United States. Outside the United States, several companies distribute the student edition, including Waterloo Maple Software, Springer-Verlag, and Brooks/Cole's International Thomson affiliates. The academic edition is available from all of these suppliers anywhere in the world. MathSoft distributes Maple primarily in the corporate world.

Waterloo Maple Software
450 Phillip Street
Waterloo, Ontario
Canada N2L 5J2
Email: `info@maplesoft.on.ca`
Telephone: 1-800-267-6583
Telephone: (519) 747-2373
Fax: (519) 747-5284

Brooks/Cole Publishing Company
511 Forest Lodge Road
Pacific Grove, CA 93950-5098
Email: `info@brookscole.com`
Telephone: (408) 373-0728
Fax: (408) 375-6414

MathSoft, Inc.
101 Main Street
Cambridge, MA 02142
Telephone: 1-800-MathCAD
Telephone: (617) 577-1017

Springer-Verlag New York, Inc.
175 Fifth Avenue
New York, NY 10010
Telephone: (212) 460-1500
Fax: (212) 473-6272

Chapter 2
User Interface

In this chapter, we describe the two main versions of Maple's user interface: the worksheet interface and the command-line interface. In Maple V Release 2 and Release 3, the worksheet interface is available on all computers with sufficient graphics capabilities. These include workstations and graphics terminals running the X Window System, PCs running MS-Windows, and NeXT and Macintosh computers. The command-line interface is used on simpler text terminals and DOS computers without MS-Windows.

1 THE WORKSHEET INTERFACE

A Maple worksheet lets you combine Maple commands and results, explanatory text, and graphics in a single document. Worksheets let you create interactive mathematical tutorials, and let you make high-quality printouts of your Maple session without translating to a document preparation system like $\TeX$. In fact, several articles published in the *Maple Technical Newsletter* have been produced by Maple worksheets.

The worksheet user interface varies somewhat on different computer platforms. This arises in part because worksheets are designed to conform to the standards or conventions of the host computer. For example, the X worksheet follows the Motif interface; the MS-Windows version looks like other Windows applications. Also, a keyboard shortcut for selecting a menu item may use the COMMAND key on a Macintosh or NeXT, the CONTROL or ALT key on an X workstation, or a function key on a PC.

This section describes the worksheet interface as it appears under the X Window System. A few nice features that are present in some other versions, notably MS-Windows, are also mentioned. The first few parts of this section describe the major components of the worksheet interface: the worksheet window, graphics windows, and help windows. The last section describes several X resources you can set to alter the appearance, behavior, and initial option settings of the worksheet interface under X.

1.1 Worksheet Window

When you begin a Maple session in a worksheet, you receive a prompt and may start entering Maple commands immediately. Results appear in a nice mathematical style below each input command. Using the mouse or the arrow keys makes it easy to correct mistakes and to copy portions of old commands into new ones. Figure 2.1 shows a sample session under the X Window System.

A worksheet is divided up into a number of *regions*. There are five different kinds of regions: input, output, text, graphics, and separator. Input regions contain Maple commands. Entering a command in an input region creates an output region holding the result of the computation. Text regions contain explanatory text and comments, and

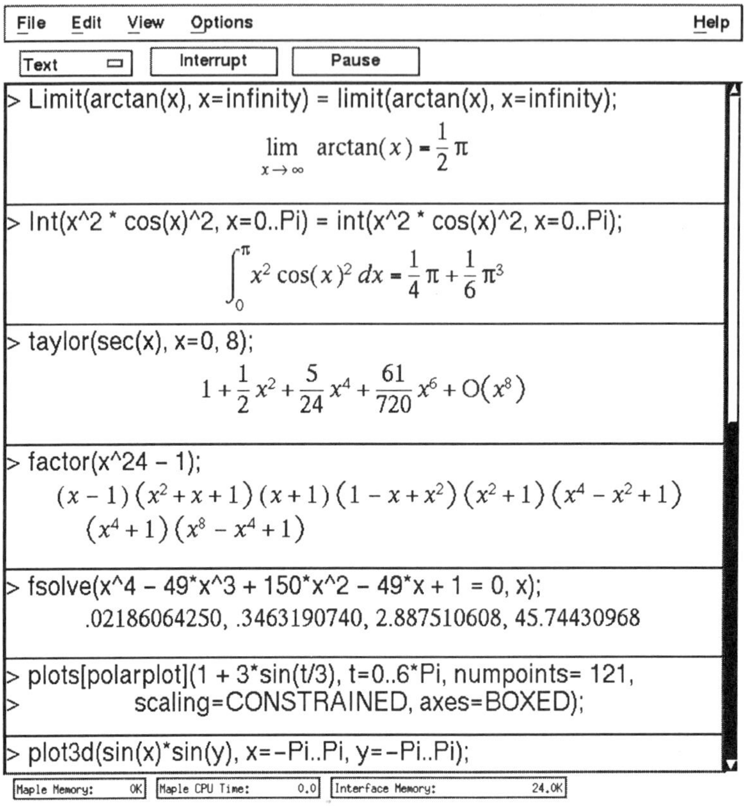

Figure 2.1: The worksheet interface

graphics regions contain plots. A separator is a thin line that divides an output region from the next input region.

To add some explanatory text to a worksheet, change the current region from an input region to a text region when you are at a new input prompt. To do this under X, click on the leftmost button in the button bar and select **Text**. Under MS-Windows, press F5, or select the **Text Region** command in the **Format** menu. To end a text region and return to input mode, find a command called **Insert Prompt** in the **Edit** menu or **Insert New Region** in the **Format** menu, then if necessary change the new region to an input region.

Graphics output in the worksheet interface appears in a separate window. To add graphics to a worksheet, select the **Copy** command in the **Edit** menu on the graphics window, then select **Paste** in the worksheet window to add it to your worksheet just below the current cursor location.

The following sections describe the various menus and buttons in the worksheet interface as they appear in the X Window System. Some commands that are present in other versions, notably MS-Windows, are also mentioned. Your interface may vary somewhat from what is presented here. Consult the "Getting Started" manual that came with your system or peruse the on-line help for more specific information on your user interface.

Buttons

There are three buttons just below the menu bar in the main worksheet window. The first indicates the type of the current region. Pressing this button lets you convert the region to another type. For example, you can convert Maple output into plain text, or plain text to Maple input.

The second button, **Interrupt**, interrupts the current calculation. You can use this as an alternative to the interrupt key.

The third button, **Pause**, pauses the display of several lines of output from a loop or a sequence of statements. Click on **Pause** again to resume printing results to the screen.

File Menu

The **File** menu handles loading and saving of worksheets, printing, and quitting.

New
: Make a new worksheet, erasing your current one entirely. Maple asks for confirmation before complying with this request.

Open
: Open a previously saved worksheet, or ".ms" file. You are given the opportunity to load the kernel state at the same time. This is assumed to be in a ".m" file with the same name as the ".ms" file loaded.

Save
: Save the worksheet. If the worksheet is new and has no name, **Save As** is executed instead.

Save As
: Prompt for a name and save the worksheet under this name. This should be a ".ms" file. You are given the opportunity to save the kernel state as a ".m" file, too.

  Under MS-Windows, you can save in the worksheet format by specifying a ".ms" name, or plain text format with a different name. Saving the kernel state is done by turning on the **Save Session State** option in the **Option** menu.

Export As LaTeX
: Save the worksheet in a form that can be processed by LaTeX. Special environments format input regions, hi-res output regions, and character-based output regions. Text regions format as paragraphs. Graphics regions are ignored, but a LaTeX macro is provided for inserting PostScript files. Style files located in the `etc` subdirectory of the main Maple directory contain definitions for the environments and macros used. This command is new in Maple V Release 3.

  Under MS-Windows, this command is an option in the **Save As** dialog box.

Export As Text
: Save the worksheet as a text file. Under MS-Windows, this command is an option in the **Save As** dialog box.

Import Text

Import a text file into the worksheet. If the file is a saved Maple worksheet in plain text form, make sure that the indicator for "Use Maple prompts to identify input" is on. If the file is a sequence of Maple commands you have created yourself, turn this indicator off.

Include

Read a text file into a region in your worksheet. If the region is a text region, the information is formatted as text; if the region is an input region, it is treated as Maple commands. In the latter case, pressing ENTER after loading the file evaluates the input commands you have read in.

Print

Save the worksheet in a format suitable for printing, such as PostScript. You may specify the dimensions of the page and the size of the margins.

Under MS-Windows, **Print** prints the worksheet directly, **Printer Setup** selects paper size and orientation, and **Page Margins** sets page margins for printouts.

Exit

Quit the session, discarding any unsaved work. Maple requests confirmation when this command is selected.

Edit Menu

Many of the commands on the **Edit** menu operate on text that has been selected with the mouse or a special keystroke. To select some part of a worksheet, click with the (left) mouse button at the beginning of the area you want to select, and drag the mouse to the end of the area, releasing the button at the end. The selected material is highlighted. You may select any portion of an input region or a text region, but you may select only an entire graphics region or "hi-res" output region. To select part of an output region, change its format to the text-based pretty print style first.

You can also select regions using the keyboard. Under X, pressing CONTROL-G selects the entire region where the cursor is currently positioned.

Cut

Delete whatever is currently selected from the worksheet and move it into a special buffer called the clipboard. The clipboard is not displayed on the screen.

Copy

Copy whatever is currently selected in the worksheet into the clipboard. The original material remains in the worksheet.

Paste

Copy whatever is currently in the clipboard into the worksheet at the current cursor position. The copied material remains in the clipboard, so you can paste the same information several times without reloading the clipboard.

Delete Selection

Delete the selected material from the worksheet without copying it to the clipboard.

Delete Cursor Region
Delete the entire region where the cursor is presently positioned. The material is not copied to the clipboard.

Delete By Type
Bring up a submenu that lets you delete every region of a given type in the worksheet: input, output, text, graphics, or separators. These regions are not copied to the clipboard.

Under MS-Windows, this command is called **Remove All** and resides in the **Format** menu.

Delete to End of Session
Delete every region below the one where the cursor is currently positioned. Again, this material is not saved in the clipboard.

Insert Prompt
Insert a prompt just below the region where the cursor is positioned. Use this command to end a text region and begin typing Maple commands again.

MS-Windows has an **Insert New Region** command in the **Format** menu for adding a new region either above or below the current region.

Insert Text
Insert a text region just below the region where the cursor is positioned.

Insert Separator
Insert a separator just below the region where the cursor is positioned.

Split Region at Cursor
Split an input or text region into two regions at the location of the cursor. Under MS-Windows, this command is in the **Format** menu.

Join Region at Cursor
Joins two adjacent input or text regions: the current region and the one immediately below it. Under MS-Windows, this command is in the **Format** menu.

View Menu

Reformat Cursor Region
Reformat the output region containing the cursor according to the formatting options currently set in the **Options** menu.

Reformat Session
Reformat all output regions according to the formatting options currently set in the **Options** menu.

Show Regions
Turning this switch on draws thin lines or dotted lines between each region in the worksheet. Turning it off removes these lines.

Execute Worksheet

Execute every command in the worksheet. This command is new in Maple V Release 3. In MS-Windows, this command is in the **Format** menu.

Options Menu

Output Mode

Pick the way Maple output is formatted in the worksheet. A submenu lets you select one of **Line Print**, **Pretty Print**, or **Hi-res Print**. These settings correspond to setting the interface variable `prettyprint` to 0, 1, or 2. For example, entering the following Maple command in Hi-res mode yields typeset output:

```
> alpha*x^3 + beta*Diff(f(x),x) + Int(c[0]/sqrt(t), t=exp(1)..Pi);
```

$$\alpha\, x^3 + \beta \left(\frac{\partial}{\partial x} \mathrm{f}(x) \right) + \int_{\mathrm{e}}^{\pi} \frac{c_0}{\sqrt{t}}\, dt$$

In Pretty Print mode, the same command gives two-dimensional output in plain text:

```
                                      Pi
                                      /
          3        / d      \        |   c[0]
  alpha x   + beta |---- f(x)| +     |   ---- dt
                   \ dx     /        |    1/2
                                    /    t
                                   exp(1)
```

In Line Print mode, the result is written on a single line:

```
alpha*x^3+beta*Diff(f(x),x)+Int(c[0]/t^(1/2),t = exp(1) .. Pi)
```

The default mode is **Hi-res Print**. After changing modes, the worksheet is not altered until you select one of the reformat commands on the **View** menu.

Output Fontsize

Select the size for "hi-res" output. The choices are **Large**, **Medium**, or **Small**. The default is **Medium**.

Under MS-Windows, the **Math Style** command on the **Format** menu performs this function as well as the **Output Mode** function. Unlike X, in MS-Windows the worksheet is reformatted automatically after changing the output mode or size of font.

Change Fonts

Change any of several fonts in the worksheet. Selecting this command brings up a submenu with five commands. The first three choices let you change the font used in every input region, text region, or output region. The fourth choice changes the font only in the region where the cursor is currently located. The fifth choice resets the font in the current region to the standard font used in regions of that type.

Whenever you change a font, you are given the opportunity to change the font used when printing the worksheet on a PostScript printer.

Most of these functions are available under MS-Windows under the **Fonts** command in the **Format** window.

Show Status Bar

A switch controlling the display of the status information at the bottom of the worksheet window. This bar contains statistics on memory usage and CPU time and is displayed by default.

Under MS-Windows, the status information appears in a separate window and is not displayed by default. The menu command for toggling its display is **Status Window**.

Automatic Separators

A switch controlling whether or not a separator is printed after each output region. The default is on. Turning this switch off does not affect separators already present. To delete these, use the **Delete By Type** command in the **Edit** menu.

Replace Mode

A switch controlling the placement of output after re-executing a command. When on (the default setting), the new output replaces the old. When off, the new output is inserted in a new region just above the old output.

The MS-Windows interface contains several additional commands in its **Options** menu that we have not yet mentioned. **Show Prompts** toggles the display of the prompt at the beginning of each input region. **Confirmation Checks** lets you avoid confirming commands like **New** and **Exit**. **Show Tool Bar** (new in Maple V Release 3) toggles the display of the icon bar at the top of the worksheet. **Fast Graphics Redraw** allows you to keep three-dimensional graphics in memory for faster redrawing, at the expense of increasing your memory usage. **Automatic Save Settings** lets you save your option settings automatically when you exit a session. **Save Settings** (on the **File** menu) saves option settings directly.

Though these commands are missing from the interface under X, you can mimic the effect of many of them by setting the proper X resources (see page 94).

Help Menu

The help menu lets you start the help topic browser and access information on the user interface and the current version of Maple.

Help Browser

Start up the help topic browser in a separate window. The topic browser organizes all Maple commands by subject and allows quick access to help text on commands. There are four main topics at the top level of the browser listed in the leftmost column: Graphics, Mathematics, Programming, and System. A topic ending with three dots has subtopics. Selecting one of these topics fills the next column with its subtopics, and so on, until you select a Maple command. Then a short synopsis of the command appears in a box near the bottom of the window, and the **Help** button becomes active. Pressing **Help** brings up the help text for this topic in a separate window. Pressing the **Close** button (or **Cancel** on some systems) quits the topic browser. See the example in Figure 2.2.

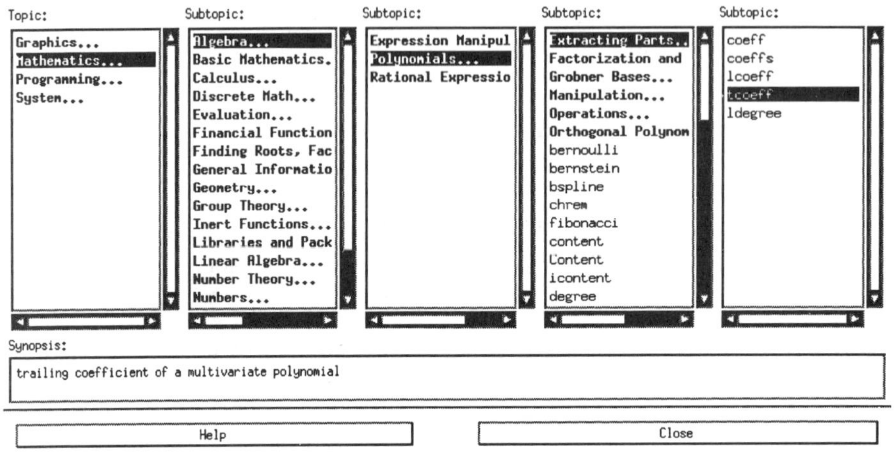

Figure 2.2: The help topic browser

Keyword Search

Search for help topics that mention a given string in their synopsis section. This command is new in many versions of Maple V Release 3. In Maple V Release 2 under DOS, pressing F2 while in the help browser starts a keyword search.

On Context

Obtain help on the worksheet user interface. Selecting this command lets you click on any part of the worksheet window to get help on a topic. For example, clicking on the name **Edit** in the menu bar after selecting **On Context** brings up help text for the commands in the **Edit** menu.

On Keys

Display help text on keys that have special meaning in the worksheet window. This is not a list of keyboard shortcuts for invoking menu commands. For example, CONTROL-G selects the entire region where the cursor is positioned, CONTROL-S pauses the display of output, and CONTROL-Q restarts paused output.

MS-Windows has an **Interface Help** command that subsumes the previous two commands in X.

On Version

Display the Maple logo, the current version number, the licensee, and the copyright notice in a separate window. Other systems may call this command **About Maple V**.

1.2 Two-Dimensional Graphics Windows

Plots

When you enter a plotting command in a Maple worksheet, the graph appears in a separate window, as in Figure 2.3. There are many operations you can perform on this graph from

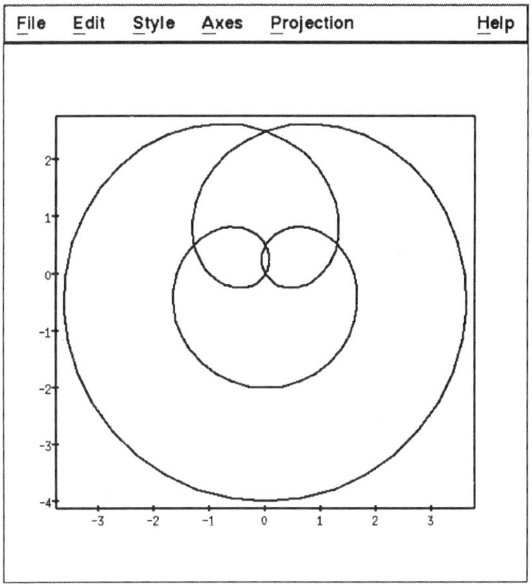

Figure 2.3: Two-dimensional plot window

this window. We discuss these operations in this section.

Clicking with the (left) mouse button inside the graphics window displays the x and y coordinates of that point in the upper-left corner of the window. (This does not work on infinity plots or log plots.)

Menu commands let you set several options for displaying a two-dimensional plot without returning to the worksheet window and altering the plotting command. The **Style** menu sets the `style`, `linestyle`, `thickness`, and `symbol` options, **Axes** sets the `axes` option, and **Projection** sets the `scaling` option. See page 26 in Chapter 1 for a description of these plotting options.

You can copy the graph into the clipboard so that it can be pasted into your worksheet. Use the **Copy** command on the **Edit** menu.

The **Print** command in the **File** menu lets you select a format and a filename for saving the graph. In some systems, like MS-Windows, you can print a graph directly by using this **Print** command.

The **Help** menu has a single command, **Plot 2D Help**, that displays help on the commands available in the two-dimensional graphics window.

Select the **Exit** command on the **File** menu to quit this window, or type a "q" inside this window.

Animation

In addition to the functions and menu commands available in the normal plotting window, the two-dimensional animation window contains a row of buttons for controlling the way the animation is presented:

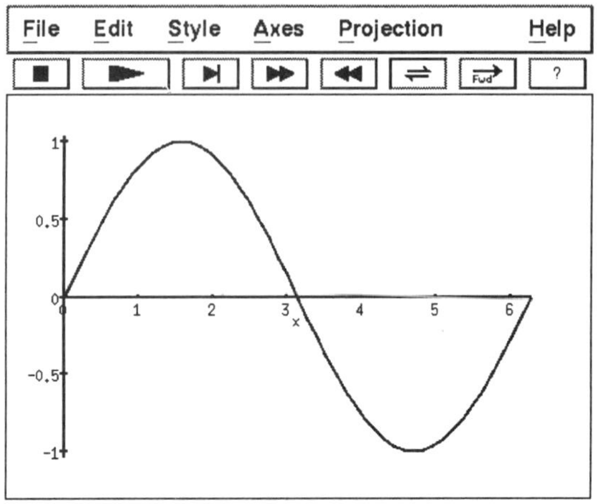

Figure 2.4: Animation window

- Play (a solid triangle or "Play") starts the animation.
- Stop (■ or "Stop") freezes the animation.
- Frame Advance (a solid triangle abutting a vertical line or "Adv") shows the next animation frame in the sequence.
- Increase Speed (two triangles pointing to the right) and Decrease Speed (two triangles pointing to the left) change the rate of drawing the animation frames. You can set this rate explicitly with an X resource (page 99).
- Playback Mode can be set to single play ($\rightleftharpoons$ or "Once") or repeat play (a circular arrow or "Loop").
- Direction lets you watch the animation in normal sequence ($\rightarrow$) or reverse sequence ($\leftarrow$).
- Show Labels (?) toggles the display of the button labels.

See Figure 2.4.

1.3 Three-Dimensional Graphics Windows

Plots

Three-dimensional plots appear in their own window. As with two-dimensional graphics, there are many operations you can perform on a three-dimensional graph from this window. See Figure 2.5.

Clicking with the (left) mouse button inside the graphics window replaces the graph with a bounding box. See Figure 2.6. This box provides a frame of reference for orienting the surface to a desired position. Dragging the mouse while the button is pressed turns this

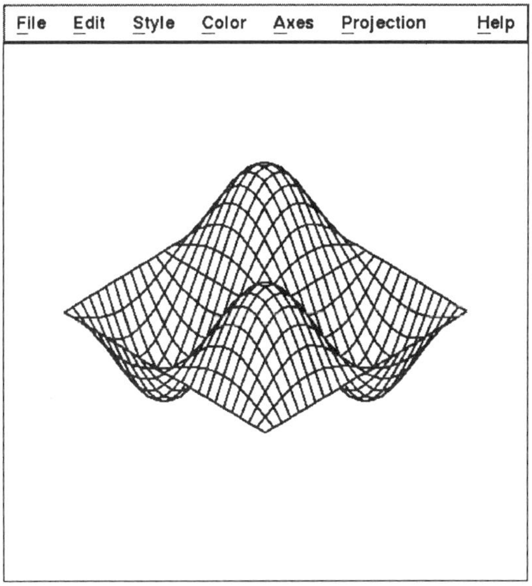

Figure 2.5: Three-dimensional plot window

box, setting the `orientation` of the graph. The current values of the spherical coordinates θ and ϕ are displayed in the upper-left corner. The value of R, the perspective, is also displayed in this corner on some systems. You can set R to a few different values (1.0, 0.8, 0.5 or 0.1) by setting **No Perspective**, **Far View**, **Medium View**, or **Near View** from the **Projection** menu.

The menus let you set several options for displaying three-dimensional graphics. In addition to setting the `perspective` option, the **Projection** menu also lets you set the `scaling` option to **Constrained** or **Unconstrained**. The **Axes** menu sets the `axes` option, **Color** sets `shading`, and **Style** sets `style`, `linestyle`, `thickness`, and `symbol`. The **Color** menu also lets you pick one of several predefined lighting schemes: particular settings of `ambientlight` and `light`. **Style** also lets you pick a triangular grid instead of the usual rectangular one (this option is not available from the `plot3d` command). See page 34 in Chapter 1 for a description of these plotting options.

Under MS-Windows, the **Color** menu allows you to select dithered patches instead of the usual solid color patches. This makes a surface appear much smoother. Dithering is not available as an option for the `plot3d` command.

After changing all the options to your liking, press the middle mouse button to redraw the graph. On some systems, you can press ENTER or "p" inside the window, or select the **Redraw** command in the **Edit** window.

The **Copy** command in the **Edit** menu copies the graph to the clipboard for pasting into the worksheet.

The **Print** command in the **File** menu saves a graph for printing, just as in the two-dimensional graphics window, and the **Plot 3D Help** command in the **Help** menu brings up help on the three-dimensional graphics window.

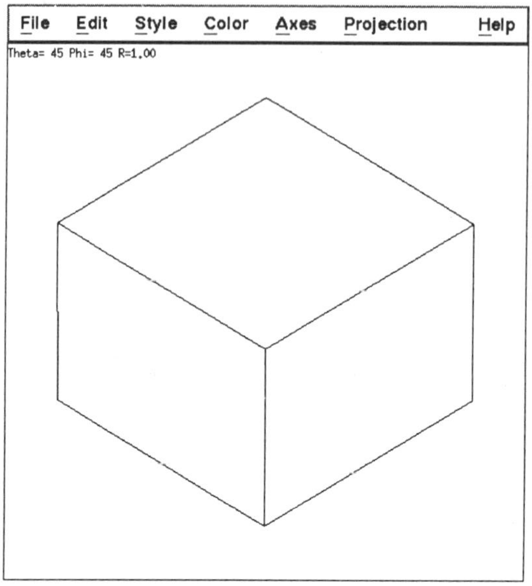

Figure 2.6: Orienting a three-dimensional plot

Quit this window by selecting **Exit** in the **File** menu or by typing "q" in the window.

Animation

The three-dimensional animation window has the same menus and commands that are in the normal three-dimensional plotting window, plus the same row of buttons present in the two-dimensional animation window.

1.4 X Resources

The X Window System lets you set *resources* to tailor applications to your own preferences. The X version of Maple's worksheet user interface allows you to set resources that specify option settings, fonts, colors, and other attributes. We discuss several of these resources in this section and record their default values. The resources are divided into seven categories: input region, text region, output region, option settings, general worksheet window, help windows and graphics windows.

Each listing starts with the resource name and the default value. For example, the following resource setting makes all input commands red:

```
MapleV.input.foreground: red
```

If you would like input commands drawn in black instead, add this line to your .Xdefaults file in your home directory:

```
MapleV.input.foreground: black
```

You may need to run `xrdb` to load these preferences into your X session. Enter the command `xrdb .Xdefaults` from a shell window inside X, or place this command in your .xinitrc file.

We list the resources that users are most likely to want to change. You can find a complete list of all the resources accepted by Maple in the X11_defaults directory inside the directory where Maple is installed on your system.

Input Region

`MapleV.input.foreground: red`
The color of input commands.

`MapleV.input.promptcolor: black`
The color of the command prompt.

`MapleV.input.font: variable`
The X11 font for displaying input commands on the screen.

`MapleV.input.psFont: Helvetica`
The PostScript font for printing input regions.

`MapleV.input.psFontSize: 12`
The size (in points) of the PostScript font for printing input regions.

`MapleV.input.leftmargin: 10`
`MapleV.input.rightmargin: 10`
The number of pixels (on the display) or points (on PostScript output) to leave as left and right margins for input regions.

Text Region

`MapleV.comment.foreground: black`
The color of the text in text regions.

`MapleV.comment.font: -adobe-courier-medium-o-normal--*-120-*`
The X11 font for displaying text on the screen.

`MapleV.comment.psFont: Helvetica-Oblique`
The PostScript font for printing text regions.

`MapleV.comment.psFontsize: 12`
The size (in points) of the PostScript font for printing text regions.

`MapleV.comment.leftmargin: 0`
`MapleV.comment.rightmargin: 0`
The number of pixels (on the display) or points (on PostScript output) to leave as left and right margins for text regions.

Output Region

`MapleV.output.foreground: black`
The color of the results in output regions.

`MapleV.output.leftmargin: 0`
`MapleV.output.rightmargin: 0`
The number of pixels (on the display) or points (on PostScript output) to leave as left and right margins for output regions in plain text format.

`MapleV.output.firstindent: 10`
The amount of indenting (in pixels or points) for the first line of hi-res output produced by a command.

`MapleV.output.subseqindent: 40`
The amount of indenting (in pixels or points) for subsequent lines of hi-res output if a command produces more than one line of output.

`MapleV.output.fontsize: medium`
The default size for formatting hi-res output. Choices are `large`, `medium`, or `small`. The actual size used depends on the fonts specified in the numberFont, variableFont, and symbolFont resources.

```
MapleV.output.numberFont1:
   -adobe-new century schoolbook-medium-r-normal--*-240-*
MapleV.output.numberFont2:
   -adobe-new century schoolbook-medium-r-normal--*-180-*
MapleV.output.numberFont3:
   -adobe-new century schoolbook-medium-r-normal--*-120-*
```
Fonts used to display numbers in hi-res output on the screen. This is a roman font by convention. Font 1 is the large font, 2 the medium, and 3 the small.

```
MapleV.output.variableFont1:
   -adobe-new century schoolbook-medium-i-normal--*-240-*
MapleV.output.variableFont2:
   -adobe-new century schoolbook-medium-i-normal--*-180-*
MapleV.output.variableFont3:
   -adobe-new century schoolbook-medium-i-normal--*-120-*
```
Fonts used to display Maple variable names in hi-res output on the screen. This is an italic font by convention. Font 1 is the large font, 2 the medium, and 3 the small.

```
MapleV.output.symbolFont1:
   -adobe-symbol-medium-r-normal--*-240-*
MapleV.output.symbolFont2:
   -adobe-symbol-medium-r-normal--*-180-*
MapleV.output.symbolFont3:
   -adobe-symbol-medium-r-normal--*-120-*
```
Fonts used to display standard mathematical symbols in hi-res output on the screen. This must be a symbol font. Font 1 is the large font, 2 the medium, and 3 the small.

`MapleV.output.psNumberFont: Times-Roman`
PostScript font family used to print numbers in hi-res output. The size of the font used depends on the value of the psFontSize resources.

`MapleV.output.psVariableFont: Times-Italic`
PostScript font family used to print variables in hi-res output. The size of the font used depends on the value of the psFontSize resources.

`MapleV.output.psSymbolFont: Symbol`
PostScript font family used to print standard mathematical symbols in hi-res output. The size of the font used depends on the value of the psFontSize resources.

`MapleV.output.psFontSize1: 24`
`MapleV.output.psFontSize2: 18`
`MapleV.output.psFontSize3: 12`
Size (in points) of the PostScript fonts used to print hi-res output. Size 1 is the large font, 2 the medium, and 3 the small.

`MapleV.output.font: fixed`
The X11 font for displaying plain text mathematical output, formatted procedures, and error and warning messages on the screen.

`MapleV.output.psFont: Courier`
The PostScript font for printing plain text mathematical output, formatted procedures, and error and warning messages.

`MapleV.output.psFontsize: 12`
The size (in points) of the PostScript font for printing plain text mathematical output, formatted procedures, and error and warning messages.

Option Settings

`MapleV*menubar*show_region_extents.set: False`
Initial setting of the **Show Regions** switch in the **View** menu.

`MapleV.output.fontsize: medium`
Initial setting for the **Output Fontsize** command in the **Options** menu.

`MapleV*menubar*show_status_bar.set: True`
Initial setting of the **Show Status Bar** switch in the **Options** menu.

`MapleV*menubar*automatic_separators.set: True`
Initial setting of the **Automatic Separators** switch in the **Options** menu.

`MapleV*menubar*replace_mode.set: True`
Initial setting of the **Replace Mode** switch in the **Options** menu.

`MapleV.psForcedPageBreaks: True`
If true, Maple treats two consecutive separators as a page break when printing the worksheet in PostScript form.

`MapleV.queryOnExit: True`
If true, Maple requests confirmation when **Exit** is selected from the **File** menu.

General Worksheet Window

`MapleV*background: grey`
The background color of all components in the worksheet window.

`MapleV*canvas.foreground: black`
`MapleV*canvas.background: white`
After setting all backgrounds to grey by default, individual components may be set to a different color. The canvas foreground includes the cursor and the separator lines, but not the color of the input nor the output, nor the color of the prompt. The canvas background refers to the background of the worksheet.

`MapleV.geometry: 600x500`
The size of the worksheet window in pixels, using the standard X11 notation for the geometry resource.

`MapleV*buttonbar*Pause.indicatorOn: False`
When true, the **Pause** button contains an indicator that shows whether or not this button is depressed.

`MapleV*buttonbar*Pause.shadowThickness: 2`
The depth of the shadow in pixels along the border of the **Pause** button in the worksheet window.

`MapleV*frame.shadowType: shadow_in`
The direction of the shadows in the worksheet window. This should be either `shadow_in` or `shadow_out`.

`MapleV*frame.shadowThickness: 3`
The depth of the shadows in pixels along borders in the worksheet window.

`MapleV.backFillOnScroll: True`
Circumvents a bug in certain X servers. Set this to false if the X server hangs while scrolling the worksheet window.

`MapleV*buttonbar*fontList: variable`
Font for displaying text in the row of buttons.

`MapleV*menubar*fontList: variable`
Font for displaying text in the menu bar.

Help Windows

`maplehelp*background: grey`
Background color of the control area of the help text window.

`maplehelp*foreground: black`
Foreground color of the control area of the help text window.

`maplehelp*helptext.background: white`
Background color of the text area in the help window.

`maplehelp*helptext.foreground: black`
Foreground color of the text in the help window.

`maplehelp*fontList: fixed`
Font used to display the help text in the help text window. Some users may want to define this to be a larger font, like `8x13` or `9x15`.

`maplebrowser*background: grey`
Background color of the help topic browser.

`maplebrowser*foreground: black`
Foreground color of the help topic browser.

`maplebrowser*XmList.fontList:`
`7x13=charset-normal, 7x13bold=charset-bold`
Fonts used in the help topic browser. Note that two fonts are specified here: one for normal text (`7x13`), and another for bold text (`7x13bold`).

Graphics Windows

Two-Dimensional Graphics

`Maple2dX11m.reverseVideo: True`
If true, two-dimensional plots are drawn on a black background; if false, on a white background.

`Maple2dX11m*background: grey`
Background color of the control area of the two-dimensional graphics window.

`Maple2dX11m*foreground: black`
Foreground color of the control area of the two-dimensional graphics window.

`Maple2dX11m*canvas.height: 500`
`Maple2dX11m*canvas.width: 500`
Size of the main area of the two-dimensional plotting window, in pixels.

`Maple2dX11m.titlefont: -adobe-helvetica-bold-r-normal--14-*`
The font used to display the title of a two-dimensional plot.

`Maple2dX11m*menubar*fontList: variable`
Font for displaying text in the menu bar.

`Maple2dX11m*animatecanvas.height: 300`
`Maple2dX11m*animatecanvas.width: 300`
Size of the main area of the two-dimensional animation window, in pixels. The default size is smaller than the normal plot window to reduce memory requirements.

`Maple2dX11m.framedelay: 100`

The delay in milliseconds between the display of successive frames in two-dimensional animations. Some platforms may not support fast frame rates.

`Maple2dX11m.fastanimation: False`

If true, prepare all frames in advance before displaying the animation. If false, display each frame as it becomes available. This resource is set to false by default in the two-dimensional case because rendering each frame is normally rather quick, and this allows many frames to be viewed without much memory use.

Three-Dimensional Graphics

`Maple3dX11m.reverseVideo: True`

If true, three-dimensional plots are drawn on a black background; if false, on a white background.

`Maple3dX11m*background: grey`

Background color of the control area of the three-dimensional graphics window.

`Maple3dX11m*foreground: black`

Foreground color of the control area of the three-dimensional graphics window.

`Maple3dX11m*canvas.height: 500`
`Maple3dX11m*canvas.width: 500`

Size of the main area of the three-dimensional plotting window, in pixels.

`Maple3dX11m.titlefont: -adobe-helvetica-bold-r-normal--14-*`

The font used to display the title of a three-dimensional plot.

`Maple3dX11m*menubar*fontList: variable`

Font for displaying text in the menu bar.

`Maple3dX11m*animatecanvas.height: 300`
`Maple3dX11m*animatecanvas.width: 300`

Size of the main area of the three-dimensional animation window, in pixels. The default size is smaller than the normal plot window to reduce memory requirements.

`Maple3dX11m.framedelay: 100`

The delay in milliseconds between the display of successive frames in three-dimensional animations. Some platforms may not support fast frame rates.

`Maple3dX11m.fastanimation: True`

If true, prepare all frames in advance before displaying the animation. If false, display each frame as it becomes available. This resource is set to true by default in the three-dimensional case because rendering each frame normally takes some time. However, this increases memory consumption, so the number of frames should be kept small under fast animation.

2 THE COMMAND-LINE INTERFACE

The command-line interface to Maple is a text-based user interface for terminals and computers without much graphics. This interface has several interesting features that make it easy to use, particularly in Maple V Release 2 and Release 3. We highlight these features in this section. We first describe the command-line editor, and how to reuse and edit previously entered commands. We then discuss several commands for changing the behavior of the command-line interface and how to save graphics in a file or switch to a special graphics mode on a terminal that supports one. Finally, we list some hints for using the command-line version to run Maple in batch mode.

2.1 Editing Commands

A facility for editing previously entered commands in the command-line interface was added to Maple V Release 2. This feature makes it easy to correct mistakes and modify parameters in commands already submitted. You can recall any of the last 100 lines you have entered.

An input line in the command-line interface can have as many as 1023 characters. When the cursor is about to move off the right edge of the screen, the input line shifts to the left by half the screen width to let you continue typing.

The editing syntax borrows from two popular editors, emacs and vi. Some editing commands can be performed by pressing the CONTROL key in combination with another letter. These commands generally follow the conventions of emacs, a popular editor from the Free Software Foundation. Other commands are typed in a command mode that you enter by pressing the ESCAPE key. These closely follow the standard Unix editor vi.

Though the editing functions of the emacs-style commands and the vi-style commands overlap considerably, there is no need to indicate to the system which set of commands you intend to use. Both styles are available simultaneously.

In addition to the commands described below, most terminals have arrow keys that can recall previous and next lines from the command-line history (the up and down keys) and move the cursor left and right on the command line (the left and right keys). Your computer may also have keys labeled Home and End that move the cursor to the beginning and end of the command line, respectively. These keys are always active.

Control Keys

These commands are available all the time when entering Maple commands, even when you are in vi command mode (except for CONTROL-I). Control keys that behave differently from standard emacs are noted.

CONTROL-P Recall the previous line from the command history. Repeatedly entering this command eventually brings you back to the most recently entered line.

CONTROL-N Recall the next line from the command history. Cycle back to the oldest line saved after the most recently entered line.

CONTROL-B Move the cursor to the left one character.

CONTROL-F Move the cursor to the right one character.

CONTROL-A Move the cursor to the beginning of the line.

CONTROL-E Move the cursor to the end of the line.

CONTROL-] Move the cursor to the matching parenthesis, bracket, or brace (not a standard emacs command).

CONTROL-W Move the cursor to the beginning of the next word (not a standard emacs command).

CONTROL-Y Move the cursor to the beginning of the current word, or to the beginning of the previous word if it is already at the beginning of a word (not a standard emacs command).

CONTROL-D Delete the character at the cursor position.

CONTROL-H *or* BACKSPACE *or* DELETE Delete the character to the left of cursor and move left one character (only DELETE is a standard emacs command).

CONTROL-K Delete to the end of the line.

CONTROL-G *or* CONTROL-X Delete the entire line (not standard emacs commands).

CONTROL-V Switch between insert mode and replace mode (not a standard emacs command). The default is insert mode.

CONTROL-I *or* TAB While inserting text, add spaces until the cursor reaches a tab stop. The tab stops are equally spaced, with the length of each determined by the interface variable `indentamount`. While replacing text, move the cursor to the next tab stop. This command does not work in vi command mode.

CONTROL-U Undo all the changes made to the line (not a standard emacs command).

CONTROL-R Search for the most recent line in the command history containing whatever is currently on the command line. Unlike emacs, the search string is typed first, then the search command.

CONTROL-L Redraw the line.

Command Completion The command-line interface includes two special control sequences for completing commands that have been only partially typed on the command line. The list of commands searched when requesting completion comes from one of two files. Under Unix, Maple first looks for a file named ".maplecmds" in your home directory. If it is not present, it looks instead for a file named "cmds" in the "src" subdirectory of the "lib" subdirectory of the main Maple area. These files must be sorted according to the order of the character set on your computer (like ASCII or EBCDIC).

Neither of these commands is a standard emacs command.

CONTROL-@ *or* CONTROL-SPACE — Attempt to complete the command that has been partially typed on the command line. If more than one command matches, complete the command as far as possible.

CONTROL-T — Display a list of all commands that match the start of the command typed so far on the command line.

Vi Command Mode

Enter command mode by pressing the ESCAPE key. Reenter text entry mode with one of the text-entry commands, like i.

k — Recall the previous line from the command history. Repeatedly entering this command eventually brings you back to the most recently entered line.

j — Recall the next line from the command history. Cycle back to the oldest line saved after the most recently entered line.

h — Move the cursor to the left one character.

l *or* SPACE — Move the cursor to the right one character.

0 — Move the cursor to the beginning of the line.

$ — Move the cursor to the end of the line.

% — Move the cursor to the matching parenthesis, bracket, or brace.

w — Move the cursor to the beginning of the next word.

b — Move the cursor to the beginning of the current word, or to the beginning of the previous word if it is already at the beginning of a word.

x — Delete the character at the cursor position.

X — Delete the character to the left of cursor and move left one character.

D — Delete to the end of the line, remaining in vi command mode.

C — Delete to the end of the line and enter insert mode.

S — Delete the entire line and enter insert mode.

U — Undo all the changes made to the line.

? — Search for the most recent line in the command history containing whatever is currently on the command line. Unlike vi, the search string is typed first, then the search command.

f — Move the cursor forward to the next occurrence of whatever character is typed next.

F Move the cursor backward to the previous occurrence of whatever character is typed next.

; Repeat the most recent f or F search.

, Repeat the most recent f or F search in the opposite direction.

i Insert text just before the current cursor position.

a Append text just after the current cursor position.

I Insert text at the beginning of the line.

A Append text to the end of the line.

R Enter replace mode beginning at the current cursor position. Whatever is typed overwrites the text on the command line.

r Replace the current character by the next character typed.

s Delete the current character and insert text in its place.

Several of the vi-style commands may be preceded by an integer to repeat the command that many times. This integer should be between 1 and 999. The commands accepting an iteration argument are `b f F , ; h j k l s w x` and `X`.

2.2 Interface Commands

A number of Maple commands affect the behavior and appearance of your Maple session. This section describes all of the interface variables and discusses several other commands that affect the user interface.

Interface Variables

The `interface` command lets you query or set several variables that affect the user interface. It is used more often in the command-line interface than in the worksheet interface, since the worksheet interface has other means of setting many of these variables.

To find the current value of an interface variable, give the variable name as an argument to the `interface` command. To set an interface variable, supply an equation. For example,

```
> interface(indentamount);
```

$$4$$

```
> interface(indentamount=2);
> interface(indentamount);
```

$$2$$

The interface variables that are common to all Maple platforms are described in this section. The variables that affect the way graphics are produced are grouped together.

`echo`

Determines when input commands are echoed to the output location. This has limited use in normal interactive sessions, but it is often desired when writing all output to a file using a function like `writeto` or when running Maple in batch mode. The `echo` variable is superseded by `quiet`: if `quiet` is true, no echoes occur. There are five possible settings for this variable.

0 Never echo.

1 Echo whenever the output is not going to the screen or whenever the input does not come from the keyboard, but do not echo when reading input commands with `read`. This is the default value.

2 Echo whenever the output is not going to the screen or whenever the input does not come from the keyboard.

3 Echo only when reading input commands with `read`.

4 Always echo.

`endcolon`

Indicates whether or not the last input statement was terminated with a colon. Returns true or false. This variable cannot be set with the `interface` command, only read.

`errorbreak`

Determines Maple's action when an error is found reading an input file. This variable was introduced in Maple V Release 2.

0 Continue reading a file even after an error has been found. This is the behavior of Maple V.

1 Stop reading an input file if a syntax error is found. This avoids displaying several errors that may all be caused by the first one. This is the default in Release 2 and Release 3.

2 Stop reading an input file if any error occurs, even an execution error.

In some systems, you can set this variable with an option on the command that starts Maple. In Unix, this option is `-e`, followed by the value.

`indentamount`

The number of spaces subsequent lines of output are indented if a command produces more than one line of output. Also, the amount of indenting used when formatting procedures and the distance between successive tab stops in Maple input. The default is 4.

`iris`

The name and version number of the user interface. This variable cannot be set with the `interface` command, only read. It is new in Maple V Release 3.

`labeling` *or* `labelling`

Enables or disables the use of labels (%1, %2, etc.) for substituting subexpressions in long output expressions. Possible values are true or false. The default is true.

`labelwidth`
The minimum length of a subexpression in a large expression that can be substituted by a label like %1. The default value is 20.

`prettyprint`
Sets the method of displaying Maple output. The possible values under Maple V Release 2 or Release 3 are:

0 Use lprint to print results in a linear style. Results printed in this style can be used as Maple input.

1 Use a two-dimensional, text-based style to print results. This is the default in the command-line interface.

2 Use the "hi-res" standard mathematical notation on a graphics device. This is the default in the worksheet interface.

Setting this variable to a larger value is permitted, but the display method remains the best one that the your device can handle.

Under Maple V, the possible values of this variable are true and false. For compatibility, these settings are still valid in later releases. False corresponds to a setting of 0, linear print; true corresponds to a setting of 1, text-based pretty-printing.

`printfile`
Shows the file named, perhaps using a browsing program available to the user interface. Intended for the help facility.

`prompt`
The string printed whenever input is expected from the user. The command-line interface inspects this variable and the `screenwidth` variable to determine how much space is available on the command line before scrolling to the right. The default is usually `'> '` in the command-line interface.

`quiet`
When true, suppresses all startup messages and status messages (logo, memory and CPU usage messages, and prompt). This setting is useful when running a Maple program in batch mode. The default value is false. In some systems, you can set this variable with an option on the command that starts Maple. In Unix, this option is `-q`.

`screenheight`
The number of lines of text on the screen.

`screenwidth`
The number of characters that fit on one line on the screen. The command-line interface uses this variable to determine the maximum length of a line of output, and (together with the `prompt` variable) to determine how much space is available on the command line before scrolling to the right. The minimum value of this variable is 10.

`verboseproc`

Determines whether procedure bodies are printed when they are read into your Maple session with `read` or `readlib` or explicitly printed with `print`. The possible values are:

0 Never print the body of any procedure. Instead, print an abbreviation of the form:

```
proc(f) ... end
```

1 Print user-defined procedures when they are read in or explicitly printed, but do not print any of the Maple library procedures. This is the default setting.

2 Print all procedures whenever they are read in or explicitly printed.

`version`

In Maple V Release 3, this is the version of Maple that is running. In Release 2, it is the name and version number of the user interface. This variable cannot be set with the `interface` command, only read.

`warnlevel`

In Maple V Release 3, an undeclared variable used in a procedure is assumed to be local if it appears on the left-hand side of an assignment statement or if it is the indexing variable of a `for` loop or a `seq` statement. This interface variable determines whether warning messages are printed when defining procedures that have undeclared variables that are implicitly defined to be local. Warning messages are printed if the value is 1 (the default); none is printed if the value is 0. This variable is new in Maple V Release 3.

`wordsize`

The size of a machine word. This variable is new in Maple V Release 3. It cannot be set with the `interface` command, only read.

Graphics The following interface variables affect the way graphics are displayed and saved in files.

`plotdevice`

The type of plotting device for which graphics are generated, whether for interactive viewing or for saving in a file. If you are using the command-line interface and your terminal supports a graphics mode like ReGIS or Tektronix, you may be able to set this variable to view graphics interactively. If you have a printer that renders graphics, set this to the printer's format. See the help text on plot[device] for the list of printing devices supported at your installation.

`plotoptions`

A string that names options to be passed to the device driver when plotting. This variable is new in Maple V Release 3.

`plotoutput`
The name of the file where graphics output is stored, in the format determined by the setting of `plotdevice`. Be sure to change the setting after each plot.

`preplot`
A list of character codes that switches the terminal into graphics mode whenever a graphics command is issued. The character codes are integer values of the character set on your computer. For example, 27 is the ASCII code for the ESCAPE key.

Some terminals require a short delay when switching modes before a plot can be drawn. A negative value in the `preplot` list specifies the delay in seconds before a graph is plotted on the screen.

For example, if the sequence of characters ESCAPE `~>` makes your terminal enter Tektronix emulation mode, set your session to switch into this mode after a two second delay whenever you request graphics with the command:

```
> interface(preplot = [-2, 27, 126, 62]);
```

See the help page on plot[setup] for some common sequences.

`postplot`
A list of character codes that switches the terminal back into text mode from graphics mode after RETURN is pressed in the graphics screen. The character codes are integer values of the character set on your computer. As in `preplot`, a negative value specifies a delay in seconds. See the help page on plot[setup] for some common sequences.

`terminal`
Set interface variables for doing Maple graphics on any of several common terminals. Setting this variable to one of the known terminal types (usually HP2393A, kd500g, vt240 and xtek) sets appropriate values for `plotdevice`, `preplot`, and `postplot`. The known terminal types are listed under the "terminal" directory in the library.

The `plotsetup` command provides a convenient way to set interface variables for controlling graphics in common situations. See page 109.

Other Interface Commands

We list some additional commands for controlling the frequency of status messages and for plotting graphs under the command-line interface.

Status Messages Under the command-line interface, Maple occasionally prints status messages showing the total amount of memory used in the session (in bytes), the total amount of memory currently allocated to your session (again in bytes), and the total CPU time used in your session. (Under the worksheet interface, this information appears in a special status area, perhaps in a separate window.) A status message looks like this:

```
bytes used=5001188, alloc=851812, time=14.78
```

Maple normally prints a status message when any of the following events occur:

1. Another n words of memory has been used by your session, where n is the value of `status[4]`.

2. A garbage collection occurs, freeing memory that is no longer being used.

3. You quit a Maple session.

Also, the same information, plus a bit more, is returned any time you ask for the value of the `status` variable.

You can change the frequency of messages you get from source number 1 with the `words` command. The command `words(`n`)` makes a status message appear each time another n words of memory are used by your session, where a word is normally four bytes. As a special case, `words(0)` turns off status messages from this source. This is the default setting on many platforms.

You can change the frequency of messages from source number 2 in a similar manner with the `gc` command. The command `gc(`n`)` makes a garbage collection occur each time another n words of memory are used by your session. The special command `gc(0)` shuts off all status messages printed during garbage collection, though it does not change the frequency of garbage collection itself.

Therefore, the following line turns off all status messages printed in the middle of your session:

```
> words(0): gc(0):
```

plotsetup The `plotsetup` command sets the `plotdevice`, `plotoptions`, `plotoutput`, `preplot`, and `postplot` interface variables for handling graphics output in many situations. For example, if you want to save some graphics as a PostScript file, the following command alters the plotting variables accordingly:

```
> plotsetup(postscript);
  Warning, plotoutput set to postscript.out by default
```

```
> interface(plotdevice);
```

$$ps$$

If you use a VT240 terminal and plan to use Tektronix emulation for displaying graphics, try this command:

```
> plotsetup(tek, vt240);
> interface(preplot);
```

$$[27, 91, 63, 51, 56, 104]$$

See the help page on `plotsetup` for a list of the device types and terminal types supported by your system.

2.3 Batch Processing

Even on computers where a worksheet interface is available, the command-line version is still very useful for running Maple programs in batch mode. In batch mode, all the commands are written ahead of time and saved in a file. These commands are then executed in sequence.

To run Maple in batch mode, write your sequence of commands in a file. Test for syntax errors by changing the parameters of the program to some small values and loading the file into an interactive Maple session. Once everything checks out, you should add a few lines for running your program without user intervention.

- Handle output explicitly. There are several ways to do this. The simplest method is executing `writeln` at the beginning of your file to direct all output to the named file. Alternatively, you can `open` an output file at the beginning of your list of commands, then explicitly `write` selected results. On some systems, such as Unix computers, you can handle output very simply by redirecting to a file any results that would normally appear on the screen.

- Set the interface variable `quiet` to `true`. This avoids printing information that is generally only useful in input mode: the startup logo, input prompts, and status messages. On some systems, you can set quiet mode with an option on the command that starts Maple. In Unix, this option is `-q`.

- Set the interface variable `echo` to control the printing of input lines. With the default setting, all input lines in batch mode are echoed to the output.

- Add a `quit` statement to the end of your program. Without this, Maple may enter interactive mode after executing your commands. Alternatively, on some systems, you can run Maple as a filter so that it exits as soon as it reaches the end of the file. Under Unix, the filter mode is indicated by the option `-f` on the command that starts Maple.

On many systems, the command for starting the command-line version of Maple is `maple`. You might start a Maple program in batch mode under Unix with a command like this:

```
maple -q -f < inputfile &
```

If output redirection is used, you might use a command like this:

```
maple -q -f < inputfile > outputfile &
```

Here, the trailing ampersand (`&`) tells Unix to run the command in the background so that you can do other things while the program is running.

Chapter 3
Subject Listing

This chapter lists all Maple commands, options, and help topics by subject. The four main categories are **Graphics**, **Mathematics**, **Programming**, and **System**. Each of these is divided into subcategories, and many of these are further subdivided.

Maple commands are printed in plain roman type. A command defined in a package is listed with the package name first, followed by the command name inside square brackets: for example, "linalg[det]". A command defined in a subpackage is similar: for instance, "stats[fit, leastsquare]". Miscellaneous library functions, which must be loaded explicitly before you can use them (see page 65), have a star ($\star$) after their names, as in "C $\star$". Most of these functions are loaded using the function name itself in a `readlib` command, such as `readlib(C)`. However, some miscellaneous library functions are defined under a different name. For instance, to use the `minpoly` command, you must `readlib(lattice)`. This case is denoted by "minpoly $\star$*lattice*" in this chapter.

Options to commands are also listed. In most cases, options are printed after the command name, with a slash (/) separating the command name and the option name, e.g., "normal/expanded", "linalg[det/sparse]", and "C/optimized $\star$". As a special case, graphics options are listed separately in the **Graphics** section, since they can be used with many different commands. A template or a sequence of choices is also given for each graphics option, as in "numpoints = *n*" or "arrows = (THIN, LINE, SLIM, THICK)". In the latter case, if there is a default value, it is listed first.

Help topics that are not simply Maple commands are listed, too. All help topics begin with a question mark (?) in this chapter.

Commands and options that are new in Maple V Release 3 are marked with a raised plus sign, as in "protect$^+$", "sqrt/symbolic$^+$", "student[Lineint]$^+$", and "piecewise$^+$ $\star$". Commands and options present in Maple V Release 2 but removed in Release 3 are marked with a slashed circle: for example, "linalg[range]$^\oslash$".

These annotations are summarized in the following table.

Listing	**Type of Command**	**Example**
fcn	Built-in or standard library function	sqrt
fcn/option	Standard function accepting an option	convert/polynom
pkg[fcn]	Command defined in a package	linalg[det]
pkg[subpkg, fcn]	Command defined in a subpackage	stats[describe, mean]
pkg[fcn/option]	Function with an option defined in a package	linalg[det/sparse]
fcn $\star$	Miscellaneous library function	FFT $\star$
fcn $\star$*name*	Miscellaneous library function defined in *name*	iFFT $\star$*FFT*
?topic	Help topic	?index
?topic[subtopic]	Help subtopic	?index[library]
command$^+$	New command in Maple V Release 3	Berlekamp$^+$
command$^\oslash$	Command removed in Maple V Release 3	linalg[range]$^\oslash$

1 GRAPHICS

1.1 Two-Dimensional

networks[draw]
plot
plots[animate]
plots[conformal]
plots[display]
plots[fieldplot]
plots[gradplot]
plots[implicitplot]
plots[loglogplot]
plots[logplot]
plots[odeplot]
plots[polarplot]
plots[polygonplot]
plots[replot]
plots[sparsematrixplot]
plots[textplot]
stats[statplots, boxplot]$^+$
stats[statplots, changecolour]$^+$
stats[statplots, histogram]$^+$
stats[statplots, notchedbox]$^+$
stats[statplots, quantile]$^+$
stats[statplots, quantile2]$^+$
stats[statplots, scatter1d]$^+$
stats[statplots, scatter2d]$^+$
stats[statplots, symmetry]$^+$
stats[statplots, xscale]$^+$
stats[statplots, xshift]$^+$
stats[statplots, xyexchange]$^+$
stats[statplot]$^\oslash$
student[leftbox]
student[middlebox]
student[rightbox]
student[showtangent]
type/PLOT
?plot[function]
?plot[infinity]
?plot[multiple]
?plot[parametric]
?plot[polar]
?plot[range]
?plot[structure]

Options

General

axes = (NORMAL, BOXED, FRAME, NONE)
axesfont$^+$= [*family*, *style*, *size*]
color *or* colour = (black, aquamarine, blue, brown, coral, cyan, gold, gray, green, grey, khaki, magenta, maroon, navy, orange, pink, plum, red, sienna, tan, turquoise, violet, wheat, white, yellow, COLOR(RGB, *r*, *g*, *b*), COLOR(HUE, *h*))
font$^+$= [*family*, *style*, *size*]
labelfont$^+$= [*family*, *style*, *size*]
scaling = (UNCONSTRAINED, CONSTRAINED)
style = (LINE, PATCH, POINT)
symbol$^+$= (BOX CIRCLE, CROSS, DEFAULT, DIAMOND, POINT)
titlefont$^+$= [*family*, *style*, *size*]

a..b
coords = polar
discont$^+$= true
labels = [str_x, str_y]
linestyle$^+$= *n*
numpoints = *n*
plots[setoptions]
resolution = *n*
thickness$^+$= *n*
tickmarks = [n_x, n_y]
title = *str*
view$^+$= [$x_1..x_2$, $y_1..y_2$]
xtickmarks = *n*
ytickmarks = *n*
?plot[options]
?plot[style]

See page 26 for details on two-dimensional plotting options.

Special

networks[draw]	Concentric($list_1$, ...), Linear($list_1$, ...)
plots[animate]	frames = *n*
plots[conformal]	grid = [n_x, n_y], numxy = [n_x, n_y]
plots[display]	insequence = true
plots[fieldplot]	grid = [n_x, n_y], arrows = (THIN, LINE, SLIM, THICK)
plots[gradplot]	grid = [n_x, n_y], arrows = (THIN, LINE, SLIM, THICK)

plots[implicitplot]	grid = $[n_x, n_y]$
plots[textplot]	align = (ABOVE, BELOW, LEFT, RIGHT)

1.2 Three-Dimensional

plot3d	plots[implicitplot3d]	plots[sphereplot]
plots[animate3d]	plots[matrixplot]	plots[surfdata]
plots[contourplot]	plots[odeplot]	plots[textplot3d]
plots[cylinderplot]	plots[pointplot]	plots[tubeplot]
plots[densityplot]	plots[polygonplot3d]	stats[statplots, changecolour]$^+$
plots[display3d]	plots[polyhedraplot]	type/PLOT3D
plots[fieldplot3d]	plots[replot]	?plot3d[structure]
plots[gradplot3d]	plots[spacecurve]	

Options

General

axes = (NONE, BOXED, FRAME, NORMAL)
axesfont$^+$= [*family*, *style*, *size*]
color *or* colour = (black, aquamarine, blue, brown, coral, cyan, gold, gray, green, grey, khaki, magenta, maroon, navy, orange, pink, plum, red, sienna, tan, turquoise, violet, wheat, white, yellow, COLOR(RGB, *r*, *g*, *b*), COLOR(HUE, *h*), *f*)
coords = (cartesian, cylindrical, spherical)
font$^+$= [*family*, *style*, *size*]
labelfont$^+$= [*family*, *style*, *size*]
projection = (ORTHOGONAL, FISHEYE, NORMAL, *p*)
scaling = (UNCONSTRAINED, CONSTRAINED)
shading = (XYZ, XY, Z, ZHUE, ZGRAYSCALE, NONE)
style = (HIDDEN, CONTOUR, LINE, PATCH, PATCHCONTOUR, PATCHNOGRID, POINT, WIREFRAME)
symbol$^+$= (BOX CIRCLE, CROSS, DEFAULT, DIAMOND, POINT)
titlefont$^+$= [*family*, *style*, *size*]
view = $z_1..z_2$ *or* $[x_1..x_2, y_1..y_2, z_1..z_2]$

ambientlight = $[r, g, b]$	linestyle$^+$= n	tickmarks = $[n_x, n_y, n_z]$
contours$^+$= c	numpoints = n	title = *str*
grid = $[n_x, n_y]$	orientation = $[\theta, \phi]$	?plot3d[colorfunc]
labels = $[str_x, str_y, str_z]$	plots[setoptions3d]	?plot3d[options]
light = $[\phi, \theta, r, g, b]$	thickness$^+$= n	

See page 34 for details on three-dimensional plotting options.

Special

plots[animate3d]	frames = n
plots[display3d]	insequence = true
plots[fieldplot3d]	grid = $[n_x, n_y, n_z]$, arrows = (THIN, LINE, SLIM, THICK)
plots[gradplot3d]	grid = $[n_x, n_y, n_z]$, arrows = (THIN, LINE, SLIM, THICK)
plots[implicitplot3d]	grid = $[n_x, n_y, n_z]$

plots[matrixplot]	heights = histogram, gap = *r*
plots[polyhedraplot]	polyscale = *r*, polytype = (tetrahedron, dodecahedron, hexahedron, icosahedron, octahedron)
plots[textplot3d]	align = (ABOVE, BELOW, LEFT, RIGHT)
plots[tubeplot]	radius = *f*, tubepoints = *n*

1.3 Differential Equations

DEtools[dfieldplot] DEtools[DEplot1] DEtools[PDEplot] plots[odeplot]
DEtools[DEplot] DEtools[DEplot2] DEtools[phaseportrait]

Options

arrows = (THIN, LINE, SLIM, THICK)
method = (rk4, backeuler, 'euler', impeuler, *f*)
label = [str_1, str_2] *or* [str_1, str_2, str_3]
scene = [var_1, var_2] *or* [var_1, var_2, var_3]
tickmarks = [n_1, n_2] *or* [n_1, n_2, n_3]
arraysize = *n* iterations = *n* mparam = [*args*] title = *str*
grid = [*n*, *m*] limitrange = (false, true) stepsize = *h* ?DEtools[options]

1.4 Graphics Devices

interface/plotdevice interface/terminal plotsetup/preplot⁺
interface/plotoptions⁺ plotsetup plotsetup/postplot⁺
interface/plotoutput plotsetup/plotdevice⁺ ?plot[device]
interface/postplot plotsetup/plotoptions⁺ ?plot[setup]⊘
interface/preplot plotsetup/plotoutput⁺

Terminals

char kd404g kd500g mac regis tek vt100 wy99gt x11

Printers and Special Formats

bmp deskjet epson9hi hplj ibm_mon laserjet paintjet postscript tiff
canon epson24 gif⁺ i300 ibm_pro ln03 pcx postscriptc toshiba
cps epson9 hpgl ibm ibmquiet oki92 pic ps unix

2 MATHEMATICS

2.1 Algebra

Algebraic Manipulation

asubs ⋆ combine/trig normal simplify/hypergeom
asubs/always ⋆ convert normal/expanded simplify/ln
collect expand Normal simplify/polar
collect/distributed Expand Normalizer simplify/power
collect/recursive expandoff op simplify/radical

combine
combine/abs$^+$
combine/exp
combine/ln
combine/power
combine/Psi
combine/radical$^+$
combine/radical/symbolic$^+$
combine/signum$^+$
expandon
factor
Factor
freeze ⋆
frontend
indets
invfunc ⋆
isolate ⋆
match
optimize ⋆
radnormal ⋆
radsimp
rationalize$^+$ ⋆
simplify
simplify/assume$^+$
simplify/atsign
simplify/Ei
simplify/GAMMA
simplify/RootOf
simplify/sqrt
simplify/trig
?simplify[siderels]
sort
subs
subsop
thaw ⋆*freeze*

Number Fields and Function Fields

AFactor
AFactors
alias
Content
convert/radical
convert/RootOf
Divide
evala
evala/Factor/lenstra
evala/Factor/linear
evala/Factor/trager
evala/Factors/lenstra
evala/Factors/linear
evala/Factors/trager
evala/independent
Expand
factor
Factor
Factors
Gcd
Gcdex
Indep
irreduc
Irreduc
maxorder ⋆
minpoly ⋆*lattice*
Norm
Normal
numtheory[factorEQ]$^+$
numtheory[sq2factor]
Prem
Primfield
Primpart
Quo
Rem
Resultant
RootOf
roots
simplify/RootOf
Sprem
sqrfree
Sqrfree
Subres
Trace
type/algext
type/algfun
type/algnum
type/algnumext
type/radext
type/radfun
type/radfunext
type/radnum
type/radnumext
type/RootOf

Finite Fields

evalgf ⋆
GF ⋆
'+' ⋆*GF*
'-' ⋆*GF*
'*' ⋆*GF*
'/' ⋆*GF*
'^' ⋆*GF*
0 ⋆*GF*
1 ⋆*GF*
ConvertIn ⋆*GF*
ConvertOut ⋆*GF*
extension ⋆*GF*
input ⋆*GF*
inverse ⋆*GF*
isPrimitiveElement ⋆*GF*
norm ⋆*GF*
order ⋆*GF*
output ⋆*GF*
PrimitiveElement ⋆*GF*
random ⋆*GF*
trace ⋆*GF*
modpol ⋆
Primitive
Randpoly
Randprime
RootOf

See also 2.1 Algebra: Polynomials: Modular Arithmetic (p. 117)
2.17 Number Theory: Modular Arithmetic (p. 130)

Gröbner Bases

grobner[finduni]
grobner[finite]
grobner[gbasis]
grobner[gbasis/plex]
grobner[gbasis/tdeg]
grobner[gsolve]
grobner[leadmon]
grobner[leadmon/plex]
grobner[leadmon/tdeg]
grobner[normalf]
grobner[normalf/plex]
grobner[normalf/tdeg]
grobner[solvable]
grobner[solvable/plex]
grobner[solvable/tdeg]
grobner[spoly]
grobner[spoly/plex]
grobner[spoly/tdeg]
?simplify[siderels]

Group Theory

galois	group[cosrep]	group[isabelian]	group[permgroup]
group[areconjugate]	group[derived]	group[isnormal]	group[permrep]
group[center]	group[DerivedS]	group[issubgroup]	group[pres]
group[centralizer]	group[grelgroup]	group[LCS]	group[RandElement]
group[convert/disjcyc]	group[groupmember]	group[mulperms]	group[subgrel]
group[convert/permlist]	group[grouporder]	group[NormalClosure]	group[Sylow]
group[core]	group[inter]	group[normalizer]	group[type/disjcyc]
group[cosets]	group[invperm]	group[orbit]	

Noncommutative Multiplication

&*	convert/c ⋆*commutat*	c ⋆*commutat*	expand/&* ⋆*commutat*
commutat ⋆	convert/&* ⋆*commutat*	expand/c ⋆*commutat*	simplify/c ⋆*commutat*

Polynomials

bernoulli	icontent	numapprox[hornerform]	resultant
bernstein ⋆	indets	numapprox[minimax]	Resultant
bspline ⋆	interp	numapprox[remez]	sign
combinat[fibonacci]	Interp	numtheory[cyclotomic]	spline ⋆
content	lcm	polynom	sprem
Content	linalg[bezout]	powseries[powpoly]	Sprem
convert/polynom	linalg[charpoly]	prem	Subres
degree	linalg[companion]	Prem	subs
discrim	linalg[hermite]	primpart	translate ⋆
divide	linalg[minpoly]	Primpart	type/expanded
Divide	linalg[smith]	quo	type/monomial
euler	linalg[sylvester]	Quo	type/polynom
fixdiv ⋆	maxnorm	recipoly ⋆	type/square
galois	minpoly ⋆*lattice*	rem	?polynomials
genpoly	norm	Rem	?polyrefs

Factoring and Finding Roots

AFactor	factors	Gcdex	RootOf
AFactors	Factors	irreduc	roots
allvalues	fsolve	Irreduc	Roots
allvalues/d	fsolve/complex	numtheory[cfracpol]	solve
Berlekamp$^+$	fsolve/fulldigits	padic[rootp]	split ⋆
convert/radical	fsolve/maxsols	ProbSplit$^+$	sqrfree
convert/RootOf	GaussInt[GIfacpoly]$^+$	proot ⋆	Sqrfree
convert/sqrfree	GaussInt[GIroots]	psqrt ⋆	student[completesquare]
DistDeg	gcd	realroot ⋆	sturmseq ⋆*sturm*
factor	Gcd	rootbound ⋆	sturm ⋆
Factor	gcdex		

See also 2.19 Solving Equations and Recurrences (p. 131)

Manipulation of Polynomials

collect	convert/horner	convert/series	sort	coeff	ldegree
collect/distributed	convert/mathorner	convert/sqrfree	sort/plex	coeffs	op
collect/recursive	convert/polynom	degree	sort/tdeg	lcoeff	tcoeff
compoly	convert/ratpoly	expand			

Modular Arithmetic

Add	Det	Gcd	modpol ⋆	Primitive	Roots
Berlekamp[+]	Diff	Gcdex	Monomial	Primpart	Shift
chrem	dinterp ⋆	GetAlgExt	msolve	ProbSplit[+]	sinterp ⋆
Chrem	Discrim	Hermite	Multiply	Quo	Smith
Coeff	DistDeg	Interp	numtheory[mipolys]	Randpoly	Sprem
Constant	Divide	Irreduc	One	Randprime	Sqrfree
Content	Embed	Lcm	Power	ratrecon ⋆	Subtract
convert/mod2	Eval	Lcoeff	Powmod	Ratrecon	Tcoeff
ConvertIn	evalgf ⋆	Ldegree	powmod ⋆	Rem	UNormal
ConvertOut	Expand	mod	Prem	Resultant	Zero
Degree	Factors	modp1			

See also 2.1 Algebra: Finite Fields (p. 115)
2.17 Number Theory: Modular Arithmetic (p. 130)

Orthogonal Polynomials

chebyshev	orthopoly[H]	orthopoly[P]	orthopoly[U]
orthopoly[G]	orthopoly[L]	orthopoly[T]	

Random Polynomials

randpoly	randpoly/degree	randpoly/expons	Randpoly
randpoly/coeffs	randpoly/dense	randpoly/terms	Randprime

Rational Functions

asympt	Normalizer	numtheory[cfrac/quotients]	sort
convert/confrac	numapprox[chebpade]	numtheory[cfrac/regular]	subs
convert/parfrac	numapprox[confracform]	numtheory[cfrac/simple]	thiele ⋆
convert/ratpoly	numapprox[hornerform]	numtheory[nthconver]	type/fraction
denom	numapprox[minimax]	rationalize[+] ⋆	type/rational
frontend	numapprox[pade]	ratrecon	type/ratpoly
normal	numapprox[remez]	Ratrecon	type/square
normal/expanded	numer	singular ⋆	?ratpolys
Normal	numtheory[cfrac]		

Student Algebra

student[changevar]	student[equate]	student[makeproc/slope]	student[powsubs]
student[combine]	student[intercept]	student[midpoint]	student[slope]
student[completesquare]	student[isolate]		

2.2 Approximations

Real Numbers

ceil
convergs ⋆
convert/binary
convert/confrac
convert/double/ibm
convert/double/maple
convert/double/mips
convert/double/vax
convert/float
convert/fraction
convert/fraction/exact
convert/octal
convert/rational
convert/rational/exact
Digits
evalf
evalhf
FFT ⋆
Float
floor
fnormal
frac
fsolve
fsolve/complex
fsolve/fulldigits
fsolve/maxsols
harmonic
iFFT ⋆*FFT*
minpoly ⋆*lattice*
numtheory[cfracpol]
numtheory[cfrac]
numtheory[cfrac/centered]
numtheory[cfrac/quotients]
numtheory[kronecker]
numtheory[minkowski]
numtheory[nearestp]
realroot ⋆
round
shake ⋆
surd$^+$ ⋆
testfloat ⋆
testfloat/digits ⋆
testfloat/model ⋆
testfloat/test ⋆
trunc
type/float
type/negative
type/nonneg
type/positive
type/realcons
?solve[floats]

Functions

asympt
bernstein ⋆
chebyshev
coeftayl ⋆
convert/confrac
convert/polynom
convert/ratpoly
convert/series
dsolve/numeric
eulermac ⋆
mtaylor ⋆
numapprox[chebpade]
numapprox[chebyshev]
numapprox[confracform]
numapprox[hornerform]
numapprox[infnorm]
numapprox[laurent]
numapprox[minimax]
numapprox[pade]
numapprox[remez]
numapprox[taylor]
numtheory[cfrac]
numtheory[cfrac/quotients]
numtheory[cfrac/regular]
numtheory[cfrac/semisimple]
numtheory[cfrac/simple]
numtheory[cfrac/simregular]
numtheory[cfrac/subdiagonal]
numtheory[cfrac/superdiagonal]
numtheory[nthconver]
O
order
Order
poisson ⋆
residue ⋆
series
taylor
type/laurent
type/series
type/taylor

Interpolation and Curve Fitting

bspline ⋆
dinterp ⋆
interp
Interp
sinterp ⋆
spline ⋆
spline/cubic ⋆
spline/linear ⋆
spline/quadratic ⋆
spline/quartic ⋆
'spline/makeproc' ⋆*spline*
stats[fit, leastsquare]$^+$
stats[linregress]$^{\oslash}$
stats[multregress]$^{\oslash}$
stats[multregress/const]$^{\oslash}$
stats[regression]$^{\oslash}$
thiele ⋆

See also 2.4 Calculus: Integration: Numerical Integration (p. 120)
2.4 Calculus: Power Series (p. 120)
2.17 Number Theory: Continued Fractions (p. 130)

2.3 Basic Mathematical Functions

Basic Arithmetic

+	/	&*	convert/'+'	igcd	irem	product	sqrt/symbolic⁺
-	^	!	convert/'*'	ilcm	max	signum	sum
*	**	abs	factorial	iquo	min	sqrt	?arithmetic

Exponentials and Logarithms

^ ** E exp gamma ln log log10 ⋆ log[*b*]

Trigonometric and Hyperbolic Functions

arccos	arccoth	arcsec	arcsinh	convert/degrees	cosh	csc	sec	sinh
arccosh	arccsc	arcsech	arctan	convert/radians	cot	csch	sech	tan
arccot	arccsch	arcsin	arctanh	cos	coth	Pi	sin	tanh

Other Mathematical Functions

Ai	binomial	erfc	Heaviside	LegendrePi	Shi ⋆
bernoulli	Chi ⋆	euler	hypergeom ⋆	LegendrePic	Si
BesselI	Ci	FresnelC	LegendreE	LegendrePic1	Ssi ⋆
BesselJ	dawson ⋆	Fresnelf ⋆	LegendreEc	Li ⋆	W
BesselK	dilog	Fresnelg ⋆	LegendreEc1	lnGAMMA ⋆	Zeta
BesselY	Dirac	FresnelS	LegendreF	MeijerG	?funcrefs
Beta	Ei	GAMMA	LegendreKc	Psi	?inifcns
Bi	erf	harmonic	LegendreKc1		

Manipulation

combine/abs⁺	convert/degrees	convert/hypergeom	simplify/hypergeom
combine/exp	convert/Ei	convert/ln	simplify/ln
combine/ln	convert/erf	convert/radians	simplify/radical
combine/Psi	convert/erfc	convert/sincos	simplify/sqrt
combine/radical⁺	convert/exp	convert/tan	simplify/trig
combine/radical/symbolic⁺	convert/expln	convert/trig	trigsubs ⋆
combine/signum⁺	convert/expsincos	radnormal ⋆	type/mathfunc
combine/trig	convert/factorial	simplify/Ei	type/trig
convert/binomial⁺	convert/GAMMA	simplify/GAMMA	

2.4 Calculus

Continuity and Limits

asympt	iscont ⋆	limit	limit/left	limit/right	student[Limit]
discont ⋆	iscont/closed ⋆	limit/complex	limit/real	Limit	

Differentiation and Extremal Problems

convert/D	extrema ⋆	numapprox[infnorm]	student[minimize]
convert/diff	maximize ⋆*minimize*	powseries[powdiff]	student[showtangent]

D	maximize/infinite^{+} ⋆*minimize*	student[D]	student[value]
diff	minimize ⋆	student[extrema]	?operators[D]
Diff	minimize/infinite^{+} ⋆	student[maximize]	

See also 2.15 Linear Programming (p. 128)
2.16 Mathematical Physics (p. 128)

Differential Equations

Ai	DEtools[DEplot2]	dsolve/type^{+}
BesselI	DEtools[dfieldplot]	fourier ⋆
BesselJ	DEtools[PDEplot]	invfourier ⋆*fourier*
BesselK	DEtools[phaseportrait]	invlaplace ⋆*laplace*
BesselY	dsolve	laplace
Bi	dsolve/explicit	mellin
DESol^{+}	dsolve/laplace	mellintable ⋆*mellin*
DEtools[Dchangevar]	dsolve/method^{+}	plots[odeplot]
DEtools[DEplot]	dsolve/numeric	powseries[powsolve]
DEtools[DEplot1]	dsolve/series	?dsolve[dverk78]$^{+}$

See also 2.16 Mathematical Physics: Lie Symmetry (p. 129)

Integration

assume	erfc	LegendreEc	Si
Chi ⋆	FresnelC	LegendreEc1	simplify/Ei
Ci	Fresnelf ⋆	LegendreF	Ssi ⋆
convert/Ei	Fresnelg ⋆	LegendreKc	student[changevar]
convert/erf	FresnelS	LegendreKc1	student[Doubleint]
convert/erfc	int	LegendrePi	student[Int]
dawson ⋆	int/CauchyPrincipalValue	LegendrePic	student[integrand]
dilog	int/continuous	LegendrePic1	student[intparts]
Ei	Int	Li ⋆	student[Lineint]$^{+}$
ellipsoid ⋆	integrate	powseries[powint]	student[Tripleint]
erf	LegendreE	Shi ⋆	?intrefs

Numerical Integration

evalf	'evalf/int'/_Dexp ⋆	student[middlebox]	student[simpson]
evalf/Int	'evalf/int'/_NCrule ⋆	student[middlesum]	student[trapezoid]
'evalf/int' ⋆	student[leftbox]	student[rightbox]	?int[numerical]
'evalf/int'/_CCquad ⋆	student[leftsum]	student[rightsum]	

Power Series

coeftayl ⋆	powseries[evalpow]	powseries[reversion]
convert/confrac	powseries[inverse]	powseries[subtract]
convert/polynom	powseries[multconst]	powseries[tpsform]
convert/ratpoly	powseries[multiply]	residue ⋆
convert/series	powseries[negative]	series
mtaylor ⋆	powseries[powcreate]	series/leadterm

numapprox[laurent]
O
order
Order
poisson ⋆
powseries[add]
powseries[compose]
powseries[powdiff]
powseries[powexp]
powseries[powint]
powseries[powlog]
powseries[powpoly]
powseries[powsolve]
powseries[quotient]
taylor
type/laurent
type/series
type/taylor
?powseries[powseries]
?solve[series]

See also 2.2 Approximations: Functions (p. 118)
2.7 Discrete Mathematics: Generating Functions and Recurrence Relations (p. 122)

Vector Calculus

linalg[curl]
linalg[curl/coords]$^+$
linalg[diverge]
linalg[diverge/coords]$^+$
linalg[grad]
linalg[grad/coords]$^+$
linalg[hessian]
linalg[jacobian]
linalg[laplacian]
linalg[laplacian/coords]$^+$
linalg[potential]
linalg[vecpotent]
linalg[Wronskian]

2.5 Complex Numbers

abs
argument
cmagdiff ⋆
combine/conjugate
conjugate
convert/polar
convert/radical
convert/RootOf
csgn
evalc
I
Im
polar ⋆
Re
residue ⋆
RootOf
signum
simplify/polar
sortcx ⋆
sortcx/abs ⋆
sortcx/lexImRe ⋆
sortcx/lexReIm ⋆
sortcx/polar ⋆
testfloat/model ⋆
type/complex
?evalc[functions]

Gaussian Integers

GaussInt[GIbasis]
GaussInt[GIchrem]$^+$
GaussInt[GIdivisor]
GaussInt[GIfacpoly]$^+$
GaussInt[GIfacset]
GaussInt[GIfactor]
GaussInt[GIfactors]
GaussInt[GIgcd]
GaussInt[GIgcdex]
GaussInt[GIhermite]$^+$
GaussInt[GIissqr]
GaussInt[GIlcm]
GaussInt[GImcmbine]
GaussInt[GInearest]
GaussInt[GInodiv]$^+$
GaussInt[GInorm]
GaussInt[GInormal]$^+$
GaussInt[GIorder]
GaussInt[GIphi]
GaussInt[GIprime]
GaussInt[GIquadres]
GaussInt[GIquo]
GaussInt[GIrem]
GaussInt[GIroots]
GaussInt[GIsieve]
GaussInt[GIsmith]$^+$
GaussInt[GIsqrfree]$^+$
GaussInt[GIsqrt]
GaussInt[GIunitnormal]$^+$
numtheory[GIgcd]
type/complex(integer)

2.6 Constants

Catalan
constants
E
FAIL
false
gamma
I
infinity
Pi
true
type/constant
type/infinity$^+$
type/realcons
?evalhf[constants]

2.7 Discrete Mathematics

Boolean Logic

convert/mod2
logic[&and]
logic[&iff]
logic[&implies]
logic[&nand]
logic[&nor]
logic[&not]
logic[&or]
logic[&xor]
logic[bequal]
logic[bsimp]
logic[canon]
logic[canon/CNF]
logic[canon/DNF]
logic[canon/MOD2]
logic[convert/frominert]
logic[convert/MOD2]
logic[convert/MOD2/expanded]
logic[convert/toinert]
logic[distrib]
logic[dual]
logic[environ]
logic[randbool]
logic[randbool/CNF]
logic[randbool/DNF]
logic[randbool/MOD2]
logic[satisfy]
logic[tautology]

Combinatorics

!
binomial
combinat[bell]
combinat[binomial]
combinat[cartprod]
combinat[character]
combinat[Chi]
combinat[choose]
combinat[composition]
combinat[conjpart]
combinat[decodepart]
combinat[encodepart]
combinat[fibonacci]
combinat[firstpart]
combinat[graycode]
combinat[inttovec]
combinat[lastpart]
combinat[multinomial]
combinat[nextpart]
combinat[numbcomb]
combinat[numbcomp]
combinat[numbpart]
combinat[numbperm]
combinat[partition]
combinat[permute]
combinat[powerset]
combinat[prevpart]
combinat[randcomb]
combinat[randpart]
combinat[randperm]
combinat[stirling1]
combinat[stirling2]
combinat[subsets]
combinat[vectoint]
convert/binomial$^+$
convert/factorial
convert/GAMMA
group[convert/disjcyc]
group[convert/permlist]
group[groupmember]
group[inter]
group[invperm]
group[mulperms]
group[permgroup]
group[type/disjcyc]
MOLS ⋆

See also 2.1 Algebra: Group Theory (p. 116)

Generating Functions and Recurrence Relations

genfunc[rgf_charseq]
genfunc[rgf_encode]
genfunc[rgf_expand]
genfunc[rgf_findrecur]
genfunc[rgf_hybrid]
genfunc[rgf_norm]
genfunc[rgf_norm/factored]
genfunc[rgf_pfrac]
genfunc[rgf_pfrac/no_RootOf]
genfunc[rgf_relate]
genfunc[rgf_sequence]
genfunc[rgf_sequence/boundary]
genfunc[rgf_sequence/coeffs]
genfunc[rgf_sequence/delta]
genfunc[rgf_sequence/first]
genfunc[rgf_sequence/firstcf]
genfunc[rgf_sequence/firstrecur]
genfunc[rgf_sequence/order]
genfunc[rgf_sequence/recur]
genfunc[rgf_simp]
genfunc[rgf_term]
genfunc[termscale]
genfunc[type/rgf_seq]
invztrans ⋆*ztrans*
rsolve
rsolve/genfunc
rsolve/makeproc
ztrans
ztrans/Delta
ztrans/Step

See also 2.4 Calculus: Power Series (p. 120)

Graph Theory

networks[acycpoly]
networks[addedge]
networks[delete]
networks[departures]
networks[incident/Out]
networks[indegree]

networks[addedge/Cycle]
networks[addedge/names]
networks[addedge/Path]
networks[addedge/weights]
networks[addvertex]
networks[addvertex/weights]
networks[adjacency]
networks[allpairs]
networks[ancestor]
networks[arrivals]
networks[bicomponents]
networks[charpoly]
networks[chrompoly]
networks[complement]
networks[complete]
networks[components]
networks[connect]
networks[connect/directed]
networks[connect/names]
networks[connect/weights]
networks[connectivity]
networks[contract]
networks[countcuts]
networks[counttrees]
networks[cube]
networks[cycle]
networks[cyclebase]
networks[daughter]
networks[degreeseq]
networks[diameter]
networks[dinic]
networks[djspantree]
networks[dodecahedron]
networks[draw]
networks[draw/Concentric]
networks[draw/Linear]
networks[duplicate]
networks[edges]
networks[edges/all]
networks[ends]
networks[eweight]
networks[flow]
networks[flow/maxflow]
networks[flowpoly]
networks[fundcyc]
networks[getlabel]
networks[girth]
networks[graph]
networks[graphical]
networks[graphical/MULTI]
networks[gsimp]
networks[gunion]
networks[gunion/SIMPLE]
networks[head]
networks[icosahedron]
networks[incidence]
networks[incident]
networks[incident/In]
networks[induce]
networks[isplanar]
networks[maxdegree]
networks[mincut]
networks[mindegree]
networks[neighbors]
networks[new]
networks[octahedron]
networks[outdegree]
networks[path]
networks[petersen]
networks[random]
networks[random/prob]
networks[rank]
networks[rankpoly]
networks[shortpathtree]
networks[show]
networks[shrink]
networks[span]
networks[spanpoly]
networks[spantree]
networks[tail]
networks[tetrahedron]
networks[tuttepoly]
networks[type/GRAPH]
networks[vdegree]
networks[vertices]
networks[void]
networks[vweight]

Sums and Products

asubs ⋆
asubs/always ⋆
bernoulli
Catalan
convert/'+'
convert/'*'
convert/hypergeom
eulermac ⋆
gamma
harmonic
hypergeom ⋆
product
Product
select
simplify/hypergeom
student[Sum]
student[value]
sum
Sum
ztrans
ztrans/Delta
ztrans/Step

2.8 Evaluation

' '
allvalues
allvalues/d
assume
convert/float
cost ⋆
Digits
evala
evalb
evalc
?evalc[functions]
evalf
'evalf/int' ⋆
evalgf ⋆
?evalhf[arrays]
?evalhf[boolean]
?evalhf[constants]
?evalhf[fcnlist]
?evalhf[fortran]
?evalhf[function]
evalm
freeze ⋆
frontend
optimize ⋆
powseries[evalpow]
student[value]
subs
subsop

Digitsp	evalhf	evaln	thaw ⋆*freeze*
eval	evalhf/Digits	padic[evalp]	value
Eval	evalhf/var	evalr ⋆	?uneval

2.9 Finance

amortization ⋆*finance*	finance ⋆	finance/interest ⋆	finance/periods ⋆
blackscholes ⋆*finance*	finance/amount ⋆	finance/payment ⋆	

2.10 Functional Transforms

Dirac	fourier ⋆	invfourier ⋆*fourier*	laplace	ztrans
dsolve/laplace	Heaviside	invlaplace ⋆*laplace*	mellin	ztrans/Delta
FFT ⋆	iFFT ⋆*FFT*	invztrans ⋆*ztrans*	mellintable ⋆*mellin*	ztrans/Step

2.11 Geometry

Student Geometry

student[distance]	student[makeproc/slope]	student[showtangent]	student[type/Point]
student[intercept]	student[midpoint]	student[slope]	

Two-Dimensional Euclidean Geometry

geometry[altitude]
geometry[Appolonius]
geometry[area]
geometry[bisector]
geometry[center]
geometry[centroid]
geometry[circle]
geometry[circle/diameter]
geometry[circumcircle]
geometry[conic]
geometry[convexhull]
geometry[coordinates]
geometry[detailf]
geometry[diameter]
geometry[distance]
geometry[ellipse]
geometry[ellipse/x_axis]
geometry[ellipse/y_axis]
geometry[Eulercircle]
geometry[Eulerline]
geometry[excircle]
geometry[find_angle]
geometry[Gergonnepoint]
geometry[harmonic]
geometry[make_square/diagonal]
geometry[median]
geometry[midpoint]
geometry[Nagelpoint]
geometry[onsegment]
geometry[orthocenter]
geometry[parallel]
geometry[perpendicular]
geometry[perpen_bisector]
geometry[point]
geometry[polar_point]
geometry[pole_line]
geometry[powerpc]
geometry[projection]
geometry[radius]
geometry[rad_axis]
geometry[rad_center]
geometry[randpoint]
geometry[reflect]
geometry[rotate]
geometry[rotate/clockwise]
geometry[rotate/counterclockwise]
geometry[sides]
geometry[similitude]

geometry[incircle]
geometry[inter]
geometry[inversion]
geometry[line]
geometry[make_square]
geometry[make_square/adjacent]
geometry[make_square/center]
geometry[Simsonline]
geometry[slope]+
geometry[square]
geometry[symmetric]
geometry[tangentpc]
geometry[tangent]
geometry[triangle]

Two-Dimensional Tests

geometry[are_collinear]
geometry[are_concurrent]
geometry[are_harmonic]
geometry[are_orthogonal]
geometry[are_parallel]
geometry[are_perpendicular]
geometry[are_similar]
geometry[are_tangent]
geometry[concyclic]
geometry[is_equilateral]
geometry[is_right]
geometry[on_circle]
geometry[on_line]
geometry[type/circle2d]
geometry[type/line2d]
geometry[type/point2d]

Three-Dimensional Euclidean Geometry

ellipsoid ⋆
geom3d[angle]
geom3d[area]
geom3d[center]
geom3d[centroid]
geom3d[coordinates]
geom3d[coplanar]
geom3d[distance]
geom3d[inter]
geom3d[line3d]
geom3d[midpoint]
geom3d[onsegment]
geom3d[parallel]
geom3d[perpendicular]
geom3d[perpendicular/line]
geom3d[perpendicular/line3d]
geom3d[plane]
geom3d[point3d]
geom3d[powerps]
geom3d[projection]
geom3d[radius]
geom3d[rad_plane]
geom3d[reflect]
geom3d[sphere]
geom3d[sphere/diameter]
geom3d[symmetric]
geom3d[tangent]
geom3d[tetrahedron]
geom3d[triangle3d]
geom3d[volume]

Three-Dimensional Tests

geom3d[are_collinear]
geom3d[are_concurrent]
geom3d[are_parallel]
geom3d[are_perpendicular]
geom3d[are_tangent]
geom3d[on_plane]
geom3d[on_sphere]
geom3d[type/line3d]
geom3d[type/plane]
geom3d[type/point3d]
geom3d[type/sphere]
geom3d[type/tetrahedron]
geom3d[type/triangle3d]

Projective Geometry

projgeom[conic]
projgeom[ctangent]
projgeom[fpconic]
projgeom[harmonic]
projgeom[inter]
projgeom[join]
projgeom[lccutc]
projgeom[lccutr]
projgeom[lccutr2p]
projgeom[linemeet]
projgeom[line]
projgeom[midpoint]
projgeom[onsegment]
projgeom[point]
projgeom[polarp]
projgeom[poleline]
projgeom[ptangent]
projgeom[rtangent]

Projective Geometry Tests

projgeom[collinear]
projgeom[concur]
projgeom[conjugate]
projgeom[tangentte]
projgeom[tharmonic]

2.12 Inert Functions

AFactor	Eigenvals	Indep	Prem	Resultant	student[Lineint]$^+$
AFactors	Eval	Int	Primfield	RootOf	student[Sum]
Ai	Expand	Interp	Primitive	Roots	student[Tripleint]
Berlekamp$^+$	Factor	Irreduc	Primpart	Signum	student[value]
Bi	Factors	Lcm	ProbSplit$^+$	Smith	Subres
Content	Gausselim	Limit	Product	Sprem	Sum
DESol$^+$	Gaussjord	Norm	Quo	Sqrfree	Svd
Det	Gcd	Normal	Randpoly	student[Doubleint]	Trace
Diff	Gcdex	Nullspace	Randprime	student[Int]	value
DistDeg	GetAlgExt	Power	Ratrecon	student[Limit]	?inert
Divide	Hermite	Powmod	Rem		

2.13 Integers

binomial	ifactor/easy	isqrt	numtheory[nthpow]
ceil	ifactor/lenstra	issqr $\star$	numtheory[phi]
chrem	ifactor/pollard	linalg[ihermite]	numtheory[sigma]
combinat[bell]	ifactor/squfof	linalg[ismith]	numtheory[tau]
combinat[fibonacci]	ifactors $\star$	mod	numtheory[thue]
combinat[multinomial]	igcd	'mod'	rand
combinat[stirling1]	igcdex	modp	round
combinat[stirling2]	ilcm	mods	trunc
convert/base	ilog10	numtheory[bigomega]$^+$	type/even
convert/binary	ilog $\star$	numtheory[divisors]	type/facint
convert/decimal	iperfpow $\star$	numtheory[ifactor]	type/integer
convert/hex	iquo	numtheory[ifactors]	type/negint
convert/octal	irem	numtheory[invphi]	type/nonnegint
euler	iroot $\star$	numtheory[isolve]	type/odd
factorial	isolve	numtheory[issqrfree]	type/posint
floor	isprime	numtheory[mcombine]	type/square
ifactor	isqrfree $\star$	numtheory[mobius]	

Primes

isprime	numtheory[fermat]	numtheory[mersenne]	numtheory[safeprime]
ithprime	numtheory[isprime]	numtheory[nextprime]	prevprime
nextprime	numtheory[ithprime]	numtheory[prevprime]	type/primeint
numtheory[factorset]			

2.14 Linear Algebra

Matrices

Creating Matrices

array	linalg[companion]	linalg[randmatrix/antisymmetric]
array/antisymmetric	linalg[concat]	linalg[randmatrix/dense]
array/diagonal	linalg[copyinto]	linalg[randmatrix/entries]

array/identity
array/sparse
array/symmetric
convert/array
convert/matrix
linalg[augment]
linalg[band]
linalg[bezout]
linalg[BlockDiagonal]
linalg[blockmatrix]
linalg[diag]
linalg[entermatrix]
linalg[fibonacci]
linalg[genmatrix]
linalg[hessian]
linalg[hilbert]
linalg[jacobian]
linalg[JordanBlock]
linalg[matrix]
linalg[randmatrix]
linalg[randmatrix/sparse]
linalg[randmatrix/symmetric]
linalg[randmatrix/unimodular]
linalg[stack]
linalg[sylvester]
linalg[toeplitz]
linalg[vandermonde]
linalg[Wronskian]
MOLS ⋆
?indexfcn

Normal Forms, Gaussian Elimination, and Solving Systems

Gausselim
GaussInt[GIhermite]+
GaussInt[GIsmith]+
Gaussjord
Hermite
linalg[backsub]
linalg[ffgausselim]
linalg[frobenius]
linalg[gausselim]
linalg[gaussjord]
linalg[hermite]
linalg[ihermite]
linalg[ismith]
linalg[jordan]
linalg[leastsqrs]
linalg[linsolve]
linalg[pivot]
linalg[ratform]
linalg[rref]
linalg[smith]
Smith

Matrix Operations

comparray ⋆
comparray/dontprint ⋆
convert/eqnlist
convert/listlist
convert/mathorner
Det
Eigenvals
evalm
linalg[add]
linalg[addcol]
linalg[addrow]
linalg[adj]
linalg[adjoint]
linalg[charmat]
linalg[charpoly]
linalg[col]
linalg[coldim]
linalg[colspace]
linalg[colspan]
linalg[cond]
linalg[cond/1]
linalg[cond/2]
linalg[cond/frobenius]
linalg[cond/infinity]
linalg[definite/negative_def]
linalg[delrows]
linalg[det]
linalg[det/sparse]
linalg[eigenvals]
linalg[eigenvals/implicit]
linalg[eigenvals/radical]
linalg[eigenvects]
linalg[eigenvects/implicit]
linalg[eigenvects/radical]
linalg[equal]
linalg[exponential]
linalg[extend]
linalg[hadamard]
linalg[htranspose]
linalg[indexfunc]
linalg[innerprod]
linalg[inverse]
linalg[iszero]
linalg[kernel]
linalg[minor]
linalg[minpoly]
linalg[mulcol]
linalg[mulrow]
linalg[multiply]
linalg[norm]
linalg[nullspace]
linalg[orthog]
linalg[permanent]
linalg[range]⊘
linalg[rank]
linalg[row]
linalg[rowdim]
linalg[rowspace]
linalg[rowspan]
linalg[scalarmul]
linalg[singularvals]
linalg[submatrix]
linalg[subvector]
linalg[swapcol]
linalg[swaprow]
linalg[trace]
linalg[transpose]
Nullspace
plots[matrixplot]
plots[sparsematrixplot]
Svd
Svd/left
Svd/right
type/array
type/matrix

linalg[definite/negative_semidef]
linalg[definite/positive_def]
linalg[definite/positive_semidef]
linalg[delcols]
linalg[norm/1]
linalg[norm/2]
linalg[norm/frobenius]
linalg[norm/infinity]
type/matrix/square
type/scalar
?linalg[references]

Vectors

comparray ⋆
comparray/dontprint ⋆
convert/list
convert/vector
lattice ⋆
lattice/integer ⋆
linalg[add]
linalg[angle]
linalg[basis]
linalg[crossprod]
linalg[curl]
linalg[diverge]
linalg[dotprod]
linalg[dotprod/orthogonal]
linalg[grad]
linalg[GramSchmidt]
linalg[innerprod]
linalg[intbasis]
linalg[laplacian]
linalg[norm/frobenius]
linalg[norm/infinity]
linalg[normalize]
linalg[potential]
linalg[randvector]
linalg[randvector/entries]
linalg[scalarmul]
linalg[sumbasis]
linalg[vecpotent]
linalg[vectdim]
linalg[vector]
type/scalar
type/vector

See also 2.16 Mathematical Physics (p. 128)
2.20 Statistics (p. 131)

2.15 Linear Programming

simplex[basis]
simplex[convert/equality]
simplex[convert/std]
simplex[convert/stdle]
simplex[convexhull]
simplex[cterm]
simplex[define_zero]$^{+}$
simplex[display]$^{+}$
simplex[dual]
simplex[feasible]
simplex[feasible/NONNEGATIVE]
simplex[feasible/UNRESTRICTED]
simplex[maximize]
simplex[maximize/NONNEGATIVE]
simplex[maximize/UNRESTRICTED]
simplex[minimize]
simplex[minimize/NONNEGATIVE]
simplex[minimize/UNRESTRICTED]
simplex[pivot]
simplex[pivoteqn]
simplex[pivotvar]
simplex[ratio]
simplex[setup]
simplex[setup/NONNEGATIVE]
simplex[standardize]
simplex[type/NONNEGATIVE]

2.16 Mathematical Physics

cartan ⋆
printcartan ⋆*cartan*
riemann ⋆*cartan*
simp1 ⋆*cartan*
simp2 ⋆*cartan*
clear ⋆*oframe*
com01 ⋆*oframe*
com02 ⋆*oframe*
com03 ⋆*oframe*
com12 ⋆*oframe*
com23 ⋆*oframe*
com31 ⋆*oframe*
const ⋆*oframe*
deriv ⋆*oframe*
e0 ⋆*oframe*
e1 ⋆*oframe*
e2 ⋆*oframe*
e3 ⋆*oframe*
id ⋆*oframe*
transfo ⋆*oframe*
petrov ⋆
nonzero ⋆*petrov*
tensor ⋆
Christoffel1 ⋆*tensor*
Christoffel2 ⋆*tensor*
d1metric ⋆*tensor*
d2metric ⋆*tensor*
display ⋆*tensor*
Einstein ⋆*tensor*
invmetric ⋆*tensor*
Ricci ⋆*tensor*
Ricciscaler ⋆*tensor*
Riemann ⋆*tensor*
Weyl ⋆*tensor*
?relativity

Differential Forms

difforms[&^]
difforms[d]
difforms[defform]
difforms[defform/const]
difforms[defform/even]
difforms[defform/form]
difforms[defform/odd]
difforms[defform/scalar]
difforms[formpart]
difforms[mixpar]
difforms[parity]
difforms[scalarpart]
difforms[simpform]
difforms[type/const]
difforms[type/form]
difforms[type/scalar]
difforms[wdegree]

Lie Symmetry

classi ⋆*bianchi*
liesymm[&^]
liesymm[&mod]
liesymm[annul]
liesymm[autosimp]
liesymm[close]
liesymm[d]
liesymm[depvars]
liesymm[determine]
liesymm[dvalue]
liesymm[Eta]
liesymm[extgen]
liesymm[extvars]
liesymm[getcoeff]
liesymm[getform]
liesymm[hasclosure]
liesymm[hook]
liesymm[indepvars]
liesymm[Lie]
liesymm[Lrank]
liesymm[makeforms]
liesymm[mixpar]
liesymm[prolong]
liesymm[reduce]
liesymm[setup]
liesymm[TD]
liesymm[translate]
liesymm[vfix]
liesymm[wcollect]
liesymm[wdegree]
liesymm[wedgeset]
liesymm[wsubs]

Newman-Penrose

curvature ⋆*debever*
npspin ⋆*debever*
printcurve ⋆*debever*
printspin ⋆*debever*
simp ⋆*debever*
NPspinor[basis]
NPspinor[checkvars]
NPspinor[conj]
NPspinor[contract]
NPspinor[D]
NPspinor[del]
NPspinor[dyad]
NPspinor[eqns]
NPspinor[findsymm]
NPspinor[makeeqn]
NPspinor[rewrite]
NPspinor[suball]
NPspinor[symm]
NPspinor[V]
NPspinor[V_D]
NPspinor[X]
NPspinor[X_D]
NPspinor[X_V]
NPspinor[Y]
NPspinor[Y_D]
NPspinor[Y_V]
NPspinor[Y_X]
np[conj]
np[D]
np[eqns]
np[suball]
np[V]
np[V_D]
np[X]
np[X_D]
np[X_V]
np[Y]
np[Y_D]
np[Y_V]
np[Y_X]

2.17 Number Theory

lattice ⋆
lattice/integer ⋆
Li ⋆
maxorder ⋆
minpoly ⋆*lattice*
numtheory[B]
numtheory[bernoulli]
numtheory[bigomega]$^+$
numtheory[cyclotomic]
numtheory[divisors]
numtheory[euler]
numtheory[F]
numtheory[factorEQ]$^+$
numtheory[factorset]
numtheory[ifactor/easy]
numtheory[ifactor/lenstra]
numtheory[ifactor/pollard]
numtheory[ifactor/squfof]
numtheory[ifactors]
numtheory[invphi]
numtheory[isolve]
numtheory[isprime]
numtheory[issqrfree]
numtheory[ithprime]
numtheory[J]
numtheory[jacobi]
numtheory[kronecker]
numtheory[L]
numtheory[mersenne]
numtheory[minkowski]
numtheory[mobius]
numtheory[nearestp]
numtheory[nextprime]
numtheory[nthpow]
numtheory[phi]
numtheory[prevprime]
numtheory[safeprime]
numtheory[sigma]
numtheory[sq2factor]
numtheory[sum2sqr]$^+$
numtheory[tau]
numtheory[thue]

numtheory[fermat]
numtheory[GIgcd]
numtheory[ifactor]
numtheory[legendre]
numtheory[M]
surd+ ⋆
Zeta

Continued Fractions

convergs ⋆
convert/confrac
numtheory[cfrac]
numtheory[cfrac/centered]
numtheory[cfrac/quotients]
numtheory[cfrac/regular]
numtheory[cfrac/semisimple]
numtheory[cfrac/simple]
numtheory[cfrac/simregular]
numtheory[cfrac/subdiagonal]
numtheory[cfrac/superdiagonal]
numtheory[cfracpol]
numtheory[nthconver]
numtheory[nthdenom]
numtheory[nthnumer]
thiele ⋆

Modular Arithmetic

chrem
convert/mod2
Det
frac
Gausselim
Gaussjord
genpoly
iratrecon ⋆
mod
'mod'
modp
mods
msolve
Nullspace
numtheory[imagunit]
numtheory[index]
numtheory[lambda]
numtheory[mcombine]
numtheory[mipolys]
numtheory[mlog]
numtheory[mroot]
numtheory[msqrt]
numtheory[order]
numtheory[pprimroot]
numtheory[primroot]
numtheory[rootsunity]

P-adic Numbers

Digitsp
genpoly
padic[arccoshp]
padic[arccosp]
padic[arccothp]
padic[arccotp]
padic[arccschp]
padic[arccscp]
padic[arcsechp]
padic[arcsecp]
padic[arcsinhp]
padic[arcsinp]
padic[arctanhp]
padic[arctanp]
padic[coshp]
padic[cosp]
padic[cothp]
padic[cotp]
padic[cschp]
padic[cscp]
padic[evalp]
padic[expp]
padic[logp]
padic[ordp]
padic[ratvaluep]
padic[rootp]
padic[sechp]
padic[secp]
padic[sinhp]
padic[sinp]
padic[sqrtp]
padic[tanhp]
padic[tanp]
padic[valuep]
?padic[functions]

See also 2.1 Algebra: Number Fields and Function Fields (p. 115)
2.1 Algebra: Finite Fields (p. 115)
2.1 Algebra: Polynomials: Modular Arithmetic (p. 117)
2.5 Complex Numbers: Gaussian Integers (p. 121)
2.13 Integers (p. 126)
2.13 Integers: Primes (p. 126)

2.18 Numbers

See 2.1 Algebra: Number Fields and Function Fields (p. 115)
2.2 Approximations: Real Numbers (p. 118)
2.5 Complex Numbers (p. 121)
2.5 Complex Numbers: Gaussian Integers (p. 121)
2.13 Integers (p. 126)
2.13 Integers: Primes (p. 126)
2.17 Number Theory: P-adic Numbers (p. 130)

2.19 Solving Equations and Recurrences

convert/eqnlist
convert/equality
convert/lessequal
convert/lessthan
DESol$^+$
dsolve
dsolve/explicit
dsolve/laplace
dsolve/method$^+$
dsolve/numeric
dsolve/series
dsolve/type$^+$
fsolve
fsolve/complex
fsolve/fulldigits
fsolve/maxsols
GaussInt[GIroots]
grobner[finite]
grobner[gsolve]
grobner[solvable]
grobner[solvable/plex]
grobner[solvable/tdeg]
invfunc ⋆
isolate ⋆
isolve
lhs
linalg[genmatrix]
linalg[leastsqrs]
linalg[linsolve]
_MaxSols
msolve
numtheory[isolve]
numtheory[thue]
padic[rootp]
powseries[powsolve]
rhs
roots
Roots
rsolve
rsolve/genfunc
rsolve/makeproc
singular ⋆
solve
solve/identity
student[equate]
student[isolate]
type/equation
?solve[floats]
?solve[functions]
?solve[ineqs]
?solve[linear]
?solve[radical]
?solve[scalar]
?solve[series]
?solve[system]

See also 2.1 Algebra: Gröbner Bases (p. 115)
2.1 Algebra: Polynomials: Factoring and Finding Roots (p. 116)
2.7 Discrete Mathematics: Generating Functions and Recurrence Relations (p. 122)

2.20 Statistics

Descriptive Statistics

stats[describe, coefficientofvariation]$^+$
stats[describe, count]$^+$
stats[describe, countmissing]$^+$
stats[describe, covariance]$^+$
stats[describe, decile]$^+$
stats[describe, geometricmean]$^+$
stats[describe, harmonicmean]$^+$
stats[describe, kurtosis]$^+$
stats[describe, linearcorrelation]$^+$
stats[describe, mean]$^+$
stats[describe, meandeviation]$^+$
stats[describe, median]$^+$
stats[describe, mode]$^+$
stats[describe, moment]$^+$
stats[describe, percentile]$^+$
stats[describe, quadraticmean]$^+$
stats[describe, quantile]$^+$
stats[describe, quartile]$^+$
stats[describe, range]$^+$
stats[describe, skewness]$^+$
stats[describe, standarddeviation]$^+$
stats[describe, variance]$^+$
?describe[gaps]$^+$
stats[average]$^\oslash$
stats[correlation]$^\oslash$
stats[covariance]$^\oslash$
stats[mean]$^\oslash$
stats[mean/continuous]$^\oslash$
stats[mean/discrete]$^\oslash$
stats[mean/geometric]$^\oslash$
stats[mean/harmonic]$^\oslash$
stats[mean/quadratic]$^\oslash$
stats[median]$^\oslash$
stats[mode]$^\oslash$
stats[projection]$^\oslash$
stats[projection/const]$^\oslash$
stats[Rsquared]$^\oslash$
stats[sdev]$^\oslash$
stats[serr]$^\oslash$
stats[variance]$^\oslash$

Distributions

stats[random]$^{+}$
stats[random/auto]$^{+}$
stats[random/builtin]$^{+}$
stats[random/default]$^{+}$
stats[random/generator]$^{+}$
stats[random/inverse]$^{+}$
?stats[distributions]

Continuous

stats[statevalf, cdf]$^{+}$
stats[statevalf, icdf]$^{+}$
stats[statevalf, pdf]$^{+}$
stats[ChiSquare]$^{\oslash}$
stats[Exponential]$^{\oslash}$
stats[Fdist]$^{\oslash}$
stats[Ftest]$^{\oslash}$
stats[N]$^{\oslash}$
stats[Q]$^{\oslash}$
stats[RandBeta]$^{\oslash}$
stats[RandChiSquare]$^{\oslash}$
stats[RandExponential]$^{\oslash}$
stats[RandFdist]$^{\oslash}$
stats[RandGamma]$^{\oslash}$
stats[RandNormal]$^{\oslash}$
stats[RandStudentsT]$^{\oslash}$
stats[RandUniform]$^{\oslash}$
stats[StudentsT]$^{\oslash}$
stats[StudentsT/area]$^{\oslash}$
stats[Uniform]$^{\oslash}$

The following continuous distributions may be used with random *or* statevalf.

beta
cauchy
chisquare
exponential
fratio
gamma
laplaced
logistic
lognormal
normald
studentst
uniform
weibull

Discrete

stats[statevalf, dcdf]$^{+}$
stats[statevalf, idcdf]$^{+}$
stats[statevalf, pf]$^{+}$
stats[RandPoisson]$^{\oslash}$

The following discrete distributions may be used with random *or* statevalf.

binomiald discreteuniform empirical hypergeometric negativebinomial poisson

Manipulation

stats[transform, apply]$^{+}$
stats[transform, classmark]$^{+}$
stats[transform, cumulativefrequency]$^{+}$
stats[transform, deletemissing]$^{+}$
stats[transform, divideby]$^{+}$
stats[transform, frequency]$^{+}$
stats[transform, moving]$^{+}$
stats[transform, multiapply]$^{+}$
stats[transform, remove]$^{+}$
stats[transform, scaleweight]$^{+}$
stats[transform, split]$^{+}$
stats[transform, standardscore]$^{+}$
stats[transform, statsort]$^{+}$
stats[transform, statvalue]$^{+}$
stats[transform, tally]$^{+}$
stats[addrecord]$^{\oslash}$
stats[evalstat]$^{\oslash}$
stats[getkey]$^{\oslash}$
stats[putkey]$^{\oslash}$
stats[removekey]$^{\oslash}$
?stats[data]
?stats[updates]$^{+}$

Regression

stats[fit, leastsquare]$^{+}$
stats[linregress]$^{\oslash}$
stats[multregress]$^{\oslash}$
stats[multregress/const]$^{\oslash}$
stats[regression]$^{\oslash}$

Statistical Plots

stats[statplots, boxplot]$^{+}$
stats[statplots, changecolour]$^{+}$
stats[statplots, histogram]$^{+}$
stats[statplots, notchedbox]$^{+}$
stats[statplots, quantile]$^{+}$
stats[statplots, quantile2]$^{+}$
stats[statplots, scatter1d]$^{+}$
stats[statplots, scatter1d/jittered]$^{+}$
stats[statplots, scatter1d/projected]$^{+}$
stats[statplots, scatter1d/stacked]$^{+}$
stats[statplots, scatter2d]$^{+}$
stats[statplots, symmetry]$^{+}$
stats[statplots, xscale]$^{+}$
stats[statplots, xshift]$^{+}$
stats[statplots, xyexchange]$^{+}$
stats[statplot]$^{\oslash}$

2.21 Student Package

student[changevar]
student[combine]
student[completesquare]
student[convert/'@']
student[convert/nested]
student[D]
student[distance]
student[Doubleint]
student[equate]
student[extrema]
student[Int]
student[integrand]
student[intercept]
student[intparts]
student[isolate]
student[leftbox]
student[leftsum]
student[Limit]
student[Lineint]$^{+}$
student[makeproc]
student[makeproc/slope]
student[maximize]
student[middlebox]
student[middlesum]
student[midpoint]
student[minimize]
student[powsubs]
student[rightbox]
student[rightsum]
student[showtangent]
student[simpson]
student[slope]
student[Sum]
student[trapezoid]
student[Tripleint]
student[type/Point]
student[value]

2.22 Units of Measure

acre
acres
bu
bushel
bushels
chain
chains
cm
convert/degrees
convert/metric
convert/radians
cord
cords
degrees
feet
foot
ft
furlong
furlongs
gal
gallon
gallons
Gals
gill
gills
gr
hr
imp
inch
inches
ins
kg
km
Lb
lbs
light_year
light_years
lt
m
mi
Mile
miles
MPG
MPH
ounce
Ounces
oz
Ozs
pint
pints
pole
poles
pound
pounds
quart
quarts
US
yard
yards
yd
yds

See also 3.5 Data Types: Conversion (p. 135)

3 PROGRAMMING

3.1 Assigning and Unassigning Variables

:=
alias
anames
assign
assigned
evaln
macro
protect^{+}
restart
type/name
type/protected^{+}
unames
unassign ⋆
unprotect^{+}
?environment
?indexed
?keywords
?strings

3.2 Assuming Properties

about	assume	OrProp	totorder[ordering]
additionally	_Envadditionally	simplify/assume+	totorder[tassume]
addproperty	is	totorder[forget]	totorder[tis]
AndProp	isgiven	totorder[init]	?asspar

Function Properties

addmul	continuous	mapping	PolynomialMap
ArithmeticOper	differentiable	monotonic	StrictlyMonotonic
binary	InfinitelyDifferentiable	operator	unary
commutative	LinearMap		

Matrix Properties

antisymmetric	matrix	PositiveDefinite	SquareMatrix
diagonal	nilpotent	PositiveSemidefinite	symmetric
ElementaryMatrix	NonSingular	RectangularMatrix	triangular
Hermitian	NonSymmetric	scalar	tridiagonal
idempotent	NonZero	ScalarMatrix	UpperTriangular
IdentityMatrix	NullMatrix	singular	vector
LowerTriangular	NullVector		

Numeric Properties

complex	GaussianInteger	integer	prime	RealRange
composite	GaussianPrime	irrational	rational	RealRange/Open
fraction	imaginary	NumeralNonZero	real	

Other Properties

BottomProp MutuallyExclusive property TopProp type

3.3 Control Statements

if	elif	fi	in	from	next	do	break	traperror	done	stop
then	else	for	by	to	while	od	ERROR	RETURN	quit	?statements

3.4 Data Structures

Array

array	comparray ⋆	convert/matrix	indices
array/antisymmetric	comparray/dontprint ⋆	convert/set	map
array/diagonal	convert/array	convert/vector	type/array
array/identity	convert/eqnlist	copy	?indexfcn
array/sparse	convert/list	entries	?index[tables]
array/symmetric	convert/listlist	evalhf[arrays]	?selection

Expression Sequence

$, NULL seq ?selection ?sequence

Heap

heap[empty] ⋆*heap*	heap[insert] ⋆*heap*	heap[new] ⋆*heap*
heap[extract] ⋆*heap*	heap[max] ⋆*heap*	heap[size] ⋆*heap*

List

[ ]	convert/list/'='	nops	sortcx/abs ⋆
(| |)	convert/list/list	select	sortcx/lexImRe ⋆
comparray ⋆	convert/listlist	sort	sortcx/lexReIm ⋆
comparray/dontprint ⋆	convert/matrix	sort/'<'	sortcx/polar ⋆
convert/'+'	convert/multiset	sort/address	type/list
convert/'*'	convert/set	sort/lexorder	type/listlist
convert/array	convert/vector	sort/numeric	zip
convert/eqnlist	map	sort/string	?lists
convert/list	member	sortcx ⋆	?selection

Queue

initialize ⋆*priqueue* insert ⋆*priqueue* extract ⋆*priqueue*

Range

.. evalr ⋆ lhs rhs shake ⋆ type/'..' type/range ?ranges

Set

{ }	convert/list	intersect	member	nops	symmdiff ⋆	union	?sets
(* *)	convert/set	map	minus	select	type/set	?selection	

Table

convert/array	convert/set	map	table/identity	?indexfcn
convert/eqnlist	copy	table	table/sparse	?index[tables]
convert/list	entries	table/antisymmetric	table/symmetric	?selection
convert/multiset	indices	table/diagonal	?indexed	

3.5 Data Types

Conversion

convert	convert/expln	convert/multiset
convert/'+'	convert/expsincos	convert/name
convert/'*'	convert/factorial	student[convert/nested]
student[convert/'@']	convert/float	convert/octal
convert/array	convert/fraction	convert/parfrac
convert/base	convert/fraction/exact	group[convert/permlist]
convert/binary	logic[convert/frominert]	convert/polar

convert/binomial$^+$
convert/confrac
convert/D
convert/decimal
convert/degrees
convert/diff
group[convert/disjcyc]
convert/double/ibm
convert/double/maple
convert/double/mips
convert/double/vax
convert/Ei
convert/eqnlist
convert/equality
simplex[convert/equality]
convert/erf
convert/erfc
convert/exp
convert/GAMMA
convert/hex
convert/horner
convert/hostfile
convert/hypergeom
convert/lessequal
convert/lessthan
convert/list
convert/list/‘=‘
convert/list/list
convert/listlist
convert/ln
convert/mathorner
convert/matrix
convert/metric
logic[convert/MOD2]
logic[convert/MOD2/expanded]
convert/mod2
convert/polynom
convert/radians
convert/radical
convert/rational
convert/rational/exact
convert/ratpoly
convert/RootOf
convert/series
convert/set
convert/sincos
convert/sqrfree
simplex[convert/std]
simplex[convert/stdle]
convert/string
logic[convert/toinert]
convert/tan
convert/trig
convert/vector

See also 2.22 Units of Measure (p. 133)

Strings

‘ ‘
.
cat
convert/name
convert/string
lexorder
parse
parse/statement
search ⋆
searchtext$^+$
SearchText$^+$
sscanf
substring
TEXT
type/string
?nullstring
?strings

Types and Type Checking

hastype
type
type/‘!‘
type/‘*‘
type/‘**‘
type/‘+‘
type/‘.‘
type/‘..‘
type/‘<‘
type/‘<=‘
type/‘<>‘
type/‘=‘
type/‘^‘
type/algebraic
type/algext
type/algfun
type/algnum
type/algnumext
type/‘and‘
type/indexed
type/infinity$^+$
type/integer
type/‘intersect‘
type/laurent
geometry[type/line2d]
geom3d[type/line3d]
type/linear
type/list
type/listlist
type/logical
type/mathfunc
type/matrix
type/matrix/square
type/‘minus‘
type/monomial
type/name
type/negative
type/negint
type/protected$^+$
type/quadratic
type/quartic
type/radext
type/radfun
type/radfunext
type/radical
type/radnum
type/radnumext
type/range
type/rational
type/ratpoly
type/realcons
type/relation
genfunc[type/rgf_seq]
type/RootOf
type/scalar
difforms[type/scalar]
type/series

type/anyfunc
type/anything
type/arctrig
type/array
type/boolean
geometry[type/circle2d]
type/complex
difforms[type/const]
type/constant
type/cubic
group[type/disjcyc]
type/equation
type/even
type/evenfunc
type/expanded
type/facint
type/float
difforms[type/form]
type/fraction
type/function
networks[type/GRAPH]
type/identical
type/nonneg
simplex[type/NONNEGATIVE]
type/nonnegint
type/'not'
type/nothing
type/numeric
type/odd
type/oddfunc
type/operator
type/'or'
geom3d[type/plane]
type/PLOT
type/PLOT3D
type/point
student[type/Point]
geometry[type/point2d]
geom3d[type/point3d]
type/polynom
type/posint
type/positive
type/primeint
type/procedure
type/set
type/specfunc
geom3d[type/sphere]
type/sqrt
type/square
type/string
type/table
type/taylor
geom3d[type/tetrahedron]
type/TEXT
geom3d[type/triangle3d]
type/trig
type/type
type/uneval
type/'union'
type/vector
whattype
?index[datatypes]
?type[argcheck]
?type[definition]
?type[structured]
?type[surface]

3.6 Debugging

debug
infolevel[all]
lasterror
printlevel
undebug
userinfo
ERROR
interface/errorbreak
mint
trace
untrace
verify ⋆
infolevel
interface/warnlevel+
option/trace
traperror

3.7 Domains of Computation (Gauss)

To access these functions, you must first load the Gauss package.

addCategory
hasCategory
show/categories
show/properties
?Gauss[domain]
defOperation
show
show/operations
?Gauss[coding]
?Gauss[example]

Categories

AbelianGroup
AbelianMonoid
CommutativeRing
DistributedMultivariatePolynomial
EuclideanDomain
ExponentVector
Field
Finite
FiniteCommutativeRing
FiniteField
FiniteRing
GcdDomain
Group
IntegralDomain
Monoid
OrderedAbelianGroup
OrderedAbelianMonoid
OrderedDomain
OrderedField
OrderedRing
OrderedSet
PartiallyOrderedSet
QuotientField
Ring
SemiGroup
Set
UniqueFactorizationDomain
UnivariatePolynomial

Domains

AlgebraicExtension
SAE
DenseExponentVector
DEV
DenseUnivariatePolynomial
DUP
ExpandedNormalForm
ENF
FactoredNormalForm
FNF
G
GaloisField
GF
LazyUnivariatePowerSeries
LUPS
MacaulayExponentVector
TEV
MapleExponentVector
MEV
MapleMultivariatePolynomial
MMP
MapleUnivariatePolynomial
MUP
OrderedUnivariatePolynomial
OUP
PrimeExponentVector
PEV
Q
QuotientField
QF
RationalFunction
RF
SparseDistributedMultivariatePolynomial
SDMP
SparseMultivariatePolynomial
SMP
SquareMatrix
SM
TableDistributedMultivariatePolynomial
TDMP
UnivariatePowerSeries
UPS
Z
Zmod

3.8 Expression Manipulation

asubs ⋆
asubs/always ⋆
freeze ⋆
length
lhs
map
nops
numboccur^{+}
op
rhs
student[powsubs]
subs
subsop
thaw ⋆*freeze*
trigsubs ⋆
?selection

3.9 Global Variables and Environment Variables

constants
Digits
Digitsp
_Envadditionally
_Envsignum0^{+}
_Env*string*
infolevel
infolevel[all]
integrate
lasterror
libname
_MaxSols
'mod'
Normalizer
NULL
Order
precision
printlevel
_seed
sharename
status
Testzero
?constants
?environment
?ininame

3.10 Input and Output

Input

convert/hostfile
interface/errorbreak
libname
linalg[entermatrix]
parse
parse/statement
read
readdata ⋆
readdata/float ⋆
readdata/integer ⋆
readlib
readline
readline/terminal
readstat
sscanf
unload ⋆
with
?files

Output

%n
appendto
appendto/terminal
close ⋆*write*
convert/hostfile
C ⋆
C/digits ⋆
C/filename ⋆
C/optimized ⋆
eqn
fortran
fortran/digits
fortran/filename
fortran/mode$^+$
fortran/optimized
precision
interface/echo
interface/endcolon
interface/indentamount
interface/labeling
interface/labelling
interface/labelwidth
interface/plotoutput
interface/prettyprint
interface/verboseproc
latex
?latex[functions]
?latex[names]
lprint
m2src$^+$
NPspinor[makeeqn]
open ⋆*write*
print
printf
printlevel
save
savelib
TEXT
updtsrc$^+$
writeln ⋆*write*
writeto
writeto/terminal
write ⋆
?files

3.11 Internal Access

addressof
assemble
disassemble
pointto

3.12 Operators

&
@
@@
combine/'@@'
combine/atatsign
define
define/antisymmetric
define/associative
define/binary
define/commutative
define/forall
define/group$^\oslash$
define/Group$^+$
define/identity
define/inverse
define/linear$^\oslash$
define/Linear$^+$
define/symmetric
define/type
define/unary
define/zero
option/angle
option/arrow
option/operator
procbody ⋆
procmake ⋆
simplify/atsign
simplify/'@'
type/operator
unapply
?arithop
?define[operators]
?index[expressions]
?neutral
?operators
?operators[binary]
?operators[examples]
?operators[functional]
?operators[nullary]
?operators[precedence]
?operators[unary]

3.13 Optimization

cost ⋆
evalhf
evalhf/var
forget ⋆
gc
optimize ⋆
'optimize/makeproc' ⋆
'optimize/makeproc'/globals ⋆
'optimize/makeproc'/locals ⋆
'optimize/makeproc'/parameters ⋆
option/remember
option/system
profile ⋆
showprofile ⋆*profile*
unprofile ⋆*profile*
showtime ⋆
off ⋆*showtime*
on ⋆*showtime*
status
time
words
?remember

3.14 Procedures and Functions

->
< >
< | >
< || >
args
nargs
'optimize/makeproc' ⋆
option
option/angle
option/arrow
options
piecewise$^+$ ⋆
proc
procbody ⋆
procmake ⋆
updtsrc$^+$
#
?comments
?functions
?index[procedures]

end	option/builtin	procname	?parameters
ERROR	option/Copyright	RETURN	?procedures
forget ⋆	option/operator	student[makeproc]	?procfile
global$^+$	option/remember	student[makeproc/slope]	?remember
local	option/system	type/procedure	?type[argcheck]
makehelp ⋆	option/trace	unapply	

3.15 Punctuation

;	' '	( )	{ }	(* *)	< | >	?quotes
:	‘ ‘	[ ]	(| |)	< >	< | | >	?separators

3.16 Tests

Arithmetic and Lexical Relations

<	=	convert/equality	lexorder	type/‘<>‘	type/relation
<=	>	convert/lessequal	type/‘<‘	type/‘=‘	?inequalities
<>	>=	convert/lessthan	type/‘<=‘	type/equation	

Boolean Evaluation

and	FAIL	not	testeq	type/boolean	?evalhf[boolean]
evalb	false	or	true	type/logical	

Numeric and Algebraic Tests

has	iscont ⋆	match	testfloat/digits ⋆	Testzero
hastype	isgiven	numtheory[issqrfree]	testfloat/model ⋆	type/algebraic
irreduc	isprime	Primitive	testfloat/test ⋆	type/numeric
Irreduc	issqr ⋆	recipoly ⋆	testfloat ⋆	verify ⋆
is				

See also

2.11 Geometry: Two-Dimensional Euclidean Geometry: Two-Dimensional Tests (p. 125)
2.11 Geometry: Three-Dimensional Euclidean Geometry: Three-Dimensional Tests (p. 125)
2.11 Geometry: Projective Geometry: Projective Geometry Tests (p. 125)
3.5 Data Types: Types and Type Checking (p. 136)

4 SYSTEM

4.1 External Functions

! helptomaple m2src$^+$ maple march mint system updtsrc$^+$?escape

4.2 Help

?	???$^+$	help	helptomaple	related$^+$	usage$^+$
??$^+$	example	info$^+$	makehelp ⋆	tutorial	

System Help Topics

?copyright	?index	?maple_group	?scg	?updates[v4.0]	?updates[v4.3]
?demo	?index[misc]	?mtn+	?support	?updates[v4.1]	?updates[v5]
?distribution	?introduction	?mtn[*n*]+	?trademark	?updates[v4.2]	?updates[v5.2]
?hardware	?maple	?references	?updates		

4.3 Libraries

libname	?index[library]	?share[combinat]+	?share[linalg]+
readlib	?index[packages]	?share[contents]	?share[numerics]+
savelib	?share	?share[contributions]	?share[numtheory]+
sharename	?share[address]	?share[engineer]+	?share[programming]+
with	?share[algebra]+	?share[geometry]+	?share[science]+
?index[external]	?share[analysis]+	?share[graphics]+	?share[statistics]+
?index[internal]	?share[autodiff]+	?share[help]	?share[system]+
?index[libmisc]	?share[calculus]+		

4.4 Packages

combinat	geom3d	liesymm	numapprox	priqueue ⋆	stats[random]+
DEtools	geometry	linalg	numtheory	projgeom	stats[statevalf]+
difforms	GF ⋆	logic	orthopoly	simplex	stats[statplots]+
Gauss	grobner	networks	padic	stats	stats[transform]+
GaussInt	group	np	plots	stats[describe]+	student
genfunc	heap ⋆	NPspinor	powseries	stats[fit]+	totorder

4.5 Resources

forget ⋆	option/system	showtime ⋆	on ⋆*showtime*	time
gc	restart	off ⋆*showtime*	status	words

See also 3.13 Optimization *(p. 139).*

4.6 User Interface

"	clear ⋆*history*	interface/plotdevice	interface/terminal
""	off ⋆*history*	interface/plotoptions+	interface/verboseproc
"""	timing ⋆*history*	interface/plotoutput	interface/version
?ditto	interface/echo	interface/postplot	interface/warnlevel+
%*n*	interface/endcolon	interface/preplot	interface/wordsize+
\	interface/errorbreak	interface/prettyprint	plotsetup
?backslash	interface/indentamount	interface/printfile	plotsetup/plotdevice+
edit ⋆	interface/iris+	interface/prompt	plotsetup/plotoptions+
?editing	interface/labeling	interface/quiet	plotsetup/plotoutput+
gc	interface/labelling	interface/screenheight	plotsetup/preplot+
history ⋆	interface/labelwidth	interface/screenwidth	plotsetup/postplot+

Quitting Maple

done quit stop

Note: more functions are available in the share library, a repository of contributed Maple programs and applications. Chapter 5 describes the share library in some detail (see page 407).

Chapter 4
Descriptions of Commands

This chapter compiles all Maple commands in one alphabetized list. Each command has a description of the following form:

function `function(`arg_1`,`arg_2`,...,`arg_n`)`
A brief description of the function ◇ Additional information on the function, its required and optional arguments, and returned value. ◇ Remarks on any Maple V Release 3 changes. ◇ One or more examples of the function. ◇ See also: other commands that are similar in capability or in name.

`Typewriter` font in the descriptions denotes fixed text that you must type exactly as shown. *Italic* font denotes variable arguments.

This chapter also includes descriptions of important global variables, symbolic constants, and package names. Headings for these entries are given in a lighter typeface: for example, Digits.

The arguments given in the templates for the commands adhere to the following conventions:

A, *B* =	matrix	*p* =	prime number
array =	array	*poly* =	polynomial
eqn =	equation	*ptr* =	pointer
expr =	expression	*r* =	real number
f =	function	*set* =	set
i, *j* =	integer	*table* =	table
K =	algebraic extension	*u*, *v* =	vector
list =	list	*var* =	variable name
m, *n* =	integer	*z* =	complex number
name =	unassigned name, or quoted name		

In addition, some packages have their own conventions for argument names. For example, *G* denotes a graph in the `networks` package. The description of a package name lists the additional conventions used for functions in that package.

!

!*command*
Invokes a command in the host operating system ◇ This operation is not supported on some platforms. The `!` should be at the beginning of the line. The `!` character is also the factorial operator. ◇ See also: `system`.

$

`$` *a*`..`*b*
expr `$` *n*
expr `$` *var* `=` *a*`..`*b*
Operator for forming an expression sequence ◇ The first form returns the expression sequence a, $a+1$, . . ., b (or up to the last value not exceeding b). The second template returns a sequence of n copies of *expr*. The third form returns the expression sequence produced by substituting for *var* in *expr* the values a, $a+1$, . . . ,b (or up to the last value not exceeding b). ◇ `z$3;` $\longrightarrow z, z, z$ ◇ `sin(i*Pi/2)$i=0..3;` $\longrightarrow 0, 1, 0, -1$ ◇ See also: `seq`.

&*

expr$_1$ `&*` *expr*$_2$
Noncommutative multiplication operator ◇ This operator is most often used to denote matrix multiplication. `&*()` denotes the matrix identity. ◇ `evalm(A &* B):` ◇ See also: `commutat`, `evalm`.

&^

expr$_1$ `&^` *expr*$_2$
Neutral exponentiation operator ◇ This operator permits efficient computation of powers modulo an integer. ◇ `237 &^ 1007 mod 641;` $\longrightarrow 96$ ◇ See also: `difforms[&^]`, `liesymm[&^]`, `mod`, `Powmod`.

?

?*topic*
?*topic*,*subtopic*
?*topic*[*subtopic*]
Obtains help on a command or other topic ◇ The last two forms are equivalent for obtaining help on a subtopic.

??

??*name*
Shows the calling sequence and descriptions of parameters for a function ◇ The information is extracted from the on-line help text. ◇ This function is new in Release 3. ◇ See also: `usage`.

???

???*name*
Shows examples of calling a function ◇ The information is extracted from the on-line help text. ◇ This function is new in Release 3. ◇ See also: `example`.

@

f `@` *g*
The function composition operator ◇ $(f@g)(x)$ is equivalent to $f(g(x))$. ◇ `(ln@cos)(0);` $\longrightarrow 0$ ◇ See also: `@@`.

@@

f `@@` *n*
The repeated composition operator ◇ This is f composed with itself n times. If n is negative, the `invfunc` table is checked for an inverse

function. ◇ `(D@@2)(sin*cos);` ⟶ −4 cos sin ◇ See also: `@`, `invfunc`.

about `about(`*name*`)`
`about(`*property*`)`
Displays information on the properties of variable names, and displays the parents and children of property names ◇ `assume(x>0): about(x);` ⟶ `Originally x, renamed x~: is assumed to be: RealRange(Open(0),infinity)` ◇ See also: `additionally`, `addproperty`, `assume`, `is`, `isgiven`.

abs `abs(`*expr*`)`
Computes the absolute value of the expression *expr* ◇ If *expr* includes a function *f*, `abs` attempts to execute the procedure `abs/f`. Thus the functionality of `abs` may be extended. ◇ `abs(4+3*I);` ⟶ 5 ◇ See also: `evalc`, `signum`.

additionally
`additionally(`*name*`, `*property*`)`
`additionally(`*relation*`)`
Adds additional assumptions to a name ◇ See also: `addproperty`, `assume`, `is`, `isgiven`.

addproperty
`addproperty(`*property*`, `*parents*`, `*children*`)`
`addproperty(`*relation*`)`
Adds a property to the property tables ◇ *parents* and *children* are sets naming the parent and children properties. ◇ See also: `about`, `additionally`, `assume`, `is`, `isgiven`.

addressof `addressof(`*expr*`)`
Obtains the address that points to an expression ◇ This function allows access to the internal representation of an expression. ◇ `addressof(a+2*b);` ⟶ 1443820 ◇ See also: `assemble`, `disassemble`, `pointto`.

AFactor `AFactor(`*poly*`)`
Inert absolute factorization ◇ When used in conjunction with `evala`, this function computes an absolute factorization of the multivariate polynomial *poly*, that is, a factorization over an algebraic closure of its coefficient field. The result is returned as an expression. ◇ `evala(AFactor(2*x^2 + 6*y^2));` ⟶ $2\,(x - \mathrm{RootOf}(\_Z^2+3)\,y)\,(x+\mathrm{RootOf}(\_Z^2+3)\,y)$ ◇ See also: `AFactors`, `evala`, `factor`, `Factor`, `RootOf`.

AFactors `AFactors(`*poly*`)`
Inert absolute factorization ◇ When used in conjunction with `evala`, this function computes an absolute factorization of the multivariate polynomial *poly*, that is, a factorization over an algebraic closure of its coefficient

field. The result is a list whose first element is the leading coefficient and whose second element is a list of factor-exponent pairs. ◇
`evala(AFactors(2*x^2+6*y^2));` ⟶
$[2,[[x - \mathrm{RootOf}(\_Z^2+3)\,y, 1], [x+\mathrm{RootOf}(\_Z^2+3)\,y, 1]]]$ ◇
See also: `AFactor, evala, factors, Factors, RootOf.`

Ai

`Ai(`*r*`)`
Airy wave function ◇ `Ai` and `Bi` are linearly independent solutions for w in the equation $w'' - wz = 0$. Use `evalf` to compute numerical values for this function. ◇ `evalf(Ai(1.));` ⟶ .1352924163 ◇ See also: `Bi, evalf.`

alias

`alias(`eqn_1`,` eqn_2`, ...)`
Defines an abbreviation or denotation ◇ Each equation has the form *newname=oldname*. Abbreviations defined with this function are used in output as well as input. The result of `alias` is a sequence of all aliases currently defined. An alias may not be defined in terms of another alias. To remove an alias, use an equation of the form *name=name*. ◇
`alias(alpha = RootOf(x^2+x+1,x));` ⟶ I, α ◇ See also: `I, macro, subs.`

allvalues

`allvalues(`*expr*`)`
`allvalues(`*expr*`, d)`
Evaluates all possible values of expressions involving `RootOfs` ◇ An expression sequence is returned. Where roots cannot be obtained exactly, `fsolve` is used to obtain numerical solutions. The second argument `d` specifies that `RootOfs` of the same equation represent the same value and they should not be evaluated independently of one another. ◇
`allvalues(RootOf(x^2-x+1, x));` ⟶ $\frac{1}{2} + \frac{1}{2}I\sqrt{3}, \frac{1}{2} - \frac{1}{2}I\sqrt{3}$ ◇
See also: `fsolve, RootOf.`

amortization

`amortization(`*amt*`,` *rate*`,` *pay*`)`
Computes an amortization schedule ◇ The loan of amount is *amt*, the interest rate per period is *rate*, and the payment per period is *pay*. The schedule is returned as an array indexed from 0 to the number of periods, where each entry in the array is a list containing the period number, periodic payment, interest, principal reduction, and new balance. ◇ You must `readlib(finance)` before using this command. ◇
`amortization(10000, .065/12, 306.49)[9];` ⟶
[9, 306.49, 43.02, 263.47, 7679.26] ◇ See also: `blackscholes, finance.`

anames

`anames()`
Assigned names ◇ This function returns an expression sequence of names that are currently assigned values other than their own name. ◇ See also: `assigned, unames.`

A B C D E F G H I J K L M N O P Q R S T U V W X Y Z

and `a and b`
Logical operator ◇ This expression is true only if both a and b are true. It is false if at least one of them is false. If one returns *FAIL* and the other is true or *FAIL*, it returns *FAIL*. b is not evaluated if a is false. ◇ See also: `evalb`, `not`, `or`.

antisymmetric
The antisymmetric indexing function for arrays and tables ◇ This function gives the result of the `symmetric` indexing function, multiplied by -1 if all indices are distinct and an odd number of transpositions are necessary to reorder the indices. The most common application is for antisymmetric matrices, where the (i, j)th element is the negative of the (j, i)th. ◇ See also: `array`, `symmetric`, `table`.

appendto `appendto(`*filename*`)`
`appendto(terminal)`
Appends all successive output to the file ◇ Under the command-line interface, prompts and input statements are also echoed to the file, and prompts are not displayed on the screen while appending to a file. The second form restores printing to the screen. If the file does not exist, it is created. ◇ See also: `save`, `write`, `writeto`.

arccos `arccos(`z`)`
The inverse cosine function ◇ See also: `cos`.

arccosh `arccosh(`z`)`
The inverse hyperbolic cosine function ◇ See also: `cosh`.

arccot `arccot(`z`)`
The inverse cotangent function ◇ See also: `cot`.

arccoth `arccoth(`z`)`
The inverse hyperbolic cotangent function ◇ See also: `coth`.

arccsc `arccsc(`z`)`
The inverse cosecant function ◇ See also: `csc`.

arccsch `arccsch(`z`)`
The inverse hyperbolic cosecant function ◇ See also: `csch`.

arcsec `arcsec(`z`)`
The inverse secant function ◇ See also: `sec`.

arcsech `arcsech(`z`)`
The inverse hyperbolic secant function ◇ See also: `sech`.

arcsin `arcsin(`z`)`
The inverse sine function ◇ See also: `sin`.

arcsinh `arcsinh(`*z*`)`
The inverse hyperbolic sine function ◇ See also: `sinh`.

arctan `arctan(`*z*`)`
`arctan(`*y*`, `*x*`)`
The inverse tangent function ◇ The second form computes the principal value of the argument of the complex number $x + I\,y$. ◇ `arctan(1,1);` $\longrightarrow \frac{1}{4}\pi$ ◇ See also: `tan`.

arctanh `arctanh(`*z*`)`
The inverse hyperbolic tangent function ◇ See also: `tanh`.

args The sequence of actual arguments passed to a procedure ◇ Within a procedure, the special name `args` has as its value the expression sequence of actual arguments with which the procedure was called. ◇ See also: `nargs, proc, procname`.

argument `argument(`*z*`)`
Complex argument function ◇ This function returns the principal value of the argument of the complex-valued expression *z*. ◇ `argument(1+I);` $\longrightarrow \frac{1}{4}\pi$ ◇ See also: `convert/polar, evalc, polar, signum`.

array `array(`*indexfcn*`, `*bounds*`, `*list*`)`
Creates an array ◇ The dimension of the array is determined by the number of bounds given. Each bound is an integer range. The elements of *list* are initial values. *indexfcn* is the indexing function for the array. Built-in indexing functions are `antisymmetric`, `diagonal`, `identity`, `sparse`, and `symmetric`. Each argument is optional. ◇ `array(1..3,[2,3,5]);` $\longrightarrow$ [2 3 5] ◇ `array([2,3,5]);` $\longrightarrow$ [2 3 5] ◇ `array(identity,1..3,1..2):` ◇ See also: `antisymmetric, copy, diagonal, entries, identity, indices, linalg[matrix], linalg[vector], sparse, symmetric, table`.

assemble `assemble(`*addrseq*`)`
Assembles a sequence of addresses into an object ◇ The first integer in the sequence represents the type of the object. ◇ `pointto( assemble( disassemble( addressof(a+2*b))));` $\longrightarrow a + 2\,b$ ◇ See also: `addressof, disassemble, pointto`.

assign `assign(`*name*`, `*expr*`)`
`assign(`*name* `=` *expr*`)`
`assign(`*eqnlist*`)`
`assign(`*eqnset*`)`
Performs assignments ◇ The first two forms assign *expr* to *name*. The last two forms perform the assignments requested by the given list or set of equations. The arguments to `assign` are fully evaluated, so a name may need to be enclosed in single right quotes if it already has a value. This function returns `NULL`. ◇ See also: `solve, unassign`.

assigned `assigned(`*name*`)`
Checks whether a name is assigned ◇ This function returns *true* if *name* has a value other than its own name, and *false* otherwise. *name* may be a name, a subscripted name, or a function call. ◇ See also: `anames`, `evaln`, `unames`.

assume `assume(`*name*`, `*property*`)`
`assume(`*relation*`)`
Associates properties with a name ◇ Properties may be assigned to a function name (e.g., `continuous`, `differentiable`), a matrix name (e.g., `singular`, `triangular`), or a numeric variable name (e.g., `integer`, `prime`). A property may also be given to a variable name by specifying a relation like $x > 0$. Lists of properties appear on page 134. A name with a property is displayed with an appended tilde (~). This function erases previous assumptions associated with a name. Use `additionally` to add additional properties. Assumptions may be eliminated by unassigning names. ◇ `assume(x>0), x;` $\longrightarrow x\sim$ ◇ See also: `about`, `additionally`, `addproperty`, `is`, `isgiven`, `unassign`.

asubs `asubs(`*eqn*`, `*expr*`)`
`asubs(`*eqn*`, `*expr*`, `*var*`)`
`asubs(`*eqn*`, `*expr*`, always)`
`asubs(`*eqn*`, `*expr*`, `*var*`, always)`
Substitutes subexpressions into a sum ◇ This function returns the expression resulting from applying the substitutions specified by the equation to the given expression. It differs from `subs` in that it is designed to work on part of a sum, and so does not rely solely on syntactic substitutions. The optional argument `always` forces substitution to occur on all sums. ◇ You must `readlib(asubs)` before using this command. ◇ `asubs(x+1=z, x+y+1);` $\longrightarrow y + z$ ◇ See also: `student[powsubs]`, `subs`, `subsop`.

asympt `asympt(`*expr*`, `*var*`)`
`asympt(`*expr*`, `*var*`, `*n*`)`
Asymptotic expansion ◇ This function computes the asymptotic expansion of *expr* with respect to the variable *var* (as *var* approaches infinity). *n* specifies the truncation order of the series; if it is absent, the value of the global variable `Order` is used. ◇ `asympt(x/(x-1), x, 3);` $\longrightarrow 1 + \frac{1}{x} + \frac{1}{x^2} + O\left(\frac{1}{x^3}\right)$ ◇ See also: `limit`, `Order`, `series`.

Berlekamp
`Berlekamp(`*poly*`, `*var*`) mod `*p*
`Berlekamp(`*poly*`, `*var*`, `*K*`) mod `*p*
Distinct degree factorization ◇ When used in conjunction with `mod`, this function factors a monic square-free univariate polynomial over a finite field using Berlekamp's algorithm. The optional argument *K* specifies an

extension field over which the factorization is performed. A set of irreducible factors is returned. ◇ This function is new in Release 3. ◇ `Berlekamp(x^5+x^4+1, x) mod 2;` ⟶ $\{x^2+x+1, x^3+x+1\}$ ◇ See also: `DistDeg, Factors, ProbSplit, Sqrfree.`

bernoulli `bernoulli(`*n*`)`
`bernoulli(`*n*`, `*x*`)`
Bernoulli numbers and polynomials ◇ This function computes the *n*th Bernoulli number or the *n*th Bernoulli polynomial in the expression *x*. The Bernoulli polynomials B(*n*, *x*) are defined by the exponential generating function:

$$\frac{te^{xt}}{(e^t-1)} = \sum_{n=0}^{\infty} \frac{B(n,x)}{n!} t^n$$

◇ `bernoulli(2);` ⟶ $\frac{1}{6}$ ◇ `bernoulli(2,x);` ⟶ $\frac{1}{6}+x^2-x$ ◇ See also: `euler, numtheory[B].`

bernstein `bernstein(`*n*`, `*f*`, `*var*`)`
Bernstein polynomial approximating a function ◇ *f* is a function of one variable. This procedure returns the *n*th degree Bernstein polynomial in *var* approximating the function *f* on the interval [0, 1]. ◇ You must `readlib(bernstein)` before using this command. ◇ `bernstein(2, x->2/(2*x+1), x);` ⟶ $2-2x+\frac{2}{3}x^2$

BesselI `BesselI(`*v*`, `*x*`)`
The modified Bessel function of the first kind ◇ v is the order of the Bessel function, and x is the argument. ◇ `BesselI(0, 1.);` ⟶ 1.266065878 ◇ See also: `BesselJ, BesselK, BesselY.`

BesselJ `BesselJ(`*v*`, `*x*`)`
The Bessel function of the first kind ◇ v is the order of the Bessel function, and x is the argument. ◇ `BesselJ(-1, 2.3);` ⟶ −.5398725326 ◇ See also: `BesselI, BesselK, BesselY.`

BesselK `BesselK(`*v*`, `*x*`)`
The modified Bessel function of the second kind ◇ v is the order of the Bessel function, and x is the argument. ◇ `BesselK(0, -1.);` ⟶ .4210244382 − 3.977463261 *I* ◇ See also: `BesselI, BesselJ, BesselY.`

BesselY `BesselY(`*v*`, `*x*`)`
The Bessel function of the second kind ◇ v is the order of the Bessel function, and x is the argument. ◇ `BesselY(1.2, 1.);` ⟶ −.9012149548 ◇ See also: `BesselI, BesselJ, BesselK.`

Beta `Beta(`z_1`, `z_2`)`
The Beta function ◇ The Beta function is defined as

$$B(z_1, z_2) = \frac{\Gamma(z_1)\Gamma(z_2)}{\Gamma(z_1+z_2)}$$

◇ `Beta(3,4);` ⟶ $\frac{1}{60}$ ◇ See also: `GAMMA`.

Bi `Bi(`*r*`)`
Airy wave function ◇ `Ai` and `Bi` are linearly independent solutions for w in the equation $w'' - wz = 0$. Use `evalf` to compute numerical values for this function. ◇ `evalf(Bi(1.));` ⟶ 1.207423595 ◇ See also: `Ai`, `evalf`.

binomial `binomial(`*n*`, `*m*`)`
Computes binomial coefficients ◇ If the arguments are both integers, this function computes the binomial coefficient $\binom{n}{m}$. For rationals or floats, the more general definition $\Gamma(n+1)/\left(\Gamma(m+1)\,\Gamma(n-m+1)\right)$ is used. ◇ `binomial(5,3);` ⟶ 10 ◇ See also: `combinat[multinomial]`.

blackscholes
`blackscholes(`*Ex*`, `*t*`, `*P*`, `*s*`, `*rf*`)`
`blackscholes(`*Ex*`, `*t*`, `*P*`, `*s*`, `*rf*`, `*name*`)`
Present value of a call option ◇ This function uses the formula derived by Black and Scholes to find the present value of a call option and the option delta or hedge ratio. *Ex* is the exercise price of the option, *t* the time to the exercise date, *P* the price of the stock now, *s* the standard deviation per period of the continuously compounded rate of return on the stock, and *rf* the continuously compounded risk-free rate of interest. If *name* is present, it is assigned the hedge ratio or option delta if it is present. ◇ You must `readlib(finance)` before using this command. ◇ See also: `amortization`, `finance`.

break Exits from the innermost repetition statement ◇ This statement must occur within a repetition statement. ◇ See also: `do`, `next`.

bspline `bspline(`*n*`, `*var*`)`
`bspline(`*n*`, `*var*`, `*list*`)`
Computes the B-spline segment polynomials ◇ This function computes the segment polynomials for the B-spline of degree *n* in the variable *var* on the knot sequence *list*. If *list* is not specified, the uniform knot sequence $[0, 1, \ldots, n+1]$ is used. A list of pairs of the form [*relation*, *expr*] is returned. ◇ You must `readlib(bspline)` before using this command. ◇ `bspline(1,t);` ⟶ $[[t < 0, 0], [t < 1, t], [t < 2, 2 - t], [2 \le t, 0]]$ ◇ See also: `interp`, `spline`.

C `C(`*expr*`, `*options*`)`
Generates C code ◇ This function generates C code for evaluating the given expression, array of expressions, or list of assignments. In the last case, each assignment is given as an equation. The option `optimized` performs common subexpression optimization on the generated code. The option `digits` specifies the number of digits to use when converting rationals or symbolic constants such as `Pi` (the default is 16). The option `filename` names a file for writing the resulting C code. By default, the

code is written to the standard output. ◇ You must `readlib(C)` before using this command. ◇ `C(sin(Pi*x^2), digits=5);` ⟶ `t0 = sin(0.31416E1*x*x);` ◇ See also: `optimize`, `eqn`, `fortran`, `latex`.

cartan

`cartan(`*h*, *hinv*, *coords*, *g*`)`
`printcartan(`*R*`)`
`riemann(`*hinv*, *gamma*, *coords*, *g*`)`
`simp1(`*name*, *f*, *gamma*`)`
`simp2(`*name*, *f*, *R*`)`
A collection of procedures for the computation of the connection coefficients and curvature components using Cartan's structure equations ◇ The arguments are:

h	an array indexed from 0 to 15
hinv	the inverse of *h*
coords	a list of four variable names
g	a constant metric with indices 0..3
var	a variable name
f	a procedure name
gamma	the connection coefficient array computed by `cartan`
R	the Riemann tensor array computed by `riemann`

`cartan` calculates the connection coefficients. `simp1` simplifies the *gamma* array and assigns the result to *var*. `printcartan` prints the nonzero entries of an array. `riemann` calculates the Riemann tensor components. `simp2` simplifies the Riemann tensor components and assigns the result to *var*. You must `readlib(cartan)` before using these functions. ◇ See also: `debever`, `tensor`.

cat

`cat(`$expr_1$, $expr_2$, `...)`
Concatenates expressions ◇ This function is equivalent to `''` . $expr_1$. $expr_2$ ... ◇ `cat(4,x,4);` ⟶ *4x4* ◇ See also: `length`, `linalg[concat]`, `substring`.

Catalan

Catalan's constant ◇

$$\text{Catalan} = \sum_{k\geq 0} \frac{(-1)^k}{(2k+1)^2}$$

It is approximately 0.915965594177219.

ceil

`ceil(`*z*`)`
`ceil(1, `*z*`)`
Returns the smallest integer greater than or equal to *z* if *z* is real ◇ If *z* is complex, the value computed is $-\text{floor}(-z)$. The second form returns the first derivative of the ceiling function wherever it is defined. ◇ `ceil(3.99-2.4*I);` ⟶ $4 - 2I$ ◇ See also: `floor`, `frac`, `round`, `trunc`.

chebyshev

`chebyshev(`*expr*`,` *var*`=`*a*`..`*b*`,` ϵ`)`
`chebyshev(`*expr*`,` *var*`,` ϵ`)`
Chebyshev series expansion ◇ This function computes the Chebyshev series expansion of the expression with respect to the variable *var* on the interval $[a, b]$, valid to accuracy ϵ. If an interval is not given, the default interval $[-1, 1]$ is used. The third argument, ϵ, is optional. If present, it may be no smaller than $10^{-\text{Digits}}$, its default value. ◇
`chebyshev(sin(x), x, .1);` $\longrightarrow$
$.8801011715\,\mathrm{T}(1, x) - .03912670797\,\mathrm{T}(3, x)$ ◇ See also: `numapprox[chebpade]`, `orthopoly[T]`, `series`, `taylor`.

Chi

`Chi(`*z*`)`
The hyperbolic cosine integral ◇

$$\mathrm{Chi}(z) = \gamma + \ln z + \int_0^z \frac{\cosh t - 1}{t}\, dt$$

◇ You must `readlib(Chi)` before using this command. ◇ See also: `Ci`, `combinat[Chi]`, `Shi`.

chrem

`chrem(`*reslist*`,` *modlist*`)`
Chinese Remainder algorithm ◇ *modlist* is a list of pairwise relatively prime moduli $[m_1, \ldots, m_n]$, and *reslist* is a list of residues $[a_1, \ldots, a_n]$. This function computes the unique positive integer $m < m_1 \cdots m_n$ satisfying $m \equiv a_i \bmod m_i$. *reslist* may be a list of integers or a list of polynomials. In the latter case, the algorithm is applied to the coefficients of each term. ◇ `chrem([1,4,5],[3,5,7]);` $\longrightarrow$ 19 ◇ See also: `GaussInt[chrem]`, `GaussInt[GImcmbine]`, `numtheory[mcombine]`.

Ci

`Ci(`*z*`)`
The cosine integral ◇

$$\mathrm{Ci}(z) = \gamma + \ln Iz - I\frac{\pi}{2} + \int_0^z \frac{\cos t - 1}{t}\, dt$$

◇ See also: `Chi`, `Si`.

classi

`classi()`
Finds the Bianchi type of any 3-dimensional Lie algebra ◇ Before invoking this function, the array `structure` must be defined as `array(`[C_{123},C_{231},C_{312},C_{112},C_{113},C_{221},C_{223},C_{331},C_{332}]`)` where the quantities C_{ijk} denote nine independent structure constants of the 3-dimensional Lie algebra, assumed to be antisymmetric in the last pair of indices. If `structure` has not been defined, the procedure prompts for the constants. This function prints the Bianchi type if the structure constants define a Lie algebra. ◇ You must `readlib(bianchi)` before using this command. ◇ See also: `liesymm`.

close `close()`
Closes the file that was opened for output by `open` ◇ You must `readlib(write)` before using this command. ◇ See also: `open`, `write`, `writeln`.

cmagdiff `cmagdiff(`*z*`)`
Complex magnitude difference ◇ This function returns the difference in magnitude between the real and complex parts of the complex number z. ◇ You must `readlib(cmagdiff)` before using this command. ◇ `cmagdiff(10+.1*I);` ⟶ 2 ◇ See also: `ilog10`.

coeff `coeff(`*poly*`, `*var*`, `*n*`)`
`coeff(`*poly*`, `*var*`^`*n*`)`
Extracts a coefficient of the multivariate polynomial ◇ This function extracts the coefficient of *var*^{n} in *poly*. The polynomial must be `collected` in the variable *var*. ◇ `coeff(x^2*y + 7*x*y^2 + 8*y^3, x);` ⟶ $7\,y^2$ ◇ See also: `Coeff`, `coeffs`, `collect`, `lcoeff`, `tcoeff`.

Coeff `Coeff(`*poly*`, `*n*`)`
Inert coefficient function ◇ When used with `modp1`, this returns the coefficient of a univariate polynomial over a given domain. ◇ See also: `coeff`, `Lcoeff`, `Tcoeff`.

coeffs `coeffs(`*poly*`)`
`coeffs(`*poly*`, `*vars*`)`
`coeffs(`*poly*`, `*vars*`, `*name*`)`
Extracts all coefficients of *var* in the multivariate polynomial ◇ This function returns an expression sequence of all the coefficients of the polynomial with respect to a single variable or a list or set of variables. If *vars* is not specified, all the indeterminates of *poly* are used. The polynomial must be `collected` with respect to the appropriate variables. If a name is given as a third argument, it is assigned an expression sequence of the terms of *poly*. ◇ `coeffs(x^2*y + 7*x*y^2 + 8*y^3,y);` ⟶ $x^2, 7\,x, 8$ ◇ See also: `coeff`, `collect`, `indets`, `lcoeff`, `tcoeff`.

coeftayl `coeftayl(`*expr*`, `*var*`=`*a*`, `*n*`)`
`coeftayl(`*expr*`, `*varlist*`=`*alist*`, `*nlist*`)`
Computes a coefficient in the Taylor series representation of *expr* without forming the series ◇ The first form computes the coefficient of $(var - a)^n$ in the (univariate) Taylor series expansion of *expr* about *var* = *a*. The second form computes the coefficient of $\prod (var_i - a_i)^{n_i}$ in the multivariate Taylor series expansion of *expr* about the a_i. ◇ You must `readlib(coeftayl)` before using this command. ◇ `coeftayl(ln(x),x=1,5);` ⟶ $\frac{1}{5}$ ◇ See also: `mtaylor`, `taylor`.

collect `collect(`*expr*`, `*vars*`)`
`collect(`*expr*`, `*vars*`, `*form*`, `*f*`)`
Collects coefficients of like powers ◇ *vars* may be a single variable name or a list or set of variable names. *form* may be `distributed` or `recursive` (the default); these yield the same result when *vars* is a single name. If a function *f* is given as a fourth argument, it is applied to the coefficients of the collected polynomial. ◇ `p := x*y + a*x*y + b*x*y^2:` ◇ `collect(p, [x,y]);` $\longrightarrow (b\,y^2 + (1+a)\,y)\,x$ ◇ `collect(p, [x,y], distributed, abs);` $\longrightarrow |1+a|\,x\,y + |b|\,x\,y^2$ ◇ See also: `coeff`, `coeffs`, `sort`.

combinat Combinatorial functions package ◇ See also: `combinat[`*function*`]`, `with`.

combinat[bell]
`bell(`*n*`)`
Computes the nth Bell number ◇ `bell(5);` $\longrightarrow 52$ ◇ See also: `binomial`.

combinat[binomial]
`binomial(`*n*`, `*m*`)`
Computes binomial coefficients ◇ Synonym for `binomial`. ◇ See `binomial`.

combinat[cartprod]
`cartprod(`*listlist*`)`
Iterates over the cartesian product of a list of lists ◇ This function returns a table with two entries: `nextvalue` and `finished`. The `nextvalue` entry is a procedure that traverses the cartesian product. Calling this function gives the next element in the cartesian product. The `finished` entry is either *true* or *false* and indicates whether the iteration is complete. ◇ See also: `combinat[subsets]`.

combinat[character]
`character(`*n*`)`
Computes character table for a symmetric group ◇ This function computes the character table for S_n, the symmetric group on n elements. ◇ See also: `combinat[Chi]`, `combinat[partition]`.

combinat[Chi]
`Chi(`λ`, `ρ`)`
Computes Chi function for partitions of symmetric group ◇ This function returns the trace of the matrices in the conjugacy class corresponding to the partition ρ in the irreducible representation corresponding to the partition λ. ρ and λ must be partitions of the same number. ◇ `Chi([2,2],[1,3]);` $\longrightarrow -1$ ◇ See also: `combinat[character]`, `combinat`, `combinat[partition]` `combinat[character]`.

A B C D E F G H I J K L M N O P Q R S T U V W X Y Z

combinat[choose]

`choose(`*n*`)`
`choose(`*list*`)`
`choose(`*n*`, `*m*`)`
`choose(`*list*`, `*m*`)`
Constructs power sets or combinations of a list ◇ The first form returns the power set of the set of the integers from 1 to n. The second form returns a list of all the possible sublists of the given list. The last two return only the sublists of length m. ◇ `choose(2);` ⟶ $\{\{\},\{1\},\{2\},\{1,2\}\}$ ◇ `choose(2,1);` ⟶ $[[1],[2]]$ ◇ See also: `combinat[numbcomb]`, `combinat[permute]`.

combinat[composition]

`composition(`*n*`, `*k*`)`
k-compositions of an integer ◇ This function returns a list containing all distinct ordered *k*-tuples of positive integers whose elements sum to *n*. ◇ `composition(4, 2);` ⟶ $\{[2,2],[1,3],[3,1]\}$ ◇ See also: `@`, `@@`, `combinat[numbcomp]`.

combinat[conjpart]

`conjpart(`*list*`)`
Computes the conjugate partition in the canonical partition sequence ◇ `conjpart([1,3]);` ⟶ $[1,1,2]$ ◇ See also: `combinat[encodepart]`, `combinat[firstpart]`, `combinat[inttovec]`, `combinat[lastpart]`, `combinat[nextpart]`, `combinat[numbpart]`, `combinat[partition]`, `combinat[prevpart]`, `combinat[randpart]`.

combinat[decodepart]

`decodepart(`*n*`, `*m*`)`
Computes canonical partition represented by an integer ◇ Given positive integers *n* and *m* between `1` and `numbpart(`*n*`)`, this function returns the partition represented by *m*. ◇ `decodepart(6,9);` ⟶ $[2,4]$ ◇ See also: `combinat[encodepart]`, `combinat[inttovec]`, `combinat[numbpart]`, `combinat[partition]`, `combinat[vectoint]`.

combinat[encodepart]

`encodepart(`*list*`)`
Computes a canonical integer representing partition ◇ Given a partition *list* of n, this function returns the integer between `1` and `numbpart(`*n*`)` uniquely representing this partition. ◇ `encodepart([2,4]);` ⟶ 9 ◇ See also: `combinat[decodepart]`.

A B C D E F G H I J K L M N O P Q R S T U V W X Y Z

combinat[fibonacci]

`fibonacci(`*n*`)`
`fibonacci(`*n*`, `*var*`)`
Computes the nth Fibonacci number, or the nth Fibonacci polynomial in the variable *var* ◇ `fibonacci(4);` ⟶ 3 ◇ `fibonacci(4,x);` ⟶ $x^3 + 2x$ ◇ See also: `linalg[fibonacci]`.

combinat[firstpart]

`firstpart(`*n*`)`
Returns the first partition in the canonical partition sequence of n ◇ `firstpart(3);` ⟶ $[1, 1, 1]$ ◇ See also: `combinat[conjpart]`, `combinat[encodepart]`, `combinat[inttovec]`, `combinat[lastpart]`, `combinat[nextpart]`, `combinat[numbpart]`, `combinat[partition]`, `combinat[prevpart]`, `combinat[randpart]`.

combinat[graycode]

`graycode(`*n*`)`
Constructs a gray code on n bits order ◇ This function returns a list containing all 2^n n-bit integers in a gray code starting at zero. ◇ `graycode(2);` ⟶ $[0, 1, 3, 2]$

combinat[inttovec]

`inttovec(`*m*`, `*n*`)`
Returns the vector corresponding to m in a canonical ordering ◇ This function provides a one-to-one correspondence between nonnegative integers and vectors of n nonnegative integers. ◇ `inttovec(1001,3);` ⟶ $[10, 3, 4]$ ◇ See also: `combinat[decodepart]`, `combinat[encodepart]`, `combinat[vectoint]`.

combinat[lastpart]

`lastpart(`*n*`)`
Returns the first partition in the canonical partition sequence of n ◇ `lastpart(3);` ⟶ $[3]$ ◇ See also: `combinat[conjpart]`, `combinat[encodepart]`, `combinat[firstpart]`, `combinat[inttovec]`, `combinat[nextpart]`, `combinat[numbpart]`, `combinat[partition]`, `combinat[prevpart]`, `combinat[randpart]`.

combinat[multinomial]

`multinomial(`*n*`, `k_1`, `k_2`, ..., `k_m`)`
Computes the multinomial coefficient $\binom{n}{k_1, k_2, \ldots, k_m}$ ◇ It is assumed that $n = k_1 + k_2 + \ldots + k_m$. ◇ `multinomial(7,3,2,2);` ⟶ 210 ◇ See also: `binomial`.

combinat[nextpart]

`nextpart(`*list*`)`
Computes the next partition in the canonical partition sequence ◇ `nextpart([2,4]);` ⟶ $[1, 5]$ ◇ See also: `combinat[conjpart]`,

combinat[encodepart], combinat[firstpart], combinat[inttovec], combinat[lastpart], combinat[numbpart], combinat[partition], combinat[prevpart], combinat[randpart].

combinat[numbcomb]

numbcomb(*n*)
numbcomb(*list*)
numbcomb(*n*, *m*)
numbcomb(*list*, *m*)
Counts combinations ◇ The first form returns the 2^n, the number of combinations of the integers from 1 to n. The second form computes the number of combinations of elements in *list*, taking into account duplicates in the list. The last two count only the sublists of length m. ◇ numbcomb(5,2); ⟶ 10 ◇ numbcomb([a,a]); ⟶ 3 ◇ See also: binomial, combinat[choose].

combinat[numbcomp]

numbcomp(*n*, *k*)
Computes the number of k-compositions of n ◇ This function counts the number of distinct ordered k-tuples of positive integers whose elements sum to n. ◇ numbcomp(4,2); ⟶ 3 ◇ See also: combinat[composition].

combinat[numbpart]

numbpart(*n*)
Computes the number of partitions of n ◇ This is the number of ways to write n as a sum without regard to order. ◇ numbpart(10); ⟶ 42 ◇ See also: combinat[partition].

combinat[numbperm]

numbperm(*n*)
numbperm(*list*)
numbperm(*n*, *m*)
numbperm(*list*, *m*)
Counts the number of permutations ◇ The first form returns the number of permutations on n objects, $n!$. The second form computes the number of permutations of the elements in *list*, taking into account duplicates in the list. The last two count the permutations of n objects or *list* elements taken m at a time. ◇ numbperm(7,3); ⟶ 210 ◇ See also: combinat[permute].

combinat[partition]

partition(*n*)
Displays all possible partitions of the integer n ◇ This function returns a list of partitions. ◇ partition(3); ⟶ [[1, 1, 1], [1, 2], [3]] ◇ See also: combinat[numbpart].

combinat[permute]

`permute(`*n*`)`
`permute(`*list*`)`
`permute(`*n*`, `*m*`)`
`permute(`*list*`, `*m*`)`
Constructs permutations ⋄ The first form returns a list of all the permutations of the integers from 1 to n. The second form computes the list of permutations of the elements in *list*. The last two list the permutations taken m at a time. ⋄ `permute([a,a,b]);` ⟶ $[[a,a,b],[a,b,a],[b,a,a]]$ ⋄ See also: `combinat[numbperm]`.

combinat[powerset]

`powerset(`*n*`)`
`powerset(`*list*`)`
`powerset(`*set*`)`
Constructs the power set ⋄ The first form returns the power set on the integers from 1 to n. The second returns the list of all sublists of *list*, and the last constructs the power set of the given set. ⋄ `powerset(2);` ⟶ $\{\{\},\{1\},\{2\},\{1,2\}\}$ ⋄ See also: `combinat[choose]`, `combinat[subsets]`.

combinat[prevpart]

`prevpart(`*list*`)`
Computes the previous partition in the canonical partition sequence ⋄ `prevpart([1,5]);` ⟶ $[2,4]$ ⋄ See also: `combinat[conjpart]`, `combinat[encodepart]`, `combinat[firstpart]`, `combinat[inttovec]`, `combinat[lastpart]`, `combinat[nextpart]`, `combinat[numbpart]`, `combinat[partition]`, `combinat[randpart]`.

combinat[randcomb]

`randcomb(`*n*`, `*m*`)`
`randcomb(`*list*`, `*m*`)`
Constructs a random combination ⋄ The first form returns a random combination of m integers among the first n positive integers. The second form returns a random combination of m elements of *list*. ⋄ `randcomb(4,2);` ⟶ $\{2,3\}$ ⋄ See also: `combinat[choose]`.

combinat[randpart]

`randpart(`*n*`)`
Constructs a random partition of n ⋄ `randpart(7);` ⟶ $[2,2,3]$ ⋄ See also: `combinat[partition]`.

combinat[randperm]

`randperm(`*n*`)`
`randperm(`*list*`)`
Constructs a random permutation ⋄ The first form returns a random permutation the first n positive integers. The second form returns a

random combination of the elements of *list*. ⋄ `randperm(4);` ⟶ $[2,4,1,3]$ ⋄ See also: `combinat[permute]`, `combinat[numbperm]`.

combinat[stirling1]

`stirling1(`*n*`,` *m*`)`

Computes Stirling numbers of the first kind ⋄ `stirling1(5,3);` ⟶ 35 ⋄ See also: `combinat[stirling2]`.

combinat[stirling2]

`stirling2(`*n*`,` *m*`)`

Computes Stirling numbers of the second kind ⋄ This equals $\frac{1}{m!}\sum_{k=0}^{m}(-1)^{m-k}\binom{m}{k}k^n$. ⋄ `stirling2(5,3);` ⟶ 25 ⋄ See also: `combinat[stirling1]`.

combinat[subsets]

`subsets(`*set*`)`

`subsets(`*list*`)`

Iterates over the power set of the set or list ⋄ This function returns a table with two entries: `nextvalue` and `finished`. The `nextvalue` entry is a procedure that traverses the power set. Calling this function gives the next element in the power set. Each returned element is a set if the original argument is a set, and a list if the original argument is a list. The `finished` entry is either *true* or *false* and indicates whether the iteration is complete. ⋄ See also: `combinat[cartprod]`, `combinat[powerset]`.

combinat[vectoint]

`vectoint(`*list*`)`

Returns the integer corresponding to *list* in a canonical ordering ⋄ This function provides a one-to-one correspondence between vectors of nonnegative integers and nonnegative integers. ⋄ `vectoint([10,3,4]);` ⟶ 1001 ⋄ See also: `combinat[encodepart]`, `combinat[decodepart]`, `combinat[inttovec]`.

combine

`combine(`*expr*`)`

`combine(`*expr*`,` *name*`)`

`combine(`*expr*`,` *namelist*`)`

Combines terms into a single term ⋄ This function applies transformations that combine terms in sums, products, and powers into a single term. It is applied to the components of lists, sets, and relations. For many functions, it is the inverse of `expand`. Subexpressions involving `Int`, `Limit`, and `Sum` are combined into one expression where possible using linearity. Other transformations are obtained by specifying a name or a list of names as a second argument. ⋄ `combine(Int(f(x),x)+Int(g(x),x));` ⟶ $\int f(x)+g(x)\,dx$ ⋄ See also: `combine/`*name*, `expand`, `factor`.

A B C D E F G H I J K L M N O P Q R S T U V W X Y Z

combine/'@@'

`combine(expr, '@@')`

Synonym for `combine/atatsign` ◇ See `combine/atatsign`.

combine/abs

`combine(expr, abs)`

Combines products of absolute values ◇ This option is new in Release 3. ◇ `combine(abs(x)*abs(y)^3, abs);` $\longrightarrow |x\, y^3|$

combine/atatsign

`combine(expr, atatsign)`

Combines expressions involving operators ◇ This function combines nested functions into function expressions using the iterated composition operator, `@@`. ◇ `combine(f(f(x)), '@@');` $\longrightarrow f^{(2)}(x)$

combine/conjugate

`combine(expr, conjugate)`

Combines complex conjugates ◇ Sums, products, and real powers of complex conjugates are combined into a single term. ◇ `combine(conjugate(a) + 3*conjugate(b), conjugate);` $\longrightarrow$ conjugate($a + 3\, b$) ◇ See also: `evalc`.

combine/exp

`combine(expr, exp)`

Combines exponentials ◇ Three transformations are applied:

$$e^x\, e^y \rightarrow e^{x+y}$$
$$(e^x)^y \rightarrow e^{x\, y}$$
$$e^{(x+n \ln y)} \rightarrow y^n\, e^x \quad (n \text{ an integer})$$

◇ `combine(exp(ln(a)), exp);` $\longrightarrow a$

combine/ln

`combine(expr, ln)`
`combine(expr, ln, type)`

Combines logarithmic terms ◇ Two transformations are applied:

$$a \ln x \rightarrow \ln x^a$$
$$\ln x + \ln y \rightarrow \ln x y$$

The first transformation is not applied if a is a logarithm. ◇ In Release 3, a type name may be given as a third argument. In this case, the first transformation is applied only if a is of the given type. The default type is `rational`. ◇ `combine(a*ln(x)+b*ln(x), [ln,power], anything);` $\longrightarrow \ln\left(x^{(a+b)}\right)$

combine/power

`combine(`*expr*`, power)`

Combines terms with powers ◇ Five transformations are applied:

$x^y\,x^z$	$\to$	x^{y+z}	
$(x^y)^z$	$\to$	$x^{y\,z}$	
$e^x\,e^y$	$\to$	e^{x+y}	
$(e^x)^y$	$\to$	$e^{x\,y}$	
$\sqrt{-n}$	$\to$	$I\,\sqrt{n}$	(n an integer)
$n^q\,m^q$	$\to$	$(n\,m)^q$	(n, m integers, q rational)

◇ `combine(5^(2/3)*7^(2/3), power);` $\longrightarrow 35^{2/3}$

combine/Psi

`combine(`*expr*`, Psi)`

Combines Psi functions ◇ The transformation applied has the form $\Psi(q \pm x) + \Psi(r \pm x) \to \Psi(x) \pm Q(x) \pm \pi\, P(\cot(\pi\, x))$, where Q is a rational function, P a univariate polynomial, and q and r are rational. ◇ `combine(Psi(x)+Psi(1-x), Psi);` $\longrightarrow 2\,\Psi(x) + \pi\cot(\pi\, x)$

combine/radical

`combine(`*expr*`, radical)`
`combine(`*expr*`, radical, symbolic)`

Combines products of radicals of the same power ◇ The main transformation applied is $x^{m/d}y^{n/d} \to (x^m y^n)^{1/d}$, where m and n are positive integers smaller than the integer d, and x and y are positive. If the `symbolic` argument is given, all radicands are assumed to be real and positive. ◇ This option is new in Release 3. ◇ `combine(sqrt(2)*sqrt(3), radical);` $\longrightarrow \sqrt{6}$ ◇ See also: `assume`.

combine/signum

`combine(`*expr*`, signum)`

Combines products of signum functions ◇ This option is new in Release 3. ◇ `combine(signum(x)*signum(y)^3, signum);` $\longrightarrow$ $\mathrm{signum}\left(x\, y^3\right)$

combine/trig

`combine(`*expr*`, trig)`

Combines trigonometric terms ◇ The transformations applied are:

$\sin(a)\sin(b)$	$\to$	$\frac{1}{2}\cos(a-b) - \frac{1}{2}\cos(a+b)$
$\sin(a)\cos(b)$	$\to$	$\frac{1}{2}\sin(a-b) + \frac{1}{2}\sin(a+b)$
$\cos(a)\cos(b)$	$\to$	$\frac{1}{2}\cos(a-b) + \frac{1}{2}\cos(a+b)$

along with analogous transformations for products and powers of `sinh` and `cosh`. ◇ `combine(sinh(x)*sinh(y), trig);` $\longrightarrow$ $\frac{1}{2}\cosh(x+y) - \frac{1}{2}\cosh(x-y)$

commutat `c(x, y)`
`commutat(c)`
`convert(expr, '&*')`
`convert(expr, c)`
`expand(expr, '&*')`
`expand(expr, c)`
`simplify(expr, c)`
Routines for manipulating and simplifying commutators ◇ `c(x, y)` stands for the commutator $xy - yx$, where the multiplication here is noncommutative. `commutat` converts a commutator to Lie bracket (list) form. The `convert` routines convert an expression from commutator notation to &* notation and back again. The `expand` and `simplify` routines provide definitions for expansions and simplifications of commutators and &* expressions. You must `readlib(commutat)` before using these functions. ◇ `commutat(c(x,y));` $\longrightarrow [x, y]$ ◇ `convert(c(x,y),'&*');` $\longrightarrow (x \mathbin{\&*} y) - (y \mathbin{\&*} x)$ ◇ See also: `&*`.

comparray `comparray(A, B)`
`comparray(A, B, dontprint)`
`comparray(A, B, ...)`
Compares the entries of two arrays or lists ◇ *A* and *B* may be arrays, lists, or lists of lists of the same dimensions. The first form returns *true* if *A* and *B* are identical and *false* otherwise. It prints an array of booleans in the same shape as *A* representing entries that agree and disagree, unless `dontprint` is present. In the third form, `testfloat` is invoked to compare *A* and *B*. Additional arguments are passed to `testfloat`. ◇ You must `readlib(comparray)` before using this command. ◇ See also: `linalg[equal]`, `testfloat`.

compoly `compoly(poly, var)`
Determines a possible composition of a polynomial ◇ This function returns a sequence of the form *expr*, *eqn* such that *poly* is `subs(eqn, expr)`. If such a pair cannot be found, it returns *FAIL*. ◇ `compoly(x^4+1, x);` $\longrightarrow x^2 + 1, x = x^2$

conjugate `conjugate(expr)`
Computes the complex conjugate of an expression ◇ If unknown variables represent real values, use `assume` or `evalc`. ◇ See also: `assume`, `evalc`, `Im`, `NPspinor[conj]`, `projgeom[conjugate]`, `Re`.

constants Global variable ◇ A sequence of names treated as system constants. You may not assign a value to any name in this sequence. This variable is initially `false`, `gamma`, `infinity`, `true`, `Catalan`, `E`, `Pi`. You can define additional protected constants by appending them to this sequence. ◇ See also: `type/constant`.

content `content(`*poly*`, `*vars*`, `*name*`)`
Content of a multivariate polynomial ◇ The content is the greatest common divisor of the coefficients of *poly* with respect to the variable or list or set of variables *vars*. If *name* is present, it is assigned the primitive part of *p*oly, that is, the polynomial divided by its content. ◇
`content(2*x+4*x*y, y);` ⟶ $2x$ ◇ See also: `coeffs, Content, gcd, icontent, primpart.`

Content `Content(`*poly*`, `*vars*`, `*name*`)`
Inert content function ◇ When used with `evala`, `evalgf`, or `mod`, this function computes the content of the polynomial in the variable or list or set of variables *vars* over the coefficient domain. If *name* is present, it is assigned the primitive part. ◇ See also: `content, evala, evalgf, mod, Primpart.`

convergs `convergs(`*list*`)`
`convergs(`*f*`)`
`convergs(`$list_1$`, `$list_2$`)`
`convergs(`$list_1$`, `$list_2$`, `*n*`)`
`convergs(`f_1`, `f_2`, `*n*`)`
Prints the convergents of a continued fraction ◇ *list* is a list of partial quotients in a simple continued fraction, or the function *f* computes these numbers as $f(1)$, $f(2)$, and so on. With two lists, $list_1$ contains the denominators in a regular continued fraction and $list_2$ contains the numerators, or the two functions compute these numbers. This function accepts one function and one list as well. The optional argument *n* controls the number of convergents displayed. This function returns `NULL`. ◇ You must `readlib(convergs)` before using this command. ◇
See also: `convert, convert/confrac.`

convert `convert(`*expr*`, `*form*`)`
`convert(`*expr*`, `*form*`, `*args*`)`
Converts an expression to another form or type ◇ Dozens of forms are known to the `convert` function. Some conversions accept additional arguments. Additional conversion routines for this function can be defined by creating procedures named `` `convert/``*f*`` ` ``, where *f* is the name of a new type or form. ◇ See also: `convert/`*name*`, type.`

convert/'+'

`convert(`*expr*``, `+`)``
Converts to a sum ◇ This function sums all the operands of *expr*. ◇
``convert([2,3,4], `+`);`` ⟶ 9 ◇ See also: ``convert/`*`.``

convert/'*'

`convert(`*expr*``, `*`)``
Converts to a product ◇ This function multiplies all the operands of *expr*.
◇ ``convert({2,3,4}, `*`);`` ⟶ 24 ◇ See also: ``convert/`+`.``

A B C D E F G H I J K L M N O P Q R S T U V W X Y Z

convert/array

`convert(`*list*`, array, `*bounds*`, `*indexfcn*`)`
`convert(`*table*`, array, `*bounds*`, `*indexfcn*`)`
`convert(`*array*`, array, `*bounds*`, `*indexfcn*`)`
Converts a list, table, or array to an array ◇ The bounds of the array and the indexing function are optional. ◇ See also: `array`, `convert/matrix`, `convert/vector`.

convert/base

`convert(`*n*`, base, `*outputbase*`)`
`convert(`*list*`, base, `*inputbase*`, `*outputbase*`)`
Converts between bases ◇ The first template converts the base 10 number *n* to base *outputbase*. The second template converts the list of digits *list* considered as a number in base *inputbase* to base *outputbase*. The lowest-order digit is first in the list. ◇ `convert([1,1,2], base,5,10);` ⟶ [6, 5] ◇ See also: `convert/binary`, `convert/decimal`, `convert/hex`, `convert/octal`.

convert/binary

`convert(`*r*`, binary)`
`convert(`*r*`, binary, `*m*`)`
Converts a decimal number to binary ◇ The number may be positive or negative and may be an integer or a floating-point number. In the case of a floating-point number, m determines the number of digits of precision in the answer (the default is `Digits`). ◇ `convert(0.75,binary,3);` ⟶ .110 ◇ See also: `convert/base`, `convert/decimal`, `convert/hex`, `convert/octal`.

convert/binomial

`convert(`*expr*`, binomial)`
Converts GAMMAs and factorials to an expression in binomial coefficients ◇ This function is new in Release 3. ◇ `convert((2*n)!/(n!)^2, binomial);` ⟶ binomial$(2n, n)$ ◇ See also: `convert/factorial`, `convert/GAMMA`.

convert/confrac

`convert(`*r*`, confrac, `*n*`, `*name*`)`
`convert(`*series*`, confrac, `*option*`)`
`convert(`*expr*`, confrac, `*var*`, `*n*`)`
Converts numbers, series, rational functions, and other algebraic objects to a continued fraction approximation ◇ The first template converts the rational or real number *r* to continued fraction form, returning a list of *n* partial quotients. The default value of *n* is 9. If *name* is present, it is assigned a list of the convergents. The second form computes a continued fraction approximation to the series to the order of the series. *option* may be `superdiagonal` (the default) or `subdiagonal`. The third form operates on rational functions or other algebraic objects. The name of the

variable *var* is required; *n* is optional. ◇ `convert(1.57,confrac,4);` ⟶ [1, 1, 1, 3] ◇ See also: `convergs`, `numtheory[cfrac]`, `numtheory[cfracpol]`.

convert/D `convert(`*expr*`, D)`
Converts expressions involving derivatives in the `diff` notation to the `D` operator notation ◇ `convert(diff(f(x),x), D);` ⟶ D(*f*)(x) ◇ See also: `convert/diff`.

convert/decimal

`convert(`*n*`, decimal, binary)`
`convert(`*string*`, decimal, hex)`
`convert(`*n*`, decimal, octal)`
Converts to base 10 ◇ This function converts a binary, hexadecimal, or octal integer to a decimal number. The binary or octal number is given as an integer; the hexadecimal number is given as a string. ◇ `convert('AB', decimal, hex);` ⟶ 171 ◇ See also: `convert/base`, `convert/binary`, `convert/hex`, `convert/octal`.

convert/degrees

`convert(`*number*`, degrees)`
Converts radians to degrees ◇ `convert(Pi, degrees);` ⟶ 180 *degrees* ◇ See also: `convert/radians`.

convert/diff

`convert(`*expr*`, diff, `*x*`)`
Replaces the `D` operator with the `diff` function wherever possible ◇ `convert(D(f)(x), diff);` ⟶ $\frac{\partial}{\partial x} f(x)$ ◇ See also: `convert/D`.

convert/double

`convert(`*r*`, double, `*outputform*`)`
`convert(`*string*`, double, maple, `*inputform*`)`
Converts a double precision floating-point number from one format to another ◇ With three arguments, *expr* is a Maple floating-point number, and *outputform* must be `ibm`, `mips`, or `vax`. With four arguments, *expr* is a hex string in *inputform*, which must be one of the same three names. ◇ `convert(0.75, double, ibm);` ⟶ *40C0000000000000* ◇ See also: `convert/float`, `convert/hex`.

convert/Ei `convert(`*expr*`, Ei)`
Converts trigonometric, hyperbolic, and logarithmic integrals to exponential integrals ◇ `convert(Li(x), Ei);` ⟶ Ei(ln(x)) ◇ See also: `Ei`, `simplify/Ei`.

convert/eqnlist

`convert(`*array*`, eqnlist)`
`convert(`*list*`, eqnlist)`
`convert(`*table*`, eqnlist)`
Converts an array, list, or table to a list of equations that define it ◇ Each item in the returned list has the form *index* = *element*. ◇ `convert([7,8],eqnlist);` ⟶ $[1 = 7, 2 = 8]$ ◇ See also: `convert/list`, `convert/listlist`.

convert/equality

`convert(`*relation*`, equality)`
Converts the relation to an equality, by replacing the relation with = ◇ `convert(x<y, equality);` ⟶ $x = y$ ◇ See also: `convert/lessequal`, `convert/lessthan`, `simplex[convert/equality]`.

convert/erf

`convert(`*expr*`, erf)`
Converts expressions involving the complementary error function or Dawson's integral to the equivalent form using the error function ◇ `convert(erfc(x), erf);` ⟶ $1 - \mathrm{erf}(x)$ ◇ See also: `convert/erfc`, `dawson`, `erf`.

convert/erfc

`convert(`*expr*`, erfc)`
Converts expressions involving the error function to the equivalent form using the complementary error function ◇ `convert(erf(x), erfc);` ⟶ $1 - \mathrm{erfc}(x)$ ◇ See also: `convert/erf`, `erfc`.

convert/exp

`convert(`*expr*`, exp)`
Converts trigonometric functions and hyperbolic trigonometric functions to exponential form ◇ `convert(2*sinh(x), exp);` ⟶ $e^x - \frac{1}{e^x}$ ◇ See also: `convert/expsincos`, `convert/expln`, `convert/ln`, `convert/sincos`, `convert/trig`.

convert/expln

`convert(`*expr*`, expln)`
Converts elementary functions to `exp` and `ln` ◇ `convert(2*cos(x)+arcsinh(x), expln);` ⟶ $e^{I\,x} + \frac{1}{e^{I\,x}} + \ln\left(x + \sqrt{x^2+1}\right)$ ◇ See also: `convert/exp`, `convert/ln`, `convert/trig`.

convert/expsincos

`convert(`*expr*`, expsincos)`
Converts trigonometric functions to `sin` and `cos`, and converts hyperbolic trigonometric functions to exponential form ◇ `convert(tan(theta),`

expsincos); ⟶ $\frac{\sin(\theta)}{\cos(\theta)}$ ◇ See also: convert/exp, convert/sincos, convert/tan, convert/trig.

convert/factorial

convert(*expr*, factorial)
convert(*expr*, factorial, *vars*)
Converts GAMMAs, binomials, and multinomial coefficients to an expression in factorials ◇ *vars* names an indeterminate or list or set of indeterminates for converting a GAMMA or binomial. ◇ convert(binomial(n,m), factorial); ⟶ $\frac{n!}{m!\,(n-m)!}$ ◇ See also: convert/binomial, convert/GAMMA.

convert/float

convert(*expr*, float)
Converts the expression to floating-point ◇ The precision is given by the value of Digits. ◇ convert(erf(1),float); ⟶ .8427007929 ◇ See also: Digits, evalf.

convert/fraction

convert(*r*, fraction)
convert(*r*, fraction, *n*)
convert(*r*, fraction, exact)
Synonym for convert/rational ◇ See convert/rational.

convert/GAMMA

convert(*expr*, GAMMA)
convert(*expr*, GAMMA, *vars*)
Converts binomials, factorials, and multinomials to an expression in the GAMMA function ◇ *vars* names an indeterminate or list or set of indeterminates for converting a binomial or factorial. ◇ convert(n!,GAMMA); ⟶ $\Gamma(n+1)$ ◇ See also: convert/binomial, convert/factorial.

convert/hex

convert(*n*, hex)
Converts to hexadecimal form ◇ This function converts a positive decimal integer to a hexadecimal number. The number is returned as a string. ◇ convert(171, hex); ⟶ *AB* ◇ See also: convert/base, convert/binary, convert/decimal, convert/double, convert/octal.

convert/horner

convert(*poly*, horner)
convert(*poly*, horner, *vars*)
Converts a polynomial to Horner (nested) form ◇ *vars* may be a single variable name or a list or set of variable names. If there is more than one variable, the conversion is applied recursively to the coefficients. A list

allows control over the order of the conversions. ◇ convert(x^2+x+1, horner); ⟶ $1+(x+1)\,x$ ◇ See also: convert/mathorner.

convert/hostfile

convert(*filename*, hostfile)
Converts the Maple file name to the actual host file name ◇ The filename returned differs from system to system. ◇ See also: read, save.

convert/hypergeom

convert(*expr*, hypergeom)
Converts summations to hypergeometrics ◇ convert(Sum(1/k^2, k=1..infinity), hypergeom); ⟶ hypergeom([1, 1, 1], [2, 2], 1) ◇ See also: hypergeom, simplify/hypergeom.

convert/lessequal

convert(*relation*, lessequal)
Converts the relation to a strong inequality, by replacing the relation with <= ◇ convert(x<>y, lessequal); ⟶ $x \leq y$ ◇ See also: convert/equality, convert/lessthan.

convert/lessthan

convert(*relation*, lessthan)
Converts the relation to a weak inequality, by replacing the relation with < ◇ convert(x=y, lessthan); ⟶ $x < y$ ◇ See also: convert/equality, convert/lessequal.

convert/list

convert(*table*, list)
convert(*array*, list)
convert(*expr*, list)
convert(*expr*, list, list)
convert(*expr*, list, '=')
Converts a one-dimensional array, table, or expression to a list ◇ An expression is converted to a list of its operands. With a third argument list, this function invokes convert/listlist on the first argument. With '=', it invokes convert/eqnlist. ◇ convert(a+b*c,list); ⟶ $[\,a, b\,c\,]$ ◇ See also: convert/eqnlist, convert/listlist, convert/set.

convert/listlist

convert(*array*, listlist)
convert(*list*, listlist)
Converts an array or a list of equations to a list of lists ◇ The list of lists formed from an array is the list that would be passed to the array function to create the array. The second form converts a list of equations defining an array to a list. ◇ convert([2=8,1=7],listlist); ⟶ $[\,7, 8\,]$ ◇ See also: array, convert/eqnlist, convert/list.

convert/ln `convert(`*expr*`, ln)`

Converts inverse trigonometric and hyperbolic trigonometric functions to logarithmic form ◇ `convert(2*arctanh(z), ln);` ⟶ $\ln(z+1)-\ln(1-z)$ ◇ See also: `convert/exp`.

convert/mathorner

`convert(`*poly*`, mathorner)`
`convert(`*poly*`, mathorner, `*var*`)`
`convert(`*poly*`, mathorner, `*A*`)`

Converts a polynomial to matrix Horner form ◇ *poly* is a polynomial in the third argument: a variable name or a matrix. ◇ `convert(x^2+x+1,mathorner);` ⟶ $1+((1+x)\,\&^*\,x)$ ◇ See also: `convert/horner`.

convert/matrix

`convert(`*array*`, matrix)`
`convert(`*listlist*`, matrix)`

Converts a two-dimensional array or a list of lists to a matrix ◇ See also: `convert/array`, `convert/vector`, `linalg[matrix]`.

convert/metric

`convert(`*expr*`, metric)`
`convert(`*expr*`, metric, imp)`
`convert(`*expr*`, metric, US)`

Converts to metric units ◇ If a third parameter, `imp` or `US`, appears, the measures specified are taken to be imperial or U.S., respectively. The following units are recognized:

acre	cord	gallon	inches	mi	oz	pounds
acres	cords	gallons	ins	Mile	Ozs	quart
bu	feet	Gals	kg	miles	pint	quarts
bushel	foot	gill	km	MPG	pints	yard
bushels	ft	gills	Lb	MPH	pole	yards
chain	furlong	gr	lbs	ounce	poles	yd
chains	furlongs	hr	light_year	Ounces	pound	yds
cm	gal	inch	light_years			

◇ `convert(1*furlong, metric);` ⟶ $201.1680\,m$

convert/mod2

`convert(`*expr*`, mod2)`

Reduces a logical expression modulo 2 ◇ The expression may contain the Boolean operators `not`, `and`, and `or`. ◇ `convert(not x^3 and y, mod2);` ⟶ $(1+x)y$ ◇ See also: `logic[convert]`, `mod`.

convert/multiset

`convert(`*expr*`, multiset)`
`convert(`*list*`, multiset)`
`convert(`*table*`, multiset)`
Converts to a multiset ◇ A multiset is represented as a list of pairs of the form $[e, m]$, where e is an element and m its multiplicity. The factors of an algebraic expression are interpreted as elements and their exponents as the associated multiplicities. For a list, the number of instances of each entry determines its multiplicity. Finally, the indices of a table are interpreted as elements, and the entries the associated multiplicities. ◇ `convert(x^4*y, multiset);` $\longrightarrow [[x,4],[y,1]]$ ◇ See also: `convert/set`.

convert/name

`convert(`*expr*`, name)`
A synonym for convert/string ◇ See `convert/string`.

convert/octal

`convert(`*r*`, octal)`
`convert(`*r*`, octal, `*m*`)`
Converts a decimal number to octal form ◇ *r* may be positive or negative and may be an integer or a floating-point number. In the case of a floating-point number, m determines the number of digits of precision in the answer (the default is `Digits`). ◇ `convert(135,octal);` $\longrightarrow 207$ ◇ See also: `convert/base`, `convert/binary`, `convert/decimal`, `convert/hex`.

convert/parfrac

`convert(`*expr*`, parfrac, `*var*`)`
`convert(`*expr*`, parfrac, `*var*`, `*factored*`)`
Converts the rational expression *expr* in the variable *var* to partial fraction form ◇ If *factored* is `true`, the denominator is already in factored form, and `normal` is not applied to *expr*. Its default value is `false`. ◇ `convert(2/(x^2-1), parfrac,x);` $\longrightarrow -\frac{1}{x+1} + \frac{1}{x-1}$ ◇ See also: `normal`.

convert/polar

`convert(`*z*`, polar)`
Converts the complex-valued expression *z* to its representation in polar coordinates ◇ The result is represented as `polar(`*r*`, `θ`)`, where *r* is the modulus and θ is the argument of *z*. ◇ `convert(1+I, polar);` $\longrightarrow$ $\mathrm{polar}\left(\sqrt{2}, \frac{1}{4}\pi\right)$ ◇ See also: `argument`, `evalc`, `polar`.

convert/polynom

`convert(`*series*`, polynom)`
Converts the series to a sum-of-products form ◇ The result may not be a polynomial if the series has terms like $1/x$ or $\ln(x)$. ◇

`convert(series(cos(x)/x, x,3), polynom);` $\longrightarrow \frac{1}{x} - \frac{1}{2}x$ ◇
See also: `convert/series`.

convert/radians

`convert(`*number*`, radians)`
Converts degrees to radians ◇ The expression must be of degree 1 in the indeterminate *degrees*. ◇ `convert(180*degrees, radians);` $\longrightarrow \pi$ ◇ See also: `convert/degrees`.

convert/radical

`convert(`*expr*`, radical)`
Converts `RootOf`s in the expression to radicals and `I` when possible ◇ `convert(RootOf(_Z^2-2), radical);` $\longrightarrow \sqrt{2}$ ◇ See also: `convert/RootOf`, `RootOf`.

convert/rational

`convert(`*r*`, rational)`
`convert(`*r*`, rational, `*n*`)`
`convert(`*r*`, rational, exact)`
Converts a floating-point number to an approximate rational ◇ The third argument determines the accuracy of the approximation. An integer n gives the number of decimal places of accuracy required. `exact` requests an exact rational equivalent. The default is the value of `Digits`. ◇ `convert(.75, rational);` $\longrightarrow \frac{3}{4}$ ◇ See also: `convert/confrac`, `Digits`.

convert/ratpoly

`convert(`*series*`, ratpoly)`
`convert(`*series*`, ratpoly, `*m*`, `*n*`)`
Converts the series to a rational polynomial ◇ If the first argument is a Taylor or Laurent series, then the result is a Pade approximation; if it is a Chebyshev series, then the result is a Chebyshev-Pade approximation. m is the desired degree of the numerator and n that of the denominator. By default, these values are chosen so that $m + n + 1 =$ order(*series*) and $m = n$ or $m = n + 1$. ◇ See also: `chebyshev`, `convert/confrac`, `numapprox`.

convert/RootOf

`convert(`*expr*`, RootOf)`
Changes all occurrences of the algebraic constant `I`, and all occurrences of radicals (whether algebraic constants or functions) to `RootOf` notation ◇ `convert(sqrt(2), RootOf);` $\longrightarrow$ RootOf($\_Z^2 - 2$) ◇ See also: `allvalues`, `convert/radical`, `RootOf`.

convert/series

`convert(`*poly*`, series)`
`convert(`*poly*`, series, `*var*`)`
Converts a polynomial to a series ◇ No terms are truncated from the polynomial on conversion. This function also accepts a series as the first

argument. ◇ `convert(1+3*x^7, series);` ⟶ $1 + 3\,x^7$ ◇ See also: `convert/polynom`, `series`.

A B C D E F G H I J K L M N O P Q R S T U V W X Y Z

convert/set

`convert(`*array*`, set)`
`convert(`*list*`, set)`
`convert(`*table*`, set)`
`convert(`*expr*`, set)`
Converts to a set ◇ This function makes a set of the entries of an array, list, table, or the operands of an expression. ◇ `convert([1,1,2], set);` ⟶ $\{1, 2\}$ ◇ See also: `convert/list`.

convert/sincos

`convert(`*expr*`, sincos)`
Converts trigonometric functions to `sin` and `cos`, and hyperbolic trigonometric functions to `sinh` and `cosh` ◇ `convert(tanh(z), sincos);` ⟶ $\frac{\sinh(z)}{\cosh(z)}$ ◇ See also: `convert/exp`, `convert/expsincos`, `convert/tan`.

convert/sqrfree

`convert(`*poly*`, sqrfree)`
`convert(`*poly*`, sqrfree, `*var*`)`
Converts the polynomial to square-free form in the variable *var* ◇ The result is an expression whose base factors are relatively prime. If *var* is not specified, a complete square-free factorization is done, recursively applying the function to the content until there are no variables left in it. ◇ `convert(x^4+4*x^3+3*x^2-4*x-4, sqrfree, x);` ⟶ $\left(x^2 - 1\right)\,(x + 2)^2$ ◇ See also: `content`, `sqrfree`.

convert/string

`convert(`*expr*`, string)`
Converts the expression to a string (name) ◇ `convert(a+1, string);` ⟶ *(a)+(1)* ◇ See also: `type/name`, `type/string`.

convert/tan

`convert(`*expr*`, tan)`
Converts trigonometric functions to `tan` ◇ `convert(sin(2*phi)/2, tan);` ⟶ $\frac{\tan(\phi)}{1+\tan(\phi)^2}$ ◇ See also: `convert/expsincos`, `convert/sincos`.

convert/trig

`convert(`*expr*`, trig)`
Converts exponentials to trigonometric and hyperbolic trigonometric functions ◇ `convert(exp(z), trig);` ⟶ $\cosh(z) + \sinh(z)$ ◇ See also: `convert/exp`, `convert/expln`, `convert/expsincos`.

convert/vector

`convert(`*array*`, vector)`
`convert(`*list*`, vector)`
Converts an array or list to a vector ◇ `convert([1,2],vector);` ⟶ [1 2] ◇ See also: `convert/array`, `convert/matrix`, `linalg[vector]`.

ConvertIn `ConvertIn(`*poly*`, `*var*`)`
Converts a univariate polynomial in the variable *var* to `modp1` representation when used with `modp1` ◇ `modp1(ConvertIn(x^2-x+7, x), 5);` ⟶ 100040002 ◇ See also: `ConvertOut`, `GF`, `modp1`.

ConvertOut

`ConvertOut(`*poly*`)`
`ConvertOut(`*poly*`, `*var*`)`
Converts a polynomial in `modp1` representation to a univariate polynomial in the variable *var* when used with `modp1` ◇ If *var* is absent, the coefficients of the polynomial are returned in a list. ◇ `modp1(ConvertOut(100040002, x), 5);` ⟶ $x^2 + 4x + 2$ ◇ See also: `ConvertIn`, `GF`, `modp1`.

copy `copy(`*array*`)`
`copy(`*table*`)`
Creates a copy of an array or table ◇ This copy may be altered without changing the original array or table.

cos `cos(`*z*`)`
The cosine function ◇ See also: `arccos`.

cosh `cosh(`*z*`)`
The hyperbolic cosine function ◇ See also: `arccosh`.

cost `cost(`*expr*$_1$`, `*expr*$_2$`, ...)`
Operation evaluation count ◇ This function counts the operations required for the numerical evaluation of the given expressions. The count is expressed as a linear combination of the names `additions`, `multiplications`, `divisions`, `functions`, `subscripts`, and `assignments`. Assigning values to these names yields a weighted cost. ◇ You must `readlib(cost)` before using this command. ◇ `cost(sin(ln(x)+y^3));` ⟶
additions + 2 *multiplications* + 2 *functions* ◇ See also: `optimize`.

cot `cot(`*z*`)`
The cotangent function ◇ See also: `arccot`.

coth `coth(`*z*`)`
The hyperbolic cotangent function ◇ See also: `arccoth`.

csc `csc(`z`)`
The cosecant function ⋄ See also: `arccsc`.

csch `csch(`z`)`
The hyperbolic cosecant function ⋄ See also: `arccsch`.

csgn `csgn(`z`)`
`csgn(1,` z`)`
`csgn(0,` z`,` z_0`)`
Sign function for real and complex expressions ⋄ This function determines in which half-plane the complex number z lies. It is 1 if z is in the right half-plane ($\mathrm{Re}(z) > 0$, or $\mathrm{Re}(z) = 0$ and $\mathrm{Im}(z) > 0$), and -1 if z is in the left half-plane ($\mathrm{Re}(z) < 0$, or $\mathrm{Re}(z) = 0$ and $\mathrm{Im}(z) < 0$). The second form returns the derivative of the `csgn` function at z. ⋄ In Release 2, csgn of 0 is 1. In Release 3, csgn of 0 is 0 by default. This value may be changed by setting the environment variable `_EnvsignumO` or by calling csgn with three arguments. In the latter case, the first argument must be 0 and the third argument names the value of csgn at 0. ⋄ `csgn(I);` $\longrightarrow 1$ ⋄ See also: `evalc`, `sign`, `signum`.

D `D(`f`)`
`D[`i`](`f`)`
`D[`i`,` j`, ...](`f`)`
Differential operator ⋄ The first form computes the derivative of a function of one argument. The second form computes the partial derivative with respect to the ith argument of a function of several variables. The third form computes derivatives of higher order. It is equivalent to `D[`i`](D[`j`, ...](`f`)`. The derivative is returned as a function. ⋄ `D(sinh);` $\longrightarrow \cosh$ ⋄ `D(sin+tan);` $\longrightarrow \cos + 1 + \tan^2$ ⋄ `D(x->x^2);` $\longrightarrow x \to 2\,x$ ⋄ `D[1,2,2]((x,y)->x^2*y^3);` $\longrightarrow (x, y) \to 12\,x\,y$ ⋄ See also: `diff`, `unapply`.

dawson `dawson(`x`)`
Dawson's integral ⋄ Dawson's integral is defined as $e^{-x^2} \int_0^x e^{t^2}\,dt$. ⋄ You must `readlib(dawson)` before using this command. ⋄ See also: `erf`, `Fresnel`.

debever `curvature(`*spincf*`,` *info*`,` *flag*`)`
`npspin(`*h*`,` *info*`,` *coords*`,` *flag*`)`
`printcurve(`*curve*`)`
`printspin(`*spincf*`)`
`simp(`*name*`,` *f*`,` *spincf*`)`
A collection of procedures for the computation of the Newman-Penrose spin coefficients and curvature components using Debever's formalism ⋄ The arguments are:

A B C D E F G H I J K L M N O P Q R S T U V W X Y Z

h	an array indexed from 0 to 15
info	an unassigned name
coords	a list of four variable names
flag	an optional argument to suppress printing
spincf	an array indexed from 0 to 11
name	an unassigned name
f	a procedure name
curve	an array indexed from 0 to 11

`npspin` calculates the spin coefficients, returning them in an array names `spincf`. `printspin` prints all the spin coefficients. `curvature` calculates the curvature coefficients, returning them in an array named `curve`. `printcurve` prints all the curvature components. `simp` simplifies the components of the array using the function *f*, assigning the new result to *name*. You must `readlib(debever)` before using these functions. ◇ See also: `cartan`, `tensor`.

debug `debug(`f_1`,` f_2`, ...)`
Turns on tracing of the named procedures for debugging ◇ When the procedures are subsequently invoked, the entry points, intermediate results calculated, and exit points are printed. Synonym for `trace`. ◇ See also: `infolevel`, `printlevel`, `profile`, `trace`, `undebug`, `untrace`, `userinfo`.

define `define(Linear(`*oper*`))`
`define(Group(`*oper*`,` *id*`,` *invoper*`))`
`define(`*oper*`,` $property_1$`,` $property_2$`, ...)`
Defines characteristics of an operator name ◇ This function defines the evaluation and simplification rules of an operator. The first template defines *oper* to be a linear operator. The second defines a group law for *oper*, with the given identity element and inverse. The third form lists specific properties for the operator. These may be `unary`, `binary`, `associative`, `commutative`, `symmetric`, `antisymmetric`, `inverse=`*invoper*, `identity=`*id*, `zero=`*zero*, or `forall(`*vars*`,` *oper*`(`$args_1$`)=`res_1`, ...)`. The `forall` expression defines special relations that hold for certain arguments of the operator. ◇ In Release 2 and earlier versions, use `linear` and `group` instead of `Linear` and `Group`. ◇ `define('&+', associative, binary, commutative, identity=0);` ◇ `define(F, forall(x, F(1/x)=F(x)));`

degree `degree(`*poly*`,` *vars*`)`
Degree of a polynomial ◇ The first argument may have negative integer exponents. *vars* is may be a single indeterminate or a list or set of such. For a set, the total degree is used; for a list, the vector degree. ◇ `degree(x^3+x^2,x);` ⟶ 3 ◇ See also: `lcoeff`, `ldegree`.

denom `denom(`*expr*`)`
Denominator of an expression ◇ If *expr* is not in normal form, it is converted to that form first. ◇ `denom(18/4);` ⟶ 2 ◇ See also: `normal`, `numer`.

DESol `DESol(`*deqns*`)`
`DESol(`*deqns*`,` *vars*`)`
`DESol(`*deqns*`,` *vars*`,` *eqns*`)`
Represents the solution of a differential equation or set of differential equations ◇ This function allows the solution of a differential equation to be manipulated symbolically without computing it first. Maple can differentiate, integrate, numerically evaluate, and generate a series expansion for a `DESol` structure. *vars* is a single variable or a set of variables; *eqns* is a single initial condition or a set of initial conditions. ◇ This function is new in Release 3. ◇ See also: `dsolve`.

Det `Det(`*A*`)`
Inert determinant ◇ This function is a placeholder for representing the determinant of the matrix `A`. It is used in conjunction with `mod` and `modp1`, which define the coefficient domain. ◇ `Det(array([[3,2],[8,7]])) mod 2;` ⟶ 1 ◇ See also: `linalg[det]`, `mod`, `modp1`.

DEtools Differential equations tools package ◇ The functions in this package must be loaded using `with` before they can be invoked. Several functions in this package display graphics. In addition to standard plotting options, these functions accept several additional options, listed on page 114. ◇
See also: `DEtools[`*function*`]`, `with`.

DEtools[Dchangevar]

`Dchangevar(`*eqnset*`,` *diffeq*`,` *constantset*`)`
Performs a change of variables on a differential equation ◇ Given one nth order differential equation and a set of equations, this function does a substitution of the equations into the differential equation. This routine functions like `subs`, but works around the fact that you cannot subs into `diff` to change the second argument. The optional third argument is a set of variables that should be treated as constant by the function. ◇ See also: `student[changevar]`, `subs`.

DEtools[DEplot]

`DEplot(`*diffeq*`,` *vars*`,` *trange*`,` *initset*`,` *options*`)`
`DEplot(`*diffeq*`,` *vars*`,` *trange*`,` *xrange*`,` *yrange*`,` *options*`)`
Plots the solution to system of differential equations ◇ Given a system of first order differential equations of the form $x' = f_1(t, x, y)$, $y' = f_2(t, x, y)$, or a higher order equation of the form $\frac{d^n}{dx^n} y(x) = f(x, y)$ and a set of initial conditions, this function plots the solution curves for the differential equation system under those initial conditions. For nonautonomous systems, by default, a three-dimensional plot of the solution curves in the first three variables is presented. For an autonomous

system of two equations, by default, the two-dimensional scene of x versus y is presented. The second argument, *vars*, may be given in the form `[`*t*`,`*y*`]` or *y*`(`*t*`)`. *trange* may have the form *a*`..`*b* or *t*=*a*`..`*b*. ◇ `DEplot(diff(y(x),x$3)+x*y = 2, [x,y], -1..3, {[1,0,0,0]});` ◇ See also: `DEtools[DEplot1]`, `DEtools[DEplot2]`, `DEtools[dfieldplot]`, `DEtools[phaseportrait]`, `dsolve`.

DEtools[DEplot1]

`DEplot1(`*diffeq*`, `*vars*`, `*trange*`, `*initset*`, `*options*`, ...)`
`DEplot1(`*diffeq*`, `*vars*`, `*trange*`, `*yrange*`, `*options*`, ...)`
Plots the solution to a first order differential equation ◇ Given a single first order differential equation, this function produces a plot of the solution curves of the differential equation for the initial conditions given, as well as a grid of arrows constituting a direction field for the equation. When no initial conditions are given, only the arrows are plotted. ◇ `DEplot1(diff(y(t),t) = cos(t-y), y(t), t=-Pi..Pi, {[0,1], [0,0], [0,-1]});` ◇ See also: `DEtools[DEplot]`, `DEtools[DEplot2]`, `DEtools[dfieldplot]`, `DEtools[phaseportrait]`, `dsolve`.

DEtools[DEplot2]

`DEplot2(`*diffeq*`, `*vars*`, `*trange*`, `*initset*`, `*options*`)`
`DEplot2(`*diffeq*`, `*vars*`, `*trange*`, `*xrange*`, `*yrange*`, `*options*`)`
Plots the solution to a two-dimensional system of differential equations ◇ Given two first order differential equations of the form $x' = f_1(t, x, y)$, $y' = f_2(t, x, y)$, and a set of initial conditions, this function plots the solution curves for the differential equation system under those initial conditions. For nonautonomous systems, by default, a three-dimensional plot of the solution curves is presented. For autonomous systems, by default, the two-dimensional scene of x versus y is presented. The direction field is included if $\frac{dy}{dx}$ is independent of t. When no initial conditions are given, only the arrows are plotted. ◇ `DEplot2([y,-x+y/7], [x,y], -5..5, {[0,1,1]}, stepsize=.1);` ◇ See also: `DEtools[DEplot]`, `DEtools[DEplot1]`, `DEtools[dfieldplot]`, `DEtools[phaseportrait]`, `dsolve`.

DEtools[dfieldplot]

`dfieldplot(`*diffeq*`, `*vars*`, `*trange*`, `*options*`)`
Plots the direction field of a one- or two-dimensional system of differential equations ◇ Given one first order differential equation or a pair of equations from which the independent variable can be eliminated, this function produces a grid of arrows constituting a direction field for the differential equation(s). The second argument, *vars*, may be given in the form `[`*t*`,`*y*`]` or *y*`(`*t*`)`. *trange* may have the form *a*`..`*b* or *t*=*a*`..`*b*. ◇ `dfieldplot(y^2*cosh(x), [x,y], -4..4);` ◇ See also:

DEtools[DEplot], DEtools[DEplot1], DEtools[DEplot2], DEtools[phaseportrait], dsolve.

DEtools[PDEplot]

PDEplot(*pdiffeq*, *vars*, *inits*, *srange*, *options*)
Plots the solution to a first order, quasi-linear partial differential equation ◇ This function produces the solution surface for a first order, quasi-linear partial differential equation of the form $P(x,y,u)\,u_x(x,y) + Q(x,y,u)\,u_y(x,y) = R(x,y,u)$. *pdiffeq* should be a list of P, Q, and R given as expressions or functions. *vars* may have the form [*x*,*y*,*u*] or *u*(*x*,*y*). *inits* is a list of length three that specifies the parametric form of a curve in three-space, in Cartesian coordinates. *srange* is the range of the initial data parameter. ◇ PDEplot([u,1,0], [x,y,u], [s,2*s,3*s], s=0..1); ◇ See also: liesymm.

DEtools[phaseportrait]

phaseportrait(*diffeq*, *vars*, *trange*, *inits*, *options*)
Plots the phase portrait to a one- or two-dimensional system of differential equations ◇ Given one or two first order differential equations of the form $x' = f_1(t,x,y)$, $y' = f_2(t,x,y)$, and a set of initial conditions, this function produces a plot of the integral curves for the differential equation system under those initial conditions. For nonautonomous systems of two equations, by default, a three-dimensional plot of the integral curves is presented. For autonomous systems, by default, the two-dimensional scene of x versus y is presented. The direction field may be included if $\frac{dy}{dx}$ is independent of t. ◇ phaseportrait([y,-x+y/5], [t,x,y], 0..15, {[1,1,1],[2,2,1]}, stepsize=.2); ◇ See also: DEtools[DEplot], DEtools[DEplot1], DEtools[DEplot2], DEtools[dfieldplot], dsolve.

diagonal — The diagonal indexing function ◇ This function gives 0 if all components of an array or table index are not identical. Defining a table or array with this indexing function prevents reassigning of off-diagonal elements. ◇ See also: array, identity, linalg[diag], table.

diff

diff(*expr*, *var$_1$*)
diff(*expr*, *var$_1$*, *var$_2$*, ...)
Partial differentiation ◇ This function computes the partial derivative of an expression with respect to the named variable or variables. diff(*expr*, *x*,*y*) is equivalent to diff(diff(*expr*, *x*),*y*). The derivative is returned as an expression. ◇ diff(sinh(a*x), x); ⟶ $\cosh(a\,x)\,a$ ◇ diff(x^2*y^3, x,y,y); ⟶ $12\,x\,y$ ◇ diff(t*ln(t)-t, t$3); ⟶ $-\frac{1}{t^2}$ ◇ See also: $, D, Diff, dsolve, int.

Diff

Diff(*expr*, *var$_1$*)
Diff(*expr*, *var$_1$*, *var$_2$*, ...)
Inert partial differentiation ◇ This function simply returns unevaluated. ◇ Diff(cos(x), x,x); ⟶ $\frac{\partial^2}{\partial x^2}\cos(x)$ ◇ See also: D, Diff, Int.

difforms Differential forms package ◇ See also: `difforms[`*function*`]`, `with`.

difforms[&^]

expr$_1$ `&^` *expr*$_2$
`&^(`*expr*$_1$`,` *expr*$_2$`, ...)`
Wedge product ◇ This operator represents the wedge product of differential forms.

difforms[d]

`d(`*expr*`)`
`d(`*expr*`,` *forms*`)`
Computes the exterior derivative of the expression ◇ When called with a list of 1-forms as a second argument, any name of type `scalar` in the expression is expanded in these 1-forms. It is assumed that the 1-forms are independent. For each 1-form, a new scalar is created to be the component for that 1-form. If the expression is a list, `d` is applied to each element in the list. ◇ See also: `D`, `diff`, `difforms[mixpar]`.

difforms[defform]

`defform(`*eqn*$_1$`,` *eqn*$_2$`, ...)`
Defines variables used in a computation ◇ Two kinds of definitions may be given: *name*=*type* and `d(`*expr*$_1$`)=`*expr*$_2$. The first declares *name* to be of type *type*, for example `const`, `even`, `form`, `nonhmg`, `odd`, or `scalar`. The second form defines the exterior derivative of *expr*$_1$. Using this function clears the remember tables of all functions in the `difforms` package, though all definitions made through `defform` are remembered throughout the session. ◇ `defform(c=const,s=scalar,f=form);` ◇ See also: `difforms[type]`.

difforms[formpart]

`formpart(`*expr*`)`
Finds the form part of the expression ◇ This function removes scalars and constants from a product, leaving only that part that is a form or is undeclared. ◇ See also: `difforms[scalarpart]`.

difforms[mixpar]

`mixpar(`*expr*`)`
Ensures equality of mixed partial derivatives ◇ This function reorders nested calls to `diff` to ensure the mixed partials are recognized as equal. ◇ See also: `diff`, `sort`.

difforms[parity]

`parity(`*expr*`)`
Extension of mod 2 ◇ This function computes the parity of an expression, assuming that unspecified exponents are integers.

difforms[scalarpart]

`scalarpart(`*expr*`)`
Finds the scalar part of the expression ◇ This function collects scalars and constants from a product, removing those parts that are forms or are undeclared. ◇ See also: `difforms[formpart]`.

difforms[simpform]

`simpform(`*expr*`)`
Simplifies an expression involving forms ◇ Simplifications include collecting like terms, simplifying wedge products, and extracting scalar factors.

difforms[type]

`type(`*expr*`, const)`
`type(`*expr*`, form)`
`type(`*expr*`, form, `*n*`)`
`type(`*expr*`, scalar)`
Checks for constants, forms, and scalars ◇ This function returns *true* or *false*. The type `const` includes anything that is of type `constant` plus those names that are defined to be `const` through `difforms[defform]`. Likewise, types `form` and `scalar` include all names defined as such. The third template tests whether *expr* is a form with `wdegree` *n*. These three types are mutually exclusive. ◇ See also: `type`, `type/constant`, `difforms[wdegree]`.

difforms[wdegree]

`wdegree(`*expr*`)`
Degree of a form ◇ This function computes the degree of an expression, considered as a form, and returns `nonhmg` when the expression is a sum of forms that have different degrees.

Digits — Environment variable ◇ The number of digits kept in floating-point calculations. Its default value is 10.

dilog — `dilog(`*x*`)`
The dilogarithm function ◇ This function is defined as $\int_1^x \frac{\ln(t)}{1-t}\,dt$ ◇
`dilog(0);` $\longrightarrow \frac{1}{6}\pi^2$

dinterp — `dinterp(`*f*`, `*n*`, `*k*`, `*d*`, `*p*`)`
Probabilistic degree interpolation ◇ Given a function *f* that evaluates a polynomial in *n* variables modulo *p*, and a degree bound *d* on the *k*th variable, this function determines probabilistically the degree of the *k*th variable in the value computed by *f*. ◇ You must `readlib(dinterp)` before using this command.

Dirac — `Dirac(`*x*`)`
`Dirac(`*n*`, `*x*`)`
The Dirac delta function ◇ The Dirac function is zero everywhere except at 0, where it has a singularity. It has the additional property that

$\int_{-\infty}^{\infty}$Dirac$(x)\,dx = 1$. The second form returns the nth derivative of the Dirac function at x. ◇ See also: `fourier`, `Heaviside`, `laplace`, `mellin`.

disassemble

`disassemble(`*ptr*`)`
Breaks the object at the given address into its components ◇ A sequence of the addresses is returned, with the first integer in the sequence representing the type of the object. This function allows access to the internal representation of an expression. ◇ `disassemble(addressof(a+2*b));` ⟶ 11, 1404964, 1240560, 1410692, 1240576 ◇ See also: `addressof`, `assemble`, `pointto`.

discont

`discont(`*expr*, *var*`)`
Computes a set of possible discontinuities for the expression in the given variable ◇ It is not necessarily certain that all the points returned are discontinuities. When a variable of the form _Z*n* appears in the result, a discontinuity is found for each assignment of an integer to that variable. Similarly, a variable of the form _NN*n* indicates a discontinuity for each assignment of a nonnegative integer. ◇ You must `readlib(discont)` before using this command. ◇ `discont(x^2+1/x,x);` ⟶ $\{0\}$ ◇ `discont(cot(x),x);` ⟶ $\{\pi\,\_Z1\}$ ◇ See also: `iscont`, `singular`.

discrim

`discrim(`*poly*, *var*`)`
Computes the discriminant of the polynomial ◇ `discrim(x^3+a*x+b, x);` ⟶ $-4\,a^3 - 27\,b^2$ ◇ See also: `Discrim`, `resultant`.

Discrim

`Discrim(`*poly*, *var*`)`
Inert discriminant of the polynomial ◇ When used in conjunction with `mod`, this function computes the discriminant of the polynomial over a given coefficient domain. ◇ `Discrim(x^2+x+1,x) mod 2;` ⟶ 1 ◇ See also: `discrim`, `Resultant`.

DistDeg

`DistDeg(`*poly*, *var*`) mod` *p*
Distinct degree factorization ◇ When used in conjunction with `mod`, this function computes the distinct degree factorization of a monic square-free univariate polynomial over a given domain. A list of pairs of the form $[q, d]$ is returned, where q is the product of the factors of *poly* of degree d. ◇ `DistDeg(x^4+x, x) mod 2;` ⟶ $[[x^2+x, 1], [x^2+x+1, 2]]$ ◇ See also: `Berlekamp`, `Factors`, `ProbSplit`, `Sqrfree`.

divide

`divide(`*poly₁*, *poly₂*, *name*`)`
Tests whether $poly_2$ divides $poly_1$ over the rationals ◇ The polynomials must have rational coefficients. This function returns *true* or *false*. If *true* is returned and a name is given, it is assigned the quotient. ◇ `divide(x^2-y^2, x+y);` ⟶ *true* ◇ See also: `Divide`, `gcd`, `quo`, `rem`.

Divide `Divide(`*poly*$_1$`, `*poly*$_2$`, `*name*`)`
Inert polynomial division test ◇ This function tests whether *poly*$_2$ divides *poly*$_1$ over a given coefficient domain. It is used in conjunction with `evala`, `evalgf`, `mod`, and `modp1`. If *true* is returned and a name is given, it is assigned the quotient. ◇ `Divide(x^2+y^2, x+y) mod 2;` ⟶ *true* ◇ See also: `divide, evala, evalgf, Gcd, mod, modp1, Quo, Rem.`

do `for` *var* `from` *expr*$_1$ `to` *expr*$_2$ `by` *expr*$_3$ `do` *statements* `od`
`for` *var* `in` *expr* `do` *statements* `od`
`while` *cond* `do` *statements* `od`
`for` *var* `...` `while` *cond* `do` *statements* `od`
`do` *statements* `od`
The repetition (looping) statement ◇ The first `for` loop executes the statements for a counted number of times, with the indexing variable *var* taking on values between *expr*$_1$ and *expr*$_2$, incremented by *expr*$_3$ after each pass through the loop. Each of the `from`, `to`, and `by` clauses is optional (the defaults are 1, infinity, and 1, respectively), and they may appear in any order. The second `for` loop has *var* assume successive operands of *expr*. The `while` loop executes the statements as long as the condition *cond* is true. The condition is tested at the beginning of each pass through the loop. The `for` and `while` forms may be combined in a single loop. A `do` statement with none of these clauses executes the loop indefinitely, or until a statement such as `break` is encountered. ◇ `s:=0: for i to 5 by 2 do s:=s+i od: s;` ⟶ 9 ◇ `s:=0: for i in [1,3,5] do s:=s+i od: s;` ⟶ 9 ◇ `t:=0: while t<5 do t:=t+2 od: t;` ⟶ 6 ◇ See also: `break, ERROR, for, if, next, op, RETURN, while.`

done Quits a Maple session ◇ See also: `quit, stop.`

dsolve `dsolve(`*deqns*`, `*vars*`)`
`dsolve(`*deqns*`, `*vars*`, `*options*`)`
Solves ordinary differential equations ◇ *deqns* is a differential equation or a set of differential equations and initial conditions. *vars* is the variable or set of variables to solve for, each of the form $y(x)$. Derivatives may be denoted with `diff` or `D`. There are three basic options in Release 3: `type`, `explicit`, and `method`. Each may be given a value by supplying an equation as an additional argument:

`type=exact`	Find an exact solution if possible (default)
`type=series`	Use a series method
`type=numeric`	Use a numerical method
`explicit=true`	Attempt to find explicit solution for dependent variable
`explicit=false`	(default)
`method=laplace`	Use Laplace transforms
`method=rkf45`	4th–5th order Runge-Kutta numerical method (default)
`method=dverk78`	7th–8th order Runge-Kutta numerical method

The `method` option should be set to `rkf45` or `dverk78` only if `type` is `numeric`. When a numerical method is used, more options may be given for controlling its parameters. ◇ In Release 2 and earlier versions, *options* may be `explicit`, `laplace`, `series`, or `numeric` (a fourth-fifth order Runge-Kutta method is used). In Release 3, this function may return a `DESol` structure if no solution is found; in earlier versions, it returns unevaluated when this happens. ◇ `dsolve(diff(y(x),x,x)=y(x), y(x));` $\longrightarrow y(x) = e^x \_C1 + \_C2\, e^{-x}$ ◇ `dsolve({(D)(y)(x)=y(x), y(0)=3}, y(x));` $\longrightarrow y(x) = 3\, e^x$ ◇ See also: `D`, `DESol`, `DEtools`, `diff`, `laplace`, `mellin`, `plots[odeplot]`, `series`.

E

Euler's number ◇ The base of natural logarithm, approximately 2.718281828459045. Identical to `exp(1)`.

edit

`edit(`*expr*`)`

Expression editor ◇ This function returns `BOX(`*expr*`)`. The following operations are available on the `BOX`:

_	the contents of BOX
child(*i*)	move BOX to the *i*th subexpression
child(*i*, *j*)	same as child(*i*); child(*j*);
down	move BOX to first operand of expression
find(*e*)	move BOX to the next occurrence of *e*
first	move BOX to first subexpression at current level
last	move BOX to last subexpression at current level
prev	move BOX to previous subexpression at current level
show	display the expression
succ	move BOX to successive subexpression at current level
top	move BOX to include the whole expression
up	move BOX to enclosing expression
use(*e*)	*e* replaces the contents of the BOX

Note: the command-line editing facility (page 101) provides a much simpler way to edit expressions. ◇ You must `readlib(edit)` before using this command.

Ei

`Ei(`*x*`)`
`Ei(`*n*`, `*z*`)`

The exponential integral ◇ The single argument case is the Cauchy Principal Value integral defined for real x as:

$$\mathrm{Ei}(x) = \int_{-\infty}^{x} \frac{e^t}{t}\, dt$$

The second case is a family of exponential integrals, indexed by a nonnegative integer n:

$$\mathrm{Ei}(n, z) = \int_{1}^{\infty} \frac{e^{-zt}}{t^n}\, dt, \quad \mathrm{Re}(z) > 0$$

◇ See also: `convert/Ei`, `simplify/Ei`.

Eigenvals `Eigenvals(`*A*`,` *name*`)`
`Eigenvals(`*A*`,` *B*`,` *name*`)`
Computes the eigenvalues/vectors of a numeric matrix ◇ This function returns an array of the eigenvalues of the real or complex square matrix A. If a name is supplied, the eigenvectors are assigned to *name* as a square matrix, where the ith column holds the eigenvector corresponding to the ith eigenvalue. This is an inert function. Use `evalf` to compute the eigenvalues and eigenvectors. The second form handles the generalized eigenvalue problem, finding eigenvalues L and eigenvectors X such that $AX = LBX$. ◇ See also: `evalf`, `linalg[eigenvals]`, `linalg[eigenvects]`.

ellipsoid `ellipsoid(`*a*`,` *b*`,` *c*`)`
Surface area of an ellipsoid ◇ This function returns the surface area of an ellipsoid whose semi-axes are a, b, and c, where $a \geq b \geq c \geq 0$. ◇ You must `readlib(ellipsoid)` before using this command. ◇ `ellipsoid(1,1,1);` $\longrightarrow 4\,\pi$

entries `entries(`*array*`)`
`entries(`*table*`)`
Entries of a table or array ◇ This function returns a sequence of lists of the entries of the table or array that have been explicitly stored. There is a one-to-one correspondence between the results of `indices` and `entries`. ◇ `entries(array([5,6,7]));` $\longrightarrow [6], [7], [5]$ ◇ See also: `array`, `indices`, `op`, `table`.

eqn `eqn(`*expr*`)`
`eqn(`*expr*`,` *filename*`)`
Produces troff/eqn output for printing the expression ◇ This function formats integrals, limits, sums, products, and matrices. If a name is given as the second parameter, the output is written to this file. ◇ See also: `C`, `fortran`, `latex`.

erf `erf(`*z*`)`
The error function ◇ The error function is defined as $\frac{2}{\sqrt{\pi}} \int_0^z \mathrm{e}^{-t^2}\, dt$ for real z and extended to the complex plane. ◇ `erf(1.);` $\longrightarrow .8427007929$ ◇ See also: `convert/erfc`, `dawson`, `erfc`, `Fresnel`.

erfc `erfc(`*z*`)`
`erfc(`*n*`,` *z*`)`
The complementary error function and its iterated integrals ◇ $\mathrm{erfc}(z) = 1 - \mathrm{erf}(z)$. Its iterated integrals are defined as $\mathrm{erfc}(-1, z) = \frac{2}{\sqrt{\pi}} \mathrm{e}^{-z^2}$ and $\mathrm{erfc}(n, z) = \int_z^\infty \mathrm{erfc}(n-1, t)\, dt$ for $n \geq 0$ and real z, and extended to the complex plane. ◇ `erfc(1.);` $\longrightarrow$ .1572992071 ◇ See also: `convert/erf`, `erf`.

ERROR `ERROR(`*expr₁*`, `*expr₂*`, ...)`
Error return from a procedure ◇ This function causes an immediate return to the point where the current procedure was invoked, and an error message is displayed with the procedure name and the sequence of expressions given. ◇ See also: `lasterror`, `proc`, `procname`, `RETURN`, `traperror`.

euler `euler(`*n*`)`
`euler(`*n*`, `*x*`)`
Euler numbers and polynomials ◇ This function computes the *n*th Euler number, or the *n*th Euler polynomial in the expression *x*. The *n*th Euler number E(n) is defined by the exponential generating function:

$$\frac{2}{(e^t + e^{-t})} = \sum_{n=0}^{\infty} \frac{E(n)}{n!} t^n$$

◇ `euler(2);` $\longrightarrow -1$ ◇ `euler(2,x);` $\longrightarrow x^2 - x$ ◇ See also: `bernoulli`.

eulermac `eulermac(`*expr*`, `*var*`)`
`eulermac(`*expr*`, `*var*`, `*n*`)`
Euler-Maclaurin summation ◇ This function computes an *n*th degree approximation to the summatory function of *expr* with respect to the variable *var* using the Euler-Maclaurin summation formula. The default value of *n* is one less than the value of `Order`. ◇ You must `readlib(eulermac)` before using this command. ◇ `eulermac(ln(n), n, 2);` $\longrightarrow n\ \ln(n) - n - \frac{1}{2}\ln(n) + \frac{1}{12}\frac{1}{n} + O\left(\frac{1}{n^3}\right)$

eval `eval(`*expr*`)`
`eval(`*expr*`, `*n*`)`
Explicit evaluation ◇ Normally, global variables are evaluated fully, and local variables and parameters one level. The first form of this function explicitly requests full evaluation, and the second requests *n* levels. ◇ `a:=b: b:=1: eval(a,1);` $\longrightarrow b$ ◇ See also: `evalf`, `evaln`, `subs`, `value`.

Eval `Eval(`*poly*`, `*var*`=`*val*`))`
`Eval(`*poly*`, {`*var₁*`=`*val₁*`, `*var₂*`=`*val₂*`,...})`
`Eval(`*poly*`, `*val*`))`
Inert polynomial evaluation ◇ When used in conjunction with `mod`, this function evaluates the polynomial in a given coefficient domain. The third form is used in conjunction with `modp1`. ◇ `Eval(x^2+x+1, x=2) mod 3;` $\longrightarrow 1$ ◇ See also: `mod`, `modp1`, `subs`.

evala `evala(`*expr*`)`
`evala(`*expr*`, independent)`
Evaluates in an algebraic number or function field ◇ If *expr* is an unevaluated function call (such as `Gcd(u,v)`), the function is performed

in the smallest algebraic number (or function) field possible. Algebraic numbers and functions must be represented with the `RootOf` notation. The option `independent` prevents checking for independence of `RootOf`s in many inert functions. The following placeholders are known to `evala`:

AFactor	Expand	Gcdex	Prem	Rem	Sqrfree
AFactors	Factor	Indep	Primfield	Resultant	Subres
Content	Factors	Norm	Primpart	Sprem	Trace
Divide	Gcd	Normal	Quo		

◇ `alias(alpha=RootOf(y^2+2,y)): evala(Factor(x^2+8, alpha));` ⟶ $(x-2\,\alpha)(x+2\,\alpha)$ ◇ See also: `alias`, `RootOf`.

evalb `evalb(`*expr*`)`
Evaluates a Boolean expression ◇ This function forces evaluation of expressions involving relational operators using a three-valued logic system. This function returns *true*, *false*, or *FAIL*. ◇ `evalb(1<2);` ⟶ *true* ◇ See also: `Boolean`.

evalc `evalc(`*expr*`)`
Evaluation over the complex field ◇ This function attempts to split the expression into its real and imaginary parts. Unknown variables are assumed to represent real-valued quantities. This function maps onto sets, lists, and equations. ◇ `evalc(exp(I*theta));` ⟶ $\cos(\theta)+I\,\sin(\theta)$ ◇ See also: `convert/polar`, `argument`, `evalf`, `Im`, `polar`, `Re`.

evalf `evalf(`*expr*`, `*n*`)`
Evaluates the expression using floating-point arithmetic ◇ *n* specifies the number of decimal digits used in the arithmetic. If it is not specified, the value of the global variable `Digits` is used. ◇ `evalf(Pi,5);` ⟶ 3.1416 ◇ See also: `Digits`, `` `evalf/int` ``, `evalhf`, `Int`.

`evalf/int` `` `evalf/int`( ``*expr*`, `*var*`=`*a*`..`*b*`)`
`` `evalf/int`( ``*expr*`, `*var*`=`*a*`..`*b*`, `*n*`, `*methodname*`)`
Numerical integration ◇ This routine is called when `evalf` is invoked on `Int`, but it may be called directly for better control of the approximation. The third parameter *n* specifies the number of digits of accuracy desired in the result (the default is given by `Digits`). The fourth parameter names the numerical method used. This name can be `_CCquad` (Clenshaw-Curtis quadrature only), `_Dexp` (double-exponential quadrature) or `_NCrule` (Newton-Cotes rule). The default is Clenshaw-Curtis, with special handling of any singularities detected. ◇ You must `` readlib(`evalf/int`) `` before using this command. ◇ `` `evalf/int`(exp(-x^2), x=0..1, 12, _NCrule); `` ⟶ .746824132812 ◇ See also: `evalf`, `int`, `Int`.

evalgf `evalgf(`*expr*`, `*p*`)`
Evaluates in an algebraic extension of a finite field ◇ If *expr* is an unevaluated function call such as `Gcd(u,v)`, the function is performed in the smallest possible algebraic extension of the integers modulo *p*. The

following inert functions are known to `evalgf`:

Content	Expand	Gcd	Normal	Primpart	Rem	Sprem
Divide	Factor	Gcdex	Prem	Quo	Resultant	

◇ You must `readlib(evalgf)` before using this command. ◇ `evalgf(Gcd(x^2+1, x^2+x*RootOf(_Z^2+1)+2), 3);` ⟶ $x + 2\,\mathrm{RootOf}\left(\_Z^2 + 1\right)$ ◇ See also: `mod`, `RootOf`, `GF`.

evalhf

`evalhf(`*expr*`)`
`evalhf(Digits)`
Evaluates an expression using the hardware floating-point arithmetic ◇ The evaluation is done in double precision. No symbolic answers are permitted. Arrays passed as parameters to `evalhf` should have all their entries evaluating to floats or unassigned. An array argument passed as `var(`*A*`)` achieves an effect similar to `var` arguments in Pascal. This function handles many standard mathematical functions and constants. The special form `evalhf(Digits)` returns the number of decimal digits used in the hardware floating-point calculations. ◇ See also: `evalf`.

evalm

`evalm(`*expr*`)`
Evaluates a matrix expression ◇ The expression may involve sums, products, and integer powers of matrices, as well as maps of functions onto matrices. This function understands 0 to denote the zero matrix or the scalar 0, and `&*()` to denote the matrix identity. ◇ See also: `&*`, `array`, `linalg[matrix]`.

evaln

`evaln(`*expr*`)`
Evaluates the expression to a name, but no further ◇ The name is itself not evaluated. This function returns an assignable object. For simple names (like a string), the same effect can be achieved with single quotes. ◇ `i:=2: evaln(a.i);` ⟶ *a2* ◇ See also: `assigned`.

evalr

`evalr(`*expr*`)`
Evaluates *expr* using range arithmetic ◇ This function evaluates an *expr* that may contain ranges written `[`*a*`..`*b*`]` with $a \leq b$, sequences of ranges written `[`*a*`..`*b*`, `*c*`..`*d*`,...]`, or bounded variables written `[`*var*`,`*a*`..`*b*`]`. A variable *var* is converted to `[`*var*`, -infinity..infinity]` by default. When called without ranges, this function evaluates `abs`, `max`, `min`, and `Signum` using range arithmetic. ◇ You must `readlib(evalr)` before using this command. ◇ `evalr(cos([1..4]));` ⟶ $[-1..cos(1)]$ ◇ `evalr(min(Pi^exp(1), exp(1)^Pi));` ⟶ $\pi^{(e)}$ ◇ See also: `shake`.

example

`example(`*name*`)`
Produces examples ◇ In Release 3, this function displays the examples section from the on-line help text. The recommended way to invoke this function in Release 3 is with the triple question mark. ◇ In Release 2 and earlier versions, this function returns an example for certain function names and types. ◇ See also: `???`, `help`, `info`, `related`, `usage`.

A B C D E F G H I J K L M N O P Q R S T U V W X Y Z

exp `exp(`*z*`)`
The exponential function ◇ This function calculates e^z, where e is approximately 2.718281828459045. `exp(`*z*`)` is preferred over `E^`*z*. ◇ `exp(I*Pi);` ⟶ -1 ◇ See also: `ln`.

expand `expand(`*expr*`)`
`expand(`*expr*`,` *subexpr₁*`,` *subexpr₂*`, ...)`
Expands an expression ◇ For polynomials, this function distributes products over sums. For rational functions, only sums in the numerator are expanded. This function also expands many mathematical functions. The optional arguments $subexpr_i$ prevent these subexpressions from being expanded. ◇ `expand(sin(2*t) + (x+y)*(x-y));` ⟶ $2\sin(t)\cos(t) + x^2 - y^2$ ◇ `expand((x+y)*(x-y), x+y);` ⟶ $(x+y)\,x - (x+y)\,y$ ◇ See also: `collect`, `combine`, `expandoff`, `expandon`, `Expand`, `factor`, `sort`.

Expand `Expand(`*expr*`)`
Inert expand function ◇ When used in conjunction with `evala`, `evalgf`, or `mod`, this function expands the expression over a given coefficient domain. ◇ `Expand((x+1)^8) mod 2;` ⟶ $x^8 + 1$ ◇ See also: `evala`, `evalgf`, `expand`, `mod`.

expandoff `expandoff(`f_1`,` f_2`, ...)`
Suppresses expansion of functions ◇ This function suppresses the expansion of the functions listed. If no arguments are passed, the expansion of all functions is suppressed. You must enter `expand(expandoff())` before using this command. ◇ `expandoff(sin): expand(sin(2*t));` ⟶ $\sin(2t)$ ◇ See also: `expand`, `expandon`.

expandon `expandon(`f_1`,` f_2`, ...)`
Unsuppresses expansion of functions ◇ This function unsuppresses the expansion of the functions listed. If no arguments are passed, expansion of all functions is reasserted. You must enter `expand(expandon())` before using this command. ◇ See also: `expand`, `expandoff`.

extrema `extrema(`*expr*`,` *constraints*`)`
`extrema(`*expr*`,` *constraints*`,` *vars*`)`
`extrema(`*expr*`,` *constraints*`,` *vars*`,` *name*`)`
Finds relative extrema of the multivariate expression subject to constraints using the method of Lagrange multipliers ◇ A set of extreme values is returned. *constraints* is a single equation or expression or a set of zero or more equations or expressions. An expression given as a constraint is set equal to zero. *vars* is a variable name or a set of variable names. If a name is given as the fourth argument, it is assigned a set of candidates for the extreme value points. Each candidate is a set of equations giving values for *vars*. ◇ You must `readlib(extrema)` before using this command. ◇

`extrema(2*x*y, x^2+y^2=1, {x,y});` ⟶ $\{-1, 1\}$ ◇ See also: `maximize`, `minimize`.

factor

`factor(`*expr*`)`
`factor(`*expr*`, `*K*`)`
Factors a multivariate polynomial with integer, rational, or algebraic number coefficients ◇ The first form factors *expr* over the field implied by its coefficients. The second form factors *expr* over the algebraic number field given by K. K may be a single `RootOf` or radical, or a list or set of `RootOf`s and radicals. If the *expr* is a rational function, it is first normalized, then the numerator and denominator of the resulting expression are factored. The result is returned as an expression. ◇ `factor(x^4-5, 5^(1/4));` ⟶ $\left(x^2 + \sqrt{5}\right)\left(x + 5^{1/4}\right)\left(x - 5^{1/4}\right)$ ◇ See also: `AFactor`, `Factor`, `factors`, `ifactor`, `irreduc`, `normal`, `roots`, `split`, `sqrfree`.

Factor

`Factor(`*expr*`)`
`Factor(`*expr*`, `*K*`)`
Inert multivariate polynomial factoring ◇ This function is used in conjunction with `evala`, `evalgf`, or `mod`. K should be a single `RootOf` with `evalgf` or `mod`, but may be a set of `RootOf`s with `evala`. Three factorization methods are available with `evala`: `lenstra`, `linear`, and `trager` (the default). The method is selected by naming it in the options for `evala`. The result is returned as an expression. ◇ `Factor(x^2+1) mod 2;` ⟶ $(x + 1)^2$ ◇ See also: `AFactor`, `evala`, `evalgf`, `factor`, `Factors`, `ifactor`, `mod`, `RootOf`.

factorial

`factorial(`*n*`)`
The factorial function ◇ This function returns the product of the integers from 1 to the nonnegative integer *n*. It may also be invoked as *n*`!` ◇ `factorial(4);` ⟶ 24 ◇ `4!;` ⟶ 24 ◇ See also: `GAMMA`.

factors

`factors(`*poly*`)`
`factors(`*poly*`, `*K*`)`
Factors a multivariate polynomial ◇ The first argument must be a polynomial. The first form factors the polynomial over the field implied by its coefficients. The second form factors the polynomial over the algebraic number field given by K. K may be a single `RootOf` or radical, or a list or set of `RootOf`s and radicals. The result is a list whose first element is a constant coefficient and whose second element is a list of irreducible factor and exponent pairs. ◇ `factors(2*x^2-4, sqrt(2));` ⟶ $[2, [[x + \sqrt{2}, 1], [x - \sqrt{2}, 1]]]$ ◇ See also: `AFactors`, `factor`, `Factors`, `GaussInt[GIfacpoly]`, `ifactors`, `irreduc`, `roots`, `sqrfree`.

Factors `Factors(`*poly*`)`
`Factors(`*poly*`, `*K*`)`
Inert multivariate polynomial factoring ◇ This function is used in conjunction with `evala`, `mod`, or `modp1`. K should be a single `RootOf` with `mod` and may be a set of `RootOf`s with `evala`. Three factorization methods are available with `evala`: `lenstra`, `linear`, and `trager` (the default). The method is selected by naming it in the options for `evala`. The result is a list whose first element is the leading coefficient and whose second element is a list of factor-exponent pairs. ◇ `Factors(x^2+1) mod 2;` ⟶ $[1,[[x+1,2]]]$ ◇ See also: `AFactors`, `Berlekamp`, `evala`, `DistDeg`, `Factor`, `factors`, `ifactors`, `Irreduc`, `mod`, `modp1`, `ProbSplit`, `Sqrfree`.

FAIL Boolean constant ◇ This name is often returned by Maple procedures to indicate that the computation failed or was abandoned.

false Boolean constant.

FFT `FFT(`*m*`, `*x-array*`, `*y-array*`)`
Fast Fourier transform ◇ *x-array* and *y-array* are one-dimensional arrays of floats indexed from 1 to 2^m. *x-array* holds the real parts of the sequence and *y-array* the imaginary parts. The fast Fourier transform is computed in place, altering the original arrays. This function returns 2^m. This procedure may be invoked with `evalhf`. ◇ You must `readlib(FFT)` before using this command. ◇ See also: `iFFT`, `evalhf`, `fourier`, `iFFT`.

finance `finance(`*eqn*$_1$`, `*eqn*$_2$`, `*eqn*$_3$`)`
`finance(amount=`*a*`, interest=`*i*`, payment=`*p*`)`
`finance(amount=`*a*`, payment=`*p*`, periods=`*n*`)`
Amount, interest, payment, or periods ◇ The procedure `finance` expects three of the following four equations as arguments: `amount=`*amt*, `interest=`*rate*, `payment=`*pay*, `periods=`*n*. It solves for the fourth equation. ◇ You must `readlib(finance)` before using this command. ◇ `finance(amount=10000, interest=.065/12, periods=36);` ⟶ *payment* = 306.4900288 ◇ See also: `amortization`, `blackscholes`.

fixdiv `fixdiv(`*poly*`, `*var*`)`
Computes the fixed divisor of a polynomial ◇ *poly* is a polynomial in *var* with integer coefficients. The fixed divisor is the largest integer that divides *poly*(n) for all integers n. ◇ You must `readlib(fixdiv)` before using this command. ◇ `fixdiv(x^2-1, x);` ⟶ 2

Float `Float(`*mantissa*`, `*exponent*`)`
An alternative way to write floating-point numbers ◇ This notation is equivalent to *mantissa*`*10^`*exponent*. It is used to represent very large or very small floating-point numbers in output, though it may denote any

floating-point number. ◇ `Float(3,1);` ⟶ 30. ◇ `2.^54321;` ⟶ Float(1779895328, 16343) ◇ See also: `Digits`, `evalf`.

floor

`floor(`*z*`)`
`floor(1,` *z*`)`
Returns the greatest integer less than or equal to z if z is real ◇ If z is complex, this function computes floor(Re(z)) + I floor(Im(z)) + X, where

$$X = \begin{cases} 0 & a+b<1 \\ 1 & a+b \geq 1 \text{ and } a \geq b \\ I & a+b \geq 1 \text{ and } a < b \end{cases}$$

and a = (Re(z)) – floor(Re(z)), b = (Im(z)) – floor(Im(z)). The second form returns the first derivative of the floor function wherever it is defined. ◇ `floor(3.99);` ⟶ 3 ◇ `floor(3.99-2.4*I);` ⟶ $4 - 3I$ ◇ See also: `ceil`, `frac`, `round`, `trunc`.

fnormal

`fnormal(`*expr*`)`
`fnormal(`*expr*, *n*`)`
`fnormal(`*expr*, *n*, ϵ`)`
Floating-point normalization ◇ This function returns a value equivalent to *expr* under the assumption that all numeric values with magnitude less than ϵ may be considered to be zero. n specifies the number of digits for floating-point evaluation (the default is the value of `Digits`). The default value of ϵ is $10^{-\text{Digits}+2}$. ◇ `fnormal(10^(-9));` ⟶ 0 ◇ See also: `evalf`, `simplify`, `testfloat`.

for

`for` *var* `from` *expr₁* `to` *expr₂* `by` *expr₃* `do` *statements* `od`
`for` *var* `in` *expr* `do` *statements* `od`
`for` *var* `...` `while` *cond* `do` *statements* `od`
Executes statements in a loop ◇ The first form executes the statements for a counted number of times, with the indexing variable *var* taking on values between $expr_1$ and $expr_2$, incremented by $expr_3$ after each pass through the loop. The `from`, `to`, and `by` clauses are optional (the defaults are 1, infinity, and 1, respectively), and they may appear in any order. The second form has *var* assume successive operands of *expr*. The `for` and `while` forms may be combined in a single loop. ◇ See `do`.

forget

`forget(`*f*`)`
`forget(`*f*, arg_1, arg_2, `...)`
Removes an entry or entries from a remember table ◇ The first form removes all entries from the remember table of *f*, and the second form removes only the entry corresponding to $f(arg_1, arg_2, ...)$. ◇ You must `readlib(forget)` before using this command. ◇ See also: `options`, `remember`.

fortran

`fortran(`*expr*`)`
`fortran(`*expr*`, ` *options*`)`
Generates FORTRAN code ◇ This function generates FORTRAN code for evaluating the given expression, array of expressions, or list of assignments. In the last case, each assignment is given as an equation. The global variable `precision` may be assigned `single` or `double` (the default is single) for single or double precision. The option `digits` specifies the number of digits to use when converting rationals or symbolic constants such as `Pi` (the default is 7 for single precision and 16 for double). The option `optimized` performs common subexpression optimization on the generated code. The option `filename` names a file for writing the resulting FORTRAN code. By default, the code is written to the standard output. ◇ In Release 3, the `mode` option determines the names of mathematical functions used in the FORTRAN code. This option may be `single`, `double`, `generic`, or `complex`. The default is determined by the setting of the `precision` variable. ◇
`fortran(sin(Pi*x^2));` ⟶ `t0 = sin(0.3141593E1*x**2)` ◇
See also: `C`, `eqn`, `latex`, `optimize`.

fourier

`fourier(`*expr*`, ` *t*`, ` *w*`)`
Fourier transform ◇ This function applies the Fourier transformation to *expr* with respect to the variable *t*, returning an expression in *w*. Expressions involving complex exponentials, polynomials, trigonometrics with linear arguments, and a variety of functions and other integral transforms can be transformed. It also recognizes the Dirac-delta function `Dirac` and Heaviside's unit step function `Heaviside`, and derivatives and integrals. This transformation is defined as:

$$F(w) = \int_{-\infty}^{\infty} (expr)\, e^{-Itw}\, dt$$

◇ You must `readlib(fourier)` before using this command. ◇
See also: `invfourier`, `laplace`, `mellin`, `ztrans`.

frac

`frac(`*z*`)`
`frac(1, ` *z*`)`
Fractional part of a number ◇ For real z, $\mathrm{frac}(z) = z - \mathrm{trunc}(z)$. For complex z, the function is applied to the real and imaginary parts. The second form returns the first derivative of the fractional part function wherever it is defined. ◇ `frac(-1.3);` ⟶ $-.3$ ◇ See also: `ceil`, `floor`, `round`, `trunc`.

freeze

`freeze(`*expr*`)`
Replaces an expression with a name ◇ You must `readlib(freeze)` before using this command. ◇ `freeze(x^2-y);` ⟶ *_R0* ◇ See also: `frontend`, `thaw`.

FresnelC `FresnelC(x)`

Fresnel cosine integral ◇ The Fresnel cosine integral is defined as $\int_0^x \cos\left(\frac{\pi}{2}t^2\right)\,dt$. ◇ See also: `erf`, `dawson`, `Fresnelf`, `Fresnelg`, `FresnelS`.

Fresnelf `Fresnelf(x)`

Fresnel auxiliary function ◇ This function is defined as

$$\left(\frac{1}{2} - \mathrm{FresnelS}(x)\right)\cos\left(\frac{\pi}{2}x^2\right) - \left(\frac{1}{2} - \mathrm{FresnelC}(x)\right)\sin\left(\frac{\pi}{2}x^2\right)$$

◇ You must `readlib(Fresnelf)` before using this command. ◇ See also: `FresnelC`, `Fresnelg`, `FresnelS`.

Fresnelg `Fresnelg(x)`

Fresnel auxiliary function ◇ This function is defined as

$$\left(\frac{1}{2} - \mathrm{FresnelC}(x)\right)\cos\left(\frac{\pi}{2}x^2\right) - \left(\frac{1}{2} - \mathrm{FresnelS}(x)\right)\sin\left(\frac{\pi}{2}x^2\right)$$

◇ You must `readlib(Fresnelg)` before using this command. ◇ See also: `FresnelC`, `Fresnelf`, `FresnelS`.

FresnelS `FresnelS(x)`

Fresnel sine integral ◇ The Fresnel sine integral is defined as $\int_0^x \sin\left(\frac{\pi}{2}t^2\right)\,dt$. ◇ See also: `erf`, `dawson`, `FresnelC`, `Fresnelf`, `Fresnelg`.

frontend `frontend(f, arglist)`
`frontend(f, arglist, [freezeset, nofreezeset])`

Processes a general expression into a rational expression ◇ This function freezes subexpressions in the list of expressions *arglist* so that they become rational functions, applies the function *f* to these arguments, then thaws the frozen expressions. This extends the domain of computation for functions that expect rational expressions as arguments. The optional third parameter names a set of expressions to freeze and a set of expressions to leave alone. This function does not work with functions *f* that assign a value to an argument. ◇ `frontend(rem, [ln(t)^4+5, ln(t)^3+1, ln(t)]);` $\longrightarrow 5 - \ln(t)$ ◇ See also: `freeze`, `thaw`.

fsolve `fsolve(eqns)`
`fsolve(eqns, vars, options)`

Solves an equation or system of equations using floating-point arithmetic ◇ *eqn* may be a single equation or expression, or a set of equations or expressions. An expression is understood to be equal to zero. *vars* is a variable name, or a set of variable names. Given a single polynomial, this function attempts to compute all its real roots. For more general expressions, a single real root is sought. The following options may be specified:

`complex`	for a polynomial, find all roots, real and complex; otherwise, search for a root in the complex plane
`fulldigits`	perform all intermediate calculations at full precision
`maxsols=`*n*	for polynomials, find only the n smallest roots
a`..`*b* or *var*`=`*a*`..`*b*	search for roots only in the given interval

◇ `fsolve(x^3+x+1);` ⟶ −.6823278038 ◇ See also: `dsolve`, `isolve`, `msolve`, `rsolve`, `solve`.

galois

`galois(`*poly*`)`
Computes the Galois group of an irreducible polynomial ◇ This function computes the Galois group of an irreducible univariate rational polynomial of degree at most 7. It returns an expression sequence with the name of the Galois group (with + if it is an even subgroup of the symmetric group), its order, and a set of generators of the group up to conjugacy. Each generator is written as a string representing an element of the symmetric group on n letters, where n is the degree of *poly*. ◇ `galois(x^4+1);` ⟶ +*V4*, 4, {(*13*)(*24*), (*12*)(*34*)} ◇ See also: `group`, `GF`.

gamma

Euler's constant ◇

$$\gamma = \lim_{n\to\infty}\left(\sum_{k=1}^{n} 1/k - \ln(n)\right)$$

It is approximately .577215664901533.

gamma

`gamma(`*n*`)`
Generalized gamma constant ◇ For any nonnegative integer *n*, this function stands for the constant

$$\lim_{m\to\infty}\left(\sum_{k=1}^{n}\frac{\ln(k)^n}{k} - \frac{\ln(m)^{n+1}}{n+1}\right)$$

`gamma(0)` equals `gamma`. ◇ `evalf(gamma(3));` ⟶ .002053834420

GAMMA

`GAMMA(`*z*`)`
`GAMMA(`*a*`, `*z*`)`
The Gamma and incomplete Gamma functions ◇ The Gamma function is defined as:

$$\Gamma(z) = \int_0^{\infty} e^{-t} t^{z-1}\, dt$$

The incomplete Gamma function is defined as:

$$\Gamma(a, z) = \int_z^{\infty} e^{-t} t^{a-1}\, dt$$

◇ `GAMMA(6);` ⟶ 120 ◇ See also: `Beta`, `factorial`, `lnGAMMA`, `Psi`.

A B C D E F G H I J K L M N O P Q R S T U V W X Y Z

Gauss The Gauss package for abstract mathematical programming ◇ Gauss supports a hierarchy of algebraic structures called categories (such as Ring, IntegralDomain and Field) and a number of abstract data types or domains (like DenseUnivariatePolynomial, QuotientField and RationalFunction). A domain is parameterized by one or more categories, for example, a polynomial domain can have coefficients in any ring. This separation allows methods for domains to be written independent of the representation of particular data structures. This way, one does not have to rewrite, for example, matrix algorithms for each possible domain of matrix entries. A list of categories and domains begins on page 137. Some Maple library code will likely be replaced by Gauss code in future versions of Maple because of the greater generality afforded by Gauss.

Gausselim

`Gausselim(`*A*`)`
`Gausselim(`*A*`, `*name₁*`)`
`Gausselim(`*A*`, `*name₁*`, `*name₂*`)`
Inert Gaussian elimination ◇ When used with `mod`, this function performs Gaussian elimination with row pivoting on *A*, a rectangular matrix of integers, modulo a positive integer. $name_1$, if present, is assigned the rank of *A*. If the rank equals the number of rows of *A*, $name_2$ is assigned the determinant of the square submatrix of *A* formed by taking the first rank(A) columns. Otherwise it is assigned zero. ◇ See also: `Gaussjord`, `linalg[gausselim]`.

GaussInt The Gaussian integer package ◇ In the descriptions of functions in this package, ξ stands for a Gaussian integer. A Gaussian integer is a complex number with integer real and imaginary parts. ◇ In Release 2, the functions in this package must first be loaded using `with` before they can be invoked. ◇ `3+4*I, -2*I` ◇ See also: `GaussInt[`*function*`]`, `with`.

GaussInt[GIbasis]

`GIbasis(`ξ_1`, `ξ_2`)`
Tests for a basis ◇ This function returns *true* if ξ_1 and ξ_2 form a basis of the Gaussian integer domain over the rational integers; *false* otherwise. ◇
`GIbasis(1+I,2+I);` ⟶ *true*

GaussInt[GIchrem]

`GIchrem(`*reslist*`, `*modlist*`)`
Chinese remainder algorithm for Gaussian integers ◇ *modlist* is a list of pairwise relatively prime Gaussian integers $[\xi_1, \ldots, \xi_n]$, and *reslist* is a list of residues $[\rho_1, \ldots, \rho_n]$. This function computes the unique Gaussian integer ξ (up to multiplication by a unit) whose norm is less than $|\xi_1 \cdots \xi_n|$ satisfying $\xi \equiv \rho_i \bmod \xi_i$. ◇ This function is new in Release 3. ◇
`GIchrem([2*I,1+I], [I+2,3+2*I]);` ⟶ $-1 + 4\,I$ ◇ See also: `chrem`, `GaussInt[mcmbine]`.

GaussInt[GIdivisor]

`GIdivisor(ξ)`

Divisors of a Gaussian integer ◇ This function returns the set of the first-quadrant representatives of the divisors of the Gaussian integer ξ. ◇ `GIdivisor(13);` $\longrightarrow \{1, 13, 3+2I, 2+3I\}$ ◇ See also: `GaussInt[GIfactor]`, `GaussInt[GIfactors]`, `GaussInt[GIfacset]`, `GaussInt[GInodiv]`.

GaussInt[GIfacpoly]

`GIfacpoly(`*poly*`)`

Factors a multivariate polynomial over the Gaussian integers ◇ The result is a list whose first element is a constant coefficient and whose second element is a list of irreducible factor and exponent pairs. ◇ This function is new in Release 3. ◇ `GIfacpoly(x^2+y^2);` $\longrightarrow$ $[1, [[x - I\,y, 1], [x + I\,y, 1]]]$ ◇ See also: `factors`.

GaussInt[GIfacset]

`GIfacset(ξ)`

Computes the set of primary prime factors of the Gaussian integer ξ ◇ A prime Gaussian integer $\xi = n + m\,I$ is primary if $n \equiv 1 \bmod 4$ and $m \equiv 0 \bmod 4$, or $n \equiv -1 \bmod 4$ and $m \equiv 2 \bmod 4$. ◇ `GIfacset(13);` $\longrightarrow \{3+2I, 3-2I\}$ ◇ See also: `GaussInt[divisor]`, `GaussInt[GIfactor]`, `GaussInt[GIfactors]`.

GaussInt[GIfactor]

`GIfactor(ξ)`

Factors a Gaussian integer ◇ This function computes the Gaussian integer factorization of ξ, returning a multiplicative expression. ◇ `GIfactor(4*I);` $\longrightarrow (-I)(1+I)^4$ ◇ See also: `GaussInt[GIfactors]`, `GaussInt[GIfacset]`, `GaussInt[GIprime]`, `ifactor`, `numtheory[factorEQ]`, `numtheory[sq2factor]`.

GaussInt[GIfactors]

`GIfactors(ξ)`

Factors a Gaussian integer ◇ This function computes the Gaussian integer factorization of ξ, returning a list of primary prime factors and exponents. ◇ `GIfactors(4*I);` $\longrightarrow [-I, [[1+I, 4]]]$ ◇ See also: `GaussInt[GIfactor]`, `GaussInt[GIfacset]`, `GaussInt[GIsqrfree]`, `ifactors`.

GaussInt[GIgcd]

`GIgcd(ξ₁, ξ₂, ..., ξₙ)`

Computes the greatest common divisor of the given Gaussian integers ◇ The result is the first quadrant associate of the gcd. ◇ `GIgcd(19+17*I,11+13*I);` $\longrightarrow 1+3I$ ◇ See also: `GaussInt[GIgcdex]`, `igcd`.

GaussInt[GIgcdex]

`GIgcdex(`ξ_1, ξ_2, *name*$_1$, *name*$_2$`)`
Extended Euclidean algorithm for Gaussian integers ◇ This function returns the gcd of ξ_1 and ξ_2, and assigns Gaussian integers to *name*$_1$ and *name*$_2$ if they are present so that the gcd equals *name*$_1$ ξ_1 + *name*$_2$ ξ_2 ◇ `GIgcdex(19+17*I,11+13*I,'s','t'): s,t;` $\longrightarrow 2+I, -3-I$ ◇ See also: `GaussInt[GIgcd]`, `igcdex`.

GaussInt[GIhermite]

`GIhermite(`A`)`
`GIhermite(`A, *name*`)`
Gaussian integer-only Hermite Normal Form ◇ This function computes the Hermite Normal Form (reduced row echelon form) of a rectangular matrix of Gaussian integers. If a name is given as a third argument, it is assigned the transformation matrix U such that the result is UA. ◇ This function is new in Release 3. ◇ See also: `GaussInt[GIsmith]`, `linalg[ihermite]`.

GaussInt[GIissqr]

`GIissqr(`ξ`)`
Tests whether a Gaussian integer is a perfect square ◇ This function returns *true* or *false*. ◇ `GIissqr(4*I-3);` $\longrightarrow$ *true* ◇ See also: `GaussInt[GIsqrt]`.

GaussInt[GIlcm]

`GIlcm(`ξ_1, ξ_2, ..., ξ_n`)`
Computes the least common multiple of the given Gaussian integers ◇ The result is the first quadrant associate of the lcm. ◇ `GIlcm(3*I-1,3-I);` $\longrightarrow 5+5I$ ◇ See also: `GaussInt[GIgcd]`, `GaussInt[GIgcdex]`, `ilcm`.

GaussInt[GImcmbine]

`GImcmbine(`ξ_1, ρ_1, ξ_2, ρ_2`)`
Chinese remainder algorithm for Gaussian integers ◇ This function computes a Gaussian integer ξ such that $\xi \equiv \rho_1 \bmod \xi_1$ and $\xi \equiv \rho_2 \bmod \xi_2$. If no such Gaussian integer exists, *FAIL* is returned. ◇ `GImcmbine(I+2,2*I,3+2*I,1+I);` $\longrightarrow -1+4I$ ◇ See also: `chrem`, `GaussInt[chrem]`, `numtheory[mcombine]`.

GaussInt[GInearest]

`GInearest(`z`)`
Finds the nearest Gaussian integer to the given complex number ◇ When there is more than one possible answer, the one with the smallest norm is chosen. ◇ `GInearest(.5+.5*I);` $\longrightarrow 0$ ◇ See also: `GaussInt[GIsqrt]`, `GaussInt[GIissqr]`.

GaussInt[GInodiv]

`GInodiv(ξ)`

Computes the number of nonassociated divisors of a Gaussian integer ◇ This function is new in Release 3. ◇ `GInodiv(13);` ⟶ 4 ◇ See also: `GaussInt[GIdivisor]`, `numtheory[tau]`.

GaussInt[GInorm]

`GInorm(ξ)`

Computes the norm of the Gaussian integer ◇ The norm of ξ is $||\xi|| = \xi\bar{\xi}$. ◇ `GInorm(1+2*I);` ⟶ 5 ◇ See also: `abs`.

GaussInt[GInormal]

`GInormal(ξ)`

Normalizes a Gaussian integer ◇ This function returns the first quadrant associate of the Gaussian integer ξ. ◇ This function is new in Release 3. ◇ `GInormal(I-1);` ⟶ $1 + I$ ◇ See also: `GaussInt[GIunitnormal]`.

GaussInt[GIorder]

`GIorder(ξ1, ξ2)`

Computes the order of a Gaussian integer ◇ This function returns the smallest positive integer i such that $\xi_1^i \equiv 1 \bmod \xi_2$. If ξ_1 and ξ_2 are not coprime, *FAIL* is returned. ◇ `GIorder(4+I,1+I);` ⟶ 16 ◇ See also: `numtheory[order]`.

GaussInt[GIphi]

`GIphi(ξ)`

Computes the number of Gaussian integers in a reduced system modulo ξ ◇ A reduced system modulo ξ is a maximal set of Gaussian integers that are pairwise incongruent modulo ξ. ◇ `GIphi(9*I-7);` ⟶ 48 ◇ See also: `numtheory[phi]`.

GaussInt[GIprime]

`GIprime(ξ)`

Tests for probable primality ◇ This function tests whether ξ is very probably a prime Gaussian integer and returns *true* or *false*. The method reduces to calling `isprime` on an integer. ◇ `GIprime(5+2*I);` ⟶ *true* ◇ See also: `isprime`.

GaussInt[GIquadres]

`GIquadres(ξ1, ξ2)`

Quadratic residue and nonresidue ◇ This function returns 1 if ξ_1 is a quadratic residue (mod ξ_2), and -1 if it is a quadratic nonresidue. ξ_1 and ξ_2 must be relatively prime. ◇ `GIquadres(1+4*I,2+3*I);` ⟶ -1 ◇ See also: `numtheory[legendre]`.

GaussInt[GIquo]

`GIquo(`ξ_1`, `ξ_2`)`
`GIquo(`ξ_1`, `ξ_2`, `*name*`)`
Computes the Gaussian integer quotient of ξ_1 divided by ξ_2 ◇ If the third argument is present, it is assigned the remainder. ◇ `GIquo(12+3*I, 2+7*I, 'r'), r;` $\longrightarrow 1 - I, 3 - 2I$ ◇ See also: `GaussInt[GIrem]`, `irem`, `iquo`.

GaussInt[GIrem]

`GIrem(`ξ_1`, `ξ_2`)`
`GIrem(`ξ_1`, `ξ_2`, `*name*`)`
Computes the Gaussian integer remainder of ξ_1 divided by ξ_2 ◇ If the third argument is present, it is assigned the quotient. ◇ `GIrem(12+3*I, 2+7*I, 'q'), q;` $\longrightarrow 3 - 2I, 1 - I$ ◇ See also: `GaussInt[GIquo]`, `irem`, `iquo`.

GaussInt[GIroots]

`GIroots(`*poly*`)`
Computes the Gaussian integer roots of the given polynomial ◇ *poly* is a univariate polynomial with Gaussian integer coefficients. A list of pairs of roots and their associated multiplicities is returned. ◇
`GIroots(x^3-(1+2*I)*x^2-2*x+2*I);` $\longrightarrow [[-1, 1], [1 + I, 2]]$

GaussInt[GIsieve]

`GIsieve(`n`)`
Gaussian prime generator ◇ This function generates a list of Gaussian primes in the first octant whose norms are less than or equal to n^2, where *n* is a positive integer. The first item in the list is the number of primes found. ◇ `GIsieve(3);` $\longrightarrow [3, [1 + I, 1 + 2I, 3]]$

GaussInt[GIsmith]

`GIsmith(`A`)`
`GIsmith(`A`, `*name₁*`, `*name₂*`)`
Gaussian integer-only Smith Normal Form ◇ This function computes the Smith normal form of a rectangular matrix of Gaussian integers. In the case of three arguments, $name_1$ and $name_2$ are assigned the transformation matrices U and V such that the result is UAV. ◇ This function is new in Release 3. ◇ See also: `GaussInt[GIhermite]`, `linalg[ismith]`.

GaussInt[GIsqrfree]

`GIsqrfree(`ξ`)`
Gaussian integer square-free factorization ◇ This function returns a list of square-free factors and exponents. ◇ This function is new in Release 3. ◇ `GIsqrfree(79-47*I);` $\longrightarrow [I, [[1 + I, 1], [1 - 8 * I, 2]]]$ ◇ See also: `isqrfree`, `GaussInt[GIfactor]`, `GaussInt[GIfactors]`.

GaussInt[GIsqrt]

`GIsqrt(`ξ`)`

Computes a Gaussian integer approximation to the square root of the given Gaussian integer ◇ The value is exact for perfect squares; otherwise the error is at most $1/\sqrt{2}$. ◇ `GIsqrt(4*I-3);` $\longrightarrow 1+2I$ ◇ See also: `GaussInt[GIissqr]`, `isqrt`.

GaussInt[GIunitnormal]

`GIunitnormal(`ξ`)`

Normalizes a Gaussian integer ◇ This function rcturns a list with three elements: the unit required to normalize ξ, the first quadrant associate of ξ, and the inverse of the normalizing unit. ◇ This function is new in Release 3. ◇ `GIunitnormal(I-1);` $\longrightarrow [-I, 1+I, I]$ ◇ See also: `GaussInt[GInormal]`.

Gaussjord

`Gaussjord(`*A*`)`
`Gaussjord(`*A*`,` $name_1$`)`
`Gaussjord(`*A*`,` $name_1$`,` $name_2$`)`

Inert reduced row echelon form ◇ *A* is a rectangular matrix with integer entries. When used with `mod`, this function computes the reduced row echelon (Gauss-Jordan) form of *A* modulo a positive integer. $name_1$, if present, is assigned the rank of *A*. If the rank equals the number of rows of *A*, $name_2$ is assigned the determinant of the square submatrix of *A* formed by taking the first rank(*A*) columns. Otherwise it is assigned zero. ◇ See also: `Gausselim`, `linalg[rref]`.

gc

`gc()`
`gc(`*n*`)`

Garbage collection ◇ Garbage collection deletes all data to which no references are made, and cleans the remember tables of procedures that have option `system` or option `builtin` by removing entries that have no other references to them. The first form causes a garbage collection to occur immediately. The second form causes automatic garbage collections to occur approximately after every *n* words of memory are used. This frequency is stored in `status[5]`. ◇ See also: `status`, `words`.

gcd

`gcd(`$poly_1$`,` $poly_2$`,` $name_1$`,` $name_2$`)`

Computes the greatest common divisor of two polynomials with rational coefficients ◇ $name_1$, if present, is assigned the cofactor of $poly_1$; likewise, $name_2$ is assigned the cofactor of $poly_2$. ◇ `gcd(x^2-1,x^3+1);` $\longrightarrow x+1$ ◇ See also: `Gcd`, `gcdex`, `igcd`, `lcm`, `numtheory[GIgcd]`.

Gcd

`Gcd(`$poly_1$`,` $poly_2$`)`
`Gcd(`$poly_1$`,` $poly_2$`,` $name_1$`,` $name_2$`)`

Inert greatest common divisor of polynomials ◇ When used with `mod` or `modp1`, this function computes the greatest common divisor of the polynomials modulo a positive integer. When used with `evala`, it computes the gcd of polynomials with algebraic coefficients. *name₁* and *name₂*, if present, are assigned the cofactors. ◇ `Gcd(x^2+1,x+1) mod 2;` ⟶ $x+1$ ◇ See also: `evala, gcd, Gcdex, Lcm.`

gcdex `gcdex(`*$poly_1$*`, `*$poly_2$*`, `*var*`, `*$name_1$*`, `*$name_2$*`)`
`gcdex(`*$poly_1$*`, `*$poly_2$*`, `*$poly_3$*`, `*var*`, `*$name_1$*`, `*$name_2$*`)`
Extended Euclidean algorithm for polynomials ◇ $poly_1$ and $poly_2$ are polynomials in the variable *var*. The first form computes polynomials s and t with degree(s) < degree($poly_2$) and degree(t) < degree($poly_1$) such that $s * poly_1 + t * poly_2 = g$ where g is the gcd of the polynomials, and returns g. The second form solves the polynomial diophantine equation $s * poly_1 + t * poly_2 = poly_3$, and returns `NULL`. s and t are stored in $name_1$ and $name_2$. ◇ See also: `gcd, Gcdex, igcdex.`

Gcdex `Gcdex(`*$poly_1$*`, `*$poly_2$*`, `*var*`, `*$name_1$*`, `*$name_2$*`)`
Inert greatest common divisor function for the extended Euclidean algorithm ◇ When used with `mod` or `modp1`, this function performs the extended Euclidean algorithm on the polynomials modulo a positive integer. When used with `evala`, it works with polynomials with algebraic coefficients. ◇ See also: `evala, Gcd, gcdex.`

genfunc The rational generating function package ◇ In the descriptions of functions in this package, *F* stands for a rational generating function in the indeterminate *gfvar*. *nvar* stands for the name of the variable that indexes the sequence encoded by the generating function. ◇ In Release 2, the functions in this package must first be loaded using `with` before they can be invoked. ◇ See also: `genfunc[`*function*`], with.`

genfunc[rgf_charseq]

`rgf_charseq(`*F*`, `*gfvar*`, `*expr*`, `*nvar*`)`
Finds the characteristic sequence of a rational generating function ◇ This function returns the characteristic sequence of the generating function *F* in the variable *gfvar*, where the *nvar*th term of the sequence is given by *expr*, an expression in *nvar*. ◇ `rgf_charseq(z/(1+z),z,a(n),n);` ⟶ $a(n+1)$ ◇ See also: `genfunc, genfunc[termscale], genfunc[rgf_encode], genfunc[rgf_expand], denom, tcoeff, ldegree.`

genfunc[rgf_encode]

`rgf_encode(`*expr*`, `*nvar*`, `*gfvar*`, `*options*`)`
Encodes a rational generating function ◇ This function computes the rational generating function in the variable *gfvar* for the sequence defined by *expr*, a valid closed form expression for a sequence with a rational generating function. An option of the form $nvar = m$ defines the mth term as the first nonzero term of the sequence. A list of equations of the form *index* = *value* defines individual values of the sequence. ◇ `rgf_encode(3^n,n,z);` ⟶ $\frac{1}{1-3z}$ ◇ See also: `type/rgf_seq,`

genfunc, genfunc[rgf_expand], genfunc[rgf_seq], ztrans, invztrans.

genfunc[rgf_expand]

rgf_expand(*F*, *gfvar*, *nvar*)

Expands the rational generating function ◇ The function finds the closed form in the variable *nvar* of the the sequence encoded by the rational generating function *F*. ◇ rgf_expand(1/(1-3*z),z,n); ⟶ 3^n ◇ See also: genfunc, genfunc[rgf_encode], RootOf, ztrans, invztrans.

genfunc[rgf_findrecur]

rgf_findrecur(*m*, *list*, *seqname*, *nvar*)

Finds a recurrence relation for the terms in a sequence ◇ This function finds the homogeneous linear recurrence of order *m* with constant coefficients satisfied by the sequence in *list*. *list* is a list of $2m$ consecutive values from the sequence. ◇ rgf_findrecur(2,[2,3,7,13],a,n); ⟶ $\mathrm{a}(n) = \mathrm{a}(n-1) + 2\,\mathrm{a}(n-2)$ ◇ See also: genfunc, genfunc[rgf_sequence], rsolve.

genfunc[rgf_hybrid]

rgf_hybrid(*gfvar*, F_1, F_2, ...)

Finds generating function of hybrid terms ◇ This function finds the generating function that results from the termwise multiplication of the sequences encoded by the generating functions $F_1, F_2, \ldots$. ◇ rgf_hybrid(z,1/(1+z),1/(1-3*z)); ⟶ $\frac{1}{1+3z}$ ◇ See also: genfunc, genfunc[rgf_expand].

genfunc[rgf_norm]

rgf_norm(*F*, *gfvar*)
rgf_norm(*F*, *gfvar*, factored)
rgf_norm(*F*, *gfvar*, factored(*K*))

Normalizes the rational generating function ◇ This function puts the generating function in normal form, expanding the numerator and denominator, and factoring the low order coefficient from the denominator. The factored option applies factor to the denominator, optionally over an algebraic extension *K*. ◇ rgf_norm(1/(1+z^2),z,factored(I)); ⟶ $\frac{1}{(I\,z+1)(-I\,z+1)}$ ◇ See also: genfunc, factor, normal.

genfunc[rgf_pfrac]

rgf_pfrac(*F*, *gfvar*)
rgf_pfrac(*F*, *gfvar*, no_RootOf)

Computes the complete partial fraction expansion of *F* over the complex numbers ◇ The option no_RootOf specifies that the denominator should be completely factored over the complex numbers without RootOf expressions. ◇ rgf_pfrac(1/(z^2+1),z); ⟶ $-\frac{1}{2}\frac{I}{z-I} + \frac{1}{2}\frac{I}{z+I}$ ◇ See also: genfunc, convert[parfrac], factor, RootOf.

genfunc[rgf_relate]

`rgf_relate(`F_1`,` *gfvar*`,` *expr*`,` *nvar*`,` F_2`)`
`rgf_relate(`F_1`,` *gfvar*$_1$`,` *expr*`,` *nvar*`,` F_2`,` *gfvar*$_2$`)`
Relates sequences with common nonzero roots in the denominators of their generating functions F_1 and F_2 ◇ *expr* is an expression in *nvar* for the *nvar*th term of the generating function F_1. The *nvar*th term of the sequence encoded by F_2 is expressed as a function of *expr*. If the generating functions have no common roots, *FAIL* is returned. ◇ `rgf_relate(1/(1-z^2),z,a(n),n,1/(1+z));` ⟶ $\mathrm{a}(n) - \mathrm{a}(n-1)$ ◇ See also: `genfunc`, `genfunc[rgf_charseq]`.

genfunc[rgf_sequence]

`rgf_sequence(`*queryname*`,` *F*`,` *gfvar*`,` *args*`)`
Extracts information about a sequence from its rational generating function ◇ The possible names for *queryname* and the information returned is summarized in the following table.

`boundary`	Boundary conditions for the sequence
`coeffs`	Coefficients of recurrence defining the sequence
`delta`	Delta terms added to the closed form
`first`	Index of first nonzero term in the sequence
`firstcf`	Minimum index where closed form determines sequence
`firstrecur`	Minimum index where recurrence determines sequence
`order`	Order of the recurrence
`recur`	The recurrence

`boundary` and `coeffs` accept a fourth argument giving the name of the sequence. `recur` requires two additional arguments: the name of the sequence and the name of the indexing variable for the sequence. ◇ `rgf_sequence(order,1/(1-3*z),z);` ⟶ 1 ◇ See also: `genfunc`, `genfunc[rgf_encode]`, `genfunc[rgf_expand]`, `rsolve`.

genfunc[rgf_simp]

`rgf_simp(`*expr*`,` *F*`,` *gfvar*`,` *seqname*`,` *nvar*`,` *options*`)`
Simplifies an expression involving a rational generating function sequence ◇ This function simplifies occurrences of *seqname* in *expr*, where *seqname* is the function name for the sequence encoded by *F*. *seqname* can have a parameter of the form *i*`*`*nvar* `+` *j*, where *i* is a positive integer and *j* is an integer. This pattern represents the index of the *nvar*th term in the sequence. If an expression in *nvar* is passed as an optional argument, the term is simplified relative to this expression. A procedure passed as an optional argument is applied to each coefficient of the collected result. ◇ `rgf_simp(a(n+2)+3*a(n+1),1/(1-3*z),z,a(n),n);` ⟶ $18\,\mathrm{a}(n)$ ◇ `rgf_simp(a(n+2)+3*a(n+1),1/(1-3*z),z,a(n),n,n+1);` ⟶ $6\,\mathrm{a}(n+1)$ ◇ `rgf_simp(a(n+2)+3*a(n+1),1/(1-3*z),z,a(n+1),n);` ⟶ $6\,\mathrm{a}(n+1)$ ◇ See also: `genfunc`, `collect`.

genfunc[rgf_term]

`rgf_term(`*F*`, `*gfvar*`, `*n*`)`
Finds terms in a sequence from its rational generating function ◇ This function finds the value of the *n*th term in the sequence encoded by the generating function *F*. ◇ `rgf_term(1/(1-3*z),z,5);` ⟶ 243 ◇ See also: `genfunc`, `genfunc[rgf_expand]`.

genfunc[termscale]

`termscale(`*poly*`, `*gfvar*`, `*expr*`, `*nvar*`)`
Determines the result of multiplying a generating function by a polynomial ◇ *expr* is an expression in *nvar* for the *nvar*th term of a sequence. This function shifts and scales *expr* to determine the sequence encoded by the product of *poly* and the generating function of *expr*. ◇ `termscale(1+z^2,z,3^n,n);` ⟶ $3^n + 3^{(n-2)}$ ◇ See also: `genfunc`, `genfunc[rgf_charseq]`, `genfunc[rgf_expand]`, `genfunc[rgf_relate]`.

genfunc[type/rgf_seq]

`type(`*expr*`, rgf_seq(`*nvar*`))`
Tests the expression to determine whether it defines a sequence encoded by a rational generating function ◇ *expr* is an expression in *nvar*. This function determines whether the sequence whose *nvar*th term is *expr* has a rational generating function. ◇ `type(3^n,rgf_seq(n));` ⟶ *true* ◇ See also: `genfunc[rgf_encode]`, `genfunc[rgf_expand]`.

genpoly

`genpoly(`*n*`, `*m*`, `*var*`)`
`genpoly(`*poly*`, `*m*`, `*var*`)`
Generates a polynomial from the integer *n* by *m*-adic expansion ◇ This function computes a polynomial in the variable *var* whose coefficients come from a symmetric set of representatives modulo m when m is positive, and the value of this polynomial at m is n. In the second template, each coefficient of the given polynomial *poly* is expanded into a polynomial in *var*. ◇ `genpoly(45,7,x);` ⟶ $3 - x + x^2$ ◇ See also: `mods`, `padic[evalp]`.

geom3d

Three-dimensional Euclidean geometry ◇ The names `x`, `y`, `z`, and `_t` are used in this package as global variables, so never use them for other purposes. ◇ In Release 2 and earlier versions, the functions in this package must first be loaded using `with` before they can be invoked. ◇ See also: `geom3d[`*function*`]`, `geometry`, `projgeom`, `with`.

geom3d[angle]

`angle(`g_1`, `g_2`)`
Finds the smallest angle between the two lines, two planes, or a line and a plane ◇ The angle returned is in radians. ◇ See also: `geom3d[distance]`.

geom3d[area]

`area(`*g*`)`

Computes the area of a triangle or the surface area of a sphere ◇ The area is always positive. ◇ See also: **geom3d[volume]**, **geometry[area]**.

geom3d[are_collinear]

`are_collinear(`pt_1`,` pt_2`,` pt_3`)`

Tests whether three points are collinear ◇ This function returns *true*, *false*, or a condition. ◇ See also: **geom3d[are_concurrent]**, geometry[are_collinear].

geom3d[are_concurrent]

`are_concurrent(`$line_1$`,` $line_2$`,` $line_3$`)`

Tests whether three lines are concurrent ◇ This function returns *true*, *false*, or a condition under which the three lines meet. ◇ See also: geom3d[are_collinear], **geometry[are_concurrent]**.

geom3d[are_parallel]

`are_parallel(`g_1`,` g_2`)`

Tests whether two planes, two lines, or a line and a plane are parallel to each other ◇ This function returns *true*, *false*, or a condition. ◇ See also: geom3d[are_perpendicular], **geometry[are_parallel]**.

geom3d[are_perpendicular]

`are_perpendicular(`g_1`,` g_2`)`

Tests whether two lines, two planes, or a line and a plane are perpendicular to each other ◇ This function returns *true*, *false*, or a condition. ◇ See also: **geom3d[are_parallel]**, **geometry[are_perpendicular]**.

geom3d[are_tangent]

`are_tangent(`g_1`,` g_2`)`

Tests whether two spheres, or one sphere and a plane or a line, are tangent to each other ◇ This function returns *true*, *false*, or a condition. ◇ See also: **geom3d[tangent]**, **geometry[are_tangent]**.

geom3d[center]

`center(`*sphere*`)`

Finds the center of the given sphere ◇ This function returns a point named `center_of_`*sphere*. ◇ See also: **geom3d[sphere]**, geom3d[point3d], **geom3d[coordinates]**, **geometry[center]**.

geom3d[centroid]

`centroid(`*g*`,` *name*`)`

`centroid(`*ptlist*`,` *name*`)`

Computes the centroid of the given tetrahedron or triangle *g*, or of a list of points *ptlist* ◇ *name* is the name given to the centroid. ◇ See also: geom3d[coordinates], **geom3d[tetrahedron]**, geometry[centroid].

geom3d[coordinates]

`coordinates(`*pt*`)`

Returns the coordinates of the given point ◇ The coordinates are returned as a list. ◇ See also: `geom3d[point3d]`, `geometry[coordinates]`.

geom3d[coplanar]

`coplanar(`pt_1`,` pt_2`,` pt_3`,` pt_4`)`
`coplanar(`$line_1$`,` $line_2$`)`

Tests whether four points or two lines are coplanar ◇ This function returns *true*, *false*, or a condition. ◇ See also: `geom3d[are_collinear]`.

geom3d[distance]

`distance(`g_1`,` g_2`)`

Finds the distance between two points, two lines, or a point and a line or a plane ◇ See also: `geom3d[angle]`, `geometry[distance]`, `student[distance]`.

geom3d[inter]

`inter(`$line_1$`,` $line_2$`,` *name*`)`
`inter(`$plane_1$`,` $plane_2$`,` *name*`)`
`inter(`*line*`,` *plane*`,` *name*`)`
`inter(`$plane_1$`,` $plane_2$`,` $plane_3$`,` *name*`)`

Finds the intersections between two lines, two planes, a line and a plane, or three planes ◇ *name* is the name given to the point, line, or plane of intersection. ◇ See also: `geom3d[coordinates]`, `geometry[inter]`, `projgeom[inter]`, `student[intersection]`.

geom3d[line3d]

`line3d(`*name*`,` *list*`)`

Defines a line in three-dimensional space ◇ *name* is the name given to the line you are defining, and *list* can contain two points or a point and a directional vector. ◇ See also: `geom3d[point3d]`, `geom3d[plane]`.

geom3d[midpoint]

`midpoint(`pt_1`,` pt_2`,` *name*`)`

Finds the midpoint of the segment joining two points ◇ *name* is the name given to the midpoint. ◇ See also: `geom3d[onsegment]`, `geom3d[symmetric]`, `geom3d[coordinates]`, `geometry[midpoint]`, `projgeom[midpoint]`, `student[midpoint]`.

geom3d[on_plane]

`on_plane(`*pt*`,` *plane*`)`
`on_plane(`*ptlist*`,` *plane*`)`
`on_plane(`*ptset*`,` *plane*`)`

Tests whether a point, list, or set of points are on a given plane ◇ This function returns *true*, *false*, or a condition. ◇ See also: `geom3d[on_sphere]`.

geom3d[onsegment]

`onsegment(`*pt*$_1$`,` *pt*$_2$`,` *r*`,` *name*`)`

Finds the point that divides the segment joining the two given points by a given ratio *r* ◇ *name* is the name given to the computed point. Selecting $r = 1$ computes the midpoint. *r* cannot be -1. ◇ See also: `geom3d[midpoint]`, `geom3d[coordinates]`, `geometry[onsegment]`, `projgeom[onsegment]`.

geom3d[on_sphere]

`on_sphere(`*pt*`,` *sphere*`)`
`on_sphere(`*ptlist*`,` *sphere*`)`
`on_sphere(`*ptset*`,` *sphere*`)`

Tests whether the point, list of points, or set of points is on the given sphere ◇ This function returns *true*, *false*, or a condition. ◇ See also: `geom3d[on_plane]`, `geom3d[sphere]`.

geom3d[parallel]

`parallel(`*pt*`,` *line*`,` *name*`)`
`parallel(`*pt*`,` *plane*`,` *name*`)`
`parallel(`*line*$_1$`,` *line*$_2$`,` *name*`)`

Finds a line or plane passing through the given point and parallel to the given line or plane, or a plane passing through a line and parallel to another line ◇ *name* is the name given to the computed line or plane. ◇ See also: `geom3d[perpendicular]`, `geometry[parallel]`.

geom3d[perpendicular]

`perpendicular(`*pt*`,` *line*`,` *name*`)`
`perpendicular(`*pt*`,` *plane*`,` *name*`)`
`perpendicular(`*pt*`,` *line*`,` *name*`, line)`
`perpendicular(`*pt*`,` *line*`,` *name*`, line3d)`

Finds a plane passing through the given point and perpendicular to the given line, or a line passing through the given point and perpendicular to the given plane, or a line passing through the given point and given line and perpendicular to the given line ◇ The last two forms have identical effects. *name* is the name given to the computed line or plane. ◇ See also: `geom3d[parallel]`, `geometry[perpendicular]`.

geom3d[plane]

`plane(`*name*`,` *list*`)`

Defines a plane ◇ *name* is the name given to the plane you are defining, and *list* can contain a linear equation (or linear polynomial) in `x`, `y`, and `z`, a point and the normal vector, or a point and two vectors. ◇ See also: `geom3d[point3d]`, `geom3d[line3d]`.

geom3d[point3d]

point3d(*name*, [*expr*$_x$, *expr*$_y$, *expr*$_z$])
point3d(*name*, *expr*$_x$, *expr*$_y$, *expr*$_z$)
Define a point in three-dimensional space ◇ *name* is the name given to the point. ◇ See also: geom3d[line3d], geom3d[plane], geom3d[coordinates].

geom3d[powerps]

powerps(*pt*, *sphere*)
Finds the power of the given point with respect to the given sphere ◇ See also: geometry[powerpc].

geom3d[projection]

projection(*pt*, *line*, *name*)
projection(*pt*, *plane*, *name*)
projection(*line*, *plane*, *name*)
Finds the projection of a point onto a line or plane, or a line onto a plane ◇ *name* is the name of the projected point or line. ◇ See also: geom3d[perpendicular], geometry[projection].

geom3d[radius]

radius(*sphere*)
Computes the radius of the given sphere ◇ This function is equivalent to the command *sphere*[radius]. ◇ See also: geom3d[sphere], geometry[radius].

geom3d[rad_plane]

rad_plane(*sphere*$_1$, *sphere*$_2$, *name*)
Finds the radical plane of the two given spheres ◇ *name* is the name of the radical plane of the two spheres.

geom3d[reflect]

reflect(*pt*, *line*, *name*)
reflect(*pt*, *plane*, *name*)
reflect(*line*, *plane*, *name*)
Finds the reflection of a point with respect to a line or plane, or the reflection of a line with respect to a plane ◇ *name* is the name given to the reflected point or line. ◇ See also: geom3d[symmetric], geometry[reflect].

geom3d[sphere]

sphere(*name*, *list*)
Defines a sphere ◇ *name* is the name given to the sphere you are defining, and *list* can contain an equation in x, y and z, the center and radius, four points, or two points and the name diameter. ◇ See also: geom3d[line3d], geom3d[point3d], geom3d[plane].

geom3d[symmetric]

`symmetric(`*pt₁*`, `*pt₂*`, `*name*`)`

Finds the symmetric point of a point with respect to another point ◇ *name* is the name given to symmetric point of pt_1 with respect to pt_2. ◇ See also: `geom3d[reflect]`, `geom3d[midpoint]`, `geom3d[coordinates]`, `geometry[symmetric]`.

geom3d[tangent]

`tangent(`*pt*`, `*sphere*`, `*name*`)`

Finds the tangent plane of the point on the given sphere ◇ *name* is the name given to the tangent plane. The point must be on the sphere. ◇ See also: `geom3d[are_tangent]`, `geometry[tangent]`.

geom3d[tetrahedron]

`tetrahedron(`*name*`, `*list*`)`

Defines a tetrahedron ◇ The list may consist of four points or four planes. ◇ See also: `geom3d[line3d]`, `geom3d[point3d]`, `geom3d[plane]`.

geom3d[triangle3d]

`triangle3d(`*name*`, `*list*`)`

Defines a triangle in three-dimensional space ◇ The list consists of three points. ◇ See also: `geom3d[line3d]`, `geom3d[point3d]`, `geom3d[tetrahedron]`, `geometry[triangle]`.

geom3d[type]

`type(`*g*`, `*typename*`)`

Checks for three-dimensional geometric types ◇ *typename* can be any of the names `point3d`, `line3d`, `plane`, `sphere`, `triangle3d`, or `tetrahedron`. This function returns *true* or *false*. ◇ See also: `geom3d[point3d]`, `geom3d[line3d]`, `geom3d[plane]`, `geometry[type]`.

geom3d[volume]

`volume(`*g*`)`

Computes the volume of a tetrahedron or a sphere ◇ The volume is always positive. ◇ See also: `geom3d[area]`.

geometry

Two-dimensional Euclidean geometry ◇ The names `x` and `y` are used in this package as global variables, so never use them for other purposes. ◇ In Release 2 and earlier versions, the functions in this package must first be loaded using `with` before they can be invoked. ◇ See also: `geometry[`*function*`]`, `geom3d`, `projgeom`, `with`.

geometry[altitude]

`altitude(`*triangle*`, `*pt*`, `*name*`)`

Finds the altitude of the given triangle at the given vertex point ◇ *name* is the name given to the line computed as the altitude of *triangle* at the vertex *pt*. ◇ See also: `geometry[median]`, `geometry[bisector]`.

geometry[Appolonius]

`Appolonius(`*circle*$_1$, *circle*$_2$, *circle*$_3$`)`
Finds the Appolonius circles of three given circles ◇ This function returns a list of the Appolonius circles. In general, there are eight circles. The coordinates of the centers and the radii of the circles must be numeric.

geometry[area]

`area(`*g*`)`
Computes the area of the given triangle, square, or circle ◇ See also: `geom3d[area]`.

geometry[are_collinear]

`are_collinear(`*pt*$_1$, *pt*$_2$, *pt*$_3$`)`
Tests whether three points are collinear ◇ This function returns *true*, *false*, or a condition. ◇ See also: `geometry[are_concurrent]`, `geom3d[are_collinear]`.

geometry[are_concurrent]

`are_concurrent(`*line*$_1$, *line*$_2$, *line*$_3$`)`
Tests whether three lines are concurrent ◇ This function returns *true*, *false*, or a condition. ◇ See also: `geometry[are_collinear]`, `geom3d[are_concurrent]`.

geometry[are_harmonic]

`are_harmonic(`*pt*$_1$, *pt*$_2$, *pt*$_3$, *pt*$_4$`)`
Tests whether the first pair of points is harmonic conjugate to the second pair of points ◇ This function returns *true*, *false*, or a condition. The given points must first be collinear. ◇ See also: `geometry[harmonic]`.

geometry[are_orthogonal]

`are_orthogonal(`*circle*$_1$, *circle*$_2$`)`
Tests whether two circles are orthogonal ◇ This function returns *true*, *false*, or a condition.

geometry[are_parallel]

`are_parallel(`*line*$_1$, *line*$_2$`)`
Tests whether two lines are parallel ◇ This function returns *true*, *false*, or a condition. ◇ See also: `geometry[parallel]`, `geometry[are_perpendicular]`, `geom3d[are_parallel]`.

geometry[are_perpendicular]

`are_perpendicular(`*line*$_1$, *line*$_2$`)`
Tests whether two lines are perpendicular ◇ This function returns *true*, *false*, or a condition. ◇ See also: `geometry[perpendicular]`, `geometry[are_parallel]`, `geom3d[are_perpendicular]`.

geometry[are_similar]

`are_similar(`*triangle*$_1$, *triangle*$_2$`)`
Tests whether two triangles are similar ◇ This function returns *true*, *false*, or a condition.

geometry[are_tangent]

`are_tangent(`g_1, g_2`)`

Tests whether a line and a circle or two circles are tangent ◇ This function returns *true*, *false*, or a condition. ◇ See also: `geometry[tangent]`, `geom3d[are_tangent]`.

geometry[bisector]

`bisector(`*triangle*`,` *pt*`,` *name*`)`

Finds the bisector of the given triangle at the given vertex point ◇ *name* is the name given to the line that bisects the angle of the triangle at the given vertex. ◇ See also: `geometry[altitude]`, `geometry[median]`.

geometry[center]

`center(`*circle*`)`

Finds the center of a given circle ◇ This function returns a point named `center_of_`*circle*. ◇ See also: `geom3d[center]`.

geometry[centroid]

`centroid(`*triangle*`,` *name*`)`
`centroid(`*ptlist*`,` *name*`)`
`centroid(`*ptset*`,` *name*`)`

Computes the centroid of the given triangle or the given list or set of points ◇ *name* is the name given to the centroid. ◇ See also: `geom3d[centroid]`.

geometry[circle]

`circle(`*name*`,` *list*`)`

Defines a circle ◇ *name* is the name given to the circle you are defining, and *list* can contain an equation for the circle, three points, the center and the radius, or two points and the name `diameter`. ◇ See also: `geometry[ellipse]`, `geometry[point]`.

geometry[circumcircle]

`circumcircle(`*triangle*`,` *name*`)`

Finds the circumcircle of the given triangle ◇ *name* is the name given to the circumcircle. ◇ See also: `geometry[excircle]`, `geometry[incircle]`.

geometry[concyclic]

`concyclic(`pt_1, pt_2, pt_3, pt_4`)`

Tests whether four points are concyclic ◇ This function returns *true*, *false*, or a condition. ◇ See also: `geometry[on_circle]`.

geometry[conic]

`conic(`*name*`,` *ptlist*`)`

Finds the conic passing through five points ◇ *name* is the name given to the conic passing through the five points in *ptlist*. ◇ See also: `geometry[ellipse]`.

geometry[convexhull]

convexhull(*ptlist*)
convexhull(*ptset*)
Finds the convex hull enclosing the given points ◇ This function returns a list of points in counterclockwise order.

geometry[coordinates]

coordinates(*pt*)
Returns the coordinates of the given point ◇ The coordinates are returned as a list. ◇ See also: geom3d[coordinates].

geometry[detailf]

detailf(*g*)
Gives information about a point, line, or circle in floating-point format ◇ This function returns the coordinates of a point, the equation of a line, or the coordinates of the center and the radius of a circle.

geometry[diameter]

diameter(*ptlist*)
diameter(*ptset*)
Computes the diameter of the given list or set of points ◇ This function returns a list of three items: the two endpoints of the diameter and the diameter. ◇ See also: geometry[convexhull].

geometry[distance]

distance(pt_1, pt_2)
distance(*pt*, *line*)
Finds the distance between two points or between a point and a line ◇ See also: geom3d[distance], student[distance].

geometry[ellipse]

ellipse(*name*, *list*)
Defines the ellipse ◇ *list* can contain an equation, five points, or the center point, the lengths of the major and minor axes, and the word x_axis or y_axis to indicate the direction of the major axis. ◇ See also: geometry[circle], geometry[conic].

geometry[Eulercircle]

Eulercircle(*triangle*, *name*)
Finds the Euler circle of the given triangle ◇ *name* is the name given to the Euler circle.

geometry[Eulerline]

Eulerline(*triangle*, *name*)
Finds the Euler line of a given triangle ◇ *name* is the name given to the Euler line.

A B C D E F G H I J K L M N O P Q R S T U V W X Y Z

geometry[excircle]

`excircle(`*triangle*`)`

Finds the three excircles of a given triangle ◇ This function returns a sequence of three circles, named `excircle_of_`*triangle_vertex-name*. ◇ See also: `geometry[circumcircle]`, `geometry[incircle]`.

geometry[find_angle]

`find_angle(`$line_1$`,` $line_2$`)`

`find_angle(`$circle_1$`,` $circle_2$`)`

Finds the angle between two lines or between two circles ◇ This function returns the smallest angle between two lines. The angle between two circles is between 0 and π. ◇ See also: `geometry[distance]`.

geometry[Gergonnepoint]

`Gergonnepoint(`*triangle*`,` *name*`)`

Finds the Gergonne point of the given triangle ◇ *name* is the name given to the Gergonne point.

geometry[harmonic]

`harmonic(`pt_1`,` pt_2`,` pt_3`,` *name*`)`

Finds a point that is harmonic conjugate to another point with respect to two other points ◇ *name* is the name given to the point that is harmonic conjugate to pt_1 with respect to pt_2 and pt_3. ◇ See also: `projgeom[harmonic]`.

geometry[incircle]

`incircle(`*triangle*`,` *name*`)`

Finds the incircle of the given triangle ◇ *name* is the name given to the incircle. ◇ See also: `geometry[circumcircle]`, `geometry[excircle]`.

geometry[inter]

`inter(`g_1`,` g_2`)`

Finds the intersections between two lines, a line and a circle, or two circles ◇ This function may return a single point named g_1`_intersect_`g_2, or a sequence of two points named g_1`_intersect1_`g_2 and g_1`_intersect2_`g_2. ◇ See also: `geom3d[inter]`, `projgeom[inter]`, `student[intersection]`.

geometry[inversion]

`inversion(`*g*`,` *circle*`,` *name*`)`

Finds the inversion of the point, line, or circle *g* with respect to the given circle ◇ *name* is the name given to the inversion.

geometry[is_equilateral]

`is_equilateral(`*triangle*`)`

Tests whether the given triangle is equilateral ◇ This function returns *true*, *false*, or a condition. ◇ See also: `geometry[is_right]`.

geometry[is_right]

`is_right(`*triangle*`)`

Tests whether the given triangle is a right triangle ◇ This function returns *true*, *false*, or a condition. ◇ See also: `geometry[is_equilateral]`.

geometry[line]

`line(`*name*`, `*list*`)`

Defines a line ◇ *name* is the name given to the line you are defining. *list* can be two points, a linear equation, or a linear polynomial in `x` and `y`. ◇ See also: `geometry[point]`, `geometry[circle]`.

geometry[make_square]

`make_square(`*name*`, `*list*`)`

Constructs a square ◇ *name* is the name given to the square you are defining. *list* can be two adjacent vertices and the name `adjacent`, two opposite vertices and the name `diagonal`, or a vertex and the center, with the center labeled `center` = *point*. ◇ See also: `geometry[square]`.

geometry[median]

`median(`*triangle*`, `*pt*`, `*name*`)`

Finds the median line of the given triangle at the given vertex point ◇ *name* is the name given to the median line. ◇ See also: `geometry[altitude]`, `geometry[bisector]`.

geometry[midpoint]

`midpoint(`pt_1`, `pt_2`, `*name*`)`

Finds the midpoint of a segment joining two points ◇ *name* is the name given to the midpoint. ◇ See also: `geom3d[midpoint]`, `projgeom[midpoint]`, `student[midpoint]`.

geometry[Nagelpoint]

`Nagelpoint(`*triangle*`, `*name*`)`

Finds the Nagel point of the given triangle ◇ *name* is the name given to the Nagel point.

geometry[on_circle]

`on_circle(`*pt*`, `*circle*`)`
`on_circle(`*ptlist*`, `*circle*`)`
`on_circle(`*ptset*`, `*circle*`)`

Tests whether a point, a list of points, or a set of points is on the given circle ◇ This function returns *true*, *false*, or a condition. ◇ See also: `geometry[concyclic]`, `geometry[on_line]`.

A B C D E F G H I J K L M N O P Q R S T U V W X Y Z

geometry[on_line]

`on_line(`*pt*`,` *line*`)`
`on_line(`*ptlist*`,` *line*`)`
`on_line(`*ptset*`,` *line*`)`
Tests whether a point, a list of points, or a set of points is on the given line ◇ This function returns *true*, *false*, or a condition. ◇ See also: `geometry[on_circle]`.

geometry[onsegment]

`onsegment(`pt_1`,` pt_2`,` *r*`,` *name*`)`
Finds the point that divides the segment joining two given points by a given ratio *r* ◇ *name* is the name given to the computed point. Selecting $r = 1$ computes the midpoint. *r* cannot be -1. ◇ See also: `geom3d[onsegment]`, `projgeom[onsegment]`.

geometry[orthocenter]

`orthocenter(`*triangle*`,` *name*`)`
Computes the orthocenter of a triangle ◇ *name* is the name given to the orthocenter.

geometry[parallel]

`parallel(`*pt*`,` *line*`,` *name*`)`
Create a line that goes through the given point and is parallel to the given line ◇ *name* is the name given to the parallel line. ◇ See also: `geometry[are_parallel]`, `geometry[perpendicular]`, `geom3d[parallel]`.

geometry[perpen_bisector]

`perpen_bisector(`pt_1`,` pt_2`,` *name*`)`
Finds the line through the midpoint of the two given points and perpendicular to the line joining them ◇ *name* is the name given to the perpendicular bisector. ◇ See also: `geometry[perpendicular]`.

geometry[perpendicular]

`perpendicular(`*pt*`,` *line*`,` *name*`)`
Finds the line passing through the given point and perpendicular to the given line ◇ *name* is the name given to the perpendicular line. ◇ See also: `geometry[are_perpendicular]`, `geometry[parallel]`, `geometry[perpen_bisector]`, `geom3d[perpendicular]`.

geometry[point]

`point(`*name*`,` $expr_x$`,` $expr_y$`)`
`point(`*name*`, [`$expr_x$`,` $expr_y$`])`
`point(`*name*`, x=`$expr_x$`, y=`$expr_y$`)`
`point(`*name*`, [x=`$expr_x$`, y=`$expr_y$`])`
Defines a point in two-dimensional space ◇ *name* is the name given to the point you are defining. ◇ See also: `geometry[line]`, `geometry[circle]`.

geometry[polar_point]

`polar_point(`*pt*, *conic*, *name*`)`
Determines the polar line of the given point with respect to given conic ◇ *name* is the name given to the polar line. ◇ See also: `geometry[pole_line]`.

geometry[pole_line]

`pole_line(`*line*, *conic*, *name*`)`
Determines the pole of the given line with respect to the given conic ◇ *name* is the name given to the pole of the line. ◇ See also: `geometry[polar_point]`.

geometry[powerpc]

`powerpc(`*pt*, *circle*`)`
Finds the power of the given point with respect to the given circle ◇ See also: `geom3d[powerps]`.

geometry[projection]

`projection(`*pt*, *line*, *name*`)`
Finds the projection of the given point on the given line ◇ *name* is the name given to the projected point. ◇ See also: `geometry[perpendicular]`, `geometry[parallel]`, `geom3d[projection]`.

geometry[rad_axis]

`rad_axis(`$circle_1$, $circle_2$, *name*`)`
Finds the radical axis of the two given circles ◇ *name* is the name given to the radical axis. ◇ See also: `geometry[rad_center]`.

geometry[rad_center]

`rad_center(`$circle_1$, $circle_2$, $circle_3$, *name*`)`
Finds the radical center of three given circles ◇ *name* is the name given to the radical center. ◇ See also: `geometry[rad_axis]`.

geometry[radius]

`radius(`*circle*`)`
Computes the radius of the given circle ◇ This function is equivalent to *circle*`[radius]`. ◇ See also: `geometry[circle]`, `geom3d[radius]`.

geometry[randpoint]

`randpoint(`*line*, $range_x$, *name*`)`
`randpoint(`*circle*, *name*`)`
`randpoint(`$range_x$, $range_y$, *name*`)`
Picks a "random" point on a line, on a circle, or in a rectangular region ◇ A random point selected on a line has x-coordinate in the given range. A rectangular region is determined by an x-range and a y-range. Consecutive calls with the same arguments yield the same point, regardless of the value of `_seed`. ◇ `randpoint(0..2,-3..1,P);` ◇ See also: `geometry[point]`.

geometry[reflect]

`reflect(`*pt*`, `*line*`, `*name*`)`

Finds the reflection of a point with respect to a line ◇ *name* is the name given to the reflected point. ◇ See also: `geometry[rotate]`, `geometry[symmetric]`, `geom3d[reflect]`.

geometry[rotate]

`rotate(`pt_1`, `θ`, `*dir*`, `*name*`, `pt_2`)`

Computes the point obtained by rotating pt_1 through an angle of θ radians in the direction *dir* with respect to the point pt_2 ◇ *name* is the name given to the computed point. *dir* must be either `clockwise` or `counterclockwise`. If pt_2 is omitted, the origin is the center of rotation. ◇ See also: `geometry[reflect]`, `geometry[symmetric]`.

geometry[sides]

`sides(`*triangle*`)`

`sides(`*square*`)`

Computes the lengths of the sides of a triangle or a square ◇ The lengths of the sides of a triangle are returned in a list. For a square, a single expression is returned.

geometry[similitude]

`similitude(`$circle_1$`, `$circle_2$`)`

Finds the insimilitude and outsimilitude of two circles ◇ This function may return one similitude or two in an expression sequence. The similitudes are named by joining the names of the given circles to the strings `in_similitude_of_` and `out_similitude_of_`.

geometry[Simsonline]

`Simsonline(`*triangle*`, `*pt*`, `*name*`)`

Finds the Simson line of the given triangle with respect to the given point on the circumcircle of the triangle ◇ *name* is the name given to the Simson line.

geometry[slope]

`slope(`*line*`)`

`slope(`pt_1`, `pt_2`)`

Computes the slope of a line ◇ The line may be given explicitly or implicitly by two points. ◇ This function is new in Release 3. ◇ See also: `geometry[line]`.

geometry[square]

`square(`*name*`, [`pt_1`, `pt_2`, `pt_3`, `pt_4`])`

Defines a square ◇ The vertices must be given in perimeter order (in either direction) in the list. ◇ See also: `geometry[make_square]`.

geometry[symmetric]

`symmetric(`pt_1`, `pt_2`, `*name*`)`
Finds the symmetric point of one point with respect to another point ◇ *name* is the name given to the symmetric point of pt_1 with respect to pt_2. ◇ See also: `geometry[rotate]`, `geometry[reflect]`, `geom3d[symmetric]`.

geometry[tangent]

`tangent(`*pt*`, `*circle*`, `$name_1$`, `$name_2$`)`
Finds the two tangents of the circle passing through the given point ◇ $name_1$ and $name_2$ are the names given to the tangent lines. ◇ See also: `geometry[are_tangent]`, `geometry[tangentpc]`, `geom3d[tangent]`.

geometry[tangentpc]

`tangentpc(`*pt*`, `*circle*`, `*name*`)`
Finds the tangent line of the given point on the given circle ◇ *name* is the name given to the tangent line. ◇ See also: `geometry[are_tangent]`, `geometry[tangent]`.

geometry[triangle]

`triangle(`*name*`, `*list*`)`
Defines a triangle in two-dimensional space ◇ *list* can consist of three points, three lines, three side lengths, or two side lengths and the angle between them. An angle is specified with an equation of the form `angle =` θ, where θ is in radians. ◇ See also: `geometry[line]`, `geometry[point]`, `geometry[circle]`, `geom3d[triangle]`.

geometry[type]

`type(`*g*`, `*typename*`)`
Checks for two-dimensional geometric types ◇ *typename* can be any of the names `point2d`, `line2d`, or `circle2d`. This function returns *true* or *false*. ◇ See also: `geometry[circle]`, `geometry[line]`, `geometry[point]`, `geom3d[type]`.

GetAlgExt `GetAlgExt(`*expr*`) mod `*n*
Returns the algebraic extension used in an expression ◇ When used in conjunction with `mod`, this function returns the `RootOf` in the given expression. ◇ See also: `RootOf`.

GF `GF(`*p*`, `*n*`)`
`GF(`*p*`, `*n*`, `*poly*`)`
Galois Field calculations ◇ This function returns a table of functions and constants for doing arithmetic in the finite field with p^n elements defined by GF(*p*)$[x]/$(*poly*), where *poly* is an irreducible polynomial in x of degree *n* over the integers mod *p*. If such a polynomial is not specified, one is chosen at random. The elements of the field are represented using the `modp1` representation. The following operations are available in the

A B C D E F G H I J K L M N O P Q R S T U V W X Y Z

table returned:

'+'	Addition in the field
'-'	Subtraction in the field
'*'	Multiplication in the field
'/'	Division in the field
'^'	Exponentiation in the field
0	Additive identity
1	Multiplicative identity
ConvertIn	Convert a polynomial to a field element
ConvertOut	Convert a field element to a polynomial
extension	Polynomial defining the extension
input	Convert an integer in $0..p^n - 1$ to a field element
inverse	Inverse of an element in the field
isPrimitiveElement	Tests whether an element is primitive
norm	Norm of an element
order	Multiplicative order of an element
output	Convert a field element to an integer
PrimitiveElement	Returns a randomly selected primitive element
random	Returns a random element from the field
trace	Trace of an element

◇ You must `readlib(GF)` before using this command. ◇ See also: `evalgf`, `modp1`, `modpol`.

global `global` *var*$_1$, *var*$_2$, `...`
Identifies global variables ◇ The `global` declaration appears at the beginning of the body of a procedure. ◇ This declaration is new in Release 3. ◇ See also: `local`, `options`, `proc`, `updtsrc`.

grobner Gröbner basis package ◇ This package is a collection of routines for doing Gröbner basis calculations. ◇ See also: `grobner[`*function*`]`, `with`.

grobner[finduni]

`finduni(`*var*, *polys*`)`
`finduni(`*var*, *polys*, *vars*`)`
Attempts to find a univariate polynomial of minimal degree in an ideal ◇ This function constructs the univariate polynomial in *var* of least degree in the ideal generated by the list or set of polynomials *polys*. A list or set of indeterminates may be given as a fourth argument. The construction makes use of the total degree Gröbner basis. `grobner[finite]` should be called first to ensure there are only finitely many solutions. ◇ `finduni(x, [y^2*z+1, x^2+y^2, x*z^2+1]);` $\longrightarrow 1 + x^3$ ◇ See also: `grobner[finite]`, `grobner[gsolve]`, `grobner[solvable]`.

grobner[finite]

`finite(`*polys*`)`
`finite(`*polys*, *vars*`)`
Decides whether a given algebraic system has only finitely many solutions ◇ Uses the total degree Gröbner basis and a criterion of Buchberger to

determine whether the list or set of polynomials *polys* in the indeterminates *vars* has only finitely many solutions. *vars* may be a list or a set. ◇ `finite([y^2*z+1, x^2+y^2, x*z^2+1]);` ⟶ *true* ◇ See also: `grobner[finduni]`, `grobner[gsolve]`, `grobner[solvable]`.

grobner[gbasis]

`gbasis(`*polys*`,` *vars*`)`
`gbasis(`*polys*`,` *vars*`,` *termorder*`)`
`gbasis(`*polys*`,` *vars*`,` *termorder*`,` *name*`)`
Computes a reduced, minimal Gröbner basis ◇ This function computes the reduced, minimal Gröbner basis of the polynomials in the list or set *polys* with respect to the variables *vars* and the given term ordering. The term ordering *termorder* may be `plex` (pure lexicographic), or `tdeg` (total degree, the default). *vars* may be a list or a set. If it is a set, the variables are ordered in a "heuristically optimal" way, and a name is required to hold this ordering. ◇ `gbasis([y^2*z+1, x^2+y^2, x*z^2+1], [x,z,y], plex);` ⟶ $[x + y^4, z - y^4, 1 + y^6]$ ◇ See also: `grobner[normalf]`.

grobner[gsolve]

`gsolve(`*polys*`)`
`gsolve(`*polys*`,` *vars*`)`
`gsolve(`*polys*`,` *vars*`,` *nonzeropolys*`)`
Prepares the given algebraic system for solving ◇ This function computes a collection of reduced (lexicographic) Gröbner bases corresponding to the list or set of polynomials *polys* in the variables *vars*. *vars* may be a list or a set. The set of polynomials *nonzeropolys* may be used to prevent certain quantities from being considered in roots. The result is a list of reduced subsystems whose roots are those of the original system, but whose variables have been successively eliminated and separated as far as possible. `solve` may be applied to each sublist of the result. ◇ In Release 2 and earlier, the order of the second and third parameters is reversed. ◇ `gsolve([y^2*z+1, x^2+y^2, x*z^2+1]);` ⟶ $[[x + 1, z - 1, y^2 + 1], [x + y^2 - 1, 1 - y^2 + z, y^4 - y^2 + 1]]$ ◇ See also: `grobner[solvable]`, `grobner[finite]`, `grobner[finduni]`, `solve`.

grobner[leadmon]

`leadmon(`*poly*`,` *varlist*`,` *termorder*`)`
`leadmon(`*poly*`,` *varlist*`)`
Computes the leading monomial of a polynomial with respect to a chosen term ordering ◇ *varlist* is a list of indeterminates used in the term ordering; *termorder* may be `plex` (pure lexicographic) or `tdeg` (total degree, the default). A list of two elements is returned: the leading coefficient and the leading term. ◇ `leadmon(2*x^2*z^4-x^3*y*z, [x,y,z], plex);` ⟶ $[-1, x^3\, y\, z]$ ◇ See also: `grobner`.

grobner[normalf]

`normalf(`*poly*`,` *polylist*`,` *varlist*`)`
`normalf(`*poly*`,` *polylist*`,` *varlist*`,` *termorder*`)`
Reduced form of a polynomial modulo an ideal ◇ This function computes the fully reduced form of *poly* with respect to the given ideal basis, indeterminates and term ordering. Normally, *polylist* is a Gröbner basis for the same list of variables and term ordering. In this case, this function computes a canonical form for *poly* modulo the ideal generated by *polylist*. The term ordering may be `plex` for pure lexicographic, or `tdeg` for total degree (the default). ◇ `normalf(x^3*y^8+x*y^4*z, [x+y^4,z-y^4,1+y^6], [x,z,y], plex);` $\longrightarrow y^2 - 1$ ◇ See also: `grobner[gbasis]`.

grobner[solvable]

`solvable(`*polys*`)`
`solvable(`*polys*`,` *vars*`)`
`solvable(`*polys*`,` *vars*`,` *termorder*`)`
Decides whether the given algebraic system is solvable ◇ This function decides, using Gröbner basis methods, whether a list or set of polynomials *polys* with respect to the indeterminates *vars* is algebraically consistent. *termorder* may be `plex` (pure lexicographic) or `tdeg` (total degree, the default). ◇ `solvable([y^2*z+1, x^2+y^2,x*z^2+1], [x,z,y], plex);` $\longrightarrow$ *true* ◇ See also: `grobner[finite]`, `grobner[finduni]`, `grobner[gsolve]`.

grobner[spoly]

`spoly(`*poly*$_1$`,` *poly*$_2$`,` *varlist*`,` *termorder*`)`
`spoly(`*poly*$_1$`,` *poly*$_2$`,` *varlist*`)`
Computes the S-polynomial of the two given polynomials with respect to the list of indeterminates and chosen term ordering ◇ *termorder* may be `plex` (pure lexicographic) or `tdeg` (total degree, the default). ◇ `spoly(x^2*z+x*y,z^2+y^3,[x,y,z]);` $\longrightarrow x\,y^4 - x^2\,z^3$ ◇ See also: `grobner[leadmon]`.

group — Permutation group and finitely presented group package ◇ In the descriptions of functions in this package, *G* stands for a group, *H* for a subgroup, *g* for an element of a group, and *perm* for a permutation. ◇ In Release 2 and earlier versions, the functions in this package must first be loaded using `with` before they can be invoked. ◇ See also: `group[`*function*`]`, `with`.

group[areconjugate]

`areconjugate(`*G*`,` g_1`,` g_2`)`
Determines whether two elements of the given permutation group *G* are conjugates to each other ◇ g_1 and g_2 must be elements of *G*. This function returns *true* or *false*. ◇ See also: `group[permgroup]`, `group[RandElement]`.

A B C D E F G H I J K L M N O P Q R S T U V W X Y Z

group[center]

`center(`*G*`)`

Finds the center of the permutation group *G* ◇ The center is the subgroup consisting of all the elements that commute with every element of *G*. This function returns a `permgroup`. ◇ `center(permgroup(3, {[[1,2,3]],[[1,2]]}));` ⟶ permgroup(3, { }) ◇ See also: `group[centralizer]`, `group[normalizer]`.

group[centralizer]

`centralizer(`*G*`,` *permset*`)`
`centralizer(`*G*`,` *perm*`)`

Finds the centralizer of a set of permutations or a single permutation ◇ *G* is a permutation group. The centralizer is the subgroup of all elements of *G* that commute with every permutation in *permset* (or with the single permutation *perm*). *perm* need not be a member of *G*. This function returns a `permgroup`. ◇ `centralizer(permgroup(3, {[[1,2,3]], [[1,2]]}), [[1,2]]);` ⟶ permgroup(3,{[[1, 2]]}) ◇ See also: `group[permgroup]`.

group[convert/disjcyc]

`convert(`*permlist*`, disjcyc)`
`convert(`*word*`, disjcyc,` *G*`)`

Converts a permutation in list notation or a word into disjoint cycle notation ◇ *permlist* is a permutation in list notation: a list whose *i*th element is the image of *i* under the permutation. The permutation is returned in disjoint cycle notation. *word* is a word in the generator names of *G*. The product represented by *word* is computed and returned in disjoint cycle notation. This function is only available after the `group` package has been loaded. ◇ `convert([3,4,1,2],disjcyc);` ⟶ [[1, 3], [2, 4]] ◇ See also: `group[convert/permlist]`, `group[permgroup]`, `group[grelgroup]`, `group[type/disjcyc]`.

group[convert/permlist]

`convert(`*perm*`, permlist,` *n*`)`

Convert a permutation in disjoint cycle notation into list notation ◇ *n* is the degree of the permutation *perm*. A list is returned whose *i*th element is the image of *i* under *perm*, for $i = 1..n$. This function is available only after the `group` package has been loaded. ◇ `convert([[1,3],[2,4]], permlist,4);` ⟶ [3, 4, 1, 2] ◇ See also: `group[convert/disjcyc]`, `group[permgroup]`, `group[type/disjcyc]`.

group[core]

`core(`*H*`,` *G*`)`

Finds the core of a subgroup of a permutation group ◇ This function finds the largest normal subgroup of the permutation group *G* contained in the subgroup *H*. It returns a `permgroup`. ◇ See also: `group[issubgroup]`, `group[NormalClosure]`.

group[cosets]

`cosets(`*H*`)`
`cosets(`*G*`, `*H*`)`
Finds a complete list of right coset representatives for a subgroup of a permutation group or a subgroup of a group given by generators and relations ◇ For groups given by generators and relations, the argument *H* should be a `subgrel`. A set of words in the generators of the group is returned. For permutation groups, *H* should be a subgroup of *G*. A set of permutations in disjoint cycle notation is returned. ◇ See also: `group[cosrep]`, `group[grelgroup]`, `group[subgrel]`, `group[permgroup]`.

group[cosrep]

`cosrep(`*g*`, `*H*`)`
Expresses a group element as a product of an element in a subgroup multiplied by a right coset representative for that subgroup ◇ This function returns a two-element list: an element of *H* and a right coset representative for *H*. If *H* is a `subgrel`, then *g* should be a word in the group generators. If *H* is a `permgroup`, then *g* should be a permutation in disjoint cycle notation. In this case, the cosets of *H* are computed in the symmetric group of the same degree as *H*. ◇ See also: `group[permgroup]`, `group[subgrel]`, `group[cosets]`.

group[derived]

`derived(`*G*`)`
Finds the derived subgroup $[G, G]$ of the given permutation group *G* ◇ This function returns a `permgroup`. ◇ `derived(permgroup(3, {[[1,2,3]],[[1,2]]}));` ⟶ permgroup(3, {[ ], [[1, 2, 3]]}) ◇ See also: `group[DerivedS]`, `group[LCS]`, `group[permgroup]`.

group[DerivedS]

`DerivedS(`*G*`)`
Finds the derived series of the given permutation group *G* ◇ The derived series determines whether *G* is solvable. This function returns a list of `permgroup`s. ◇ `DerivedS(permgroup(3,{[[1,2,3]],[[1,2]]}));` ⟶ [permgroup(3, {[[1, 2, 3]], [[1, 2]]}), permgroup(3, {[ ], [[1, 2, 3]]}), permgroup(3, {[ ]})] ◇ See also: `group[derived]`, `group[LCS]`, `group[permgroup]`.

group[grelgroup]

`grelgroup(`*genset*`, `*relset*`)`
Represents a group by generators and relations ◇ *genset* is a set of names, the generators of the group. *relset* is a set of words in the generators, the relations defining the group. A word is represented as a list of generators and inverses of generators. The inverse of a generator `a` is represented by `1/a`. An empty list represents the empty word. This function returns the `grelgroup` call. ◇ `grelgroup({a,b}, {[a,a,a], [b,b],`

[a,b,a,1/b]}): ◇ See also: group[convert/disjcyc], group[permgroup], group[subgrel].

group[groupmember]

groupmember(*perm*, *G*)
Tests whether a permutation is in the permutation group *G* ◇ This function returns *true* or *false*. ◇ groupmember([[1,2]], permgroup(4,{[[1,2,3]],[[1,2],[3,4]]})); ⟶ *false* ◇ See also: group[permgroup].

group[grouporder]

grouporder(*G*)
Computes the order of a group ◇ This function returns the number of elements in *G*, a permgroup or a grelgroup. ◇ grouporder(permgroup(4,{[[1,2,3]],[[1,2],[3,4]]})); ⟶ 12 ◇ See also: group[permgroup], group[grelgroup].

group[inter]

inter(G_1, G_2)
Finds the intersection of two permutation groups ◇ The groups must be of the same degree. This function returns a permgroup. ◇ See also: group[permgroup].

group[invperm]

invperm(*perm*)
Finds the inverse of a permutation ◇ This returns the inverse permutation in disjoint cycle notation. ◇ invperm([[1,2,3],[4,5]]); ⟶ [[1, 3, 2], [4, 5]] ◇ See also: group[mulperms], group[permgroup].

group[isabelian]

isabelian(*G*)
Determines whether the permutation group *G* is abelian ◇ This function returns *true* or *false*. ◇ isabelian(permgroup(4, {[[1,2,3]], [[1,2], [3,4]]})); ⟶ *false* ◇ See also: group[permgroup].

group[isnormal]

isnormal(*H*)
isnormal(*G*, *H*)
Determines whether a subgroup is normal ◇ The first form determines whether the subgrl *H* is normal, the second form determines whether the permutation group *H* is normal in the permutation group *G*. *G* and *H* must be of the same degree. This function returns *true* or *false*. ◇ isnormal(subgrel({x=[a]}, grelgroup({a,b}, {[a,a,a], [b,b], [a,b,a,b]}))); ⟶ *true* ◇ See also: group[issubgroup], group[permgroup], group[subgrel], group[grelgroup].

group[issubgroup]

`issubgroup(`*H*`, `*G*`)`
Determines whether the permutation group *H* is a subgroup of the permutation group *G* ◇ *G* and *H* must be of the same degree. This function returns *true* or *false*. ◇ `issubgroup(permgroup(4, {[[1,2], [3,4]]}), permgroup(4, {[[1,2]], [[1,2,3,4]]}));` ⟶ *true* ◇ See also: `group[permgroup]`, `group[isnormal]`.

group[LCS]

`LCS(`*G*`)`
Finds the lower central series of the permutation group *G* ◇ This series determines whether *G* is nilpotent and its class of nilpotency. This function returns a list of `permgroups`. ◇ `LCS(permgroup(3, {[[1,2,3]]}));` ⟶ [permgroup(3,{[[1, 2, 3]]}), permgroup(3,{[ ]})] ◇ See also: `group[derived]`, `group[DerivedS]`.

group[mulperms]

`mulperms(`$perm_1$`, `$perm_2$`)`
Multiplies two permutations in disjoint cycle notation ◇ This function returns the product in disjoint cycle notation. Permutations act on the right, so $(perm_1 * perm_2)(i) = perm_2(perm_1(i))$ ◇ `mulperms([[1,2,4]],[[1,2],[3,4]]);` ⟶ [[2, 3, 4]] ◇ See also: `group[invperm]`, `group[permgroup]`.

group[NormalClosure]

`NormalClosure(`*H*`, `*G*`)`
Finds the normal closure of a subgroup of a permutation group ◇ This function finds the smallest normal subgroup of the permutation group *G* containing *H*. It returns a `permgroup`. ◇ `NormalClosure(permgroup(4, {[[1,2]]}), permgroup(4, {[[1,2]], [[1,2,3,4]]}));` ⟶ permgroup(4, {[[3, 4]], [[2, 3]], [[1, 2]]}) ◇ See also: `group[issubgroup]`, `group[core]`.

group[normalizer]

`normalizer(`*G*`, `*H*`)`
Finds the normalizer of the subgroup *H* of the permutation group *G* ◇ This function finds the largest subgroup of *G* in which *H* is a normal subgroup. This function returns a `permgroup`. ◇ See also: `group[NormalClosure]`, `group[permgroup]`.

group[orbit]

`orbit(`*G*`, `*i*`)`
`orbit(`*permset*`, `*i*`)`
Computes the orbit of a point under the action of a permutation group or a set of permutations *G* ◇ The orbit is returned as a set of points. ◇ `orbit({[[1,2],[3,4]],[[1,2,3]]},3);` ⟶ {1, 2, 3, 4} ◇ See also: `group[permgroup]`.

group[permgroup]

`permgroup(`*n*`, `*permset*`)`
Represents a permutation group ◇ *n* is the degree of the permutation group, and *permset* is a set of named or unnamed permutations generating the group. A named permutation has the form *name* = *perm*. A permutation is represented in disjoint cycle notation: a cycle is a list of distinct integers between 1 and *n*; a permutation is a list of pairwise disjoint cycles. The identity permutation is represented by the empty list. Permutations act on the right. This function returns the `permgroup` call. ◇ `permgroup(4, {[[1,2],[3,4]], [[2,3,4]]}):` ◇ See also: `group[convert/disjcyc]`, `group[convert/permlist]`, `group[grelgroup]`, `group[type/disjcyc]`.

group[permrep]

`permrep(`*H*`)`
Finds a permutation representation of a subgroup *H* of a group given by generators and relations ◇ *H* is a `subgrl`. This function finds all the right cosets of *H* in its parent group *G*, assigns consecutive integers to these cosets, constructs a permutation on these coset numbers for each group generator, and returns the permutation group generated by these permutations. This group is a homomorphic image of *G*, but not necessarily isomorphic to *G*. This function returns a `permgroup` whose generators are named the same as the original group generators. ◇ `permrep(subgrel({x=[a]}, grelgroup({a,b}, {[a,a,a], [b,b], [a,b,a,b]})));` $\longrightarrow$ permgroup(2,$\{a = [\,], b = [[1,2]]\}$) ◇ See also: `group[permgroup]`, `group[grelgroup]`, `group[subgrel]`.

group[pres]

`pres(`*H*`)`
Finds a presentation for a subgroup *H* of a group given by generators and relations ◇ This function computes a set of relations among the names of the generators of *H* sufficient to define the subgroup. This function returns a `grelgroup`. ◇ `pres(subgrel({x=[a]}, grelgroup({a,b}, {[a,a,a], [b,b], [a,b,a,b]})));` $\longrightarrow$ grelgroup($\{x\}, \{[x,x,x]\}$) ◇ See also: `group[grelgroup]`, `group[subgrel]`.

group[RandElement]

`RandElement(`*G*`)`
Selects a random element of the group *G* ◇ *G* may be a `permgroup` or a `grelgroup`. ◇ `RandElement(grelgroup({a,b}, {[a,a,a], [b,b,b], [b,a,b,a]}));` $\longrightarrow \left[\frac{1}{a}, \frac{1}{a}, \frac{1}{a}, b\right]$ ◇ See also: `group[groupmember]`.

group[subgrel]

`subgrel(`*eqnset*`, `*G*`)`
Represents a subgroup of a group ◇ *eqnset* is a set of equations of the form *name* = *word*. *name* is the name of a subgroup generator, and *word* is

a word in the generators of the group G. This function returns the `subgrel` call. ◇ `subgrel({x=[b,1/a]}, grelgroup({a,b}, {[a,a,a], [b,b,b], [b,a,b,a]})):` ◇ See also: `group[grelgroup]`.

group[Sylow]

`Sylow(`G`,` p`)`
Finds the Sylow p-subgroup of the permutation group G ◇ p is a prime number that divides the order of G. This function finds a maximal p-group contained in G. It returns a `permgroup`. ◇ `Sylow(permgroup(6, {[[1,2]], [[1,2,3,4,5,6]]}), 3);` ⟶ permgroup(4, {[[3, 4]], [[1, 3, 2, 4]]}) ◇ See also: `group[grouporder]`, `group[issubgroup]`, `group[permgroup]`.

group[type/disjcyc]

`type(`*expr*`, disjcyc(`n`))`
Checks for a permutation in disjoint cycle notation ◇ This function returns *true* if *expr* is a valid permutation of degree n in disjoint cycle notation and *false* otherwise. It is available only after the `group` package has been loaded. ◇ `type([[3,4],[1,2]],disjcyc(4));` ⟶ *true* ◇ See also: `group[convert/disjcyc]`, `group[convert/permlist]`, `group[permgroup]`.

harmonic `harmonic(`z`)`
The harmonic function ◇ For a positive integer n, the harmonic function is $\sum_{k=1}^{n} 1/k$. In general, for complex z, it is $\Psi(z+1)+\gamma$. You must use `evalf` to compute the value of this function. ◇ `evalf(harmonic(100));` ⟶ 5.187377518 ◇ See also: `gamma`, `Psi`.

has `has(`*expr*`,` *subexpr*`)`
`has(`*expr*`,` *list*`)`
`has(`*expr*`,` *set*`)`
Tests for a specified subexpression ◇ In the first form, the function returns *true* if *expr* contains the subexpression *subexpr*; otherwise it returns *false*. In the other forms, it returns *true* if and only if the expression contains at least one of the subexpressions in the given list or set. ◇ `has(a+b,a);` ⟶ *true* ◇ See also: `member`, `numboccur`, `op`.

hastype `hastype(`*expr*`,` *type*`)`
Tests for a specified type ◇ This function returns *true* if and only if *expr* has any subexpressions of type *type*. ◇ `hastype(a+2*b,'*');` ⟶ *true* ◇ See also: `has`, `type`, `whattype`.

heap Heap data structure routines ◇ You must `readlib(heap)` before using these commands.

heap[empty]

`heap[empty](`h`)`
Tests whether the heap h is empty ◇ This function returns *true* or *false*. ◇ You must `readlib(heap)` before using this command.

heap[extract]

`heap[extract](`*h*`)`
Removes the maximal element from the heap *h* and returns it ◇ You must `readlib(heap)` before using this command.

heap[insert]

`heap[insert](`*elem*`,` *h*`)`
Inserts *elem* into the heap *h* ◇ You must `readlib(heap)` before using this command.

heap[max]

`heap[max](`*h*`)`
Returns the maximal element from the heap *h* ◇ The element is not removed from the heap. ◇ You must `readlib(heap)` before using this command.

heap[new]

`heap[new](`*f*`)`
`heap[new](`*f*`,` $expr_1$`, ...,` $expr_n$`)`
Create a new heap ◇ Returns a new heap using the boolean-valued function *f* as the total ordering function. The optional arguments $expr_i$ are elements initially inserted into the heap. ◇ You must `readlib(heap)` before using this command.

heap[size]

`heap[size](`*h*`)`
Returns the number of elements in the heap ◇ You must `readlib(heap)` before using this command.

Heaviside

`Heaviside(`*x*`)`
The Heaviside step function ◇ This function is 0 for $x < 0$ and 1 for $x \geq 0$. ◇ See also: `Dirac`, `fourier`, `laplace`, `mellin`.

help

`help(`*topic*`)`
`help(`*topic*`,` *subtopic*`)`
`help(`*topic*`[`*subtopic*`])`
Help command ◇ Maple keywords must be enclosed in back quotes. The recommended way to invoke help is with the question mark. ◇ See also: `?`, `??`, `???`, `info`, `example`, `related`, `usage`.

Hermite

`Hermite(`*A*`,` *var*`)`
`Hermite(`*A*`,` *var*`,` *name*`)`
Inert Hermite Normal Form ◇ When used with `mod`, this function computes the Hermite Normal Form (reduced row echelon form) of a rectangular matrix of univariate polynomials in *var*. If a name is given as a third argument, it is assigned the transformation matrix U such that the result is UA. ◇ See also: `linalg[hermite]`, `Smith`.

history `history()`
Maintains a history of all values computed ◇ This command starts saving previous commands. Output of successive statements is saved in the variables `O1`, `O2`, and so on. The command `off` turns this feature off. A timing facility is available when history is on: use `timing(`*expr*`)` to time the execution of *expr*. When history is on, "" and """ are not available. ◇ You must `readlib(history)` before using this command. ◇ See also: `showtime`, `time`.

hypergeom

`hypergeom([`n_1`, ..., `n_j`], [`d_1`, ..., `d_k`], `*var*`)`
Computes the generalized hypergeometric function in the given variable ◇ This function is defined as

$$\sum_{m=0}^{\infty} \frac{var^m \prod_{i=1}^{j} \Gamma(n_i+m)/\Gamma(n_i)}{m!\prod_{i=1}^{k} \Gamma(d_i+m)/\Gamma(d_i)}$$

This function is also known as ${}_jF_k$. ◇ You must `readlib(hypergeom)` before using this command. ◇ `hypergeom([1],[2],z);` $\longrightarrow \frac{e^z-1}{z}$ ◇ See also: `convert/hypergeom`, `GAMMA`.

I Square root of -1 ◇ `I` is an alias for `(-1)^(1/2)`. ◇ See also: `alias`.

icontent `icontent(`*poly*`)`
Integer content of a polynomial with rational coefficients ◇ The content is the gcd of the numerators of the coefficients, divided by the lcm of the denominators. The polynomial should be in expanded form. ◇ `icontent(15/2*x+6*y);` $\longrightarrow \frac{3}{2}$ ◇ See also: `content`.

identity The identity indexing function ◇ This function gives 1 if all components of an array or table index are identical and 0 otherwise. ◇ `array(identity,1..3,1..3):` ◇ See also: `array`, `diagonal`, `table`.

if

`if` *cond* `then` *stmts* `fi`
`if` *cond* `then` *stmts* `else` *stmts* `fi`
`if` *cond* `then` *stmts* `elif` *cond* `then ... fi`

The selection (conditional) statement ◇ A conditional expression is a Boolean expression formed using the relational operators (`<`, `<=`, `>`, `>=`, `=`, `<>`), the logical operators (`and`, `or`, `not`), and the logical names (`true`, `false`, `FAIL`). The `elif` ... `then` ... construct may be repeated any number of times. ◇ See also: `do`.

ifactor
`ifactor(`n`)`
`ifactor(`n`, easy)`
`ifactor(`n`, lenstra)`
`ifactor(`n`, pollard)`
`ifactor(`n`, pollard, `m`)`
`ifactor(`n`, squfof)`
Integer factorization ◇ This function computes the complete integer factorization of n, returning a multiplicative expression. It first tries some "easy" tests, then calls one of several methods to extract any remaining large factors. The default method is the Morrison-Brillhart algorithm. The option `lenstra` selects Lenstra's elliptic curve method; `pollard`, Pollard's rho method; and `squfof`, Shanks's square forms method. The option `easy` does no extra work beyond the initial easy tests. In this case a partial factorization is returned, including a symbol that indicates the number of digits in the remaining unfactored part (e.g., *_c47*). The `pollard` method accepts an additional option m that improves efficiency when one of the factors is congruent to 1 modulo m. `ifactor` calls `isprime` on the remaining factor before invoking the other methods. This function also operates on fractions (factoring numerator and denominator), and lists or sets of integers and fractions. ◇ `ifactor(7007);` ⟶ $(7)^2(11)(13)$ ◇ See also: `ifactors`, `isprime`, `factor`, `type[facint]`.

ifactors
`ifactors(`n`)`
Integer factorization ◇ This function returns the complete integer factorization of the integer n and returns a list of prime factors and exponents. The factorization is performed using the default strategy in `ifactor`. ◇ You must `readlib(ifactors)` before using this command. ◇ `ifactors(-7007);` ⟶ $[-1,[[7,2],[11,1],[13,1]]]$ ◇ See also: `ifactor`, `factor`, `factors`, `isqrfree`.

iFFT
`iFFT(`m`, `*x-array*`, `*y-array*`)`
Computes the inverse fast Fourier transform of a complex sequence ◇ *x-array* and *y-array* are one-dimensional arrays of floats indexed from 1 to 2^m. *x-array* holds the real parts of the sequence and *y-array* the imaginary parts. The inverse fast Fourier transform is computed in place, altering the original arrays. This function returns 2^m. This procedure may be invoked with `evalhf`. ◇ You must `readlib(FFT)` before using this command. ◇ See also: `FFT`, `evalhf`, `FFT`, `invfourier`.

igcd
`igcd(`n_1`, `n_2`, ...)`
Computes the greatest common divisor of the given integers ◇ `igcd(21,49);` ⟶ 7 ◇ See also: `gcd`, `ilcm`.

igcdex
`igcdex(`n_1`, `n_2`, `*name*$_1$`, `*name*$_2$`)`
Extended Euclidean algorithm for integers ◇ This function returns the gcd g of n_1 and n_2 and computes integers s and t such that $g = s\,n_1 + t\,n_2$. s and t are stored in *name*$_1$ and *name*$_2$. ◇ See also: `gcdex`, `igcd`.

ilcm `ilcm(`n_1`, `n_2`, ...)`
Computes the least common multiple of the given integers ◇ This function computes the least common multiple of an arbitrary number of integers. ◇ `ilcm(21,49);` ⟶ 147 ◇ See also: `igcd, lcm.`

ilog `ilog(`*z*`)`
`ilog[`*b*`](`*z*`)`
Integer base *b* logarithm ◇ For real *z*, this function returns the greatest integer less than or equal to the base *b* logarithm of *z*. For complex *z*, it returns the maximum of the same function evaluated on the real and imaginary parts of *z*. The default base is exp(1). ◇ You must `readlib(ilog)` before using this command. ◇ `ilog[3](82);` ⟶ 4 ◇ See also: `ilog10, log.`

ilog10 `ilog10(`*z*`)`
Integer base 10 (common) logarithm ◇ For real *z*, this function returns the greatest integer less than or equal to the base 10 logarithm of *z*. For complex *z*, it returns the maximum of `ilog10` evaluated on the real and imaginary parts of *z*. ◇ `ilog10(101);` ⟶ 2 ◇ See also: `ilog, log10.`

Im `Im(`*expr*`)`
Extracts the imaginary part of the expression ◇ Use `assume` or `evalc` if unknown variables should represent real values. If *expr* includes a function *f*, `Im` tries to execute the procedure `'Im/`*f*`'`. ◇ `Im(ln(-1));` ⟶ π ◇ See also: `assume, evalc, Re.`

Indep `Indep(`*K*`)`
`Indep(`*K*`, `*name*`)`
Inert independence checking ◇ When used with `evala`, this function checks whether the given `RootOf` or set of `RootOf`s is independent and returns *true* or *false*. If *false*, *name* is assigned the set of relations found, if this argument is present. ◇ See also: `evala, RootOf.`

indets `indets(`*expr*`)`
`indets(`*expr*`, `*typename*`)`
Finds the indeterminates in the expression ◇ With only one argument, this function returns a set containing all the indeterminates of *expr*. *expr* is viewed as a rational expression, so subexpressions like `sqrt(x)` and `f(x,y)` are treated as indeterminates. In the second case, this function returns a set containing all subexpressions of type `typename`. ◇ `indets(sqrt(x)+2*y);` ⟶ $\{x, y, \sqrt{x}\}$

indices `indices(`*array*`)`
`indices(`*table*`)`
Indices of a table or array ◇ This function returns a sequence of lists of the indices (or keys) of the table or array corresponding to the entries that have been explicitly stored. There is a one-to-one correspondence between

the results of `indices` and `entries`. ◇ `indices(array([5,6,7]));` ⟶ [2],[3],[1] ◇ See also: `array, entries, op, table`.

infinity — ∞ ◇ Real or complex infinity, depending on context.

info — `info(`*name*`)`
Shows the brief description of a function ◇ The information is extracted from the on-line help text. Maple keywords must be enclosed in back quotes. ◇ This function is new in Release 3. ◇ `info(dsolve);` ⟶
`FUNCTION: dsolve - solve ordinary differential equations`
◇ See also: `example, help, related, usage`.

infolevel — `infolevel[`*f*`] :=` *n*
`infolevel[all] :=` *n*
Causes information of level n or below to be displayed when invoking the function f ◇ The information is encoded as `userinfo` statements inside the procedure. n is typically an integer between 1 and 5. If the name `all` is given, information of level n or below is displayed for every function. ◇ See also: `printlevel, userinfo`.

int — `int(`*expr*`,` *var*`)`
`int(`*expr*`,` *var*`=`*a*`..`*b*`,` *options*`)`
Definite and indefinite integration ◇ The first form computes the indefinite integral of *expr* with respect to the variable *var*. No constant of integration appears in the result. The second form computes the definite integral on the interval given by the range *a..b*. If a closed form for the integral cannot be found, the function call itself is returned. By default, `int` checks for discontinuities in the given interval. The option `continuous` instructs `int` not to look for discontinuities. The option `CauchyPrincipalValue` instructs `int` to consider the limits from the left and right at each interior discontinuity as a single limit, so that the independent variable approaches the discontinuity at the same rate from the left and the right. ◇ `int(cos(x),x);` ⟶ $\sin(x)$ ◇ `int(sin(x),x=0..Pi);` ⟶ 2 ◇ See also: `diff, Int, iscont, limit, series, student[Int]`.

Int — `Int(`*expr*`,` *var*`)`
`Int(`*expr*`,` *var*`=`*a*`..`*b*`)`
Inert definite and indefinite integration ◇ This function is a placeholder for integration. When the second form is used with `evalf`, a numerical approximation to the definite integral is calculated. ◇ `evalf(Int(exp(-x^2), x=0..1));` ⟶ .7468241328 ◇ See also: `evalf, 'evalf/int', int, student[Doubleint], student[Lineint], student[Tripleint], value`.

integrate — Global variable ◇ Its value is a procedure for integrating expressions. It is initially set to `int`. ◇ See also: `int`.

interface `interface(`*name*`=`*value*`)`
`interface(`*name*`)`
Sets or queries user interface variables ◇ The first form specifies a setting for the named variable. The second form returns the current value of the named variable. The accepted names are:

echo	labeling	plotoutput	prompt	verboseproc
endcolon	labelling	postplot	quiet	version
errorbreak	labelwidth	preplot	screenheight	warnlevel
indentamount	plotdevice	prettyprint	screenwidth	wordsize
iris	plotoptions	printfile	terminal	

Descriptions of all interface variables appear in Chapter 2 beginning on page 104. ◇ See also: `plotsetup`.

interp `interp(`*xvals*`, `*yvals*`, `*var*`)`
Polynomial interpolation ◇ *xvals* and *yvals* are lists or vectors of length n. This function computes the polynomial of degree $\leq n-1$ in the variable *var* that interpolates the points ($xvals_i$, $yvals_i$) for each i. All x values must be distinct. ◇ `interp([-1,0,1],[1,0,1],t);` $\longrightarrow t^2$ ◇
See also: `Interp`, `sinterp`.

Interp `Interp(`*xvals*`, `*yvals*`, `*var*`)`
`Interp(`*xvals*`, `*yvals*`)`
Inert polynomial interpolation function ◇ *xvals* and *yvals* are lists or vectors of length n. When used with `mod`, this function computes the polynomial that interpolates the points ($xvals_i$, $yvals_i$). With `modp1`, the interpolation polynomial is computed in the `modp1` representation, and *xvals* and *yvals* must be lists. All x values must be distinct in the coefficient domain. ◇ `Interp([0,1,2],[0,3,2],x) mod 5;` $\longrightarrow$ $3\,x^2$ ◇ See also: `interp`, `mod`, `modp1`.

intersect set_1 `intersect` set_2
`'intersect'(`set_1`, `set_2`, ...)`
The set intersection operator ◇ This function returns the set of elements that are in all of the given sets. ◇ `{a,b} intersect {b,c};` $\longrightarrow \{b\}$ ◇ See also: `minus`, `symmdiff`, `union`.

invfourier `invfourier(`*expr*`, `*t*`, `*w*`)`
Inverse fourier transform ◇ This function applies the inverse Fourier transformation to *expr* with respect to the variable *t*, returning an expression in *w*. Expressions involving complex exponentials, polynomials, trigonometrics with linear arguments, and a variety of functions and other integral transforms can be transformed. It also recognizes the Dirac-delta function `Dirac` and Heaviside's unit step function `Heaviside`, and derivatives and integrals. This transformation is defined as:

$$F(w) = \frac{1}{2\pi}\int_{-\infty}^{\infty} (expr)\, e^{Itw}\, dt$$

◇ You must `readlib(fourier)` before using this command. ◇
See also: `fourier`.

invfunc `invfunc[`*f*`]`
Inverse function table ◇ This table is used by `solve`, `simplify`, and the `'@@'` operator to replace (explicit or implicit) occurrences of expressions of the form *f*`@@(-1)` with the corresponding inverse function. Entries may be explicitly added to or deleted from this table. You must `readlib(invfunc)` before using this table. ◇ See also: `@@`, `solve`, `simplify`.

invlaplace `invlaplace(`*expr*`, `*s*`, `*t*`)`
Inverse Laplace transform ◇ This function applies the inverse Laplace transformation to *expr* with respect to the variable *s*, returning an expression in *t*. Sums of rational functions of polynomials can be transformed, as well as transforms of sums of terms involving exponentials, polynomials, and Bessel functions with linear arguments. It also recognizes the Dirac-delta function `Dirac` and Heaviside's unit step function `Heaviside`. ◇ You must `readlib(laplace)` before using this command. ◇ `invlaplace(1/(s-1),s,t);` $\longrightarrow e^t$ ◇ See also: `Dirac`, `dsolve`, `laplace`.

invztrans `invztrans(`*expr*`, `*z*`, `*n*`)`
Inverse Z transform ◇ This function finds the inverse Z transformation of an expression *expr* in the variable *z*, yielding an expression in *n*. It can find the inverse Z transform of rational polynomials and a few special functions. ◇ You must `readlib(ztrans)` before using this command. ◇ `invztrans(z/(z-a),z,n);` $\longrightarrow a^n$ ◇ See also: `rsolve`, `ztrans`.

iperfpow `iperfpow(`*n*`)`
`iperfpow(`*n*`, `*name*`)`
Determines whether a positive integer is a perfect power of another integer ◇ If $n = m^k$, this function returns m. It returns *FAIL* if it is not able to establish that *n* is a perfect power. If a name is given as a second argument, it is assigned k. ◇ You must `readlib(iperfpow)` before using this command. ◇ `iperfpow(343);` $\longrightarrow 7$

iquo `iquo(`*m*`, `*n*`)`
`iquo(`*m*`, `*n*`, `*name*`)`
Computes the integer quotient of *m* over *n* ◇ If the third argument is present, it is assigned the remainder. ◇ `iquo(19,4);` $\longrightarrow 4$ ◇ See also: `irem`, `quo`.

iratrecon `iratrecon(`n`, `m`, `N`, `D`, `$name_N$`, `$name_D$`)`
Reconstructs a signed rational number from its image n mod m ◇ *N* and *D* are integers representing upper bounds for the magnitude of the numerator and denominator computed. The solution is unique if $2ND < m$. This function returns *true* if a solution is found and assigns the

numerator to $name_N$ and the denominator to $name_D$; otherwise it returns *false*. ◇ You must `readlib(iratrecon)` before using this command. ◇ `iratrecon(28,41,4,5,'n','d'),n/d;` ⟶ *true*, $\frac{2}{3}$ ◇ See also: `mod`.

irem

`irem(`*m*`, `*n*`)`
`irem(`*m*`, `*n*`, `*name*`)`
Computes the integer remainder of *m* over *n* ◇ If the third argument is present, it is assigned the quotient. ◇ `irem(19,4);` ⟶ 3 ◇ See also: `iquo`, `rem`.

iroot

`iroot(`*m*`, `*n*`)`
Integer nth root ◇ This function computes an integer approximation to the nth root of m. The answer is exact for a perfect nth power; otherwise the absolute error is at most 1/2. ◇ You must `readlib(iroot)` before using this command. ◇ `iroot(126,3);` ⟶ 5 ◇ See also: `iperfpow`, `isqrt`, `proot`, `numtheory[mroot]`.

irreduc

`irreduc(`*poly*`)`
`irreduc(`*poly*`, `*K*`)`
Polynomial irreducibility test ◇ The first form tests whether a multivariate polynomial is irreducible over the field implied by the coefficients of *poly*. The second form test for irreducibility over the algebraic number field determined by *K*. It returns *true* if *poly* is irreducible; *false* otherwise. *K* may be a single `RootOf`, a list or set of `RootOf`s, a single radical, or a list or set of radicals. ◇ `irreduc(x^2+1,I);` ⟶ *false* ◇ See also: `factor`, `Irreduc`, `RootOf`.

Irreduc

`Irreduc(`*poly*`)`
`Irreduc(`*poly*`, `*K*`)`
Inert irreducibility test ◇ When used with `mod` or `modp1`, this tests the irreducibility of the multivariate polynomial *poly* over the given finite field. The second form test for irreducibility over the algebraic extension determined by *K*, a `RootOf`. ◇ `Irreduc(x^2+x+1) mod 2;` ⟶ *true* ◇ See also: `irreduc`, `mod`, `modp1`, `RootOf`, `Factor`.

is

`is(`*name*`, `*property*`)`
`is(`*relation*`)`
Determines whether the name has the property, or whether the relation is satisfied ◇ This function returns *true*, *false*, or *FAIL*. ◇ `assume(x,real): is(x^2>=0);` ⟶ *true* ◇ See also: `about`, `additionally`, `addproperty`, `assume`, `isgiven`.

iscont

`iscont(`*expr*`, `*var*`=`*a*`..`*b*`)`
`iscont(`*expr*`, `*var*`=`*a*`..`*b*`, closed)`
Tests continuity on an interval ◇ This function returns *true* if the expression in the variable *var* is continuous on the open interval (a, b); *false* otherwise. It assumes that any symbols in the expression are real. If this function cannot determine the result, it returns *FAIL*. The option

`closed` tests for continuity on the closed interval $[a, b]$. ◇ You must `readlib(iscont)` before using this command. ◇ `iscont(1/x,x=0..1);` ⟶ *true* ◇ See also: `discont`.

isgiven `isgiven(`*name*`, `*property*`)`
`isgiven(`*relation*`)`
Checks whether the name has been given a property ◇ This function returns *true* or *false*. This is a weaker test than `is`. ◇ `assume(x,real):` `isgiven(x^2>=0);` ⟶ *false* ◇ See also: `about`, `additionally`, `addproperty`, `assume`, `is`.

isolate `isolate(`*eqn*`, `*expr*`)`
`isolate(`*eqn*`, `*expr*`, `*n*`)`
Isolates a subexpression to left side of an equation ◇ If the first argument is not an equation, then the equation *eqn* = 0 is assumed. The optional third argument is the maximum number of transformation steps the function performs. If an expression is given as the first argument, it is assumed to be equal to 0. ◇ You must `readlib(isolate)` before using this command. ◇ `isolate(x*ln(x)=1,ln(x));` ⟶ $\ln(x) = \frac{1}{x}$ ◇ See also: `solve`.

isolve `isolve(`*eqns*`)`
`isolve(`*eqns*`, `*vars*`)`
Solves equations for integer solutions ◇ This function solves the single equation or set of equations *eqns* over the integers. The second argument names variables to be used in parameterizing a family of solutions. It may be a set or a single name. `NULL` is returned if no solutions are found. ◇ `isolve(x^2-y^2=1);` ⟶ $\{y = 0, x = 1\}, \{y = 0, x = -1\}$ ◇ See also: `numtheory[thue]`, `solve`.

isprime `isprime(`*n*`)`
`isprime(`*n*`, `*m*`)`
Probable prime test ◇ This function returns *false* if *n* is shown to be composite and *true* otherwise. There is a very small chance that `isprime` will report a composite number as prime, although it will never report a prime number as composite. In Release 3, one modified Miller–Rabin test is followed by one Lucas test. No composite number that satisfies both tests has ever been found, although such a counterexample may exist. ◇ In Release 2 and earlier versions, *m* iterations of the Miller–Rabin test are used. *m* may be given as a second argument. Its default value is 5 and its maximum value is 25. The chance of an incorrect result decreases with larger *m*. ◇ `isprime(211);` ⟶ *true* ◇ See also: `ithprime`, `nextprime`, `prevprime`, `numtheory[safeprime]`, `type/primeint`.

isqrfree `isqrfree(`*n*`)`
Computes the square-free integer factorization of *n* ◇ This function returns a list of square-free factors and exponents. ◇ You must `readlib(isqrfree)` before using this command. ◇ `isqrfree(150);`

$\longrightarrow$ [1, [[6, 1], [5, 2]]] ◇ See also: `ifactor`, `ifactors`, `sqrfree`, `Sqrfree`.

isqrt `isqrt(`*m*`)`
Integer square root ◇ This function computes an integer approximation to the square root of *m*. The answer is exact for a perfect square; otherwise the absolute error is at most 1/2. ◇ `isqrt(126);` $\longrightarrow$ 11 ◇ See also: `iroot`, `issqr`, `numtheory[msqrt]`, `psqrt`, `sqrt`.

issqr `issqr(`*n*`)`
Tests whether an integer is a perfect square ◇ This function returns *true* or *false*. ◇ You must `readlib(issqr)` before using this command. ◇ `issqr(15625);` $\longrightarrow$ *true* ◇ See also: `isqrt`.

ithprime `ithprime(`*i*`)`
Determines the *i*th prime number ◇ The first prime number is 2. This function uses `isprime` to test for large primes. ◇ `ithprime(100);` $\longrightarrow$ 541 ◇ See also: `isprime`, `nextprime`, `prevprime`.

laplace `laplace(`*expr*`, `*t*`, `*s*`)`
Laplace transform ◇ This function applies the Laplace transformation to *expr* with respect to the variable *t*, returning an expression in *s*. Expressions involving exponentials, polynomials, trigonometrics with linear arguments, and Bessel functions with linear arguments can be transformed. It also recognizes the Dirac-delta function `Dirac` and Heaviside's unit step function `Heaviside`, and derivatives and integrals. ◇ `laplace(exp(t),t,s);` $\longrightarrow \frac{1}{s-1}$ ◇ See also: `BesselI`, `BesselJ`, `Dirac`, `dsolve`, `invlaplace`.

lasterror Global variable containing the text of the most recent error message ◇ See also: `ERROR`, `traperror`.

latex `latex(`*expr*`)`
`latex(`*expr*`, `*filename*`)`
Produces LaTeX output for printing the expression ◇ This function can format integrals, limits, sums, products, and matrices. If a name is given as the second parameter, the output is written to this file. ◇ See also: `C`, `eqn`, `fortran`.

lattice `lattice(`*list*`)`
`lattice(`*list*`, integer)`
Finds a reduced basis of a lattice ◇ *list* is a list of lists or a list of vectors specifying a basis for a lattice. This function uses the LLL algorithm to compute the reduced basis. The option `integer` specifies that only integer arithmetic should be used (instead of rational arithmetic). ◇ You must `readlib(lattice)` before using this command. ◇ `lattice([[3,6,4],[6,9,7]]);` $\longrightarrow$ [[0, −3, −1], [3, 0, 2]] ◇ See also: `minpoly`.

lcm `lcm(`*poly₁*`, `*poly₂*`, ...)`
Computes the least common multiple of polynomials over the rationals ◇ An arbitrary number of multivariate polynomials may be given. ◇ `lcm(x^2+1,x+1);` ⟶ $x^3 + x^2 + x + 1$ ◇ See also: `gcd`, `ilcm`, `Lcm`.

Lcm `Lcm(`*poly₁*`, `*poly₂*`, ...)`
Inert least common multiple of polynomials ◇ When used with `mod` or `modp1`, this function computes the least common multiple of the polynomials modulo a positive integer. ◇ `Lcm(x^2+1,x+1) mod 2;` ⟶ $x^2 + 1$ ◇ See also: `Gcd`, `lcm`.

lcoeff `lcoeff(`*poly*`)`
`lcoeff(`*poly*`, `*vars*`)`
`lcoeff(`*poly*`, `*vars*`, `*name*`)`
Leading coefficient of a multivariate polynomial ◇ *vars* may be a single indeterminate, or a list or set of indeterminates. If *vars* is not specified, all the indeterminates of *poly* are used. The polynomial must be `collected` with respect to the appropriate variables. If a name is given as a third argument, it is assigned the leading term of *poly*. ◇ `lcoeff(2*x^3-4);` ⟶ 2 ◇ See also: `coeff`, `coeffs`, `collect`, `degree`, `indets`, `Lcoeff`, `tcoeff`.

Lcoeff `Lcoeff(`*poly*`)`
Inert leading coefficient ◇ When used with `modp1`, this computes the leading coefficient of a univariate polynomial over a given domain. ◇ See also: `Coeff`, `lcoeff`, `Tcoeff`.

ldegree `ldegree(`*poly*`, `*vars*`)`
Low degree of a polynomial ◇ The first argument may have negative integer exponents. *vars* is may be a single indeterminate, or a list or set of such. For a set, the total degree is used; for a list, the vector degree. ◇ `ldegree(x^3+x^2,x);` ⟶ 2 ◇ See also: `degree`, `tcoeff`.

LegendreE
`LegendreE(`*x*`, `*k*`)`
Legendre's canonical incomplete elliptic integral of the second kind ◇ `x` is the variable limit of integration and `k` is the modulus.

LegendreEc
`LegendreEc(`*k*`)`
Legendre's complete elliptic integral of the second kind ◇ `k` is the modulus.

LegendreEc1
`LegendreEc1(`*k*`)`
Associated complete elliptic integral of the second kind using the complementary modulus ◇ `k` is the modulus.

LegendreF

`LegendreF(`*x*`,` *k*`)`
Legendre's canonical incomplete elliptic integral of the first kind ◇ `x` is the variable limit of integration and `k` is the modulus.

LegendreKc

`LegendreKc(`*k*`)`
Legendre's complete elliptic integral of the first kind ◇ `k` is the modulus.

LegendreKc1

`LegendreKc1(`*k*`)`
Associated complete elliptic integral of the first kind using the complementary modulus ◇ `k` is the modulus.

LegendrePi

`LegendrePi(`*x*`,` *alphasq*`,` *k*`)`
Legendre's canonical incomplete elliptic integral of the third kind ◇ `x` is the variable limit of integration, `alphasq` is the characteristic, and `k` is the modulus.

LegendrePic

`LegendrePic(`*alphasq*`,` *k*`)`
Legendre's complete elliptic integral of the third kind ◇ `alphasq` is the characteristic and `k` is the modulus.

LegendrePic1

`LegendrePic1(`*alphasq*`,` *k*`)`
Associated complete elliptic integral of the third kind using the complementary modulus ◇ `alphasq` is the characteristic and `k` is the modulus.

length `length(`*expr*`)`
Length of an object ◇ The length of an integer is number of decimal digits. The length of an string is number of characters in the string. For other objects, the length is the sum of the lengths of the operands plus the number of words used to represent the expression. ◇ See also: `nops`.

lexorder `lexorder(`$name_1$`,` $name_2$`)`
Lexicographical ordering test ◇ This function returns *true* if $name_1$ occurs before $name_2$ in lexicographical order or if the two names are equal; otherwise it returns *false*. ◇ `lexorder(ab,ba);` ⟶ *true* ◇ See also: `sort`.

lhs `lhs(`*expr*`)`
Returns the left-hand side of an expression ◇ The expression may be an equation, an inequality, or a range. ◇ `lhs(3..4);` ⟶ 3 ◇ See also: `op, rhs`.

Li `Li(`x`)`
The logarithmic integral ◇ This function is defined as $\text{Li}(x) = \int_0^x \frac{1}{\ln t}\,dt$ where we take the Cauchy Principal Value of the integral. `Li(`x`)` provides an approximation to the number of primes less than x. ◇ You must `readlib(Li)` before using this command. ◇ `Li(1000.);` ⟶ 177.6096580

libname Global variable ◇ A sequence of directories containing Maple libraries that are to be searched each time a `readlib` or `with` command is executed.

liesymm The Lie symmetries package ◇ This is an implementation of the Harrison-Estabrook procedure (as outlined in the "Journal of Mathematical Physics", vol. 12, American Institute of Physics, New York, 1971, pp. 653–665). It obtains the determining equations leading to the similarity solutions of a system of partial differential equations using a number of refinements and extensions as developed by J. Carminati. ◇ In Release 2 and earlier versions, the functions in this package must first be loaded using `with` before they can be invoked. ◇ See also: `liesymm[`*function*`]`, `with`.

liesymm[&^]

expr₁ `&^` *expr₂*
`&^(`*expr₁*`,` *expr₂*`, ...)`
Wedge product ◇ This function computes the wedge product of differential forms relative to the coordinates defined by `setup`. ◇ `setup(x,y): a*d(x) &^ (b*d(y));` ⟶ $a\,b\,(\mathrm{d}(x)\,\&\hat{}\ \mathrm{d}(y))$ ◇ See also: `liesymm[getcoeff]`, `liesymm[getform]`, `liesymm[setup]`, `liesymm[wcollect]`, `liesymm[wdegree]`, `liesymm[wedgeset]`, `liesymm[wsubs]`.

liesymm[&mod]

form `&mod` *formlist*
Reduces a form modulo an exterior ideal ◇ This function reduces the form *form* modulo the exterior ideal generated by the list or set of differential forms *formlist*. Use a closed *formlist* to specify a differential ideal. ◇ See also: `liesymm[&^]`, `liesymm[close]`, `liesymm[d]`, `liesymm[hasclosure]`.

liesymm[annul]

`annul(`*forms*`)`
`annul(`*forms*`,` *vlist*`)`
Annuls a list or set of differential forms ◇ Given a list or set of differential forms and a list of coordinates to be treated as independent, this function sections the forms and sets them equal to 0. The result is a set of partial differential equations corresponding to the differential forms. If *vlist* is absent, the annulling process is done with respect to the coordinates as indicated by `indepvars`. ◇ See also: `liesymm[determine]`, `liesymm[indepvars]`, `liesymm[makeforms]`.

liesymm[autosimp]

`autosimp(`*eqnset*`)`

Simplifies a set of differential forms ◇ The system of pde's characterizing the symmetries of a system of pde's often includes a number of equations that are easy to integrate. This function carries out many of these simple integrations. The result is returned in the form *remaining* &where *rels*, where *remaining* is the set of remaining equations and *rels* is a set of algebraic equations indicating the relationship between the new functions in the set of simplified equations and the original functions. ◇ See also: `liesymm[determine]`, `liesymm[reduce]`.

liesymm[close]

`close(`*forms*`)`

Computes the closure of the list or set of differential forms ◇ The closure with respect to the exterior derivative `d` is computed by adding additional forms to the original set. ◇ See also: `liesymm[&mod]`, `liesymm[d]`, `liesymm[hasclosure]`, `liesymm[Lrank]`.

liesymm[d]

`d(`*form*`)`

The exterior derivative ◇ This function computes the exterior derivative of the differential form `form` with respect to the coordinates defined by `setup`. ◇ `setup(x,y): d(a*x+b*y);` ⟶ $a\,\mathrm{d}(x) + b\,\mathrm{d}(y)$ ◇ See also: `liesymm[&^]`, `liesymm[hook]`, `liesymm[Lie]`, `liesymm[setup]`, `liesymm[wedgeset]`.

liesymm[depvars]

`depvars()`

`depvars(`x_1`, ..., `x_n`)`

Dependent variables of the differential forms ◇ With no argument, the current setting for dependent variables is reported. With one or more arguments, these become the dependent variables. ◇ See also: `liesymm[extvars]`, `liesymm[indepvars]`, `liesymm[makeforms]`.

liesymm[determine]

`determine(`*forms*`, `*name*`)`

`determine(`*deqns*`, `*name*`, `*fcns*`, `*rootname*`)`

`determine(`*deqns*`, `*name*`, `*fcns*`, `*extdlist*`)`

Finds the determining equations for the isovectors of a pde ◇ Given a list or set of differential forms and a name *name*, this function constructs the determining equations for the isovectors that are generators of the invariance group (isogroup) of the differential equations. The forms must be closed. If a list or set differential equations are given then the required differential forms are constructed automatically using `makeforms`. Additional arguments identify the dependent and independent variables (e.g., `u(t,x)`, or a list of such function names) and name the extended variables. ◇ See also: `liesymm[close]`, `liesymm[hasclosure]`, `liesymm[makeforms]`.

A B C D E F G H I J K L M N O P Q R S T U V W X Y Z

liesymm[dvalue]

`dvalue(`*expr*`)`

Forces evaluation of derivatives in an expression involving `Diff` ◇ This function forces evaluation of the expression to take place as if `diff` had been used and then rewrites the final result in terms of the inert `Diff`. ◇ See also: `Diff`, `diff`, `liesymm[vfix]`.

liesymm[Eta]

`Eta(`*f*`,` *x*`)`
`Eta[1](`*f*`,` *x*`)`
`Eta[2](`*f*`,` *x*`,` *y*`)`
`Eta[3](`*f*`,` *x*`,` *y*`,` *z*`)`

Computes the coefficients of the generator of a finite point transformation ◇ *f* is a named partial in the sense of `depvars`, and *x*, *y*, and *z* are independent variables in the sense of `indepvars`. This is a special differential operator defined in terms of `TD`. The result is an inert expression reported in terms of the `Diff` procedure. ◇ See also: `liesymm[depvars]`, `liesymm[indepvars]`, `liesymm[TD]`.

liesymm[extgen]

`extgen(`*expr*`)`
`extgen(`*expr*`,` *name*`)`
`extgen[`*i*`](`*expr*`)`
`extgen[`*i*`](`*expr*`,` *name*`)`

The extended differential operator ◇ The expression is regarded as a function of the dependent variables, the independent variables, and named partial derivatives of order less than *i*. The default value of *i* is 1. *name* gives the root name of the covariant vector. The default name is `V`. The extended derivative is constructed using `Diff`. ◇ See also: `liesymm[depvars]`, `liesymm[indepvars]`, `liesymm[TD]`.

liesymm[extvars]

`extvars()`
`extvars(`*f*`,` *x*`)`
`extvars(`*f*`,` *x*`,` *y*`)`

Extended variables of the differential forms ◇ During the investigation of a system of pde's, names are dynamically created for the partials involving some of the dependent and independent variables. This function reports on the names used for those partials. The result is the name used for the partial of the given dependent variable *f* with respect to the given independent variables. With no argument, a table indicating all the names is returned. ◇ See also: `liesymm[depvars]`, `liesymm[indepvars]`, `liesymm[makeforms]`, `liesymm[translate]`.

liesymm[getcoeff]

`getcoeff(`*expr*`)`

Extracts the coefficient of a basis wedge product ◇ *expr* is an expression involving a single wedge product. ◇ `setup(x,y):`

`getcoeff(d(y)+a*d(y));` ⟶ $1+a$ ◇ See also: `liesymm[&^]`, `liesymm[d]`, `liesymm[getform]`, `liesymm[setup]`.

liesymm[getform]

`getform(`*expr*`)`

Extracts the basis element of a single wedge product ◇ *expr* is an expression involving a single wedge product. ◇ `setup(x,y):` `getform(d(y)+a*d(y));` ⟶ $\mathrm{d}(y)$ ◇ See also: `liesymm[&^]`, `liesymm[d]`, `liesymm[getcoeff]`, `liesymm[setup]`.

liesymm[hasclosure]

`hasclosure(`*forms*`)`

Tests for closure of the list or set of differential forms ◇ This function tests *forms* for closure with respect to the exterior derivative `d`. ◇ See also: `liesymm[&mod]`, `liesymm[close]`, `liesymm[d]`.

liesymm[hook]

`hook(`*expr*`,` *v*`)`
`hook(`*expr*`,` *list*`)`
`hook(`*expr*`,` *name*`)`

Inner product (hook) ◇ *expr* is an expression involving differential forms. This function computes the inner product of the expression with respect to a vector or a list or a name of a vector. ◇ See also: `liesymm[&^]`, `liesymm[d]`, `liesymm[Lie]`, `liesymm[setup]`, `liesymm[wedgeset]`.

liesymm[indepvars]

`indepvars()`
`indepvars(`x_1`,` x_2`,` x_3`)`

Independent variables of the differential forms ◇ With no argument, the current setting for independent variables is reported. With one or more arguments, these become the independent variables. ◇ See also: `liesymm[depvars]`, `liesymm[extvars]`, `liesymm[makeforms]`.

liesymm[Lie]

`Lie(`*form*`,` *isovector*`)`
`Lie(`*form*`,` *name*`)`

The Lie derivative ◇ The Lie derivative of the differential form *form* is constructed with respect to the coordinates of *isovector*, or with respect to symbolic coordinates based on the given *name*. ◇ See also: `liesymm[&^]`, `liesymm[d]`, `liesymm[hook]`, `liesymm[setup]`.

liesymm[Lrank]

`Lrank(`*forms*`)`

The Lie rank of a list or set of forms ◇ This function removes forms that are redundant with respect to the generation of the determining equations. ◇ See also: `liesymm[close]`, `liesymm[Lie]`, `liesymm[makeforms]`.

liesymm[makeforms]

`makeforms(`*eqns*`,` *fcns*`,` *rootname*`)`
`makeforms(`*eqns*`,` *fcns*`,` *extdlist*`)`
Constructs a set of differential forms from a pde ◇ Given a partial differential equation or several in a list or a set, this function constructs a set of differential forms, which after closure is equivalent to the original system of equations in the sense of Cartan. The second argument is a single function name specifying the dependent and independent variables (e.g., `h(t,x)`), or a list of such function names. The third argument is used to construct the names of the extended variables. It is either a base name or a list of names. ◇ `makeforms(Diff(h(t,x),x,x) = Diff(h(t,x),t,t),h(t,x),w);` ⟶ $[\mathrm{d}(h) - w1\,\mathrm{d}(t) - w2\,\mathrm{d}(x), -\&\hat{}(\mathrm{d}(t), \mathrm{d}(w2))+\&\hat{}(\mathrm{d}(w1), \mathrm{d}(x))]$ ◇ See also: `liesymm[annul]`, `liesymm[close]`, `liesymm[depvars]`, `liesymm[determine]`, `liesymm[indepvars]`, `liesymm[setup]`.

liesymm[mixpar]

`mixpar(`*expr*`)`
Recognize mixed partials as equivalent ◇ Given an expression involving `Diff` or `diff`, this function reorders the partials to a standard ordering so that mixed partials can be recognized as equivalent. ◇ `mixpar(Diff(Diff(f(x,y),y),x));` ⟶ $\frac{\partial^2}{\partial y\,\partial x}\mathrm{f}(x, y)$ ◇ See also: `Diff`, `diff`, `liesymm[d]`, `liesymm[Lie]`.

liesymm[prolong]

`prolong(`*isovector*`)`
`prolong[1](`*isovector*`)`
`prolong[2](`*isovector*`)`
`prolong[3](`*isovector*`)`
Makes substitutions for components of the extended isovector in terms of partials of the original isovector ◇ *isovector* may be a list or a name. This function constructs equations that can be used to rewrite the components of the isovector corresponding to the extended variables in terms of partials of isovector components corresponding to the independent and dependent variables. This routine is used by `reduce` to help simplify the determining equations of the contact symmetries. ◇ See also: `liesymm[reduce]`.

liesymm[reduce]

`reduce(`*eqnset*`)`
Reduces a set of differential forms ◇ *eqnset* is a set of determining equations for the symmetries of a given system of pde's. This function forces the isovector components to be independent of any extended variables that have been introduced during the construction of the determining equations. This allows simplification of the system of determining equations produced by `determine`. ◇ See also: `liesymm[autosimp]`, `liesymm[determine]`, `liesymm[prolong]`.

liesymm[setup]

`setup()`
`setup(`*names*`)`
`setup(`*namelist*`)`
`setup(`*nameset*`)`

Defines the coordinates ◇ This function returns a list of the coordinates to be used in constructing wedge products and defining the actions of functions such as `d`. If no argument is given, all existing coordinates are removed. It clears the remember tables of many functions, including `d`, `&^`, `Lie`, `hook`, `makeforms`, and `wcollect` This function should be used between problems to keep the underlying coordinate list to a minimum size. ◇ See also: `liesymm[&^]`, `liesymm[d]`, `liesymm[hook]`, `liesymm[Lie]`, `liesymm[makeforms]`, `liesymm[wcollect]`.

liesymm[TD]

`TD(`*expr*`, `*x*`)`
`TD[`*i*`](`*expr*`, `*x*`)`

Constructs the total derivative of *expr* with respect to *x* ◇ *x* must be one of the independent variables. The expression is regarded as a function of the dependent variables, the independent variables, and named partial derivatives of order less than *i*. The default value of *i* is 1. The total derivative is constructed using `Diff`. ◇ See also: `liesymm[depvars]`, `liesymm[Eta]`, `liesymm[extgen]`, `liesymm[indepvars]`.

liesymm[translate]

`translate()`
`translate(`*name*`)`

Translate names of partial derivatives ◇ This function indicates which partial derivative is indicated by the name *name*, or returns the entire name translation table if no name is given. ◇ See also: `liesymm[depvars]`, `liesymm[extvars]`, `liesymm[indepvars]`.

liesymm[vfix]

`vfix(`*expr*`, `*varlist*`, `*name*`)`

Changes variable dependencies in unevaluated derivatives ◇ This function changes the variable dependencies for functions appearing in an expression involving `Diff`. The function name *name* is declared to depend only on the variables in *varlist*. ◇ See also: `Diff`, `liesymm[dvalue]`.

liesymm[wcollect]

`wcollect(`*expr*`)`

Regroups the terms as a sum of products ◇ This function returns the expression rewritten in a sum of products form, with each distinct wedge product occurring exactly once. ◇ See also: `liesymm[&^]`, `liesymm[d]`, `liesymm[setup]`.

liesymm[wdegree]

`wdegree(`*form*`)`
Computes the wedge degree of the differential form ◇ The wedge degree is defined relative to a given coordinate system, so using `setup` may change its value. ◇ `setup(x,y): wdegree(d(x));` ⟶ 1 ◇ See also: `liesymm[&^]`, `liesymm[d]`, `liesymm[setup]`.

liesymm[wedgeset]

`wedgeset(`*n*`)`
Forms of degree *n* ◇ All forms of degree *n* over the underlying coordinates are returned in an expression sequence. ◇ `setup(x,y,z): wedgeset(3);` ⟶ &^(d(y), d(z), d(x)) ◇ See also: `liesymm[&^]`, `liesymm[d]`, `liesymm[setup]`.

liesymm[wsubs]

`wsubs(`*eqn*`, `*expr*`)`
`wsubs(`*eqns*`, `*expr*`)`
Replaces part of a wedge product ◇ *eqn* has the form *wedgeprod* = *value*. It specifies a substitution for occurrences of *wedgeprod* in *expr*. Two or more substitutions may be given in a list or a set *eqns*. They are completed in the order specified, even if they are given in a set. ◇ See also: `liesymm[&^]`, `liesymm[d]`.

limit

`limit(`*expr*`, `*var*`=`*a*`);`
`limit(`*expr*`, `*var*`=`*a*`, `*dir*`)`
`limit(`*expr*`, `*set*`)`
`limit(`*expr*`, `*set*`, `*dir*`)`
Computes the limiting value of *expr* as *var* approaches *a* ◇ The first two forms compute the limiting value of *expr* as *var* approaches *a*. Limits at `infinity` and `-infinity` are supported. The optional argument *dir* may be `real` for the bidirectional real limit (the default, except for limits at infinity), `complex` for an omnidirectional complex limit (in this case, `infinity` stands for complex infinity), `left` for approaching from the left (the default for *a*=`infinity`), or `right` for approaching from the right (the default for *a*=`-infinity`). If the limit is not defined, a numeric range is returned if the value of the expression is known to lie in that range when restricted to a neighborhood of the limit point. Otherwise, `undefined` is returned. If a closed form cannot be found, the function call itself is returned. The last two forms compute a limit in higher-dimensional space. *set* is a set of equations of the form *var*=*a*. The optional direction *dir* is applied to each variable. ◇ `limit(1/x,x=0);` ⟶ *undefined* ◇ `limit(1/x,x=0,left);` ⟶ $-\infty$ ◇ `limit(1/x,x=0,complex);` ⟶ ∞ ◇ `limit(sin(1/x),x=0);` ⟶ $-1\,..\,1$ ◇ `limit((x^2-y^2)/(x-y),{x=1,y=1});` ⟶ 2 ◇ See also: `Limit`.

Limit

`Limit(`*expr*`,` *var*`=`*a*`)`
`Limit(`*expr*`,` *var*`=`*a*`,` *dir*`)`
`Limit(`*expr*`,` *set*`)`
`Limit(`*expr*`,` *set*`,` *dir*`)`
Inert limit function ◇ See also: `limit`, `value`.

linalg

Linear algebra package ◇ For the functions in this package, *A* and *B* denote matrices, *u* and *v* designate vectors, *row* stands for a row index of a matrix, and *col* represents a column index of a matrix. ◇ See also: `&*`, `evalm`, `linalg[`*function*`]`, `with`.

linalg[add]

`add(`*A*`,` *B*`)`
`add(`*A*`,` *B*`,` $expr_1$`,` $expr_2$`)`
`add(`*u*`,` *v*`)`
`add(`*u*`,` *v*`,` $expr_1$`,` $expr_2$`)`
Matrix or vector addition ◇ If the scalar expressions are present, this function computes the weighted sum $expr_1 A + expr_2 B$ (or $expr_1 u + expr_2 v$). ◇ See also: `evalm`.

linalg[addcol]

`addcol(`*A*`,` col_1`,` col_2`,` *expr*`)`
Forms linear combinations of matrix columns ◇ This function returns a copy of the matrix *A* where *expr* times col_1 is added to col_2. ◇ See also: `linalg[addrow]`.

linalg[addrow]

`addrow(`*A*`,` row_1`,` row_2`,` *expr*`)`
Forms linear combinations of matrix rows ◇ This function returns a copy of the matrix *A* where *expr* times row_1 is added to row_2. ◇ See also: `linalg[addcol]`.

linalg[adjoint]

`adjoint(`*A*`)`
Computes the adjoint of the matrix *A* ◇ The product of *A* and its adjoint is the product of `det(`*A*`)` and the identity matrix. ◇ See also: `linalg[det]`, `linalg[minor]`.

linalg[angle]

`angle(`*u*`,` *v*`)`
Computes the angle between vectors ◇ This function computes the angle θ as $\cos\theta = \left(\frac{u \cdot v}{|u||v|}\right)^{1/2}$ using the Euclidean inner product and norm. ◇ `angle(vector([1,0]), vector([1,1]));` $\longrightarrow \frac{1}{4}\pi$ ◇ See also: `linalg[norm]`, `linalg[dotprod]`, `linalg[innerprod]`.

linalg[augment]

`augment(`*A*`, `*B*`, ...)`
`augment(`*u*`, `*v*`, ...)`
Joins two or more matrices or vectors together horizontally ◇ Matrices must have the same number of rows; vectors must have the same length. A vector is interpreted as a column vector. ◇ See also: `linalg[concat]`, `linalg[stack]`.

linalg[backsub]

`backsub(`*A*`)`
Back substitution on a matrix ◇ If *A* is the result of applying forward Gaussian elimination to the augmented matrix of a system of linear equations, this function completes the solution by back substitution. If a solution exists, it is returned as a vector. If the solution is not unique, it is parameterized in terms of global symbols. ◇ See also: `linalg[gausselim]`, `linalg[gaussjord]`, `linalg[linsolve]`.

linalg[band]

`band(`*list*`, `*n*`)`
`band(`*v*`, `*n*`)`
Creates a band matrix ◇ This function creates an $n \times n$ band matrix using the elements of the given list or vector, starting with the subdiagonals. The list or vector must have an odd number of elements not exceeding *n*. ◇ `band([1,2,3],5);` ◇ See also: `linalg[diag]`.

linalg[basis]

`basis(`*v*`)`
`basis(`*vlist*`)`
`basis(`*vset*`)`
Finds a basis for a vector space ◇ This function returns a list or set of vectors among the given vectors forming a basis for the vector space spanned by the given vectors. In the case of a single vector, the same vector is returned. ◇ See also: `linalg[intbasis]`, `linalg[sumbasis]`.

linalg[bezout]

`bezout(`$poly_1$`, `$poly_2$`, `*var*`)`
Creates the Bezout matrix of two polynomials ◇ $poly_1$ and $poly_2$ are polynomials in the variable *var*. The determinant of this matrix is equal to the resultant of the polynomials. ◇ See also: `linalg[det]`, `linalg[sylvester]`, `resultant`.

linalg[BlockDiagonal]

`diag(`B_1`, `B_2`, ..., `B_n`)`
A synonym for `linalg[diag]` ◇ See `linalg[diag]`.

linalg[blockmatrix]

`blockmatrix(`m`, `n`, `B_{11}`, `B_{12}`, ..., `B_{mn}`)`
Creates a block matrix ◇ The $m \times n$ block matrix is filled in row major order.

linalg[charmat]

`charmat(`A`, `λ`)`
Constructs the characteristic matrix ◇ This function returns the characteristic matrix $\lambda I - A$, where I is the identity matrix. λ may be a name or an algebraic expression. ◇ See also: `linalg[charpoly]`, `linalg[eigenvals]`, `linalg[nullspace]`.

linalg[charpoly]

`charpoly(`A`, `λ`)`
Computes the characteristic polynomial of a matrix ◇ The characteristic polynomial is the determinant of the characteristic matrix. λ may be a name or an algebraic expression. ◇
`charpoly(matrix([[2,3],[3,4]]),x);` $\longrightarrow -1 - 6x + x^2$ ◇
See also: `linalg[charmat]`, `linalg[eigenvals]`, `linalg[minpoly]`.

linalg[col]

`col(`A`, `*col*`)`
`col(`A`, `col_1`..`col_2`)`
Extracts columns of a matrix ◇ The first form extracts the *col*th column of the matrix and returns the result as a vector. The second form extracts a range of columns and returns a sequence of vectors. ◇ See also: `linalg[addcol]`, `linalg[coldim]`, `linalg[delcols]`, `linalg[mulcol]`, `linalg[row]`, `linalg[swapcol]`.

linalg[coldim]

`coldim(`A`)`
Returns the number of columns in the matrix A ◇ See also: `linalg[colspace]`, `linalg[rowdim]`, `linalg[vectdim]`.

linalg[colspace]

`colspace(`A`)`
`colspace(`A`, `*name*`)`
Computes a basis for the column space ◇ This function returns a set of vectors that form a basis for the vector space spanned by the columns of the matrix A. The vectors are returned in a canonical form with leading entries 1. If a name is given as a second parameter, it is assigned the rank of A. ◇ See also: `linalg[coldim]`, `linalg[colspan]`, `linalg[nullspace]`, `linalg[rank]`, `linalg[rowspace]`.

linalg[colspan]

`colspan(`A`)`
`colspan(`A`, `*name*`)`
Computes spanning vectors for the column space ◇ This function computes a spanning set of vectors for the column space of A, a matrix of

multivariate polynomials over the rationals. Unlike `colspace`, this function uses "fraction-free" Gaussian elimination to avoid introduction of rational expressions during the elimination. If a name is given as a second parameter, it is assigned the rank of A. ⋄ See also: `linalg[colspace]`, `linalg[ffgausselim]`, `linalg[rowspan]`.

linalg[companion]

`companion(`*poly*`, `*var*`)`

Returns the companion matrix associated with a polynomial ⋄ *poly* is a polynomial in the variable *var*, in expanded form.

linalg[concat]

`concat(`*A*`, `*B*`, ...)`
`concat(`*u*`, `*v*`, ...)`

A synonym for linalg[augment] ⋄ Joins two or more matrices or vectors together horizontally. ⋄ See `linalg[augment]`.

linalg[cond]

`cond(`*A*`)`
`cond(`*A*`, 1)`
`cond(`*A*`, 2)`
`cond(`*A*`, frobenius)`
`cond(`*A*`, infinity)`

Condition number of a matrix with respect to a norm ⋄ This function computes the norm of A times the norm of A^{-1}. The second parameter names the norm: see `linalg[norm]` for descriptions. The default norm is `infinity`. ⋄ `cond(matrix([[2,3],[3,4]]),1);` ⟶ 49 ⋄ See also: `linalg[norm]`.

linalg[copyinto]

`copyinto(`*A*`, `*B*`, `*m*`, `*n*`)`

Moves entries from one matrix into another ⋄ This function copies the entries of the matrix A into the matrix B beginning at index position $[m, n]$. Thus $B[m, n]$ is assigned $A[1, 1]$. ⋄ See also: `linalg[extend]`, `linalg[submatrix]`.

linalg[crossprod]

`crossprod(`*u*`, `*v*`)`
`crossprod(`*list*`, `*list*`)`

Computes the vector cross product of the two given vectors or lists ⋄ The lists or vectors must have exactly three elements. ⋄ `crossprod([1,0,0],[0,1,0]);` ⟶ [0 0 1] ⋄ See also: `linalg[dotprod]`, `linalg[innerprod]`.

linalg[curl]

curl(*u*, *v*)
curl(*u*, *v*, coords=*coordtype*)
curl(*u*, *v*, *list*)
Computes the curl of the vector-valued expression *u* with respect to the variables in *v* ◇ The first argument is a list or vector of three expressions in three variables named in the second argument, which may also be a list or a vector. ◇ In Release 3, an optional third argument specifies the coordinate system. *coordtype* may be **cartesian** (the default), **cylindrical**, or **spherical**. A list of three elements giving the scale factors requests computation in the corresponding orthogonally curvilinear coordinate system. ◇ **curl([-z^2/2,x*z,0],[x,y,z]);** ⟶ $[-x \;\; -z \;\; z]$ ◇ See also: **linalg[grad]**, **linalg[diverge]**, linalg[vecpotent].

linalg[definite]

definite(*A*, negative_def)
definite(*A*, negative_semidef)
definite(*A*, positive_def)
definite(*A*, positive_semidef)
Tests whether a square symmetric matrix is positive or negative (semi-)definite ◇ For numerical matrices, this function returns *true* or *false*. For matrices with symbols, this function returns a conjunction of Boolean expressions that must be satisfied for the property to hold.

linalg[delcols]

delcols(*A*, col_1..col_2)
Deletes columns of a matrix ◇ This function returns the submatrix of *A* obtained by deleting the indicated range of columns. ◇ See also: linalg[col], linalg[delrows], linalg[submatrix].

linalg[delrows]

delrows(*A*, row_1..row_2)
Deletes rows of a matrix ◇ This function returns the submatrix of *A* obtained by deleting the indicated range of rows. ◇ See also: linalg[delcols], linalg[row], linalg[submatrix].

linalg[det]

det(*A*)
det(*A*, sparse)
Computes the determinant of a square matrix ◇ The **sparse** option indicates that the determinant should be computed by minor expansion. ◇ det(matrix([[2,3],[3,4]])); ⟶ −1 ◇ See also: linalg[permanent].

linalg[diag]

`diag(`B_1`, `B_2`, ..., `B_n`)`
Creates a block diagonal matrix ◇ This function returns a matrix on whose diagonal are the matrix blocks B_1, B_2, ..., B_n. ◇ See also: `linalg[BlockDiagonal]`, `linalg[JordanBlock]`.

linalg[diverge]

`diverge(`*u*`, `*v*`)`
`diverge(`*u*`, `*v*`, coords=`*coordtype*`)`
`diverge(`*u*`, `*v*`, `*list*`)`
Computes the divergence of the vector-valued expression *u* with respect to the variables in *v* ◇ The first argument is a list or vector of expressions in the variables named in the second argument, which may also be a list or a vector. *u* and *v* must have the same length. ◇ In Release 3, an optional third argument specifies the coordinate system. *coordtype* may be `cartesian` (the default), `cylindrical`, or `spherical`. A list of three elements giving the scale factors requests computation in the corresponding orthogonally curvilinear coordinate system. ◇ `diverge([x*y,y*z,x*z], [x,y,z]);` $\longrightarrow y+z+x$ ◇ See also: `linalg[curl]`, `linalg[grad]`, `linalg[vecpotent]`.

linalg[dotprod]

`dotprod(`*u*`, `*v*`)`
`dotprod(`*u*`, `*v*`, orthogonal)`
Computes the dot product of two vectors or lists ◇ The `orthogonal` option computes the dot product without taking the complex conjugate of the elements in *v*. ◇ `dotprod([1,2,1],[2,-1,2]);` $\longrightarrow 2$ ◇ See also: `linalg[crossprod]`, `linalg[innerprod]`.

linalg[eigenvals]

`eigenvals(`*A*`)`
`eigenvals(`*A*`, implicit)`
`eigenvals(`*A*`, radical)`
`eigenvals(`*A*`, `*B*`)`
Computes the eigenvalues of a matrix ◇ This function returns a sequence of the eigenvalues of *A*. If *A* contains floating-point numbers, a numerical method is used. Otherwise, with the option `implicit`, the eigenvalues are expressed in `RootOf` notation. With the option `radical` (the default), Maple attempts to express the eigenvalues in terms of exact radicals. The last template finds the roots of the polynomial $\det(\lambda B - A)$. ◇ `eigenvals(matrix([[2,3],[3,4]]));` $\longrightarrow 3+\sqrt{10}, 3-\sqrt{10}$ ◇ See also: `Eigenvals`, `linalg[charpoly]`, `linalg[eigenvects]`, `RootOf`.

linalg[eigenvects]

`eigenvects(A)`
`eigenvects(A, implicit)`
`eigenvects(A, radical)`

Finds the eigenvectors of a matrix ◇ *A* may be a matrix of rationals, polynomials, algebraic numbers, or algebraic functions. The result is a sequence of lists. Each list contains an eigenvalue, its multiplicity, and a set containing a basis for the eigenspace corresponding to the eigenvalue. If *A* contains floating-point numbers, a numerical method is used. Otherwise, with the option `implicit` (the default), the eigenvalues are expressed in `RootOf` notation. With the option `radical`, Maple attempts to express the eigenvalues in terms of exact radicals. ◇ See also: `Eigenvals`, `linalg[eigenvals]`, `linalg[charpoly]`, `RootOf`.

linalg[entermatrix]

`entermatrix(A)`

Interactive matrix entry ◇ This function prompts for values to be inserted into the matrix *A*. The values are entered in row major order. Each value must be terminated with a semicolon.

linalg[equal]

`equal(A, B)`

Determines whether two matrices are equal ◇ This function returns *true* or *false*. ◇ See also: `linalg[iszero]`.

linalg[exponential]

`exponential(A)`
`exponential(A, t)`

Computes the matrix exponential ◇ *A* must be a square matrix, and *t* is an optional parameter name. The matrix exponential is defined as $\exp(tA) = \sum_{k\geq 0} A^k t^k / k!$. If a variable name is not given, the first indeterminate (if any) in *A* is removed and used as a parameter.

linalg[extend]

`extend(A, m, n, expr)`
`extend(A, m, n)`

Enlarges a matrix ◇ This function returns a new matrix that is a copy of the matrix *A* with *m* additional rows and *n* additional columns. Each new entry is initialized to *expr* if this argument is present. ◇ See also: `linalg[augment]`, `linalg[copyinto]`, `linalg[stack]`.

linalg[ffgausselim]

`ffgausselim(A)`
`ffgausselim(A, name_1)`
`ffgausselim(A, name_1, name_2)`
`ffgausselim(A, col)`

Fraction-free Gaussian elimination on a matrix ◇ This function performs fraction-free Gaussian elimination with row pivoting on *A*, a rectangular matrix of multivariate polynomials over the rationals. *$name_1$*, if present, is assigned the rank of A. If the rank equals the number of rows of A, *$name_2$* is assigned the determinant of the square submatrix of A formed by taking the first rank(A) columns. Otherwise it is assigned zero. Elimination is terminated at column position *col* if this argument is present. ◇ See also: `linalg[gausselim]`, `linalg[rank]`, `linalg[rref]`.

linalg[fibonacci]

`fibonacci(`*n*`)`
Computes the *n*th fibonacci matrix ◇ See also: `combinat[fibonacci]`.

linalg[frobenius]

`frobenius(`*A*`)`
`frobenius(`*A*`, `*name*`)`
Computes the Frobenius form of a matrix ◇ This is also known as the Rational Canonical form of a matrix. A must be a square matrix. If a name is given as the second argument, it is assigned the transformation matrix P such that $P^{-1}AP$ is the Frobenius form. ◇ See also: `linalg[ratform]`.

linalg[gausselim]

`gausselim(`*A*`)`
`gausselim(`*A*`, `*$name_1$*`)`
`gausselim(`*A*`, `*$name_1$*`, `*$name_2$*`)`
`gausselim(`*A*`, `*col*`)`
Gaussian elimination on a matrix ◇ This function performs Gaussian elimination with row pivoting on A, a rectangular matrix of rationals, complex rationals, or rational functions with these coefficients. *$name_1$*, if present, is assigned the rank of A. If the rank equals the number of rows of A, *$name_2$* is assigned the determinant of the square submatrix of A formed by taking the first rank(A) columns. Otherwise it is assigned zero. Elimination is terminated at column position *col* if this argument is present. ◇ See also: `Gausselim`, `linalg[ffgausselim]`, `linalg[rank]`, `linalg[rref]`.

linalg[gaussjord]

`gaussjord(`*A*`)`
`gaussjord(`*A*`, `*$name_1$*`)`
`gaussjord(`*A*`, `*$name_1$*`, `*$name_2$*`)`
`gaussjord(`*A*`, `*col*`)`
A synonym for `linalg[rref]` ◇ Gauss-Jordan elimination ◇ See `linalg[rref]`.

linalg[genmatrix]

`genmatrix(`*eqns*`, `*vars*`)`
`genmatrix(`*eqns*`, `*vars*`, `*anything*`)`
Generates the coefficient matrix from equations ◇ This function generates the coefficient matrix from the list or set of linear equations *eqns* in the list or set of unknowns *vars*. If a third argument is present, the negative of the

"right-hand side" vector is included as the last column of the matrix. ◇ See also: `linalg[linsolve]`.

linalg[grad]

`grad(`*expr*`, `*v*`)`
`grad(`*expr*`, `*v*`, coords=`*coordtype*`)`
`grad(`*expr*`, `*v*`, `*list*`)`
Computes the gradient of the expression with respect to the variables in *v* ◇ *v* may be a list or a vector. ◇ In Release 3, an optional third argument specifies the coordinate system. *coordtype* may be `cartesian` (the default), `cylindrical`, or `spherical`. A list of three elements giving the scale factors requests computation in the corresponding orthogonally curvilinear coordinate system. ◇ `grad(x*y*z,[x,y,z]);` ⟶ $[\, yz \; xz \; yx \,]$ ◇ See also: `linalg[curl]`, `linalg[diverge]`, `linalg[hessian]`, `linalg[laplacian]`.

linalg[GramSchmidt]

`GramSchmidt([`v_1`, `v_2`, ..., `v_n`])`
`GramSchmidt(`*vectors*`)`
Computes a list or set of orthogonal vectors from the given linearly independent vectors, using the Gram-Schmidt orthogonalization process ◇ *vectors* is a list or set of linearly independent vectors. The result is given as a list or a set, depending on the form of the input. The vectors returned are not normalized. ◇ See also: `linalg[basis]`, `linalg[norm]`.

linalg[hadamard]

`hadamard(`*A*`)`
Bound on the determinant of a matrix ◇ The result is an upper bound on the absolute value of the determinant of a numeric matrix *A*, or more generally on the `maxnorm` of the determinant of a polynomial matrix. *A* must be a square matrix. ◇ See also: `linalg[det]`, `maxnorm`.

linalg[hermite]

`hermite(`*A*`, `*var*`)`
`hermite(`*A*`, `*var*`, `*name*`)`
Hermite Normal Form ◇ This function computes the Hermite Normal Form (reduced row echelon form) of a rectangular matrix of univariate polynomials in *var* over the field of rational numbers **Q**, or rational expressions over **Q**. If a name is given as a third argument, it is assigned the transformation matrix U such that the result is UA. ◇ See also: `GaussInt[GIhermite]`, `Hermite`, `linalg[ihermite]`, `linalg[smith]`.

linalg[hessian]

`hessian(`*expr*`, `*v*`)`
Computes the Hessian matrix of an expression ◇ *v* is a list or vector of variables. The result is an $n \times n$ matrix, where n is the length of *v*. The

(i, j)th entry of the matrix result is $\frac{\partial^2}{\partial v_i \partial v_j} expr$. ◇ See also: `linalg[grad]`, `linalg[jacobian]`.

linalg[hilbert]

`hilbert(`*n*`)`

`hilbert(`*n*`, `*expr*`)`

Creates a generalized Hilbert matrix ◇ This matrix has $1/(i + j - expr)$ as its (i, j)th entry. The default value of *expr* is 1.

linalg[htranspose]

`htranspose(`*A*`)`

`htranspose(`*v*`)`

Computes the Hermitian transpose of a matrix ◇ The Hermitian transpose is the conjugate transpose. The result inherits the indexing function of *A*. A vector is treated as a matrix with one column. ◇ See also: `linalg[transpose]`.

linalg[ihermite]

`ihermite(`*A*`)`

`ihermite(`*A*`, `*name*`)`

Integer-only Hermite Normal Form ◇ This function computes the Hermite Normal Form (reduced row echelon form) of a rectangular matrix of integers. If a name is given as a third argument, it is assigned the transformation matrix U such that the result is UA. ◇ See also: `GaussInt[GIhermite]`, `Hermite`, `linalg[hermite]`, `linalg[ismith]`.

linalg[indexfunc]

`indexfunc(`*A*`)`

Determines the indexing function of an array ◇ This could be one of the standard indexing functions (`antisymmetric`, `diagonal`, `identity`, `sparse`, or `symmetric`), or the name of a user-defined indexing function. `NULL` is returned if *A* has no indexing function. ◇ See also: `indexfcn`.

linalg[innerprod]

`innerprod(`u`, `A_1`, `A_2`, ..., `A_n`, `v`)`

Calculates the inner product of a sequence of vectors and matrices ◇ The dimension of each matrix and vector must be compatible for multiplication in the order given. ◇ See also: `linalg[crossprod]`, `linalg[dotprod]`, `linalg[multiply]`.

linalg[intbasis]

`intbasis(`S_1`, `S_2`, ..., `S_n`)`

Determines a basis for the intersection of vector spaces ◇ Each S_i may be a single vector, a list of vectors, or a set of vectors. A set of vectors forming a basis for the intersection is returned. ◇ See also: `linalg[basis]`, `linalg[sumbasis]`.

linalg[inverse]

`inverse(A)`

Computes the inverse of a matrix ◇ An error occurs if the matrix is singular.

linalg[ismith]

`ismith(A)`

`ismith(A, name1, name2)`

Integer-only Smith Normal Form ◇ This function computes the Smith normal form of a rectangular matrix of integers. In the case of three arguments, $name_1$ and $name_2$ are assigned the transformation matrices U and V such that the result is UAV. ◇ See also: `GaussInt[GIsmith]`, `linalg[ihermite]`, `linalg[smith]`, `Smith`.

linalg[iszero]

`iszero(A)`

Determines whether a matrix is the zero matrix ◇ This function returns *true* if all of the entries of the matrix are zero and *false*. ◇ See also: `linalg[equal]`.

linalg[jacobian]

`jacobian(u, v)`

Computes the Jacobian matrix of a vector function ◇ The first argument is a list or vector of expressions in the variables named in the second argument, which may also be a list or a vector. *u* and *v* need not have the same length. The (i, j)th entry of the matrix result is $\frac{\partial u_i}{\partial v_j}$. ◇ See also: `linalg[hessian]`.

linalg[jordan]

`jordan(A)`

`jordan(A, name)`

Computes the Jordan form of a matrix ◇ The Jordan form is a block diagonal matrix of Jordan block matrices, where the diagonal entries of the Jordan blocks are the eigenvalues of *A*. If a name is given, it is assigned the transformation matrix P such that the Jordan form is $P^{-1}JP$. ◇ See also: `linalg[diag]`, `linalg[eigenvals]`, `linalg[JordanBlock]`.

linalg[JordanBlock]

`JordanBlock(expr, n)`

Returns a Jordan block matrix ◇ This function returns an $n \times n$ matrix with *expr* on the main diagonal, 1 on the superdiagonal, and 0 elsewhere. ◇ See also: `linalg[diag]`, `linalg[jordan]`.

linalg[kernel]

`kernel(A)`

`kernel(A, name)`

Computes a basis for the kernel (null space) of the linear transformation defined by the matrix *A* ◇ The result is a set of vectors. If a name is given, it is assigned the dimension of the kernel. ◇ See also: `linalg[colspace]`, `linalg[nullspace]`, `linalg[rowspace]`, `Nullspace`.

linalg[laplacian]

`laplacian(`*expr*`, `*v*`)`
`laplacian(`*expr*`, `*v*`, coords=`*coordtype*`)`
`laplacian(`*expr*`, `*v*`, `*list*`)`
Computes the Laplacian of an expression with respect to the variables in *v* ◇ *v* may be a list or a vector. ◇ In Release 3, an optional third argument specifies the coordinate system. *coordtype* may be `cartesian` (the default), `cylindrical`, or `spherical`. A list of three elements giving the scale factors requests computation in the corresponding orthogonally curvilinear coordinate system. ◇ `laplacian(x^2*y^2*z,[x,y,z]);` $\longrightarrow 2\,y^2 z + 2\,x^2 z$ ◇ See also: `linalg[grad]`.

linalg[leastsqrs]

`leastsqrs(`*A*`, `*v*`)`
`leastsqrs(`*set*`, `*varset*`)`
Least-squares solution of equations ◇ The first form returns the vector x that best satisfies $Ax = v$ in the least-squares sense. The second form finds the values for the variables in the set *varset* that minimize the set of expressions or equations *set* in the least-squares sense. It returns a set of equations for the variables. ◇ See also: `linalg[linsolve]`.

linalg[linsolve]

`linsolve(`*A*`, `*v*`)`
`linsolve(`*A*`, `*v*`, `$name_1$`, `$name_2$`)`
`linsolve(`*A*`, `*B*`)`
`linsolve(`*A*`, `*B*`, `*name*`)`
Solution of linear equations ◇ The first form finds the vector x that satisfies the matrix equation $Ax = v$. The length of *v* must equal the number of rows of *A*. `NULL` is returned if no solution exists. If there are many solutions, global names are employed to describe them. The latter form finds the matrix X that satisfies $AX = B$. `NULL` is returned if there is not a unique solution. A third argument is assigned the rank of *A*. A fourth argument lets you specify the base name of the global names used to parameterize solutions. ◇ See also: `linalg[genmatrix]`, `linalg[leastsqrs]`, `solve`.

linalg[matrix]

`matrix(`*list*`)`
`matrix(`*m*`, `*n*`)`
`matrix(`*m*`, `*n*`, `*list*`)`
`matrix(`*m*`, `*n*`, `*v*`)`
`matrix(`*m*`, `*n*`, `*f*`)`
Creates a matrix ◇ The first form creates a matrix with elements from *list*, a list of lists or vectors. The dimensions of the matrix are determined by

the structure of *list*. The second form creates an $m \times n$ matrix with unspecified elements. The third form makes an $m \times n$ matrix with elements from *list*. Here *list* may be a list of lists or a list of vectors, or a list of elements, in which case the matrix is initialized in row major order. The fourth form initializes the matrix in row major order with the elements of the vector. In the last form, the function f determines the matrix elements: $A_{ij} = f(i,j)$. ◇ `matrix([[2,3],[3,4]]):` ◇ `matrix(2,2,[2,3,3,4]):` ◇ `matrix(2,2,(i,j)->i+j):` ◇ See also: `array`, `linalg[vector]`.

linalg[minor]

`minor(`*A*`, `*row*`, `*col*`)`
Computes a minor of a matrix ◇ This function returns the minor of *A* obtained by removing the given row and column. ◇ See also: `linalg[delrows]`, `linalg[delcols]`, `linalg[submatrix]`.

linalg[minpoly]

`minpoly(`*A*`, `*var*`)`
Computes the minimum polynomial of the matrix ◇ The minimum polynomial is the polynomial of lowest degree that annihilates *A*. The polynomial is given in the variable *var*. ◇ `minpoly(matrix([[2,3],[3,4]]),x);` $\longrightarrow -1 - 6\,x + x^2$ ◇ See also: `linalg[charpoly]`.

linalg[mulcol]

`mulcol(`*A*`, `*col*`, `*expr*`)`
Multiplies a column of a matrix by an expression ◇ This function returns a copy of the matrix *A* where the given column number is multiplied by *expr*. ◇ See also: `linalg[addcol]`, `linalg[mulrow]`.

linalg[mulrow]

`mulrow(`*A*`, `*row*`, `*expr*`)`
Multiplies a row of a matrix by an expression ◇ This function returns a copy of the matrix *A* where the given row number is multiplied by *expr*. ◇ See also: `linalg[addrow]`, `linalg[mulcol]`.

linalg[multiply]

`multiply(`*A*`, `*B*`, ...)`
`multiply(`*A*`, `*v*`)`
Matrix-matrix or matrix-vector multiplication ◇ A vector is interpreted as a column vector. ◇ See also: `evalm`, `linalg[innerprod]`.

linalg[norm]

`norm(`A`)`
`norm(`A`, 1)`
`norm(`A`, 2)`
`norm(`A`, frobenius)`
`norm(`A`, infinity)`
`norm(`v`)`
`norm(`v`, `n`)`
`norm(`v`, frobenius)`
`norm(`v`, infinity)`

Computes the norm of a matrix or a vector ◇ Matrix norms (sums are sums of absolute values):

1	Maximum column sum
2	Square root of maximum eigenvalue of AA^t
frobenius	Square root of sum of squares of matrix elements
infinity	Maximum row sum

Vector norms (again sums are sums of absolute values):

n	nth root of sum of nth powers of elements
frobenius	Same as $n = 2$
infinity	Maximum magnitude of elements

The default norm is `infinity` throughout the `linalg` package. ◇ `norm(matrix([[2,3],[3,4]]), frobenius);` $\longrightarrow \sqrt{38}$ ◇ See also: `linalg[cond]`, `linalg[normalize]`, `norm`, `Norm`.

linalg[normalize]

`normalize(`v`)`

Normalizes a vector ◇ This function normalizes the specified vector with the 2-norm. ◇ `normalize(vector([6,8]));` $\longrightarrow \left[\frac{3}{5}\ \frac{4}{5}\right]$ ◇ See also: `linalg[norm]`, `linalg[GramSchmidt]`.

linalg[nullspace]

`nullspace(`A`)`
`nullspace(`A`, `*name*`)`

A synonym for `linalg[kernel]` ◇ See `linalg[kernel]`.

linalg[orthog]

`orthog(`A`)`

Tests for an orthogonal matrix ◇ A matrix is orthogonal if the inner product of any column with itself is 1, and the inner product of any column with any other column is 0. This function returns *true* if it can show that the matrix is orthogonal, *false* if it can show it is not orthogonal, and *FAIL* otherwise. ◇ See also: `testeq`.

linalg[permanent]

`permanent(`A`)`

Computes the permanent of a matrix ◇ The permanent of a matrix is computed in a way similar to the determinant, but there is no alternating sign in the minor expansions. ◇ `permanent(matrix([[2,3], [3,4]]));` ⟶ 17 ◇ See also: `linalg[det]`.

linalg[pivot]

`pivot(`*A*`, `*i*`, `*j*`)`
`pivot(`*A*`, `*i*`, `*j*`, `*row₁*`..`*row₂*`)`
Pivots about the matrix entry $A[i,j]$ ◇ $A[i,j]$ must be nonzero. The first form adds multiples of row i to every other row so that the jth column becomes zero in each. The second form clears the jth column in only the specified range of rows. ◇ See also: `linalg[gausselim]`.

linalg[potential]

`potential(`*list*`, `*varlist*`, `*name*`)`
`potential(`*v*`, `*varlist*`, `*name*`)`
Computes the potential of a vector field ◇ The first argument is a list or vector of expressions representing a vector field, and the second argument is a list of variables. This function returns *true* if the vector field has a scalar potential, and it assigns a scalar potential to the third argument in this case. Otherwise it returns *false*. ◇ `potential([y*z,x*z,x*y], [x,y,z],'ans'), ans;` ⟶ *true*, $x\,y\,z$ ◇ See also: `linalg[grad]`, `linalg[vecpotent]`.

linalg[randmatrix]

`randmatrix(`*m*`, `*n*`, `*options*`)`
Random matrix generator ◇ This function generates a random $m \times n$ matrix. The `entries = `*f* option, where *f* is a nullary function (default `rand(-99..99)`), determines the entries of the matrix. Options also specify the structure of the matrix: `antisymmetric`, `dense` (the default), `sparse`, `symmetric` and `unimodular`. ◇ See also: `linalg[matrix]`, `linalg[randvector]`, `rand`, `_seed`.

linalg[randvector]

`randvector(`*n*`)`
`randvector(`*n*`, entries=`*f*`)`
Random vector generator ◇ This function generates a random vector of dimension *n*. The `entries = `*f* option, where *f* is a nullary function (default `rand(-99..99)`), determines the entries of the vector. ◇ See also: `linalg[vector]`, `linalg[randmatrix]`.

linalg[range]

`range(`*A*`)`
`range(`*A*`, `*name*`)`
A synonym for `linalg[colspace]` ◇ This function has been removed in Release 3. ◇ See `linalg[colspace]`.

linalg[rank]

`rank(`*A*`)`

Rank of a matrix ◇ This function computes the rank of a matrix by performing Gaussian elimination on the rows of the given matrix. ◇ See also: `linalg[colspace]`, `linalg[rowspace]`.

linalg[ratform]

`ratform(`*A*`)`
`ratform(`*A*`,` *name*`)`

A synonym for `linalg[frobenius]` ◇ This function computes the rational canonical form of the square matrix *A*. ◇ See `linalg[frobenius]`.

linalg[row]

`row(`*A*`,` *row*`)`
`row(`*A*`,` row_1`..`row_2`)`

Extracts rows of a matrix ◇ The first form extracts the *row*th row of the matrix and returns the result as a vector. The second form extracts a range of rows and returns a sequence of vectors. ◇ See also: `linalg[addrow]`, `linalg[col]`, `linalg[delrows]`, `linalg[mulrow]`, `linalg[rowdim]`, `linalg[swaprow]`.

linalg[rowdim]

`rowdim(`*A*`)`

Returns the number of rows in the matrix *A* ◇ See also: `linalg[coldim]`, `linalg[rowspace]`, `linalg[vectdim]`.

linalg[rowspace]

`rowspace(`*A*`)`
`rowspace(`*A*`,` *name*`)`

Computes a basis for the row space ◇ This function returns a set of vectors that form a basis for the vector space spanned by the rows of the matrix *A*. The vectors are returned in a canonical form with leading entries 1. If a name is given as a second parameter, it is assigned the rank of *A*. ◇ See also: `linalg[colspace]`, `linalg[nullspace]`, `linalg[rank]`, `linalg[rowdim]`, `linalg[rowspan]`.

linalg[rowspan]

`rowspan(`*A*`)`
`rowspan(`*A*`,` *name*`)`

Computes spanning vectors for the row space ◇ This function computes a spanning set of vectors for the row space of *A*, a matrix of multivariate polynomials over the rationals. Unlike `rowspace`, this function uses "fraction-free" Gaussian elimination to avoid introduction of rational expressions during the elimination. If a name is given as a second parameter, it is assigned the rank of *A*. ◇ See also: `linalg[colspan]`, `linalg[ffgausselim]`, `linalg[rowspace]`.

linalg[rref] `rref(`A`)`

`rref(`A, *$name_1$*`)`

`rref(`A, *$name_1$*, *$name_2$*`)`

`rref(`A, *col*`)`

Reduced row echelon form ◇ A is a rectangular matrix of rational numbers or rational functions with rational coefficients. Elementary row operations are performed on A to reduce it to row Echelon (Gauss-Jordan) form. $name_1$, if present, is assigned the rank of A. If the rank equals the number of rows of A, $name_2$ is assigned the determinant of the square submatrix of A formed by taking the first rank(A) columns. Otherwise it is assigned zero. Elimination is terminated at column position *col* if this argument is present. ◇ See also: `Gaussjord`, `linalg[backsub]`, `linalg[gausselim]`, `linalg[gaussjord]`, `linalg[hermite]`, `linalg[pivot]`, `linalg[rank]`.

linalg[scalarmul]

`scalarmul(`A, *expr*`)`

`scalarmul(`v, *expr*`)`

Multiplies each element of a matrix or vector by a scalar expression ◇ See also: `evalm`, `linalg[mulcol]`, `linalg[mulrow]`, `linalg[multiply]`.

linalg[singularvals]

`singularvals(`A`)`

Computes the singular values of a matrix ◇ This function returns a list of the singular values of the square matrix A. The singular values of a matrix are the square roots of the eigenvalues of A^tA. ◇

`singularvals(matrix([[2,3],[3,4]]));` ⟶

$\left[\sqrt{19+6\sqrt{10}}, \sqrt{19-6\sqrt{10}}\right]$ ◇ See also: `linalg[eigenvals]`, `Svd`.

linalg[smith]

`smith(`A, *var*`)`

`smith(`A, *var*, *$name_1$*, *$name_2$*`)`

Computes the Smith Normal Form of a matrix ◇ The entries of A are univariate polynomials in the variable *var* over a field. In the case of four arguments, the two names are assigned the transformation matrices U and V such that the result is UAV. ◇ See also: `GaussInt[GIsmith]`, `linalg[hermite]`, `linalg[ismith]`, `Smith`.

linalg[stack]

`stack(`A, B, `...)`

`stack(`u, v, `...)`

Joins two or more matrices or vectors together vertically ◇ Matrices must have the same number of columns; vectors must have the same length. A vector is interpreted as a row vector. ◇ See also: `linalg[augment]`, `linalg[extend]`.

linalg[submatrix]

`submatrix(`*A*`,` *rows*`,` *cols*`)`

Extracts a specified submatrix from a matrix ◇ This function returns the submatrix of *A* determined by the given row and column indices. *rows* and *cols* may be ranges or lists of indices. ◇ See also: `linalg[subvector]`, `linalg[row]`, `linalg[col]`.

linalg[subvector]

`subvector(`*A*`,` *rows*`,` *cols*`)`

Extracts a specified vector from a matrix ◇ This function returns the subvector of *A* determined by the given row and column indices. One of *rows* or *cols* must be a single index; the other may be a range or a list of indices. ◇ See also: `linalg[submatrix]`.

linalg[sumbasis]

`sumbasis(`S_1`,` S_2`, ...,` S_n`)`

Determines a basis for the sum of vector spaces ◇ Each S_i may be a single vector, a list of vectors, or a set of vectors. A set of vectors forming a basis for the sum of the vector spaces is returned. ◇ See also: `linalg[intbasis]`, `linalg[basis]`.

linalg[swapcol]

`swapcol(`*A*`,` col_1`,` col_2`)`

Swaps two columns in a matrix ◇ This function creates a new matrix in which column col_1 has been swapped with column col_2. ◇ See also: `linalg[swaprow]`.

linalg[swaprow]

`swaprow(`*A*`,` row_1`,` row_2`)`

Swaps two rows in a matrix ◇ This function creates a new matrix in which row row_1 has been swapped with row row_2. ◇ See also: `linalg[swapcol]`.

linalg[sylvester]

`sylvester(`$poly_1$`,` $poly_2$`,` *var*`)`

Creates the Sylvester matrix of two polynomials ◇ $poly_1$ and $poly_2$ are polynomials in the variable *var*. The determinant of this matrix is equal to the resultant of the polynomials. ◇ See also: `linalg[bezout]`, `linalg[det]`, `resultant`.

linalg[toeplitz]

`toeplitz(`*list*`)`

Creates a Toeplitz matrix from the expressions in the list ◇ The result is a symmetric square matrix with as many rows as there are elements in the list. The first element of *list* is placed along the diagonal, and successive elements are placed along successive subdiagonals and superdiagonals.

linalg[trace]

`trace(`*A*`)`

Computes the trace of a square matrix ◇ The trace of *A* is the sum of the diagonal elements of *A*. ◇ See also: `trace`, `Trace`.

linalg[transpose]

`transpose(`*A*`)`
`transpose(`*v*`)`
`transpose(`*list*`)`

Computes the transpose of a matrix ◇ The result inherits the indexing function of *A*. A vector or list is treated as a matrix with one column. ◇ See also: `linalg[htranspose]`.

linalg[vandermonde]

`vandermonde(`*list*`)`

Creates a Vandermonde matrix from the elements of the list ◇ This function returns a square matrix with as many rows as there are elements in the list. The (i, j)th entry is $list_i^{j-1}$.

linalg[vecpotent]

`vecpotent(`*list*`, `*varlist*`, `*name*`)`
`vecpotent(`*v*`, `*varlist*`, `*name*`)`

Computes the vector potential of a vector field ◇ The first argument is a list or vector of three expressions representing a vector field, and the second argument is a list of three variables. This function returns *true* if the vector field has a vector potential, and it assigns a vector potential field to the third argument in this case. Otherwise it returns *false*. The vector potential exists if and only if the divergence of the vector field is zero. ◇ `vecpotent([-x,-z,z],[x,y,z],'ans'), eval(ans);` ⟶ *true*, $\left[-\frac{1}{2}z^2 \; xz \; 0\right]$ ◇ See also: `linalg[curl]`, `linalg[diverge]`, `linalg[potential]`.

linalg[vectdim]

`vectdim(`*v*`)`
`vectdim(`*list*`)`

Determines the dimension of a vector ◇ The dimension of a vector or a list is the number of elements in the vector or list. ◇ `vectdim([2,3,3,7]);` ⟶ 4 ◇ See also: `linalg[coldim]`, `linalg[rowdim]`.

linalg[vector]

`vector(`*list*`)`
`vector(`*n*`)`
`vector(`*n*`, `*list*`)`
`vector(`*n*`, `*f*`)`

Creates a vector ◇ The first form creates a vector with elements from *list*, with the dimension of the vector determined by the length of the list. The second form creates a vector of length *n* with unspecified elements. The third form makes a vector of length *n* with elements from *list*. In the last form, the function f determines the vector elements: $v_k = f(k)$. ◇ `vector([2,4,6]):` ◇ `vector(3,[2,4,6]):` ◇ `vector(3,k->2*k):` ◇ See also: `array`, `linalg[matrix]`.

linalg[Wronskian]

`Wronskian(`*v*`, `*var*`)`
`Wronskian(`*list*`, `*var*`)`
Computes the Wronskian matrix of a vector or list of expressions ◇ The first argument is a vector or list of expressions in the variable *var*. The (i, j)th entry of the result is the $(i - 1)$th derivative of the jth expression in the vector or list.

ln `ln(`*z*`)`
The natural logarithm ◇ This logarithm has base exp(1). ◇ `ln(exp(1));` $\longrightarrow$ 1 ◇ See also: `exp`, `ilog`, `log`, `log10`.

lnGAMMA `lnGAMMA(`*z*`)`
The natural logarithm of the Gamma function ◇ You must `readlib(lnGAMMA)` before using this command. ◇ `lnGAMMA(6);` $\longrightarrow$ ln(120) ◇ See also: `GAMMA`.

local `local `*var₁*`, `*var₂*`, ...`
Identifies local variables ◇ The `local` declaration appears at the beginning of the body of a procedure. ◇ See also: `global`, `options`, `proc`.

log `log(`*z*`)`
`log[b](`*z*`)`
The logarithm with base b ◇ If the index b is omitted, the natural logarithm is used. ◇ See also: `ilog`, `ln`, `log10`.

log10 `log10(`*z*`)`
The base 10 (common) logarithm ◇ You must `readlib(log10)` before using this command. ◇ See also: `ilog10`, `log`.

logic Boolean
logic package ◇ The following Boolean operators are used in this package:

`&and`	`&implies`	`&nor`	`&or`
`&iff`	`&nand`	`&not`	`&xor`

In the descriptions of functions in this package, b stands for a Boolean expression. ◇ In Release 2 and earlier versions, the functions in this package must first be loaded using `with` before they can be invoked. ◇ See also: `logic[`*function*`]`, `with`.

logic[bequal]

`bequal(`b_1`, `b_2`)`
`bequal(`b_1`, `b_2`, `*name*`)`
Tests whether the two given Boolean expressions are logically equivalent ◇ If *name* is present and the two expressions are not logically equivalent,

a valuation is assigned to this name in which the two expressions have different results. This valuation is a set of equations in the symbols of *b*. If the two expressions are equivalent, *name* is assigned `NULL`. This function returns *true* or *false*. ◇ `bequal(&not a &or b, a &implies b);` ⟶ *true* ◇ See also: `logic`.

logic[bsimp]

`bsimp(`*b*`)`
Boolean expression simplification ◇ This function returns a minimal sum of products expansion of the Boolean expression *b*. ◇ `bsimp(a &and (a &or b));` ⟶ *a* ◇ See also: `logic`, `logic[distrib]`, `logic[environ]`.

logic[canon]

`canon(`*b*`,` *namelist*`)`
`canon(`*b*`,` *namelist*`,` *formname*`)`
Canonical representation of expressions ◇ This function converts the boolean expression *b* to a canonical form with respect to the symbol names given in *namelist*. *formname* selects the canonical form. It may be `CNF` (conjunctive normal form), `DNF` (disjunctive normal form, the default), or `MOD2` (modulo 2 canonical form). ◇ `canon(a &iff b,{a,b},MOD2);` ⟶ $a + b + 1$ ◇ See also: `logic`, `logic[convert]`.

logic[convert]

`convert(`*b*`, frominert)`
`convert(`*b*`, toinert)`
`convert(`*b*`, MOD2)`
`convert(`*b*`, MOD2, expanded)`
Converts a Boolean expression to another form ◇ The `frominert` option replaces the inert operators `&and`, `&or`, and `&not` with the system operators `and`, `or`, and `not`. The `toinert` option performs the reverse conversion. The `MOD2` option converts *b* to its equivalent modulo 2 format. If `expanded` also appears, then the modulo 2 representation is returned fully expanded. ◇ `convert(a &and b,MOD2);` ⟶ $a\,b$ ◇ `convert(a and b,toinert);` ⟶ *a* &and *b* ◇ See also: `logic`, `convert/mod2`.

logic[distrib]

`distrib(`*b*`)`
`distrib(`*b*`,` *nameset*`)`
Expands the Boolean expression *b* into a sum of products (a disjunction of conjunctions) ◇ The expanded expression is not guaranteed to be in minimized form or canonical form. If *nameset* is present, the expansion should be the canonical disjunctive normal form of the expression with respect to the symbols in this set. ◇ `distrib(&not(a &or b));` ⟶ &not(*a*)&and&not(*b*) ◇ See also: `logic`, `logic[canon]`, `logic[bsimp]`, `expand`.

logic[dual] `dual(`*b*`)`

Constructs the dual of a Boolean expression ◇ This function returns the dual of the given Boolean expression: all occurrences of `true`, `false`, `&and`, and `&or` are replaced by `false`, `true`, `&or`, and `&and`, respectively. Other Boolean operators are left unchanged. ◇ `dual(a &or b);` ⟶ *a* &and *b* ◇ See also: `logic`, `logic[environ]`.

logic[environ]

`environ(`*n*`)`

Sets the level of automatic logical simplification ◇ This function sets the type of automatic simplification of Boolean expressions. There are four possibilities for *n*:

0 No simplifications (the default)

1 Apply associativity to remove redundant parentheses, and rewrite the expression in terms of `&and`, `&or`, and `&not`

2 The simplifications above, plus *a* `&and` $a \to a$, *a* `&or` $a \to a$, and simplifications with `true` and `false`

3 Convert to the unexpanded modulo 2 form

◇ See also: `logic`, `logic[bsimp]`, `logic[convert]`.

logic[randbool]

`randbool(`*names*`,` *formname*`)`

Constructs a random Boolean expression in the given canonical form ◇ *names* is a list or a set of symbols in which the boolean expression should be expressed. The name of the canonical form *formname* may be `CNF` (conjunctive normal form), `DNF` (disjunctive normal form), or `MOD2` (modulo 2 normal form). ◇ `randbool([a,b],MOD2);` ⟶ $a\,b + 1 + b$ ◇ See also: `logic`.

logic[satisfy]

`satisfy(`*b*`)`

`satisfy(`*b*`,` *nameset*`)`

Returns an evaluation satisfying a Boolean expression ◇ This function returns a set of equations giving values for each of the variables of *b* that make the Boolean expression *true*. If *nameset* is present, an equation is given for each of the symbols in this set as well. If *b* is not satisfiable, `NULL` is returned. ◇ `satisfy(a &implies b);` ⟶ $\{a = \mathit{false}, b = \mathit{false}\}$ ◇ See also: `logic`.

logic[tautology]

`tautology(`*b*`)`

`tautology(`*b*`,` *name*`)`

Tests for a tautology ◇ This function tests whether the Boolean expression *b* is a tautology. It returns *true* or *false*. If *name* is present and *b* is not a tautology, a valuation that gives a negative result is assigned to this name. This valuation is a set of equations in the symbols of *b*. If *b* is a tautology, `NULL` is assigned to *name*. ◇ `tautology((a &xor b) &or (a &iff b));` ⟶ *true* ◇ See also: `logic`.

lprint `lprint(`*expr*$_1$`, `*expr*$_2$`, ...)`
Prints the expressions in a linear format ◇ In general, an expression printed by this function is valid input to Maple. Multiple expressions are printed on the same line, each separated from the others by three blanks. ◇ See also: `interface`, `print`.

m2src `m2src` *filename*`.m`
`m2src` *options filename*`.m`
Converts files in internal format to text form ◇ This external command converts Maple V Release 2 ".m" files to equivalent text files. The `-p` option prints the full text of library procedures so that the text produced may be read by Maple. The `-t` option saves the output in the file *filename*`.t` instead of sending it to the standard output. ◇ This command is new in Release 3. ◇ See also: `updtsrc`.

macro `macro(`*eqn*$_1$`, `*eqn*$_2$`, ..., `*eqn*$_n$`)`
An abbreviation facility ◇ If the macro $f = g$ is in effect, Maple translates all occurrences of f (except for parameters and local variables) to g when reading input from the terminal or a file. ◇ See also: `alias`, `subs`.

makehelp `makehelp(`*topic*`, `*textfile*`)`
`makehelp(`*topic*`, `*textfile*`, `*library*`)`
Converts a text file into a help file ◇ This function reads a text file, assumed to contain help for the specified topic, and creates an internal help object for that topic. The topic can be of the form *name* or *name*$_1$/*name*$_2$. The resulting help object has the name `help/text/`*name* or `help/`*name*$_1$`/text/`*name*$_2$. ◇ You must `readlib(makehelp)` before using this command.

map `map(`*f*`, `*expr*`, `*arg*$_2$`, ..., `*arg*$_n$`)`
Applies a function to each operand of an expression ◇ This function applies the function *f* to each operands of *expr*. Additional arguments are passed to *f*. The function returns an expression where each operand is replaced by its image under *f*. ◇ `map(f,[a,b,c]);` $\longrightarrow$ $[f(a), f(b), f(c)]$ ◇ `map(f,a+b*c);` $\longrightarrow f(a) + f(b\,c)$ ◇ See also: `select`, `zip`.

march `march -a` *archivedir filename*$_1$ *indexname*$_1$ ...
`march -c` *archivedir n*
`march -d` *archivedir indexname*
`march -l` *archivedir*
`march -p` *archivedir*
`march -u` *archivedir filename*$_1$ *indexname*$_1$ ...
`march -x` *archivedir indexname*$_1$ *filename*$_1$ ...
Maple library archive manager ◇ This program is an external command for managing an archive of Maple ".m" files. *archivedir* names the location of the archive. Each option is described below.

`-a` Adds the specified ".m" files using the given index names
`-c` Creates a new archive for holding approximately *n* files
`-d` Deletes the file at the given index (new in Release 3)
`-l` Lists the contents of the archive
`-p` Packs the archive, purging old versions of files
`-u` Updates the specified ".m" files at the given index names
`-x` Extracts $indexname_i$ and stores it in $filename_i$

◇ The `-d` option is new in Release 3.

match `match(`*expr* `=` *pattern*`,` *var*`,` *name*`)`
Pattern matching ◇ *expr* is an expression in the main variable *var*. This function returns *true* if it is able to match *expr* to *pattern*, and *false* otherwise. If a match is made, *name* is assigned a set of substitutions that turn the pattern into the expression. ◇ `match(2*x = a*x+b, x,s),s;` ⟶ *true*, $\{b = 0, a = 2\}$ ◇ See also: `solve`, `subs`.

max `max(`$expr_1$`,` $expr_2$`, ...)`
Computes the maximum of the arguments ◇ It may help to increase `Digits` if this function returns unevaluated. ◇ `max(exp(Pi), Pi^exp(1));` ⟶ e^{π} ◇ See also: `maximize`, `min`.

maximize `maximize(`*expr*`)`
`maximize(`*expr*`,` *vars*`)`
`maximize(`*expr*`,` *vars*`, infinite)`
`maximize(`*expr*`,` *vars*`,` *ranges*`)`
Computes the maximum ◇ This function returns the maximum value of *expr* or an expression sequence with all the values that are candidates for the maximum value. In Release 3, *vars* may be a single variable or a set of variables. In Release 2 and earlier versions, *vars* must be a set of variables. If *vars* is not given, the expression is maximized with respect to all its indeterminates. ◇ A third argument is supported in Release 3. `infinite` requests the unconstrained maximum of the expression, including its limits at infinity (this is the behavior in earlier versions). *ranges* is a single range or a set of ranges that restricts variables to the given closed sets. ◇ You must `readlib(minimize)` before using this command. ◇ `maximize(sin(t));` ⟶ 1 ◇ `maximize(x^2-3, x, 1..2);` ⟶ 1 ◇ See also: `extrema`, `minimize`.

maxnorm `maxnorm(`*poly*`)`
Infinity norm of a polynomial ◇ Given an expanded polynomial, this function returns the maximum of the absolute values of the coefficients. The coefficients of the polynomial must be real. ◇ `maxnorm(x^2-3*y);` ⟶ 3 ◇ See also: `norm`.

maxorder `maxorder(`*K*`)`
`maxorder(`*K*`,` *var*`)`
Computes an integral basis for a number field or function field ◇ *K* is a `RootOf` expression or a set of `RootOf` expressions that define the field

extension. The variable name *var* can be supplied in the case of a function field extension. ◇ You must `readlib(maxorder)` before using this command. ◇ `maxorder(RootOf(y^2-5,y));` ⟶ $[1, -\frac{1}{2} + \frac{1}{2}\text{RootOf}(\_Z^2 - 5)]$ ◇ `maxorder(RootOf(y^2-x^2+x,y),x);` ⟶ $[1, \text{RootOf}(\_Z^2 - x^2 + x)]$ ◇ See also: `RootOf`.

MeijerG

`MeijerG(`*m*`,` *a*`,` *z*`)`
Special case of the general Meijer G function ◇ *m* is an integer greater than or equal to 2, *a* is a real expression, and *z* is a complex expression. ◇ `MeijerG(2,a,z);` ⟶ $\frac{\Gamma(a,z)}{z}$ ◇ See also: `GAMMA`.

mellin

`mellin(`*expr*`,` *x*`,` *s*`)`
Mellin transform ◇ This function applies the Mellin transformation to *expr* with respect to the variable *x*, returning an expression in *s*. This is defined as $\int_0^\infty expr\, x^{s-1}\, dx$. Some expressions involving exponentials, polynomials, algebraic functions, trigonometrics, and various special functions can be transformed. ◇ `mellin(exp(-x),x,s);` ⟶ $\Gamma(s)$ ◇ See also: `mellintable`.

mellintable

`mellintable(`*expr*`,` *ftransform*`,` *x*`,` *s*`)`
Adds an entry to the basic table of Mellin transforms ◇ This function creates an entry for *expr* in an internal table of basic Mellin transforms. *expr* is an expression in the variable *x*, and *ftransform* is its Mellin transform in the variable *s*. From this entry, the function `mellin` can deduce the Mellin transforms of expressions of the form $a\ln(x)^k t^b f(cx^d)$, where *expr* is $f(x)$, *a*, *b*, *c*, and *d* are real, *k* is a positive integer, and *c* is positive. ◇ You must `readlib(mellin)` before using this command.

member

`member(`*expr*`,` *list*`,` *name*`)`
`member(`*expr*`,` *set*`,` *name*`)`
Determines whether *expr* is a member of the given list or set ◇ This function returns *true* or *false*. The third argument is optional. If it is present and the function returns *true*, this name is assigned the position of the first occurrence of *expr* in the list or set. ◇ `member(b,[a,b,c]);` ⟶ *true* ◇ See also: `has`.

min

`min(`*expr₁*`,` *expr₂*`, ...)`
Computes the minimum of the arguments ◇ It may help to increase `Digits` if this function returns unevaluated. ◇ `min(exp(Pi),Pi^exp(1));` ⟶ $\pi^{(e)}$ ◇ See also: `max`, `minimize`.

minimize

`minimize(`*expr*`)`
`minimize(`*expr*`,` *vars*`)`
`minimize(`*expr*`,` *vars*`, infinite)`
`minimize(`*expr*`,` *vars*`,` *ranges*`)`
Computes the minimum ◇ This function returns the minimum value of *expr* or an expression sequence with all the values that are candidates for

the minimum value. In Release 3, *vars* may be a single variable or a set of variables. In Release 2 and earlier versions, *vars* must be a set of variables. If *vars* is not given, the expression is minimized with respect to all its indeterminates. ◇ A third argument is supported in Release 3. `infinite` requests the unconstrained minimum of the expression, including its limits at infinity (this is the behavior in earlier versions). *ranges* is a single range or a set of ranges that restricts variables to the given closed sets. ◇ You must `readlib(minimize)` before using this command. ◇ `minimize(x^2+y^2);` ⟶ 0 ◇ `minimize(sin(t), t, {t=1..2});` ⟶ sin(1) ◇ See also: `extrema`, `maximize`.

minpoly

`minpoly(`*r*`,` *n*`)`
`minpoly(`*r*`,` *n*`,` *acc*`)`
Finds the minimum polynomial with an approximate root ◇ This function uses the lattice reduction algorithm to find a polynomial of degree *n* or less with small integer coefficients which has the given approximation *r* of a real algebraic number as one of its roots. The third argument controls accuracy and has a default value of 10^{d-2}, where *d* is the current value of `Digits`. ◇ You must `readlib(lattice)` before using this command. ◇ `minpoly(1.414214,2);` ⟶ $-2+\_X^2$ ◇ See also: `lattice`, `linalg[minpoly]`.

mint

`mint` *filename*
`mint -i` *n filename*
The `mint` command ◇ This program is an external command that checks for syntax errors and common semantic errors in a Maple program file. It is patterned after lint, the syntax checker for C. The option `-i` *n* controls the detail of information produced. *n* may be 1, 2 (the default), 3, or 4. $n \geq 1$ reports severe errors, $n \geq 2$ reports serious errors, $n \geq 3$ reports warnings, and $n = 4$ gives a variable usage report.

minus

set_1 `minus` set_2
`` `minus`( ``set_1, set_2`)`
The set difference operator ◇ This function returns the set of elements in set_1 that are not in set_2. ◇ `{a,b} minus {b,c}` ⟶ $\{a\}$ ◇ See also: `intersect`, `symmdiff`, `union`.

mod

expr `mod` *m*
Computation over the integers modulo m ◇ This binary operator evaluates *expr* over the integers modulo m, using a system of representatives determined by the value of the environment variable '`mod`' The following functions for polynomial and matrix arithmetic over finite rings and fields are known to `mod`.

Berlekamp	Expand	GetAlgExt	Nullspace	ProbSplit	Resultant
Content	Factor	Hermite	Power	Quo	RootOf
Det	Factors	Interp	Powmod	Randpoly	Roots
Discrim	Gausselim	Irreduc	Prem	Randprime	Smith
DistDeg	Gaussjord	Lcm	Primitive	Ratrecon	Sprem
Divide	Gcd	Normal	Primpart	Rem	Sqrfree
Eval	Gcdex				

◇ `50 mod 11;` $\longrightarrow 6$ ◇ `Factor(x^2+1) mod 2;` $\longrightarrow (x+1)^2$ ◇
See also: `&^`, `'mod'`, `modp`, `modp1`, `mods`.

'mod'

Environment variable ◇ Its value is a procedure that chooses a system of representatives for performing arithmetic modulo an integer. Supported choices are `modp` for nonnegative representatives and `mods` for symmetric representatives. The default value is `modp`. ◇ See also: `mod`, `modp`, `mods`.

modp

`modp(`*expr*`,` *m*`)`
Computation over the integers modulo m: nonnegative representatives ◇ This function evaluates *expr* over the integers modulo m, using the system of representatives $\{0, 1, \ldots, |m| - 1\}$. ◇ `modp(50,11);` $\longrightarrow 6$ ◇
See also: `mod`, `'mod'`, `mods`.

modp1

`modp1(`*expr*`,` *m*`)`
Univariate polynomial arithmetic modulo m ◇ This function makes available efficient arithmetic and other operations in domain of univariate polynomials over the integers modulo *m*. The following functions are known to `modp1`.

Add	Det	Gcdex	Multiply	Randpoly	Smith
Chrem	Diff	Interp	One	Randprime	Sqrfree
Coeff	Divide	Irreduc	Power	Rem	Subtract
Constant	Embed	Lcm	Powmod	Resultant	Tcoeff
ConvertIn	Eval	Lcoeff	Prem	Roots	UNormal
ConvertOut	Factors	Ldegree	Quo	Shift	Zero
Degree	Gcd	Monomial			

◇ See also: `ConvertIn`, `ConvertOut`, `evalgf`, `GF`, `mod`.

modpol

`modpol(`*expr*`,` *poly*`,` *var*`,` *p*`)`
Expression evaluation in a quotient field ◇ *expr* is a rational expression in *var*, and *poly* is a polynomial in *var*. *p* is a prime number. This function finds a canonical representative for *expr* in GF(p)$[x]/$(*poly*). ◇ You must `readlib(modpol)` before using this command. ◇
`modpol(x/(x+1),x^2+x+1,x,2);` $\longrightarrow x+1$ ◇ See also: `GF`.

mods

`mods(`*expr*`,` *m*`)`
Computation over the integers modulo m: symmetric representatives ◇ This function evaluates *expr* over the integers modulo m, using the system of representatives $\left\{-\left\lfloor\frac{|m|-1}{2}\right\rfloor, \ldots, \left\lfloor\frac{|m|}{2}\right\rfloor\right\}$. ◇ `mods(50,11);` $\longrightarrow -5$
◇ See also: `mod`, `'mod'`, `modp`.

MOLS `MOLS(`*p*`, `*m*`, `*n*`)`
Mutually orthogonal Latin squares ◇ This function returns a list of n mutually orthogonal Latin squares of size p^m, if $n < p^m$. Two Latin squares A and B are orthogonal if the collection of ordered pairs (A_{ij}, B_{ij}) has no repeated elements. ◇ You must `readlib(MOLS)` before using this command.

msolve `msolve(`*eqns*`, `*vars*`, `*m*`)`
`msolve(`*eqns*`, `*m*`)`
Solves equations over the integers modulo m ◇ *eqns* is a single equation or a set of equations. By default at most 5 solutions are found. This number may be changed by assigning to the environment variable `_MaxSols`. *vars* names variables to be used in parameterizing a family of solutions. It may be a set or a single name. ◇ `msolve(x^3=11,41);` $\longrightarrow \{x = 6\}$ ◇ See also: `mod`, `isolve`, `numtheory[mroot]`, `numtheory[msqrt]`, `Roots`, `solve`.

mtaylor `mtaylor(`*expr*`, `*vars*`)`
`mtaylor(`*expr*`, `*vars*`, `*n*`)`
`mtaylor(`*expr*`, `*vars*`, `*n*`, `*list*`)`
Computes a multivariate Taylor series expansion of the input expression *expr*, with respect to the variables *vars*, to order *n*, using the variable weights *w* ◇ *vars* may be a list or set of names or equations. A name is short for the equation *name* = 0. *n* specifies the order of truncation (total degree) for the series. The fourth argument specifies weights (positive integers) for the variables: a weight of 2 halves the order to which the series is computed in the corresponding variable. The result is a polynomial. ◇ You must `readlib(mtaylor)` before using this command. ◇ `mtaylor(sin(x)*sin(y),[x,y],5);` $\longrightarrow$ $x\,y - \frac{1}{6}x\,y^3 - \frac{1}{6}x^3\,y$ ◇ See also: `coeftayl`, `poisson`, `series`, `taylor`.

nargs The number of arguments passed to a procedure ◇ Within a procedure, the special name `nargs` has as its value the number of elements in the expression sequence `args`. ◇ See also: `args`, `proc`, `procname`.

networks The networks package ◇ In the descriptions of functions in this package, *G* and *H* denote graphs, *v* stands for a vertex, and *e* represents an edge. A graph is represented by a special procedure of type `GRAPH`. ◇ In Release 2, the functions in this package must first be loaded using `with` before they can be invoked. ◇ See also: `networks[`*function*`]`, `type/GRAPH`, `with`.

networks[acycpoly]
`acycpoly(`*G*`, `*p*`)`
Acyclicity polynomial of an undirected graph ◇ This function gives the probability that *G* is acyclic when each edge operates with probability *p*. ◇ See also: `networks[chrompoly]`, `networks[flowpoly]`,

A B C D E F G H I J K L M N O P Q R S T U V W X Y Z

networks[rankpoly], networks[spanpoly], networks[tuttepoly].

networks[addedge]

addedge({v_1, v_2}, G)
addedge([v_1, v_2], G)
addedge(v_1, v_2, names=*enames*, weights=*eweights*, G)
addedge(Cycle(v_1, ..., v_n), G)
addedge(Path(v_1, ..., v_n), G)

Adds one or more edges to the graph ◇ An undirected edge is indicated by a set of vertices. A directed edge is represented by a list of two vertices: the tail is the first vertex, and the head is the second vertex. A list or set of such vertex collections may be given to add more than one edge at once. The names option gives a single name or a list of names to use for the new edges created. The weights option gives a single weight or a list of weights for the created edges (the default weight is 1). The Cycle form adds a cycle of edges connecting the named vertices, and Path adds edges connecting the named vertices in a path. An expression sequence of the names of the new edges is returned. ◇ See also: networks[addvertex], networks[edges], networks[ends], networks[eweight], networks[head], networks[tail].

networks[addvertex]

addvertex(*vertices*, G)
addvertex(*vertices*, weights=*vweights*, G)

Adds one or more vertices to the graph ◇ *vertices* may be a single vertex or a sequence, list, or set of vertices. The weights option gives a single weight or a list of weights for the new vertices (the default weight is 0). An expression sequence of the names of the new vertices is returned. ◇ See also: networks[addedge], networks[edges], networks[vertices], networks[vweight].

networks[adjacency]

adjacency(G)

Constructs the adjacency matrix of a graph ◇ This function constructs a matrix whose rows and columns are indexed by the vertices (in address order) and whose i, jth entry is the number of edges from vertex i to vertex j. ◇ See also: networks[charpoly], networks[incidence].

networks[allpairs]

allpairs(G)
allpairs(G, *name*)

Shortest distances in a graph ◇ This function is an implementation of Floyd's algorithm for finding the shortest distances between all pairs of vertices. The result is a table recording the shortest distance between each pair of vertices. A name given as a second parameter is assigned a table of ancestors. ◇ See also: networks[ancestor], networks[daughter], `networks[eweight]`, `networks[shortpathtree]`, `networks[spantree]`, `networks[vweight]`.

networks[ancestor]

`ancestor(`*v*`,` *G*`)`
`ancestor(`*vset*`,` *G*`)`
`ancestor(`*G*`)`
Finds ancestors in a directed tree ◇ The first form returns the set of ancestors of *v* in the graph. The second form returns the set of ancestors of the subgraph induced by the given set of vertices in *G*. With a single argument, this function returns the ancestor table of *G*, indexed by vertex. ◇ See also: `networks[daughter]`, `networks[path]`, `networks[shortpathtree]`, `networks[spantree]`.

networks[arrivals]

`arrivals(`*v*`,` *G*`)`
`arrivals(`*vset*`,` *G*`)`
`arrivals(`*G*`)`
Neighboring vertices found along incoming edges ◇ The first form returns the set of vertices that are at the tail of edges directed into *v*. Undirected edges are treated as bidirectional. The second form computes the arrival into the subgraph induced by the given set of vertices. The third form returns a table of arrivals indexed by vertex. ◇ See also: `networks[departures]`, `networks[head]`, `networks[neighbors]`, `networks[tail]`.

networks[bicomponents]

`bicomponents(`*G*`)`
Computes the biconnected components of a graph ◇ A connected graph can be decomposed into two connected components possibly connected by bridges. This function returns two sets in a list: the first indicates the bridges and the second indicates the biconnected components. Bridges are specified as edges. ◇ See also: `networks[addvertex]`, `networks[components]`, `networks[edges]`, `networks[ends]`, `networks[head]`, `networks[tail]`.

networks[charpoly]

`charpoly(`*G*`,` *x*`)`
Characteristic polynomial of an undirected graph ◇ This routine computes the characteristic polynomial of the adjacency matrix of *G* expressed as a polynomial in *var*. ◇ See also: `networks[adjacency]`.

networks[chrompoly]

`chrompoly(`*G*`,` λ`)`
Computes the chromatic polynomial of the graph *G* as a polynomial in λ ◇ The value of this polynomial gives the number of proper vertex-colorings of *G* using λ colors. ◇ See also: `networks[acycpoly]`,

networks[flowpoly], networks[rankpoly], networks[spanpoly], networks[tuttepoly].

networks[complement]

complement(*G*)
complement(*G*, *H*)
Finds the complement of a graph ◇ In the first form, the complement is relative to the complete graph on the same number of vertices. In the second form, the complement is relative to *H*. ◇ See also: networks[complete], networks[induce].

networks[complete]

complete(*n*)
complete(*vset*)
complete(*m*, *n*)
complete(m_1, ..., m_k)
Generates a complete graph ◇ The number of arguments indicates the number of parts. Each part is specified by an integer indicating the number of vertices in that part; for example, the third form generates a complete bipartite graph. In the case of one part, a set of vertex names may be given. ◇ See also: networks[cycle], networks[petersen], networks[void].

networks[components]

components(*G*)
Finds the connected components of a graph ◇ This function returns the components of the graph *G* as a set of sets of vertices. ◇ See also: networks[bicomponents], networks[induce], networks[vertices].

networks[connect]

connect(vs_1, vs_2, *G*)
connect(vs_1, vs_2, weights=$list_1$, *G*)
connect(vs_1, vs_2, names=$list_2$, *G*)
connect(vs_1, vs_2, directed, *G*)
Connects two sets of vertices in a graph ◇ Each *vs* is a list or set of vertices from *G*. This function connects each vertex in vs_1 to each vertex in vs_2 with an edge. The names of the new edges are returned in an expression sequence. The option directed makes directed edges from vs_1 to vs_2 (the default is undirected). Edge weights and names can be indicated in lists associated with the weights and names options. ◇ See also: networks[addedge], networks[edges].

A B C D E F G H I J K L M N O P Q R S T U V W X Y Z

networks[connectivity]

`connectivity(`*G*`)`
`connectivity(`*G*`,` *n*`)`
`connectivity(`*G*`,` *n*`,` $name_1$`,` $name_2$`)`
Computes the edge connectivity of the graph *G* ◇ *n* is a lower bound for the connectivity. $name_1$, if present, is assigned a set of vertices known to contain a minimum edge cut. $name_2$, if present, is assigned a set of vertices known to be supersaturated. ◇ See also: `networks[components]`, `networks[countcuts]`, `networks[flow]`.

networks[contract]

`contract(`*edges*`,` *G*`)`
`contract(`*vpairs*`,` *G*`)`
Contracts edges in a graph ◇ *edges* is a single edge or a list or set of edges. *vpairs* is a connected vertex pair uniquely identifying an edge or a list or set of connected vertex pairs. This function contracts the edges of *G* specified by the first argument. It returns the name of the vertex surviving the contraction of the specified edge or a set of surviving names if there is more than one contracting edge. ◇ See also: `networks[delete]`, `networks[edges]`, `networks[ends]`, `networks[shrink]`.

networks[countcuts]

`countcuts(`*G*`)`
Counts minimum network cuts of an undirected graph ◇ This function computes the number of network cutsets of minimum cardinality in an undirected multigraph. ◇ See also: `networks[connectivity]`, `networks[counttrees]`.

networks[counttrees]

`counttrees(`*G*`)`
Counts spanning trees of an undirected graph ◇ This function uses the Kirchoff Matrix-Tree theorem to count spanning trees of *G*. ◇ See also: `networks[spantree]`.

networks[cube]

`cube(`*n*`)`
`cube(`*vertices*`)`
`cube()`
Creates a hypercube ◇ The first form creates the *n*th hypercube, a graph with 2^n vertices. The second form names the vertices used in the graph. *vertices* is a list or a set with length a power of 2. The third form is equivalent to `cube(3)`. ◇ See also: `networks[dodecahedron]`, `networks[icosahedron]`, `networks[octahedron]`, `networks[petersen]`, `networks[tetrahedron]`.

networks[cycle]

`cycle(`*n*`)`

Creates a cycle ◇ This function returns a graph that is an undirected cycle containing *n* vertices and edges. ◇ See also: `networks[complete]`, `networks[void]`.

networks[cyclebase]

`cyclebase(`*G*`)`

Finds a cycle basis in an undirected graph ◇ This function finds all the fundamental cycles with respect to a spanning tree. It returns a set of cycles, each cycle represented by a set of edges. ◇ See also: `networks[components]`, `networks[fundcyc]`, `networks[spantree]`.

networks[daughter]

`daughter(`*v*`, `*G*`)`
`daughter(`*G*`)`

Finds daughters in a directed tree ◇ The first form returns the set of daughters of *v* in the graph. The second form returns the set of daughters of the subgraph induced by the given set of vertices in *G*. With a single argument, this function returns the daughter table of *G*, indexed by vertex. ◇ See also: `networks[ancestor]`, `networks[path]`, `networks[shortpathtree]`, `networks[spantree]`.

networks[degreeseq]

`degreeseq(`*G*`)`

Finds the degree sequence of a graph ◇ This function returns a list of vertex degrees sorted in ascending order. ◇ See also: `networks[indegree]`, `networks[maxdegree]`, `networks[mindegree]`, `networks[outdegree]`, `networks[vdegree]`.

networks[delete]

`delete(`*edges*`, `*G*`)`
`delete(`*vertices*`, `*G*`)`

Deletes edges or vertices from a graph ◇ This function modifies the given graph *G* by deleting the indicated edges or vertices. The first argument may be a single edge or vertex or a list or set of edges or vertices. The modified graph is also returned by the procedure. ◇ See also: `networks[complement]`, `networks[contract]`.

networks[departures]

`departures(`*v*`, `*G*`)`
`departures(`*vset*`, `*G*`)`
`departures(`*G*`)`

Neighboring vertices found along outgoing edges ◇ The first form returns the set of vertices that are at the head of edges directed out from *v*. Undirected edges are treated as bidirectional. The second form computes the departure from the subgraph induced by the given set of vertices. The third form returns a table of departures indexed by vertex. ◇ See also: `networks[arrivals]`, `networks[head]`, `networks[neighbors]`, `networks[tail]`.

networks[diameter]

`diameter(`*G*`)`

Computes the diameter of the graph ◇ The result is infinity if the graph is disconnected. Edge weights are assumed to be lengths or distances, and so must be nonnegative. ◇ See also: `networks[allpairs]`, `networks[eweight]`, `networks[shortpathtree]`, `networks[spantree]`.

networks[dinic]

`dinic(`*G*`,` v_1`,` v_2`)`
`dinic(`*G*`,` v_1`,` v_2`,` $name_1$`,` $name_2$`)`
`dinic(`*G*`,` v_1`,` v_2`,` $name_1$`,` $name_2$`,` *n*`)`

Augmenting-path flow algorithm ◇ This function returns the maximum flow from the source vertex v_1 to the sink vertex v_2. Edge weights of *G* are interpreted as capacities. $name_1$, if specified, is assigned the set of saturated edges; $name_2$, the set of vertices among these edges. If a nonnegative integer upper bound *n* is specified, the routine terminates after a flow of *n* is found, even if greater flows are possible. This function is normally called by `networks[flow]`. ◇ See also: `networks[connectivity]`, `networks[flow]`, `networks[spantree]`, `networks[shortpathtree]`.

networks[djspantree]

`djspantree(`*G*`)`

Edge-disjoint spanning trees of a graph ◇ This function uses Edmonds's matroid partitioning algorithm to find a partition of *G* into the minimum number of forests. As many as possible of the forests in the final partition are spanning trees. It returns a table of edge sets. ◇ See also: `networks[components]`, `networks[countcuts]`, `networks[flow]`, `networks[spantree]`.

networks[dodecahedron]

`dodecahedron()`

Creates a dodecahedron graph ◇ See also: `networks[cube]`, `networks[icosahedron]`, `networks[octahedron]`, `networks[petersen]`, `networks[tetrahedron]`.

networks[draw]

`draw(`*G*`)`
`draw(Concentric(`$vlist_1$`,` $vlist_2$`, ...),` *G*`)`
`draw(Linear(`$vlist_1$`,` $vlist_2$`, ...),` *G*`)`

Draws a graph ◇ This function displays the edges and vertices of a graph. The first form places the vertices at equal intervals around a circle. The

`Linear` option draws the vertex groups specified by the lists in parallel lines. The `Concentric` option draws the vertex groups specified by the lists in concentric circles, with the first list forming the innermost circle. ◇ See also: `networks[show]`.

networks[duplicate]

`duplicate(`*G*`)`

Copies a graph ◇ See also: `networks[complement]`, `networks[induce]`.

networks[edges]

`edges(`*G*`)`
`edges([`v_1`, `v_2`], `*G*`)`
`edges({`v_1`, `v_2`}, `*G*`)`
`edges(`*vpair*`, `*G*`, all)`

Finds edges in a graph ◇ The first form returns a set of all the edges in *G*. The second form returns a set of all directed edges from v_1 to v_2. The third form returns a set of all undirected edges connecting v_1 and v_2. In the last template, the set of all edges connecting the two vertices in *vpair* is returned, regardless of direction. *vpair* may be a set or list. ◇ See also: `networks[addedge]`, `networks[ends]`, `networks[head]`, `networks[tail]`, `networks[vertices]`.

networks[ends]

`ends(`*e*`, `*G*`)`
`ends(`*edges*`, `*G*`)`
`ends(`*G*`)`

Finds the ends of an edge in a graph ◇ Given a single edge as the first argument, this function returns the pair of vertices at the ends of the edge. If the edge is directed, this pair is a list with the tail vertex first; otherwise this pair is a set. Given a list or set of edges as the first argument, it returns a list or set of vertex pairs. The last form is equivalent to `ends(edges(`*G*`), `*G*`)` ◇ See also: `networks[edges]`, `networks[head]`, `networks[tail]`.

networks[eweight]

`eweight(`*e*`, `*G*`)`
`eweight(`*elist*`, `*G*`)`
`eweight(`*G*`)`

Finds edge weights in a graph ◇ This function returns the weight associated with the edge *e*, or the list of weights associated with the list of edges *elist*. The third form returns the edge weight table of *G*. ◇ See also: `networks[edges]`, `networks[vweight]`.

networks[flow]

`flow(`G`,` v_1`,` v_2`)`
`flow(`G`,` v_1`,` v_2`, maxflow=`n`)`
`flow(`G`,` v_1`,` v_2`,` *name*$_1$`,` *name*$_2$`)`
`flow(`G`,` v_1`,` v_2`,` *name*$_1$`,` *name*$_2$`, maxflow=`n`)`

Finds the maximum flow in a network ◇ This function returns the maximum flow from the source vertex v_1 to the sink vertex v_2. Edge weights of G are interpreted as capacities. *name*$_1$, if specified, is assigned the set of saturated edges; *name*$_2$, the set of vertices among these edges. The `maxflow` option sets n to be the maximum flow found. ◇ See also: `networks[connectivity]`, `networks[dinic]`, `networks[mincut]`, `networks[shortpathtree]`, `networks[spantree]`.

networks[flowpoly]

`flowpoly(`G`,` *var*`)`

Computes the flow polynomial of an undirected graph ◇ When *var* is an integer m, this is the number of nowhere-zero flows on G with edge labels chosen from integers modulo m. ◇ See also: `networks[acycpoly]`, `networks[chrompoly]`, `networks[rankpoly]`, `networks[spanpoly]`, `networks[tuttepoly]`.

networks[fundcyc]

`fundcyc(`*eset*`,` G`)`

Finds the cycle in a unicyclic undirected graph ◇ Given a subset of edges forming a unicyclic subgraph of G, this function returns the edges forming the unique cycle as a set. ◇ See also: `networks[components]`, `networks[cyclebase]`, `networks[rank]`, `networks[span]`.

networks[getlabel]

`getlabel(`G`)`

Retrieves the unique internal label of the graph ◇ Each graph has an internal label so that copies of the same graph can be distinguished from one another. ◇ See also: `networks[duplicate]`, `networks[new]`.

networks[girth]

`girth(`G`)`
`girth(`G`,` *name*`)`

Finds the length of the shortest cycle in the undirected graph G ◇ Infinity is returned if the graph has no cycle. If a name is given, it is assigned the set of edges forming a shortest cycle. ◇ See also: `networks[cycle]`.

networks[graph]

`graph(`*vset*`,` *eset*`)`

Generates a graph from a specified set of vertices and edges ◇ *eset* is a set of pairs of vertices. Each pair may be a list or a set: a list indicates a directed edge; a set, an undirected edge. ◇ See also: `networks[induce]`, `networks[new]`.

networks[graphical]

`graphical(`*intlist*`)`
`graphical(`*intlist*`, MULTI)`
Tests whether a list of integers is graphical ⋄ This function tests whether the list of integers is the degree sequence of a simple graph or a multigraph with no loops if `MULTI` is specified. If so, it returns a list of edges for one realization of the list as a graph. Otherwise, it returns *FAIL*. ⋄ `graphical([1,2,1]);` $\longrightarrow [\{2,3\},\{1,2\}]$ ⋄ See also: `networks[degreeseq]`, `networks[maxdegree]`, `networks[mindegree]`, `networks[vdegree]`.

networks[gsimp]

`gsimp(`*G*`)`
Creates a simple graph from a multigraph ⋄ This function deletes loops, replaces directed edges with undirected edges, and replaces multiple edges with a single edge of capacity equal to the combined capacities of the old edges. The modified graph *G* is returned. ⋄ See also: `networks[delete]`.

networks[gunion]

`gunion(`*G*`, `*H*`)`
`gunion(`*G*`, `*H*`, SIMPLE)`
Union of two graphs ⋄ This function creates a graph whose vertex set is the union of the vertex sets of *G* and *H*, and that has an edge for each edge of *G* and an edge for each edge of *H*. Multiple edges with the same endpoints are retained unless `SIMPLE` is specified, in which case they are reduced to a single edge. ⋄ See also: `networks[addedge]`, `networks[addvertex]`.

networks[head]

`head(`*e*`, `*G*`)`
`head(`*edges*`, `*G*`)`
`head(`*G*`)`
Finds the head of directed edges ⋄ If *e* is undirected, `NULL` is returned. If a list or set of edges *edges* is supplied, a list or set of head vertices is returned. If only the graph *G* is given, this function returns a table recording the head vertices of all the edges. ⋄ See also: `networks[ends]`, `networks[tail]`.

networks[icosahedron]

`icosahedron()`
Creates an icosahedron graph ⋄ See also: `networks[cube]`, `networks[dodecahedron]`, `networks[octahedron]`, `networks[petersen]`, `networks[tetrahedron]`.

networks[incidence]

`incidence(`*G*`)`

Makes an incidence matrix from a graph ◇ The rows of the incidence matrix represent vertices and columns represent edges. An outgoing edge at a vertex yields a -1 in the matrix; an incoming or undirected edge yields a 1. All other entries are 0. ◇ See also: `networks[adjacency]`, `networks[edges]`, `networks[ends]`, `networks[incident]`.

networks[incident]

`incident(`*v*`, `*G*`, `*option*`)`
`incident(`*vset*`, `*G*`, `*option*`)`

Finds the edges incident to a vertex set ◇ The first form returns the set of edges incident with the vertex *v*. Given a set of vertices, this function returns the set of cut edges to the subgraph induced by that set of vertices. *option* may be `In` or `Out` to specify which kind of directed edges should be considered. Undirected edges are considered bidirectional. ◇ See also: `networks[arrivals]`, `networks[departures]`, `networks[neighbors]`.

networks[indegree]

`indegree(`*v*`, `*G*`)`

Finds the in-degree of the vertex *v* ◇ This function considers only directed edges. ◇ See also: `networks[maxdegree]`, `networks[mindegree]`, `networks[neighbors]`, `networks[outdegree]`, `networks[vdegree]`.

networks[induce]

`induce(`*eset*`, `*G*`)`
`induce(`*vset*`, `*G*`)`

Induces a subgraph in a graph ◇ This function creates a subgraph of *G* from either a set of vertices and their interconnecting edges or a set of edges and their ends. ◇ See also: `networks[complement]`, `networks[edges]`, `networks[ends]`, `networks[incident]`, `networks[vertices]`.

networks[isplanar]

`isplanar(`*G*`)`

Checks whether the graph is planar ◇ This function returns *true* if the graph has a planar embedding and *false* otherwise. ◇ See also: `networks[bicomponents]`.

networks[maxdegree]

`maxdegree(`*G*`)`
`maxdegree(`*G*`, `*name*`)`

Finds the maximum vertex degree in an undirected graph ◇ This function computes the total number of undirected edges (including loops) incident at each vertex and returns the maximum. If a name is given, it is assigned a vertex with maximum degree. ◇ See also: `networks[indegree]`, networks[mindegree], networks[outdegree], networks[vdegree].

networks[mincut]

mincut(*G*, v_1, v_2)
mincut(*G*, v_1, v_2, *name*)
Finds the minimum cut in a network flow problem ◇ This function returns the smallest set of edges whose deletion disconnects all directed paths from the source node v_1 to the sink node v_2. If a name is given, it is assigned the value of the cut. ◇ See also: networks[connectivity], networks[countcuts], networks[edges], networks[flow].

networks[mindegree]

mindegree(*G*)
mindegree(*G*, *name*)
Finds the minimum vertex degree in an undirected graph ◇ This function computes the total number of undirected edges (including loops) incident at each vertex and returns the minimum. If a name is given, it is assigned a vertex with minimum degree. ◇ See also: networks[indegree], networks[maxdegree], networks[outdegree], networks[vdegree].

networks[neighbors]

neighbors(*v*, *G*)
neighbors(*vset*, *G*)
neighbors(*G*)
Finds neighboring vertices, treating all edges as undirected ◇ The first form returns the set of vertices that are at the ends of any edges incident with *v*, independent of direction. The second form computes the neighbors of the subgraph induced by the given set of vertices. The third form returns a table of neighbors indexed by vertex. ◇ See also: networks[arrivals], networks[departures].

networks[new]

new()
new(*name*)
Creates a new graph with no vertices or edges ◇ If a name is given, it is assigned the new graph. This function returns the new graph. ◇ See also: networks[addedge], networks[addvertex], networks[void].

networks[octahedron]

octahedron()
Creates an octahedron graph ◇ See also: networks[cube], networks[dodecahedron], networks[icosahedron], networks[petersen], networks[tetrahedron].

networks[outdegree]

`outdegree(`*v*`,` *G*`)`

Finds the out-degree of the vertex *v* ◇ This function considers only directed edges. ◇ See also: `networks[indegree]`, `networks[maxdegree]`, `networks[mindegree]`, `networks[neighbors]`, `networks[vdegree]`.

networks[path]

`path([`v_1`,` v_2`],` *G*`)`

Finds a path in a directed tree ◇ If there is a directed path from v_1 to v_2, this function returns a list of the vertices on the path. *FAIL* is returned when no such path can be found. ◇ See also: `networks[ancestor]`, `networks[daughter]`, `networks[shortpathtree]`.

networks[petersen]

`petersen()`

Creates the petersen graph ◇ See also: `networks[cube]`, `networks[dodecahedron]`, `networks[icosahedron]`, `networks[octahedron]`, `networks[tetrahedron]`.

networks[random]

`random(`*n*`)`

`random(`*n*`,` *m*`)`

`random(`*n*`, prob=`*p*`)`

Creates a random graph ◇ With a single argument *n*, this function returns a random graph with *n* vertices, with each edge appearing with probability $\frac{1}{2}$. Additional arguments specify the number of edges *m* or the probability *p* with which each edge occurs. ◇ See also: `networks[complete]`, `networks[cycle]`, `networks[new]`.

networks[rank]

`rank(`*e*`,` *G*`)`

`rank(`*eset*`,` *G*`)`

Finds the rank of an edge set ◇ The first argument may be a single edge or a set of edges. The rank is the number of vertices of *G* minus the number of components of the subgraph induced by the edge set. ◇ See also: `networks[components]`, `networks[induce]`, `networks[span]`.

networks[rankpoly]

`rankpoly(`*G*`,` *x*`,` *y*`)`

(Whitney) rank polynomial of an undirected graph ◇ The coefficient of $x^i y^j$ in the rank polynomial is the number of spanning subgraphs of *G* having i more components than *G* and having a cycle space of dimension j. ◇ See also: `networks[acycpoly]`, `networks[chrompoly]`, `networks[flowpoly]`, `networks[spanpoly]`, `networks[tuttepoly]`.

networks[shortpathtree]

`shortpathtree(`*G*`,` *v*`)`

Constructs a shortest path spanning tree ◇ This function implements Dijkstra's algorithm for a shortest path spanning tree rooted at *v*. Edge weights in *G* are assumed to be lengths or distances, and so must be nonnegative. The final distances are recorded as the vertex weights in the returned graph. ◇ See also: `networks[allpairs]`, `networks[diameter]`, `networks[eweight]`, `networks[spantree]`, `networks[vweight]`.

networks[show]

`show(`*G*`)`

Shows a table of information about a graph ◇ See also: `networks[draw]`, `networks[edges]`, `networks[eweight]`, `networks[neighbors]`, `networks[vertices]`, `networks[vweight]`.

networks[shrink]

`shrink(`*vset*`,` *G*`)`
`shrink(`*vset*`,` *G*`,` *vname*`)`

Identifies a group of vertices as a single vertex ◇ This function replaces the given set of vertices with a single new vertex in *G* and returns the name of this new vertex. The name of the shrunken vertex may be given as the third argument. ◇ See also: `networks[contract]`.

networks[span]

`span(`*e*`,` *G*`)`
`span(`*eset*`,` *G*`)`

Finds the span of an edge set ◇ This function returns the set of edges spanned by the given edge or set of edges in *G*. An edge is spanned by a given edge set if both its ends are in the set of ends of edges in the edge set. ◇ See also: `networks[fundcyc]`, `networks[induce]`, `networks[rank]`.

networks[spanpoly]

`spanpoly(`*G*`,` *p*`)`

Span polynomial of an undirected graph ◇ This function gives the probability that *G* is spanning when each edge operates with probability *p*. ◇ See also: `networks[acycpoly]`, `networks[chrompoly]`, `networks[flowpoly]`, `networks[rankpoly]`, `networks[tuttepoly]`.

networks[spantree]

`spantree(`*G*`)`
`spantree(`*G*`,` *v*`)`
`spantree(`*G*`,` *v*`,` *name*`)`

Finds a minimum weight spanning tree ◇ The chosen tree, rooted at *v*, has edges that minimize the total edge weight of the tree. If *name* is given, it is assigned the total weight. The result is returned as a new graph. ◇ See also: `networks[shortpathtree]`.

networks[tail]

`tail(`*e*`, `*G*`)`
`tail(`*edges*`, `*G*`)`
`tail(`*G*`)`

Finds the tail of directed edges ◇ If *e* is undirected, `NULL` is returned. If a list or set of edges *edges* is supplied, a list or set of tail vertices is returned. If only the graph *G* is given, this function returns a table recording the tail vertices of all the edges. ◇ See also: `networks[ends]`, `networks[head]`.

networks[tetrahedron]

`tetrahedron()`

Creates a tetrahedron graph ◇ See also: `networks[cube]`, `networks[dodecahedron]`, `networks[icosahedron]`, `networks[octahedron]`, `networks[petersen]`.

networks[tuttepoly]

`tuttepoly(`*G*`, `*t*`, `*z*`)`

Tutte polynomial of an undirected graph ◇ The Tutte polynomial is a sum over all maximal forests H of G of $t^{\mathrm{ia}(H)}z^{\mathrm{ea}(H)}$, where $\mathrm{ia}(H)$ is the internal activity of H and $\mathrm{ea}(H)$ is the external activity of H. ◇ See also: `networks[acycpoly]`, `networks[chrompoly]`, `networks[flowpoly]`, `networks[rankpoly]`, `networks[spanpoly]`.

networks[type/GRAPH]

`type(`*G*`, GRAPH)`

Checks for type GRAPH ◇ A graph is represented by a special procedure of type `GRAPH`. This type check is only available after the `networks` package has been loaded. ◇ See also: `networks`.

networks[vdegree]

`vdegree(`*v*`, `*G*`)`

Finds the vertex degree of *v* ◇ This function considers only undirected edges. ◇ See also: `networks[indegree]`, `networks[maxdegree]`, `networks[mindegree]`, `networks[neighbors]`, `networks[outdegree]`.

networks[vertices]

`vertices(`*G*`)`

Returns the vertex set of *G* ◇ This function returns the vertex set of *G*. ◇ See also: `networks[addvertex]`, `networks[edges]`.

networks[void]

`void(`*n*`)`
`void(`*vset*`)`
Generates a void graph ◇ This function generates a graph with *n* vertices and no edges. The vertex names may be provided in a set. ◇ See also: `networks[complete]`, `networks[cycle]`, `networks[petersen]`.

networks[vweight]

`vweight(`*v*`, `*G*`)`
`vweight(`*vlist*`, `*G*`)`
`vweight(`*G*`)`
Finds vertex weights in a graph ◇ This function returns the weight associated with the vertex *v* or the list of weights associated with the list of vertices *vlist*. The third form returns the vertex weight table of *G*. ◇ See also: `networks[eweight]`, `networks[vertices]`.

next Proceeds immediately with the next iteration of the surrounding `do` loop ◇ This statement must occur within a repetition statement. ◇ See also: `break`, `do`.

nextprime `nextprime(`*n*`)`
Computes the smallest probable prime larger than *n* ◇ This function uses `isprime` to compute the next probable prime. ◇ `nextprime(100);` ⟶ 101 ◇ See also: `isprime`, `ithprime`, `numtheory[safeprime]`, `prevprime`.

nops `nops(`*expr*`)`
Returns the number of operands in the expression ◇ For a list or a set, this is the number of elements in it. ◇ `nops([4,5,6]);` ⟶ 3 ◇ See also: `op`.

norm `norm(`*poly*`, `*r*`, `*vars*`)`
`norm(`*poly*`, infinity, `*vars*`)`
Norm of a polynomial ◇ For a given $r \geq 1$, this function computes $\left(\sum |c|^r\right)^{1/r}$ for c ranging over the coefficients of *poly* in the variables *vars*. If the second argument is `infinity`, the maximum of the $|c|$ is returned. *vars* may be a single variable name or a list or a set of variable names. If it is absent, `indets(`*poly*`)` is used. ◇ `norm(x^2+3,2);` ⟶ $\sqrt{10}$ ◇ See also: `coeffs`, `indets`, `linalg[norm]`, `maxnorm`, `Norm`.

Norm `Norm(`*expr*`)`
`Norm(`*expr*`, `*L*`, `*K*`)`
Inert norm function ◇ When used with `evala`, this function computes the norm of an algebraic number or algebraic function. In the first form, if *expr* is an algebraic number, the norm over the rationals is computed. If *expr* is an algebraic function, the norm over the smallest possible transcendental extension of the rationals is computed. In the second form, the norm is computed over the algebraic number or function field K, with *expr* viewed

as an element of *L(expr)*. K and L are sets of `RootOf`s, and L must be a subset of K. ◇ See also: `evala`, `linalg[norm]`, `norm`, `RootOf`.

normal `normal(`*expr*`)`
`normal(`*expr*`, expanded)`
Normalizes a rational expression ◇ The expression is converted into "factored normal form": a sum of terms is combined over a common denominator, and the resulting rational function is normalized so that the numerator and denominator are relatively prime factored polynomials with integer coefficients. If the option `expanded` is given, the numerator and denominator are expanded polynomials. ◇ `normal(x/y - x/(x+y));` $\longrightarrow \frac{x^2}{y\,(x+y)}$ ◇ See also: `denom`, `expand`, `factor`, `Normal`, `numer`, `simplify`.

Normal `Normal(`*expr*`)`
Inert normal function ◇ When used with `mod`, this function computes a normal form for the expression over a finite field. When used with `evala`, the normal form is computed over a field determined by the `RootOf`s in the coefficients of the expression. ◇ `Normal((x^3+1)/(x^2+x)) mod 2;` $\longrightarrow \frac{x^2+x+1}{x}$ ◇ See also: `evala`, `Expand`, `mod`, `normal`, `RootOf`.

Normalizer Environment variable ◇ Its value is a procedure for normalizing expressions. It is initially set to `normal`.

not `not` *a*
Logical operator ◇ This expression is true if *a* is false and false if *a* is true. It returns *FAIL* if *a* returns *FAIL*. ◇ See also: `and`, `evalb`, `or`.

np Newman-Penrose package ◇ The functions available are:

conj	eqns	V	X	X_V	Y_D	Y_X
D	suball	V_D	X_D	Y	Y_V	

This package has been superseded by the `NPspinor` package, which contains all of these functions. ◇ In Release 2 and earlier versions, the functions in this package must first be loaded using `with` before they can be invoked. ◇ See also: `NPspinor`, `NPspinor[`*function*`]`.

NPspinor NPspinor package ◇ This package is an extension of the obsolete `np` package. The functions in this package must be loaded using `with` before they can be invoked. This package uses global names for the various NP quantities. For a list of all global names used, display the value of `` `NPspinor/globals` `` after loading the package. ◇ See also: `NPspinor[`*function*`]`, `with`.

NPspinor[basis]
`basis(`*expr*`)`
`basis(`*expr*`, `*indexlist*`)`
Computes the set of fundamental spinors (i.e., basis spinors or spin metrics) present in the given expression ◇ If a list of spinor indices is

A B C D E F G H I J K L M N O P Q R S T U V W X Y Z

given as a second argument, then only those basis elements in the given indices will be found.

NPspinor[checkvars]

`checkvars()`

Checks for assignments to package variables ◇ This function checks the global variables used for the NP spin coefficients and curvature components and the spinors `i` and `o`. A warning is issued if any of these have values.

NPspinor[conj]

`conj(`*expr*`)`

Computes the complex conjugate of an object in the Newman-Penrose/spinor formalism according to the usual rules ◇ The value of `conj(`*name*`)` is output as *name*`c`.

NPspinor[contract]

`contract(`*expr*`)`

Evaluates contractions between basis spinors ◇ The procedure NPspinor[dyad] must be used to expand other objects in the spinor basis before contractions are evaluated. ◇ See also: `NPspinor[dyad]`.

NPspinor[D]

`D(`*expr*`)`

Newman-Penrose Pfaffian operator associated with the null tetrad vector l ◇ The global variable `const` allows names to be marked constant.

NPspinor[del]

`del(`*expr*`,` $index_1$`,` $index_2$`)`

Spinor covariant differentiation ◇ This function computes the covariant derivative (using the Newman-Penrose/spinor formalism) of `object` explicitly for scalars, some unevaluated procedure calls, and basis 1-spinors. Exactly one of the free spinor indices (i.e., names) must be "dotted" (i.e., end in the letter "c"). The global variable `const` contains the set of quantities with covariant derivative 0. ◇ See also: `NPspinor[basis]`, `NPspinor[D]`, `NPspinor[dyad]`.

NPspinor[dyad]

`dyad(`*expr*`)`

Expands expression in spinor dyad ◇ This function fully expands any forms of the curvature spinors (and their complex conjugates) with respect to dyad spinors `o` and `i`. ◇ See also: `NPspinor[basis]`, `NPspinor[contract]`, `NPspinor[del]`.

NPspinor[eqns]

`eqns()`

`eqns(`*n*`,` *m*`)`

Initialization and display of NP equations ◇ The first form initializes the "basic" equations (the Newman-Penrose Ricci and Bianchi identities) `eq1` through eq29 and assigns 29 to the variable `neqs`, or displays all `neqs` equations if they have already been initialized. The second form displays eq*n* through eq*m* if the equations have been initialized. The user must reset `neqs` when the set of basic equations is extended. ◇ See also: `NPspinor[suball]`.

NPspinor[findsymm]

`findsymm(`*expr*`,` *indexlist*`,` *flag*`)`
`findsymm(`*expr*`,` *indexlist*`)`
Rewrites symmetric subexpressions of an object ◇ This function rewrites any symmetrized collections of basis spinors in the free indices name in *indexlist* using the unevaluated function `Symm`. If a third argument appears, a nicer-looking (but henceforth inert) dummy notation for symmetrizations is used instead. ◇ See also: `NPspinor[rewrite]`, `NPspinor[symm]`.

NPspinor[makeeqn]

`makeeqn(`*expr*`)`
Converts expression to eqn/troff form ◇ The behavior is almost identical to the library function `eqn`, except with Newman-Penrose quantities and tables that may represent spinors. ◇ See also: `eqn`.

NPspinor[rewrite]

`rewrite(`*expr*`,` *newname*`,` *oldname*`)`
`rewrite(`*expr*`,` *newname*`,` *oldname*`,` *indexlist*`)`
Rewrites a spinor expression in terms of symbolic coefficients ◇ This function rewrites the given spinor condition in terms of symbolic coefficients (*newname*`[1]`, ...); the old coefficients are stored as an array in `oldname`. If a list of names is given as a fourth argument, the expression is treated as an expression in only those indices. ◇ See also: `NPspinor[symm]`.

NPspinor[suball]

`suball(`*expr*`)`
`suball(`*expr*`,` *n*`,` *m*`)`
Substitution utility ◇ This function makes substitutions from the set of basic Newman-Penrose equations (the Ricci and Bianchi identities) and their conjugates. The first form uses all `neqs` equations; the second form uses only equations eq*n* through eq*m*. ◇ See also: `NPspinor[eqns]`.

NPspinor[symm]

`symm(`*expr*`,` *indexlist*`)`
Symmetrizes an object in the given list of free indices ◇ The indices given must be all dotted or all undotted. ◇ See also: `NPspinor[findsymm]`, `NPspinor[rewrite]`.

NPspinor[V]

`V(`*expr*`)`
Newman-Penrose Pfaffian operator associated with the null tetrad vector n ◇ The global variable `const` allows names to be marked constant.

NPspinor[V_D]

`V_D(`*expr*`)`

Corresponds to the commutation relation associated with the Lie bracket $[V, D]$ ⋄ This function returns an equation whose left-hand side is the Lie bracket and whose right-hand side is its known linear expansion in terms of the NP derivatives.

NPspinor[X]

`X(`*expr*`)`

Newman-Penrose Pfaffian operator associated with the null tetrad vector m ⋄ The global variable `const` allows names to be marked constant.

NPspinor[X_D]

`X_D(`*expr*`)`

Corresponds to the commutation relation associated with the Lie bracket $[X, D]$ ⋄ This function returns an equation whose left-hand side is the Lie bracket and whose right-hand side is its known linear expansion in terms of the NP derivatives.

NPspinor[X_V]

`X_V(`*expr*`)`

Corresponds to the commutation relation associated with the Lie bracket $[X, V]$ ⋄ This function returns an equation whose left-hand side is the Lie bracket and whose right-hand side is its known linear expansion in terms of the NP derivatives.

NPspinor[Y]

`Y(`*expr*`)`

Newman-Penrose Pfaffian operator associated with the null tetrad vector $\overline{m}$ ⋄ The global variable `const` allows names to be marked constant.

NPspinor[Y_D]

`Y_D(`*expr*`)`

Corresponds to the commutation relation associated with the Lie bracket $[Y, D]$ ⋄ This function returns an equation whose left-hand side is the Lie bracket and whose right-hand side is its known linear expansion in terms of the NP derivatives.

NPspinor[Y_V]

`Y_V(`*expr*`)`

Corresponds to the commutation relation associated with the Lie bracket $[Y, V]$ ⋄ This function returns an equation whose left-hand side is the Lie bracket and whose right-hand side is its known linear expansion in terms of the NP derivatives.

NPspinor[Y_X]

`Y_X(`*expr*`)`

Corresponds to the commutation relation associated with the Lie bracket $[Y, X]$ ⋄ This function returns an equation whose left-hand side is the Lie bracket and whose right-hand side is its known linear expansion in terms of the NP derivatives.

NULL The empty expression sequence.

Nullspace `Nullspace(`*A*`)`
Inert nullspace function ◇ When used with `mod`, this function computes the nullspace of the linear transformation defined by the matrix *A* modulo a positive integer. The result is a set of vectors. ◇ See also: `linalg[kernel]`.

numapprox
Procedures for the numerical approximation of functions ◇ In Release 2, the functions in this package must first be loaded using `with` before they can be invoked. ◇ See also: `numapprox[`*function*`]`, `with`.

numapprox[chebpade]
`chebpade(`*expr*`, `*var*`=`*a*`..`*b*`, [`*m*`, `*n*`])`
`chebpade(`*expr*`, `*var*`, [`*m*`, `*n*`])`
`chebpade(`*f*`, `*a*`..`*b*`, [`*m*`, `*n*`])`
Computes a Chebyshev-Pade approximation of degree (m, n), where m is the desired degree of the numerator and n that of the denominator ◇ The first argument may be an expression in the given variable *var* or a function. This is expanded in a Chebyshev series on the interval *a..b* ($-1..1$ is used by default). The numerator and denominator are expressed in terms of the Chebyshev polynomials $\mathtt{T}(n, x)$. The third argument may be a single integer m, in which case n is taken to be 0. ◇ See also: `chebyshev`, `convert[ratpoly]`, `numapprox[minimax]`, `orthopoly[T]`.

numapprox[chebyshev]
`chebyshev(`*expr*`, `*var*`=`*a*`..`*b*`, `ϵ`)`
`chebyshev(`*expr*`, `*var*`, `ϵ`)`
Chebyshev series expansion ◇ See `chebyshev`.

numapprox[confracform]
`confracform(`*f*`)`
`confracform(`*expr*`)`
`confracform(`*expr*`, `*var*`)`
Converts a rational function or expression to continued fraction form ◇ The continued fraction form minimizes the number of operations required to evaluate the rational function or expression. If the first argument is a function, a function is returned. Otherwise it must be a rational expression, and an expression is returned. The variable in which *expr* is a rational expression may be given as the second argument. ◇ See also: `convert[confrac]`, `numapprox[hornerform]`, `numtheory[cfrac]`.

numapprox[hornerform]

`hornerform(`*f*`)`
`hornerform(`*expr*`)`
`hornerform(`*expr*`, ` *var*`)`
Converts the numerator and denominator of a rational expression to Horner form ⋄ For a polynomial, this form minimizes the number of arithmetic operations required for evaluation. If the first argument is a function, a function is returned. Otherwise it is a rational expression, and an expression is returned. A variable may be given as the second argument in this case. ⋄ See also: `convert[horner]`, `numapprox[confracform]`.

numapprox[infnorm]

`infnorm(`*expr*`, ` *var*`=`*a*`..`*b*`, ` *name*`)`
`infnorm(`*f*`, ` *a*`..`*b*`, ` *name*`)`
Computes the L-infinity norm $[a, b]$ ⋄ This function estimates the maximum value of the function or expression on the given interval. The first argument may be a function or an expression. In the latter case, *expr* is an expression in the variable *var*. If a name is given as a third argument, it is assigned the value of *var* where the maximum occurs. ⋄ `infnorm(sin,0..2);` ⟶ 1.0 ⋄ See also: `maxnorm`, `norm`.

numapprox[laurent]

`laurent(`*expr*`, ` *var*`=`*a*`, ` *n*`)`
`laurent(`*expr*`, ` *var*`, ` *n*`)`
Computes the laurent series expansion of the expression in the variable *var* up to order *n* ⋄ The expansion is computed about the point *a*, which is taken to be zero in the second form. The result must have finite principal part. ⋄ `laurent(1/(x*exp(x)),x,3);` ⟶
$x^{-1} - 1 + \frac{1}{2}x - \frac{1}{6}x^2 + \mathrm{O}\left(x^3\right)$ ⋄ See also: `series`, `taylor`, `type/laurent`.

numapprox[minimax]

`minimax(`*expr*`, ` *var*`=`*a*`..`*b*`, [`*m*`, ` *n*`], ` *weight*`, name)`
`minimax(`*f*`, ` *a*`..`*b*`, [`*m*`, ` *n*`], ` *weight*`, name)`
Computes the best minimax rational approximation of degree (m, n) for the given real function or expression on the interval $[a, b]$ ⋄ This function uses the Remez algorithm. *m* is the desired degree of the numerator; *n* that of the denominator. The third argument may simply be given as *m* if a polynomial approximation is desired. The first argument may be an expression in the variable *var* or a function. The optional fourth argument *weight* is a procedure or expression specifying the positive weight function (the default is 1). If a name is given as the fifth argument, it is assigned the maximum error of the approximation on $[a, b]$ with respect to the weight function. ⋄ `minimax(sinh(x),x=0..1,2);` ⟶
$.00596984887 + (.8958620986 + .2673993974\,x)\,x$ ⋄ See also: `numapprox[remez]`.

numapprox[pade]

`pade(`*expr*`,` *var*`=`*a*`, [`*m*`,` *n*`])`
`pade(`*expr*`,` *var*`, [`*m*`,` *n*`])`
Computes a Pade approximation of degree (*m*,*n*) for the given expression with respect to the variable *var* ◇ *m* is the desired degree of the numerator; *n* that of the denominator. The expression is expanded in a Laurent series about *var* = *a* (*var* = 0 in the second form) to order $m + n + 1$, then the Pade rational approximation is computed. The third argument may simply be given as *m* if a polynomial approximation is desired. ◇ `pade(sinh(x), x=0, [3,3]);` $\longrightarrow \frac{\frac{7}{60}x^3+x}{1-\frac{1}{20}x^2}$ ◇ See also: `convert[ratpoly]`, `numapprox[chebpade]`, `numapprox[laurent]`.

numapprox[remez]

`remez(`*weight*`,` *f*`,` *a*`,` *b*`,` *m*`,` *n*`,` *array*`,` *name*`)`
Remez algorithm for minimax rational approximation ◇ This is not usually invoked as a user-level routine. This function computes the best minimax rational approximation of degree (m, n) for a given real function *f* on the interval $[a, b]$ with respect to the positive weight function *weight*. *array* is an array indexed $1..m + n + 2$ containing an initial estimate of the critical set (the extremal points of the error curve). *name* is assigned the maximum X error of the approximation on $[a, b]$ with respect to the weight function. ◇ See also: `numapprox[minimax]`.

numapprox[taylor]

`taylor(`*expr*`,` *var*`=`*a*`,` *n*`)`
`taylor(`*expr*`,` *var*`,` *n*`)`
Taylor series expansion ◇ See `taylor`.

numboccur

`numboccur(`*expr*`,` *subexpr*`)`
Counts the number of times the subexpression is found in the expression ◇ This function is new in Release 3. ◇ `numboccur((s+t)^2*sin(s+t) + ln(s+t+1), s+t);` $\longrightarrow 2$ ◇ See also: `has`.

numer `numer(`*expr*`)`
Numerator of an expression ◇ If *expr* is not in normal form, it is converted to that form first. ◇ `numer(x+1/x);` $\longrightarrow x^2 + 1$ ◇ See also: `denom`, `normal`.

numtheory Number theory package ◇ See also: `numtheory[`*function*`]`, `with`.

numtheory[B]

`B(`*n*`)`
`B(`*n*`,` *x*`)`
Synonym for `bernoulli` and `numtheory[bernoulli]` ◇ Computes Bernoulli numbers and polynomials. ◇ See `bernoulli`.

numtheory[bernoulli]

`bernoulli(`*n*`)`
`bernoulli(`*n*`, `*x*`)`
Synonym for `bernoulli` ◇ Computes Bernoulli numbers and polynomials. ◇ See `bernoulli`.

numtheory[bigomega]

`bigomega(`*n*`)`
Number of prime divisors of *n* counted with multiplicity ◇ This function is new in Release 3. ◇ `bigomega(50);` $\longrightarrow 3$ ◇ See also: `numtheory[divisors]`, `numtheory[sigma]`, `numtheory[tau]`.

numtheory[cfrac]

`cfrac(`*confrac*`)`
`cfrac(`*z*`, `*n*`, `$name_c$`, `$name_d$`, `*options*`)`
`cfrac(`*rat*`, `*n*`, `*var*`, `*form*`, `*option*`)`
`cfrac(`*expr*`, `*n*`, `*var*`, `*diag*`, `*form*`, `*option*`)`
Continued fraction expansion of numbers, rational functions, series, and other algebraic objects expansion ◇ Only the first argument is required. *n* controls the number of partial quotients computed. Its default value is 10. The option `quotients` makes the function return a list of partial quotients rather than print them in fraction form. The first template computes the last convergent of the given continued fraction *confrac*. In the second template, *z* can be any number: rational, real, or complex. $name_c$ is a name that is assigned a list of the convergents, and $name_d$ is a list of the denominators of the convergents. The option `centered` in the numeric case computes the centered form of the continued fraction (where the numerators are allowed to be ± 1). The third template puts rational functions in continued fraction form. *rat* is a rational function in the variable *var*, and *form* can be `simple` (the default) or `regular`. The fourth template is for a series or other algebraic expressions in the given variable *var*. *diag* can be `superdiagonal` (the default) or `subdiagonal`. *form* can be `simple`, `semisimple` (the default), or `simregular`. ◇ `cfrac(E-1,5,quotients);` $\longrightarrow [1,1,2,1,1,4,\ldots]$ ◇ `cfrac((x^3+1)/(x^2+1),quotients);` $\longrightarrow \left[x, -x-1, -\frac{1}{2}x+\frac{1}{2}\right]$ ◇ `cfrac(exp(x),4,x,semisimple,quotients);` $\longrightarrow$ $[1,[x,1],[-x,2],[x,3],[-x,2],\ldots]$ ◇ See also: `convergs`, `convert/confrac`, `numtheory[cfracpol]`, `numtheory[nthconver]`, `thiele`.

numtheory[cfracpol]

`cfracpol(`*poly*`)`
`cfracpol(`*poly*`, `*n*`)`
Computes simple continued fraction expansions for the real roots of the given polynomial for all real roots of a rational polynomial ◇ This function returns a sequence of lists of partial quotients. *n* controls the number of partial quotients computed for each root. Its default value is 10.

◇ `cfracpol(x^2-2,4);` ⟶ $[-2, 1, 1, 2, 2, \ldots], [1, 2, 2, 2, 2, \ldots]$ ◇ See also: `convert[confrac]`, `numtheory[cfrac]`, `numtheory[nthconver]`.

numtheory[cyclotomic]

`cyclotomic(`*n*`, `*var*`)`
Calculates the *n*th cyclotomic polynomial in the variable *var* ◇ `cyclotomic(4,x);` ⟶ $x^2 + 1$

numtheory[divisors]

`divisors(`*n*`)`
Computes the set of positive divisors of *n* ◇ `divisors(6);` ⟶ $\{1, 2, 3, 6\}$ ◇ See also: `ifactor`, `numtheory[bigomega]`, `numtheory[factorset]`, `numtheory[sigma]`, `numtheory[tau]`.

numtheory[euler]

`euler(`*n*`)`
`euler(`*n*`, `*x*`)`
Synonym for `euler` ◇ Computes Euler numbers and polynomials. ◇ See `euler`.

numtheory[F]

`F()`
`F(`*n*`)`
`F(`*n*`, `*name*`)`
Synonym for `numtheory[fermat]` ◇ See `numtheory[fermat]`.

numtheory[factorEQ]

`factorEQ(`α`, `*d*`)`
`factorEQ(`*list*`, `*d*`)`
`factorEQ(`*set*`, `*d*`)`
Factorization in the ring of integers of the quadratic number field $\mathbf{Q}(\sqrt{d})$ ◇ The ring of integers must be Euclidean, so *d* must be one of the following integers: −1, −2, −3, −7, −11, 2, 3, 5, 6, 7, 11, 13, 17, 19, 21, 29, 33, 37, 41, 57, 73. The first argument may a single integer or a list or set of integers in $\mathbf{Q}(\sqrt{d})$. ◇ This function is new in Release 3. ◇ `factorEQ(5-sqrt(7),7);` ⟶ $\left(3 + \sqrt{7}\right) \left(2 - \sqrt{7}\right)^2$ ◇ See also: `GaussInt[GIfactor]`, `ifactor`, `numtheory[sq2factor]`.

numtheory[factorset]

`factorset(`*n*`)`
Computes the set of prime factors of *n* ◇ −1 is included if *n* is negative. ◇ `factorset(36);` ⟶ $\{2, 3\}$ ◇ See also: `ifactor`, `ifactors`, `numtheory[divisors]`.

numtheory[fermat]

`fermat()`
`fermat(`*n*`)`
`fermat(`*n*`, `*name*`)`
Information on Fermat numbers ◇ This function returns the *n*th Fermat number for *n* less than 20. The *n*th Fermat number is $2^{2^n}+1$. If a name is given as the second argument, known information on the factors and primality of this number is assigned to it. With no arguments, a list of the *n* for which the primality character of `fermat(`*n*`)` is known is returned. ◇ `fermat(4);` ⟶ 65537 ◇ See also: `numtheory[F]`.

numtheory[GIgcd]

`GIgcd(`ξ_1`, `ξ_2`, ..., `ξ_n`)`
Computes the greatest common divisor of the given Gaussian integers ◇ The result is the first quadrant associate of the gcd. ◇ `GIgcd(19+17*I,11+13*I);` ⟶ $1+3I$ ◇ See also: `GaussInt[GIgcd]`, `igcd`.

numtheory[ifactor]

`ifactor(`*n*`)`
`ifactor(`*n*`, `*methodname*`)`
Synonym for `ifactor` ◇ Integer factorization. ◇ See `ifactor`.

numtheory[ifactors]

`ifactors(`*n*`)`
Synonym for `ifactors` ◇ Integer factorization. ◇ See `ifactors`.

numtheory[imagunit]

`imagunit(`*n*`)`
Computes the square root of -1 mod n ◇ If there is no such square root, the function returns *FAIL*. ◇ `imagunit(101);` ⟶ 91 ◇ See also: `msolve`, `numtheory[mroot]`, `numtheory[msqrt]`.

numtheory[index]

`index(`*m*`, `*b*`, `*n*`)`
`index(`*m*`, `*b*`, `*n*`, `*name*`)`
Computes the index (or discrete logarithm) of *m* to the base *b* modulo *n* ◇ This function returns an integer `i` such that $b^i \equiv m \bmod n$ if it exists, otherwise it returns *FAIL*. If *name* is present, it is assigned the period of the solution. ◇ `index(17,2,47);` ⟶ 6 ◇ See also: `numtheory[mlog]`.

numtheory[invphi]

`invphi(`*n*`)`
Inverse of the totient function ◇ This function returns a list of all positive integers that map to *n* under the totient function. ◇ `invphi(6);` ⟶ [9, 18, 7, 14] ◇ See also: `numtheory[phi]`.

numtheory[isolve]

`isolve(`*eqns*`)`
`isolve(`*eqns*`, `*vars*`)`
Synonym for `isolve` ◇ Solves equations for integer solutions. ◇ See `isolve`.

numtheory[isprime]

`isprime(`*n*`, `*m*`)`
Synonym for `isprime` ◇ Probable prime test. ◇ See `isprime`.

numtheory[issqrfree]

`issqrfree(`*n*`)`
Tests whether an integer is squarefree ◇ An integer is squarefree if it is not divisible by a perfect square. This function returns *true* or *false*. ◇ `issqrfree(50);` ⟶ *false* ◇ See also: `ifactor`, `numtheory[nthpow]`.

numtheory[ithprime]

`ithprime(`*i*`)`
Synonym for `ithprime` ◇ Determines the *i*th prime number. The first prime number is 2. This function uses `isprime` to test for large primes. ◇ See also: `isprime`, `ithprime`, `nextprime`, `prevprime`.

numtheory[J]

`J(`*m*`, `*n*`)`
Synonym for `numtheory[jacobi]` ◇ See `numtheory[jacobi]`.

numtheory[jacobi]

`jacobi(`*m*`, `*n*`)`
Jacobi symbol ◇ This function computes the Jacobi symbol $\left(\frac{m}{n}\right)$, for *n* an odd positive integer relatively prime to *m*. ◇ `jacobi(15,1001);` ⟶ -1 ◇ See also: `numtheory[J]`, `numtheory[legendre]`.

numtheory[kronecker]

`kronecker(`*ineqset*`, `*xvarset*`, `*yvarset*`)`
`kronecker(`*coeflist*`, `*alphalist*`, `*errlist*`)`
Inhomogeneous Diophantine approximation ◇ This function finds an integer solution to the inequalities:

$$\begin{aligned}
|a_{11}\,x_1 + \cdots + a_{1n}\,x_n - \alpha_1 - y_1| &\leq \mathrm{err}_1 \\
|a_{21}\,x_1 + \cdots + a_{2n}\,x_n - \alpha_2 - y_2| &\leq \mathrm{err}_2 \\
&\vdots \\
|a_{n1}\,x_1 + \cdots + a_{nn}\,x_n - \alpha_n - y_n| &\leq \mathrm{err}_n
\end{aligned}$$

where | | is the ordinary absolute value or a *p*-adic valuation. The inequalities may be given explicitly in a set, as in the first template, with the x and y variables also named explicitly in sets (if there is only one inequality or variable, the set delimiters may be dropped). Or, the a_{ij} may be given as a list of lists, with the α_i and the err_i also given in lists, as in the second template. A list of x-values and a list of y-values are returned. ◇ `kronecker([[Pi,log(2)]],[E],[1/100]);` ⟶ [−15, −6], [−54] ◇ See also: `isolve`, `numtheory[minkowski]`, `padic`.

numtheory[L]

`L(`*m*`, `*n*`)`

Synonym for `numtheory[legendre]` ◇ See `numtheory[legendre]`.

numtheory[lambda]

`lambda(`*n*`)`

Carmichael's lambda function ◇ This computes the size of the largest cyclic subgroup of the group of units modulo n. This is the smallest positive integer i such that $a^i \equiv 1 \bmod n$ for all a relatively prime to n. ◇ `lambda(1001);` ⟶ 60 ◇ See also: `numtheory[phi]`.

numtheory[legendre]

`legendre(`*m*`, `*n*`)`

Legendre symbol ◇ This function computes the Legendre symbol $\left(\frac{m}{n}\right)$, which is defined to be 1 if m is a quadratic residue (mod n) and −1 if m is a quadratic nonresidue (mod n). ◇ `legendre(100,17);` ⟶ 1 ◇ See also: `numtheory[jacobi]`, `numtheory[L]`.

numtheory[M]

`M(`*n*`)`

`M([`*i*`])`

Synonym for `numtheory[mersenne]` ◇ See `numtheory[mersenne]`.

numtheory[mcombine]

`mcombine(`n_1`, `r_1`, `n_2`, `r_2`)`

Chinese remainder algorithm ◇ This function computes the unique positive integer n less than n_1n_2 such that $n \equiv r_1 \bmod n_1$ and $n \equiv r_2 \bmod n_2$. If no such integer exists, *FAIL* is returned. ◇ `mcombine(8,3,7,1);` ⟶ 43 ◇ See also: `chrem`.

numtheory[mersenne]

`mersenne(`*n*`)`

`mersenne([`*i*`])`

Mersenne primes ◇ If $2^n - 1$ is known to be composite, this function returns *false*. If it is known to be prime, then the Mersenne prime $2^n - 1$ is returned. If its primality character is unknown, this function returns *FAIL*. In the second template, this function returns the *i*th Mersenne prime. ◇ `mersenne(7);` ⟶ 127

numtheory[minkowski]

`minkowski(`*ineqset*`, `*xvarset*`, `*yvarset*`)`
`minkowski(`*coeflist*`, `*errlist*`)`
Homogeneous diophantine approximation (Minkowski's linear forms) ◇ This function finds an integer solution to the inequalities:

$$\begin{aligned} |a_{11}\,x_1 + \cdots + a_{1n}\,x_n - y_1| &\leq \text{err}_1 \\ |a_{21}\,x_1 + \cdots + a_{2n}\,x_n - y_2| &\leq \text{err}_2 \\ &\vdots \\ |a_{n1}\,x_1 + \cdots + a_{nn}\,x_n - y_n| &\leq \text{err}_n \end{aligned}$$

where | | is the ordinary absolute value or a *p*-adic valuation. The inequalities may be given explicitly in a set, as in the first template, with the x and y variables also named explicitly in sets (if there is only one inequality or variable, the set delimiters may be dropped). Or, the a_{ij} may be given as a list of lists, with the err_i also given in a list, as in the second template. A list of x-values and a list of y-values are returned. ◇ `minkowski([[Pi,log(2)]],[1/100]);` ⟶ [9, 40], [56] ◇ See also: `isolve`, `numtheory[kronecker]`.

numtheory[mipolys]

`mipolys(`*n*`, `*p*`)`
`mipolys(`*n*`, `*p*`, `*m*`)`
Number of monic irreducible univariate polynomials ◇ This function computes the number of monic irreducible univariate polynomials of degree *n* over the field with *p* elements or over the field with p^m elements. ◇ `mipolys(3,2);` ⟶ 2

numtheory[mlog]

`mlog(`*m*`, `*b*`, `*n*`)`
`mlog(`*m*`, `*b*`, `*n*`, `*name*`)`
Computes the discrete logarithm (or index) of *m* to the base *b* modulo *n* ◇ This function returns an integer i such that $b^i \equiv m \bmod n$ if it exists, otherwise it returns *FAIL*. If *name* is present, it is assigned the period of the solution. ◇ `mlog(17,2,47);` ⟶ 6 ◇ See also: `numtheory[index]`.

numtheory[mobius]

`mobius(`*n*`)`
Computes the Möbius function of *n* ◇ The Möbius function is 0 if *n* is not squarefree and $(-1)^{\omega(n)}$ otherwise, where $\omega(n)$ is the number of distinct prime divisors of *n*. ◇ `mobius(1001);` ⟶ -1 ◇ See also: `numtheory[issqrfree]`.

numtheory[mroot]

`mroot(`*m*`, `*p*`, `*n*`)`
Computes the *p*th root of *m* modulo *n* ◇ If a root does not exist, *FAIL* is returned. ◇ `mroot(2,3,11);` ⟶ 7 ◇ See also: `msolve`,

`numtheory[msqrt]`, `numtheory[primroot]`, `numtheory[rootsunity]`.

numtheory[msqrt]

`msqrt(`*m*`,` *n*`)`

Computes the square root of *m* modulo *n* ◇ If a square root does not exist, *FAIL* is returned. ◇ `msqrt(2,17);` ⟶ 6 ◇ See also: `msolve`, `numtheory[imagunit]`, `numtheory[mroot]`.

numtheory[nearestp]

`nearestp(`*basislist*`,` *pt*`)`

Computes the lattice point closest to the given point in the given lattice ◇ *basislist* is a list $[b_1, \ldots, b_n]$, where each b_i is a list of n real numbers. It represents a basis for a lattice in $\mathbf{R}^n$. *pt* is a list of n real numbers representing a point in $\mathbf{R}^n$. The function returns a list of integers $[m_1, \ldots, m_n]$ such that *pt* is closest to the lattice point $m_1 b_1 + \cdots + m_n b_n$. ◇ `nearestp([[1,0],[1/2,sqrt(3)/2]],[E,Pi]);` ⟶ [1,4] ◇ See also: `GaussInt[nearest]`, `numtheory[kronecker]`, `numtheory[minkowski]`.

numtheory[nextprime]

`nextprime(`*n*`)`

Synonym for `nextprime` ◇ Computes the smallest probable prime larger than *n*. This function uses `isprime` to compute the next probable prime. ◇ `nextprime(100);` ⟶ 101 ◇ See also: `isprime`, `nextprime`, `numtheory[prevprime]`, `numtheory[safeprime]`.

numtheory[nthconver]

`nthconver(`*confrac*`,` *n*`)`

Computes the *n*th convergent of a simple or regular continued fraction ◇ *confrac* may be in list or fraction form. ◇ `nthconver([1,2,2,2],3);` ⟶ $\frac{17}{12}$ ◇ See also: `convert[confrac]`, `numtheory[cfrac]`, `numtheory[cfracpol]`, `numtheory[nthdenom]`, `numtheory[nthnumer]`.

numtheory[nthdenom]

`nthdenom(`*confrac*`,` *n*`)`

Computes the denominator of the *n*th convergent of a simple or regular continued fraction ◇ `nthdenom([1,2,2,2],3);` ⟶ 12 ◇ See also: `numtheory[cfrac]`, `numtheory[nthconver]`, `numtheory[nthnumer]`.

numtheory[nthnumer]

`nthnumer(`*confrac*`,` *n*`)`

Computes the numerator of the *n*th convergent of a simple or regular continued fraction ◇ `nthnumer([1,2,2,2],3);` ⟶ 17 ◇ See also: `numtheory[cfrac]`, `numtheory[nthconver]`, `numtheory[nthdenom]`.

numtheory[nthpow]

`nthpow(m, n)`

Finds the largest *n*th power factor of an integer *m* ◇ `nthpow(5760,2);` ⟶ $(24)^2$ ◇ See also: `ifactor`, `numtheory[issqrfree]`.

numtheory[order]

`order(n, m)`

Computes the order of *n* modulo *m* ◇ This is the smallest integer *i* such that $n^i \equiv 1 \bmod m$. If *n* and *m* have a common factor, *FAIL* is returned. ◇ `order(2,7);` ⟶ 3 ◇ See also: `numtheory[pprimroot]`, `numtheory[primroot]`.

numtheory[phi]

`phi(n)`

Totient function ◇ This function computes the number of positive integers less than *n* and relatively prime to *n*. This is the size of the group of units modulo *n*. ◇ `phi(14);` ⟶ 6 ◇ See also: `numtheory[invphi]`, `numtheory[lambda]`.

numtheory[pprimroot]

`pprimroot(m, n)`
`pprimroot(n)`

Computes the smallest pseudo-primitive root larger than *m* in the group of units modulo *n* ◇ A pseudo-primitive root is an element of order `lambda(n)` in the group of units modulo *n*. If there are no pseudo-primitive roots larger than *m*, *FAIL* is returned. If *m* is absent, it is assumed to be 0. ◇ `pprimroot(1001);` ⟶ 2 ◇ See also: `numtheory[lambda]`, `numtheory[order]`, `numtheory[phi]`, `numtheory[primroot]`.

numtheory[prevprime]

`prevprime(n)`

Synonym for `prevprime` ◇ Computes the largest probable prime smaller than *n*. This function uses `isprime` to compute the previous probable prime. ◇ `prevprime(100);` ⟶ 97 ◇ See also: `isprime`, `prevprime`, `numtheory[nextprime]`.

numtheory[primroot]

`primroot(m, n)`
`primroot(n)`

Computes the smallest primitive root larger than *m* in the group of units modulo *n* ◇ A primitive root is an element of order `phi(n)` in the group of units modulo *n*. If there are no primitive roots larger than *m*, *FAIL* is returned. If *m* is absent, it is assumed to be 0. ◇ `primroot(17);` ⟶ 3 ◇ See also: `numtheory[lambda]`, `numtheory[order]`, `numtheory[phi]`, `numtheory[pprimroot]`.

numtheory[rootsunity]

`rootsunity(`p`, `n`)`

Roots of unity ◇ This function calculates all the pth roots of unity mod n, returning an expression sequence. ◇ `rootsunity(3,7);` ⟶ $1, 2, 4$ ◇ See also: `numtheory[order]`.

numtheory[safeprime]

`safeprime(`n`)`

Computes the smallest safe probable prime larger than n ◇ A prime p is safe if $(p-1)/2$ is also prime. In Release 3, `isprime` is used to test primality. ◇ In Release 2 and earlier versions, a Fermat–Euler test with bases 2, 3, 5, 7, and 11 is used to test primality. This is weaker than the test in `isprime`. ◇ `safeprime(100);` ⟶ 107 ◇ See also: `isprime`, `numtheory[nextprime]`.

numtheory[sigma]

`sigma(`n`)`
`sigma[`k`](`n`)`

Computes the sum of the positive divisors of n ◇ The second form computes the sum of the kth powers of the positive divisors of n, where k is any nonnegative integer. ◇ `sigma(70);` ⟶ 144 ◇ See also: `ifactor`, `numtheory[bigomega]`, `numtheory[divisors]`, `numtheory[tau]`.

numtheory[sq2factor]

`sq2factor(`α`)`
`sq2factor(`*list*`)`
`sq2factor(`*set*`)`

Factorization in $\mathbf{Z}\left[\sqrt{2}\right]$ ◇ The argument is either a single number α of the form $n + m\sqrt{2}$, where n and m are integers, or a list or set of such numbers. It returns a multiplicative expression. ◇ `sq2factor(sqrt(2)-11);` ⟶ $\left(1-2\sqrt{2}\right)\left(1+3\sqrt{2}\right)$ ◇ See also: `GaussInt[GIfactor]`, `ifactor`, `numtheory[factorEQ]`.

numtheory[sum2sqr]

`sum2sqr(`n`)`

Finds the ways to write n as a sum of two squares ◇ A list of solutions is returned, where each solution is a list of two integers. ◇ This function is new in Release 3. ◇ `sum2sqr(65);` ⟶ $[[4,7],[1,8]]$ ◇ See also: `isolve`.

numtheory[tau]

`tau(`n`)`

Computes the number of positive divisors of n ◇ `tau(70);` ⟶ 8 ◇ See also: `ifactor`, `numtheory[bigomega]`, `numtheory[divisors]`, `numtheory[sigma]`.

numtheory[thue]

`thue(`*eqn*`, [`*var₁*`, `*var₂*`], `*n*`, `*name*`)`
`thue(`*ineq*`, [`*var₁*`, `*var₂*`], `*n*`, `*name*`)`
Computes all solutions to a Thue equation or inequality up to a given bound ◇ A Thue equation has the form $f(x, y) = m$, where $f(x, y)$ is a homogeneous irreducible polynomial in x and y with integer coefficients, and m is an integer. A Thue inequality has the form $|f(x, y)| \leq m$. This function computes all integer solutions of the given Thue equation or inequality in the variables var_1 and var_2, subject to the constraint $|var_2| \leq 10^n$. Only the first two arguments are required. If n is omitted, it defaults to 10. When *name* is present with a Thue equation, if no solutions are found, then *name* is assigned a sequence of all integers m' with $m' \leq 2|m|$ for which a solution was found. With an inequality, *name* is assigned a sequence of integers $m' \leq m$ for which equality holds among the solutions. ◇ `thue(x^3-3*x*y^2-y^3=37,[x,y],3);` ⟶ $[x = -5, y = -3], [x = -3, y = 8], [x = 8, y = -5]$ ◇ See also: `isolve`.

O — The order term function for series ◇ `series(exp(x), x=0, 2);` ⟶ $1 + x + O\left(x^2\right)$ ◇ See also: `order`, `Order`.

oframe — Orthonormal tetrads in the Ellis-MacCallum formalism ◇ This collection of procedures allows formal manipulation of the variables in the Ellis-MacCallum formalism:

com01	com03	com23	clear	e0	e2	deriv	transfo
com02	com12	com31	const	e1	e3	id	

The `com`*ij* give the commutation relations. The `e`*i* are the directional derivative operators, and `deriv` is the derivative operator. `const` defines constants and `id` tests whether an equation is an identity. The global variables `j1` through `j16` contain the Jacobi identities, and `ein00` through `ein33` contain the Einstein field equations. You must `readlib(oframe)` before using these functions. ◇ See also: `NPspinor`.

op

`op(`*i*`, `*expr*`)`
`op(`*i*`..`*j*`, `*expr*`)`
`op(`*expr*`)`
Extracts operands from an expression ◇ This first form extracts the ith component of an expression. The second returns the sequence of the ith through jth components, and the last form returns the sequence of all the components of the expression. ◇ `op(2,a+b);` ⟶ b ◇ See also: `nops`.

open

`open(`*filename*`)`
Opens the file for output ◇ Only one file may be open at a time. ◇ You must `readlib(write)` before using this command. ◇ See also: `close`, `write`, `writeln`.

optimize `optimize(`*expr*`)`
`optimize(`*A*`)`
Common subexpression optimization ◇ This function returns a sequence of equations representing an optimized computation sequence for the input expression. This function optimizes only syntactically identical subexpressions. ◇ You must `readlib(optimize)` before using this command. ◇ `optimize(x^3);` $\longrightarrow t1 = x^2, t2 = t1\,x$ ◇ See also: `C`, `cost`, `fortran`, `'optimize/makeproc'`.

'optimize/makeproc'
`'optimize/makeproc'(`*eqnlist*`,` *options*`)`
Generates a procedure from a computation sequence ◇ The returned procedure executes the assignments represented by the equations. Which names are to be considered parameters, locals, and globals can be indicated by specifying options of the form `parameters=`*list*, `locals=`*list*, `globals=`*list*. ◇ You must `readlib('optimize/makeproc')` before using this command. ◇ See also: `C`, `cost`, `fortran`, `optimize`.

options `option` *option₁*`,` *option₂*`, ...`
`options` *option₁*`,` *option₂*`, ...`
Procedure options ◇ The `option` declaration appears immediately after the `local` declaration at the beginning of a procedure. Recognized options are `angle`, `arrow`, `builtin`, `Copyright`, `operator`, `remember`, `system`, and `trace`. ◇ See also: `global`, `local`, `proc`, `remember`, `trace`.

or *a* `or` *b*
Logical operator ◇ This expression is true if at least one of *a* and *b* are true, and false if both *a* and *b* are false. If one returns *FAIL* and the other is false or *FAIL*, it returns *FAIL*. *b* is not evaluated if *a* is true. ◇ See also: `and`, `evalb`, `not`.

order `order(`*series*`)`
Determines the truncation order of a series ◇ This function returns the degree of the indeterminate in the `O( )` term of a series. ◇ See also: `numtheory[order]`, `Order`.

Order Environment variable ◇ The default order of truncation for series. Its initial value is 6. ◇ See also: `O`.

orthopoly Orthogonal polynomial package ◇ See also: `orthopoly[`*function*`]`, `with`.

orthopoly[G]
`G(`*n*`,` *a*`,` *x*`)`
Computes the *n*th Gegenbauer (ultraspherical) polynomial, with parameter *a*, evaluated at *x* ◇ `G(2,1,x);` $\longrightarrow 4\,x^2 - 1$

orthopoly[H]

`H(`*n*`,` *x*`)`

Computes the *n*th Hermite polynomial, evaluated at *x* ◇ `H(3,y);` ⟶ $8\,y^3 - 12\,y$ ◇ See also: `linalg[hermite]`.

orthopoly[L]

`L(`*n*`,` *a*`,` *x*`)`
`L(`*n*`,` *x*`)`

Computes the *n*th generalized Laguerre polynomial with parameter *a*, evaluated at *x* ◇ The two argument case takes $a = 0$. ◇ `L(2,2,x);` ⟶ $6 - 4\,x + \frac{1}{2}x^2$

orthopoly[P]

`P(`*n*`,` *a*`,` *b*`,` *x*`)`
`P(`*n*`,` *x*`)`

Jacobi and Legendre polynomials ◇ The first form computes the *n*th Jacobi polynomial, with parameters *a* and *b*, evaluated at *x*. With two arguments, this function computes the *n*th Legendre (spherical) polynomial evaluated at *x*. It is equivalent to the first form with $a = 0$ and $b = 0$. ◇ `P(2,2,2,s);` ⟶ $7\,s^2 - 1$ ◇ See also: `numtheory[jacobi]`, `numtheory[legendre]`.

orthopoly[T]

`T(`*n*`,` *x*`)`

Computes the *n*th Chebyshev polynomial of the first kind, evaluated at *x* ◇ `T(5,x);` ⟶ $16\,x^5 - 20\,x^3 + 5\,x$ ◇ See also: `chebyshev`, `numapprox[chebpade]`.

orthopoly[U]

`U(`*n*`,` *x*`)`

Computes the *n*th Chebyshev polynomial of the second kind, evaluated at *x* ◇ `U(5,x);` ⟶ $32\,x^5 - 32\,x^3 + 6\,x$

padic — The *p*-adic number package ◇ Suggested references on *p*-adic numbers: *p-adic Numbers: An Introduction*, F. Q. Gouvêa (Springer-Verlag 1993); *p-adic Numbers, p-adic Analysis, and Zeta-Functions*, N. Koblitz (Springer-Verlag 1977); *Introduction to p-adic Numbers and Their Functions*, K. Mahler (Cambridge 1973); *Ultrametric Calculus: An Introduction to p-adic Analysis*, W. H. Schikhof (Cambridge 1984). ◇ In Release 2, the functions in this package must first be loaded using `with` before they can be invoked. ◇ See also: `padic[`*function*`]`, `with`.

padic[arccoshp]

`arccoshp(`*expr*`,` *p*`,` *s*`)`
`arccoshp(`*expr*`,` *p*`)`
`arccoshp(`*expr*`)`

p-adic arccosh ◇ This function computes the *p*-adic hyperbolic inverse cosine of the expression *expr* and prints *s* terms of the result. If *s* is missing, the current value of `Digitsp` is assumed. If the result is not p-adically convergent, *FAIL* is returned. `arccoshp` is a short form for `evalp@arccosh`. ◇ `arccoshp(10,5,4);` $\longrightarrow 5+2\,5^2+4\,5^3+\mathrm{O}\left(5^5\right)$ ◇ See also: `padic[coshp]`.

padic[arccosp]

`arccosp(`*expr*`,` *p*`,` *s*`)`
`arccosp(`*expr*`,` *p*`)`
`arccosp(`*expr*`)`
p-adic arccos ◇ This function computes the p-adic inverse cosine of the expression *expr* and prints s terms of the result. If s is missing, the current value of `Digitsp` is assumed. If the result is not p-adically convergent, *FAIL* is returned. `arccosp` is a short form for `evalp@arccos`. ◇ `arccosp(97,7,4);` $\longrightarrow 2\,7+5\,7^3+\mathrm{O}\left(7^4\right)$ ◇ See also: `padic[cosp]`.

padic[arccothp]

`arccothp(`*expr*`,` *p*`,` *s*`)`
`arccothp(`*expr*`,` *p*`)`
`arccothp(`*expr*`)`
p-adic arccoth ◇ This function computes the p-adic hyperbolic inverse cotangent of the expression *expr* and prints s terms of the result. If s is missing, the current value of `Digitsp` is assumed. If the result is not p-adically convergent, *FAIL* is returned. `arccothp` is a short form for `evalp@arccoth`. ◇ `arccothp(3/4,2,6);` $\longrightarrow 2^2+2^3+\mathrm{O}\left(2^8\right)$ ◇ See also: `padic[cothp]`.

padic[arccotp]

`arccotp(`*expr*`,` *p*`,` *s*`)`
`arccotp(`*expr*`,` *p*`)`
`arccotp(`*expr*`)`
p-adic arccot ◇ This function computes the p-adic inverse cotangent of the expression *expr* and prints s terms of the result. If s is missing, the current value of `Digitsp` is assumed. If the result is not p-adically convergent, *FAIL* is returned. `arccotp` is a short form for `evalp@arccot`. ◇ `arccotp(64/3,3,6);` $\longrightarrow 3+2\,3^2+3^3+\mathrm{O}\left(3^6\right)$ ◇ See also: `padic[cotp]`.

padic[arccschp]

`arccschp(`*expr*`,` *p*`,` *s*`)`
`arccschp(`*expr*`,` *p*`)`
`arccschp(`*expr*`)`
p-adic arccsch ◇ This function computes the p-adic hyperbolic inverse cosecant of the expression *expr* and prints s terms of the result. If s is missing, the current value of `Digitsp` is assumed. If the result is not p-adically convergent, *FAIL* is returned. `arccschp` is a short form for `evalp@arccsch`. ◇ `arccschp(4/121,11,4);` $\longrightarrow 3\,11^2+11^4+\mathrm{O}\left(11^5\right)$ ◇ See also: `padic[cschp]`.

padic[arccscp]

`arccscp(`*expr*`, `*p*`, `*s*`)`
`arccscp(`*expr*`, `*p*`)`
`arccscp(`*expr*`)`
p-adic arccsc ◇ This function computes the p-adic inverse cosecant of the expression *expr* and prints s terms of the result. If s is missing, the current value of `Digitsp` is assumed. If the result is not p-adically convergent, *FAIL* is returned. `arccscp` is a short form for `evalp@arccsc`. ◇ `arccscp(7/4,2,4);` $\longrightarrow 2^2+2^3+2^4+O\left(2^5\right)$ ◇ See also: `padic[cscp]`.

padic[arcsechp]

`arcsechp(`*expr*`, `*p*`, `*s*`)`
`arcsechp(`*expr*`, `*p*`)`
`arcsechp(`*expr*`)`
p-adic arcsech ◇ This function computes the p-adic hyperbolic inverse secant of the expression *expr* and prints s terms of the result. If s is missing, the current value of `Digitsp` is assumed. If the result is not p-adically convergent, *FAIL* is returned. `arcsechp` is a short form for `evalp@arcsech`. ◇ `arcsechp(6,17,3);` $\longrightarrow 2\,17+4\,17^2+O\left(17^3\right)$ ◇ See also: `padic[sechp]`.

padic[arcsecp]

`arcsecp(`*expr*`, `*p*`, `*s*`)`
`arcsecp(`*expr*`, `*p*`)`
`arcsecp(`*expr*`)`
p-adic arcsec ◇ This function computes the p-adic inverse secant of the expression *expr* and prints s terms of the result. If s is missing, the current value of `Digitsp` is assumed. If the result is not p-adically convergent, *FAIL* is returned. `arcsecp` is a short form for `evalp@arcsec`. ◇ `arcsecp(49,5,4);` $\longrightarrow 5+2\,5^2+O\left(5^4\right)$ ◇ See also: `padic[secp]`.

padic[arcsinhp]

`arcsinhp(`*expr*`, `*p*`, `*s*`)`
`arcsinhp(`*expr*`, `*p*`)`
`arcsinhp(`*expr*`)`
p-adic arcsinh ◇ This function computes the p-adic hyperbolic inverse sine of the expression *expr* and prints s terms of the result. If s is missing, the current value of `Digitsp` is assumed. If the result is not p-adically convergent, *FAIL* is returned. `arcsinhp` is a short form for `evalp@arcsinh`. ◇ `arcsinhp(1,7,3);` $\longrightarrow 5\,7+4\,7^2+O\left(7^3\right)$ ◇ See also: `padic[sinhp]`.

padic[arcsinp]

`arcsinp(`*expr*`, `*p*`, `*s*`)`
`arcsinp(`*expr*`, `*p*`)`
`arcsinp(`*expr*`)`
p-adic arcsin ◇ This function computes the p-adic inverse sine of the expression *expr* and prints s terms of the result. If s is missing, the current value of `Digitsp` is assumed. If the result is not p-adically convergent, *FAIL* is returned. `arcsinp` is a short form for `evalp@arcsin`. ◇ `arcsinp(39/4,13,3);` $\longrightarrow 4\,13 + 3\,13^2 + O\left(13^3\right)$ ◇ See also: `padic[sinp]`.

padic[arctanhp]

`arctanhp(`*expr*`, `*p*`, `*s*`)`
`arctanhp(`*expr*`, `*p*`)`
`arctanhp(`*expr*`)`
p-adic arctanh ◇ This function computes the p-adic hyperbolic inverse tangent of the expression *expr* and prints s terms of the result. If s is missing, the current value of `Digitsp` is assumed. If the result is not p-adically convergent, *FAIL* is returned. `arctanhp` is a short form for `evalp@arctanh`. ◇ `arctanhp(2/3,11,5);` $\longrightarrow 2\,11 + 6\,11^4 + O\left(11^5\right)$ ◇ See also: `padic[tanhp]`.

padic[arctanp]

`arctanp(`*expr*`, `*p*`, `*s*`)`
`arctanp(`*expr*`, `*p*`)`
`arctanp(`*expr*`)`
p-adic arctan ◇ This function computes the p-adic inverse tangent of the expression *expr* and prints s terms of the result. If s is missing, the current value of `Digitsp` is assumed. If the result is not p-adically convergent, *FAIL* is returned. `arctanp` is a short form for `evalp@arctan`. ◇ `arctanp(27/8,3,3);` $\longrightarrow 2\,3^3 + 2\,3^4 + O\left(3^6\right)$ ◇ See also: `padic[tanp]`.

padic[coshp]

`coshp(`*expr*`, `*p*`, `*s*`)`
`coshp(`*expr*`, `*p*`)`
`coshp(`*expr*`)`
p-adic cosh ◇ This function computes the p-adic hyperbolic cosine of the expression *expr* and prints s terms of the result. If s is missing, the current value of `Digitsp` is assumed. If the result is not p-adically convergent, *FAIL* is returned. `coshp` is a short form for `evalp@cosh`. ◇ `coshp(95,19,2);` $\longrightarrow 1 + O\left(19^2\right)$ ◇ See also: `padic[arccoshp]`.

padic[cosp]

`cosp(`*expr*`, `*p*`, `*s*`)`
`cosp(`*expr*`, `*p*`)`
`cosp(`*expr*`)`
p-adic cosine ◇ This function computes the *p*-adic cosine of the expression *expr* and prints *s* terms of the result. If *s* is missing, the current value of `Digitsp` is assumed. If the result is not *p*-adically convergent, *FAIL* is returned. `cosp` is a short form for `evalp@cos`. ◇
`cosp(343,7,12,8);` $\longrightarrow 1 + 3\,7^6 + O\left(7^7\right)$ ◇ See also: `padic[arccosp]`.

padic[cothp]

`cothp(`*expr*`, `*p*`, `*s*`)`
`cothp(`*expr*`, `*p*`)`
`cothp(`*expr*`)`
p-adic coth ◇ This function computes the *p*-adic hyperbolic cotangent of the expression *expr* and prints *s* terms of the result. If *s* is missing, the current value of `Digitsp` is assumed. If the result is not *p*-adically convergent, *FAIL* is returned. `cothp` is a short form for `evalp@coth`. ◇
`cothp(rootp(x^3+3*x-2,2));` $\longrightarrow 2^{-1} + 1 + 2^2 + 2^6 + O\left(2^7\right)$ ◇
See also: `padic[arccothp]`.

padic[cotp]

`cotp(`*expr*`, `*p*`, `*s*`)`
`cotp(`*expr*`, `*p*`)`
`cotp(`*expr*`)`
p-adic cot ◇ This function computes the *p*-adic cotangent of the expression *expr* and prints *s* terms of the result. If *s* is missing, the current value of `Digitsp` is assumed. If the result is not *p*-adically convergent, *FAIL* is returned. `cotp` is a short form for `evalp@cot`. ◇
`cotp(116,29,4);` $\longrightarrow 22\,29^{-1} + 21 + O(29)$ ◇ See also: `padic[arccotp]`.

padic[cschp]

`cschp(`*expr*`, `*p*`, `*s*`)`
`cschp(`*expr*`, `*p*`)`
`cschp(`*expr*`)`
p-adic csch ◇ This function computes the *p*-adic hyperbolic cosecant of the expression *expr* and prints *s* terms of the result. If *s* is missing, the current value of `Digitsp` is assumed. If the result is not *p*-adically convergent, *FAIL* is returned. `cschp` is a short form for `evalp@csch`. ◇
`cschp(18/5,3,5);` $\longrightarrow 3^{-2} + 2\,3^{-1} + 1 + O(3)$ ◇ See also: `padic[arccschp]`.

padic[cscp]

`cscp(`*expr*`, `*p*`, `*s*`)`
`cscp(`*expr*`, `*p*`)`
`cscp(`*expr*`)`

p-adic csc ◇ This function computes the p-adic cosecant of the expression *expr* and prints s terms of the result. If s is missing, the current value of `Digitsp` is assumed. If the result is not p-adically convergent, *FAIL* is returned. `cscp` is a short form for `evalp@csc`. ◇ `cscp(4,2,4);` ⟶ $2^{-2} + O(2)$ ◇ See also: `padic[arccscp]`.

padic[evalp]

`evalp(`*expr*`, `*p*`, `*s*`)`
`evalp(`*expr*`, `*p*`)`
`evalp(`*expr*`)`

p-adic evaluation ◇ This function computes the p-adic value of the expression *expr*. The parameter s determines the number of terms printed; if omitted, the value of `Digitsp` is used. `Digitsp` is initially assigned 10. If the result of the computation is not p-adically convergent, *FAIL* is returned. ◇ `evalp(942/10,5);` ⟶ $5^{-1} + 4 + 3\,5 + 3\,5^2 + O\left(5^9\right)$ ◇ `evalp(RootOf(x^2-2),7);` ⟶ $3 + 7 + O\left(7^2\right), 4 + 5\,7 + O\left(7^2\right)$ ◇ See also: `padic[ordp]`, `padic[ratvaluep]`, `padic[rootp]`.

padic[expp]

`expp(`*expr*`, `*p*`, `*s*`)`
`expp(`*expr*`, `*p*`)`
`expp(`*expr*`)`

p-adic exp ◇ This function computes the p-adic exponential of the expression *expr* and prints s terms of the result. If s is missing, the current value of `Digitsp` is assumed. If the result is not p-adically convergent, *FAIL* is returned. `expp` is a short form for `evalp@exp`. ◇ `expp(85,17,3);` ⟶ $1 + 5\,17 + O\left(17^2\right)$ ◇ See also: `padic[logp]`.

padic[logp]

`logp(`*expr*`, `*p*`, `*s*`)`
`logp(`*expr*`, `*p*`)`
`logp(`*expr*`)`

p-adic log ◇ This function computes the p-adic logarithm of the expression *expr* and prints s terms of the result. If s is missing, the current value of `Digitsp` is assumed. If the result is not p-adically convergent, *FAIL* is returned. `logp` is a short form for `evalp@log`. ◇ `logp(expp(15,3));` ⟶ $2\,3 + 3^2 + O\left(3^{10}\right)$ ◇ See also: `padic[expp]`.

padic[ordp]

`ordp(`*expr*`, `*p*`)`
`ordp(`*expr*`)`

Computes the order of a p-adic number ◇ This function computes the p-adic order of the expression *expr*, that is, the degree of the leading term of *expr* (or of `evalp(expr,p)` in the case of two arguments). ◇ `ordp(744653,7);` $\longrightarrow 3$ ◇ See also: `padic[evalp]`, `padic[valuep]`.

padic[ratvaluep]

`ratvaluep(expr, n)`
Approximates the rational value of a p-adic number ◇ This function returns a rational number that is the sum of the first n terms of the p-adic number *expr*. ◇ `map(ratvaluep,[rootp(x^2-2,7)],6);` $\longrightarrow$ $[38181, 79468]$ ◇ See also: `padic[evalp]`, `padic[ordp]`.

padic[rootp]

`rootp(poly, p, s)`
`rootp(poly, p)`
Finds all roots of a polynomial with rational coefficients in a p-adic field ◇ This function computes all roots of the polynomial *poly* in the field of p-adic numbers. The parameter s determines the number of terms printed; if omitted, the value of `Digitsp` is used. ◇ `rootp(x^2-2,7,3);` $\longrightarrow$ $3+7+O\left(7^2\right), 4+5\,7+O\left(7^2\right)$ ◇ See also: `padic[evalp]`, `padic[sqrtp]`.

padic[sechp]

`sechp(expr, p, s)`
`sechp(expr, p)`
`sechp(expr)`
p-adic sech ◇ This function computes the p-adic hyperbolic secant of the expression *expr* and prints s terms of the result. If s is missing, the current value of `Digitsp` is assumed. If the result is not p-adically convergent, *FAIL* is returned. `sechp` is a short form for `evalp@sech`. ◇ `sechp(33,11,5);` $\longrightarrow 1+11+O\left(11^3\right)$ ◇ See also: `padic[arcsechp]`.

padic[secp]

`secp(expr, p, s)`
`secp(expr, p)`
`secp(expr)`
p-adic secant ◇ This function computes the p-adic secant of the expression *expr* and prints s terms of the result. If s is missing, the current value of `Digitsp` is assumed. If the result is not p-adically convergent, *FAIL* is returned. `secp` is a short form for `evalp@sec`. ◇ `secp(logp(23,5,9));` $\longrightarrow 1+2\,5^2+3\,5^6+O\left(5^7\right)$ ◇ See also: `padic[arcsecp]`.

padic[sinhp]

`sinhp(expr, p, s)`
`sinhp(expr, p)`
`sinhp(expr)`
p-adic sinh ◇ This function computes the p-adic hyperbolic sine of the expression *expr* and prints s terms of the result. If s is missing, the current

value of `Digitsp` is assumed. If the result is not p-adically convergent, *FAIL* is returned. `sinhp` is a short form for `evalp@sinh`. ◇
`sinhp(4,2,5);` $\longrightarrow 2^2+2^5+\mathrm{O}\left(2^6\right)$ ◇ See also: `padic[arcsinhp]`.

padic[sinp]

`sinp(`*expr*`, `*p*`, `*s*`)`
`sinp(`*expr*`, `*p*`)`
`sinp(`*expr*`)`
p-adic sine ◇ This function computes the p-adic sine of the expression *expr* and prints s terms of the result. If s is missing, the current value of `Digitsp` is assumed. If the result is not p-adically convergent, *FAIL* is returned. `sinp` is a short form for `evalp@sin`. ◇
`sinp(sqrtp(2,7)+4,7,5);` $\longrightarrow 2\,7+2\,7^2+\mathrm{O}\left(7^5\right)$ ◇ See also: `padic[arcsinp]`.

padic[sqrtp]

`sqrtp(`*expr*`, `*p*`, `*s*`)`
`sqrtp(`*expr*`, `*p*`)`
`sqrtp(`*expr*`)`
p-adic square root ◇ This function computes the p-adic square root of the expression *expr* and prints s terms of the result. If s is missing, the current value of `Digitsp` is assumed. If the result is not p-adically convergent, *FAIL* is returned. `sqrtp` is a short form for `evalp@sqrt`. ◇
`sqrtp(-1,41,4);` $\longrightarrow 9+9\,41+34\,41^2+\mathrm{O}\left(41^3\right)$ ◇ See also: `padic[rootp]`.

padic[tanhp]

`tanhp(`*expr*`, `*p*`, `*s*`)`
`tanhp(`*expr*`, `*p*`)`
`tanhp(`*expr*`)`
p-adic tanh ◇ This function computes the p-adic hyperbolic tangent of the expression *expr* and prints s terms of the result. If s is missing, the current value of `Digitsp` is assumed. If the result is not p-adically convergent, *FAIL* is returned. `tanhp` is a short form for `evalp@tanh`. ◇
`tanhp(5,2);` $\longrightarrow$ *FAIL* ◇ See also: `padic[arctanhp]`.

padic[tanp]

`tanp(`*expr*`, `*p*`, `*s*`)`
`tanp(`*expr*`, `*p*`)`
`tanp(`*expr*`)`
p-adic tangent ◇ This function computes the p-adic tangent of the expression *expr* and prints s terms of the result. If s is missing, the current value of `Digitsp` is assumed. If the result is not p-adically convergent, *FAIL* is returned. `tanp` is a short form for `evalp@tan`. ◇
`tanp(arctanp(25,5));` $\longrightarrow 5^2+\mathrm{O}\left(5^{12}\right)$ ◇ See also: `padic[arctanp]`.

padic[valuep]

`valuep(`*expr*`, `*p*`)`
`valuep(`*expr*`)`
p-adic valuation ◇ This function computes the p-adic valuation of the number *expr*. This is $1/p$^(p-adic order of *expr*). ◇ `valuep(744653,7);` ⟶ 7^{-3} ◇ See also: `padic[ordp]`.

parse

`parse(`*string*`)`
`parse(`*string*`, statement)`
Parses a string as an expression or statement ◇ The first form parses the given string as an expression and evaluates this expression, returning the result. With the `statement` option, the string is parsed as a Maple statement. Note that statements are a superset of expressions. If the string ends in a colon, the result is not displayed. ◇ `parse('1+1');` ⟶ 2 ◇ See also: `readline, readstat, sscanf`.

petrov

`petrov(`*Psi*`)`
`petrov(`*Psi*`, `*flag*`)`
`nonzero()`
Determines the Petrov classification of the given Weyl tensor in Newman-Penrose form ◇ *Psi* is an array indexed from 0 to 4. If a second argument is present, the components of *Psi* are simplified using an internal routine. `petrov` returns a string naming the Petrov type. `nonzero` displays expressions that were determined to be nonzero. ◇ You must `readlib(petrov)` before using this command. ◇ `petrov(array(0..4, [x,y,0,0,0]));` ⟶ *type_III* ◇ See also: `array, debever`.

Pi

π ◇ The ratio of the circumference to the diameter of a circle, approximately 3.141592653589793.

piecewise

`piecewise(`rel_1`, `$expr_1$`, ..., `rel_n`, `$expr_n$`)`
`piecewise(`m`, `rel_1`, `$expr_1$`, ..., `rel_n`, `$expr_n$`)`
`piecewise(`rel_1`, `$expr_1$`, ..., `rel_n`, `$expr_n$`, `$expr_{n+1}$`)`
`piecewise(`m`, `rel_1`, `$expr_1$`, ..., `rel_n`, `$expr_n$`, `$expr_{n+1}$`)`
Defines a function piecewise ◇ The function is defined by $expr_i$ whenever the relation rel_i is true. The relations are tested in the given order, so the first true relation determines the value of the function. If none of the relations is true, the value of the function is given by $expr_{n+1}$. The default for $expr_{n+1}$ is `undefined`. The optional parameter m specifies the order of smoothness at the joins of the function. Its default value is 0. A value of -1 indicates that the function is not continuous at the joins. ◇ This function is new in Release 3. ◇ You must `readlib(piecewise)` before using this command. ◇ `piecewise(x<0,-1,1): eval(subs(x=2,"));` ⟶ 1

plot

plot(*expr*, *var*=*a*..*b*, *options*)
plot(*f*, *a*..*b*, *options*)
plot([$expr_x$, $expr_y$, *var*=*a*..*b*], *options*)
plot([$expr_r$, $expr_\theta$, *var*=*a*..*b*], coords=polar, *options*)
plot([x_1, y_1, x_2, y_2, ...], style=POINT, *options*)
plot([[x_1, y_1], [x_2, y_2], ...], style=POINT, *options*)
plot({$expr_1$, $expr_2$, ...}, *var*=*a*..*b*, *options*)

Creates a two-dimensional plot ◇ The first form graphs the expression in the given variable *var* on the real interval $[a, b]$. The second form graphs the function *f* on the interval $[a, b]$. All other templates involving an *expr* have a corresponding functional form. The third template creates a parametric plot; the fourth, a parametric plot in polar coordinates. The next two templates plot points in the plane. The range is optional in this case. The last template plots all of the expressions on the same graph. In all cases, *a* may be `-infinity`, and *b* may be `infinity`. If no interval is given, $[-10, 10]$ is used by default, except for plotting points explicitly, where the range is determined by the given points. An adaptive sampling algorithm is used when plotting an expression or function, but not for parametric plots. Options for `plot` are listed in brief on page 112 and described in detail beginning on page 26. ◇ `plot(sin(x),x=0..Pi);` ◇ `plot(sin,0..Pi);` ◇ `plot([cos(t),sin(t)],t=0..Pi);` ◇ `plot({cos(x),sin(x)},x=0..Pi);` ◇ `plot(tan(x),x=0..10, -5..5);` ◇ See also: `DEtools`, `plot3d`, `plots[animate]`, `plots[conformal]`, `plots[display]`, `plots[implicitplot]`, `plots[loglogplot]`, `plots[logplot]`, `plots[polarplot]`, `plots[setoptions]`, `type/PLOT`.

plot3d

plot3d(*expr*, *x*=*a*..*b*, *y*=*c*..*d*, *options*)
plot3d(*f*, *a*..*b*, *c*..*d*, *options*)
plot3d([$expr_x$, $expr_y$, $expr_z$], *s*=*a*..*b*, *t*=*c*..*d*, *options*)
plot3d(*expr*, θ=*a*..*b*, *z*=*c*..*d*, coords=cylindrical);
plot3d(*expr*, θ=*a*..*b*, ϕ=*c*..*d*, coords=spherical);
plot3d({$expr_1$, $expr_2$, ...}, *x*=*a*..*b*, *y*=*c*..*d*, *options*)

Creates a three-dimensional plot ◇ The first form plots the surface defined by the expression on the rectangular region defined by the two ranges. *a* and *b* must be real constants; *c* and *d* may be real constants or expressions in the first variable. In the second form, *f* must be a function of two arguments. *a* and *b* must be real constants; *c* and *d* may be real constants or functions in one variable. The other templates have a corresponding functional form. The third template defines a parametric surface. The fourth defines a surface in cylindrical coordinates, with *expr* determining r from θ and z. The fifth defines a surface in spherical coordinates, with *expr* determining r from θ and ϕ. The last form plots several surfaces on the same graph. Options for `plot3d` are listed in brief on page 113 and described in detail beginning on page 34. ◇ `plot3d(x+y,x=0..1,y=0..1);` ◇ `plot3d([s*cos(t), s*sin(t),` t], s=1..2,t=0..4*Pi); ◇ See also: plot, plots[animate3d], plots[contourplot], plots[cylinderplot], plots[densityplot], plots[implicitplot3d], plots[pointplot], plots[setoptions3d], plots[spacecurve], plots[sphereplot], plots[tubeplot], type/PLOT3D.

plots Two- and three-dimensional plotting package ◇ See also: DEtools, plot, plot3d, plots[*function*], with.

plots[animate]

animate(*expr*, *x*=*a*..*b*, *t*=*c*..*d*, *options*)
animate(*f*, *a*..*b*, *c*..*d*, *options*)
animate([$expr_x$, $expr_y$, *s*=*a*..*b*], *t*=*c*..*d*, *options*)
animate({$expr_1$, $expr_2$, ...}, *x*=*a*..*b*, *t*=*c*..*d*, *options*)
Creates an animation of 2d plots ◇ The first form plots *expr* against *x* for values of *t* ranging from *c* to *d*, making an animation. The second template is similar, with *f* a function of two variables. The third form animates a parametric plot, and the last template animates a sequence of curves simultaneously. In addition to the standard options for 2d plots, this function accepts the option frames=*n*, where *n* is the number of frames produced in the animation. Its default value is 16. ◇ animate(x^t, x=0..1, t=1..4); ◇ animate([sin(s), t*cos(s), s=0..2*Pi], t=1..3); ◇ See also: plot, plots[animate3d], plots[display].

plots[animate3d]

animate3d(*expr*, *x*=*a*..*b*, *y*=*c*..*d*, *t*=*q*..*r*)
animate3d(*f*, *a*..*b*, *c*..*d*, *q*..*r*)
animate([$expr_x$, $expr_y$, $expr_z$], *s*=*a*..*b*, *u*=*c*..*d*, *t*=*q*..*r*)
animate({$expr_1$, $expr_2$, ...}, *x*=*a*..*b*, *y*=*c*..*d*, *t*=*q*..*r*)
Creates an animation of 3d plots ◇ The first form plots *expr* against *x* and *y* for values of *t* ranging from *q* to *r*, making an animation. The second template is similar, with *f* a function of three variables. The third form animates a parametric plot, and the last template animates a sequence of surfaces simultaneously. In addition to the standard options for 3d plots, this function accepts the option frames=*n*, where *n* is the number of frames produced in the animation. Its default value is 8. ◇ animate3d(x^2+t*y^2, x=-1..1, y=-1..1, t=1..3); ◇ animate3d([s*cos(t*u), s*sin(t*u), u], s=1..2, u=0..2*Pi, t=1..2); ◇ See also: plot3d, plots[animate], plots[display].

plots[conformal]

conformal(*expr*, *z*=*a*..*b*, *c*..*d*, *options*)
conformal(*f*, *a*..*b*, *c*..*d*, *options*)
Conformal plot of a complex function ◇ This function draws the image of a grid in the complex plane under the given expression or function. *a* and *b* are complex numbers giving the lower-left and upper-right corners of the

original grid. The second range is optional and describes the complex coordinates of the image window. The option `grid=[m,n]` sets the number of lines in the grid in each direction: the default is $m = 11$ and $n = 11$. The option `numxy=[m,n]` sets the number of points plotted on each gridline: the default is $m = 15$ and $n = 15$. ◇ `conformal(z^4,z=0..1+I);`

plots[contourplot]

`contourplot(`*expr*`, `*x*`=`*a*`..`*b*`, `*y*`=`*c*`..`*d*`)`
`contourplot(`*f*`, `*a*`..`*b*`, `*c*`..`*d*`)`
`contourplot([`$expr_x$`, `$expr_y$`, `$expr_z$`], `*s*`=`*a*`..`*b*`, `*t*`=`*c*`..`*d*`)`
`contourplot({`$expr_1$`, `$expr_2$`, ...}, `*x*`=`*a*`..`*b*`, `*y*`=`*c*`..`*d*`)`
Draws a contour plot of a surface or surfaces ◇ This first two calling sequences describe contour plots in Cartesian coordinates. The third describes a contour plot of a surface given parametrically, and the fourth draws multiple contour plots at once. The last two templates also have corresponding forms where the expressions are replaced by function names and no variable name is supplied in the ranges, as in the second template. This function accepts the standard 3d plotting options. ◇ `contourplot(x*cos(y),x=1..3,y=-1..1);` ◇ See also: `plot3d`, `plots[densityplot]`.

plots[cylinderplot]

`cylinderplot(`*expr*`, `θ`=`*a*`..`*b*`, `*z*`=`*c*`..`*d*`)`
`cylinderplot(`*f*`, `*a*`..`*b*`, `*c*`..`*d*`)`
`cylinderplot([`$expr_r$`, `$expr_\theta$`, `$expr_z$`], `*s*`=`*a*`..`*b*`, `*t*`=`*c*`..`*d*`)`
`cylinderplot({`$expr_1$`, `$expr_2$`, ...}, `θ`=`*a*`..`*b*`, `*z*`=`*c*`..`*d*`)`
Plots a surface in cylindrical coordinates ◇ In the first calling sequence, *expr* is an expression that determines r from θ and z. In the second template, f is a function of two variables that gives r from θ and z. The first range is the range for θ; the second for z. The last two templates also have corresponding functional forms. The third template plots a surface whose cylindrical coordinates are given parametrically. The last form plots multiple cylindrical surfaces on the same graph. This function accepts the standard 3d plotting options. ◇ `cylinderplot({1,2}, theta=0..2*Pi, z=0..2);` ◇ See also: `plot3d`, `plots[sphereplot]`.

plots[densityplot]

`densityplot(`*expr*`, `*x*`=`*a*`..`*b*`, `*y*`=`*c*`..`*d*`)`
`densityplot(`*f*`, `*a*`..`*b*`, `*c*`..`*d*`)`
`densityplot([`$expr_x$`, `$expr_y$`, `$expr_z$`], `*s*`=`*a*`..`*b*`, `*t*`=`*c*`..`*d*`)`
`densityplot({`$expr_1$`, `$expr_2$`, ...}, `*x*`=`*a*`..`*b*`, `*y*`=`*c*`..`*d*`)`
Draws a density plot of a surface or surfaces ◇ The first two calling sequences describe density plots in Cartesian coordinates. The third describes a density plot of a surface given parametrically, and the fourth draws multiple density plots at once. The last two templates also have

corresponding forms where the expressions are replaced by function names and no variable name is supplied in the ranges, as in the second template. This function accepts the standard 2d plotting options, plus the option `grid=[`*m*`,`*n*`]` for setting the mesh of the grid used to make the density plot. Its default value is $m = 25$, $n = 25$. ◇ In Release 3, this function also accepts the option `style=PATCHNOGRID` for drawing the density plot without the grid lines. ◇ `densityplot(x*cos(y),x=1..3,y=-1..1);` ◇ See also: `plot3d`, `plots[contourplot]`.

plots[display]

`display(`*plot*`,` *options*`)`
`display([`$plot_1$`,` $plot_2$`, ...],` *options*`)`
`display({`$plot_1$`,` $plot_2$`, ...},` *options*`)`
`display([`$plot_1$`,` $plot_2$`, ...], insequence=true)`
Displays a list or set of plots ◇ By default, this function renders all the given plots on the same graph. The plots may be two-dimensional or three-dimensional, but these types may not be mixed together. If the option `insequence=true` appears, the graphs are presented in sequence as an animation. Standard plotting options that determine cosmetic features of the graph, such as `axes` or `title`, may be given, but options that control the way a graph is generated, such as `numpoints`, are ignored. ◇ See also: `plots[animate]`, `plots[animate3d]`.

plots[display3d]

A synonym for `plots[display]` ◇ See `plots[display]`.

plots[fieldplot]

`fieldplot([`$expr_1$`,` $expr_2$`],` *x*`=`*a*`..`*b*`,` *y*`=`*c*`..`*d*`,` *options*`)`
`fieldplot([`f_1`,` f_2`],` *a*`..`*b*`,` *c*`..`*d*`,` *options*`)`
Plots a 2d vector field ◇ This function plots the vector field determined by the two expressions or functions. In addition to standard 2d plotting options, this function accepts the options `grid=[`*n*`,`*m*`]` and `arrows=`*style*. The `grid` option sets the mesh of the grid used to make the vector field (default $n = 20$, $m = 20$), and the arrow style may be `LINE`, `SLIM`, `THICK`, or `THIN` (the default). ◇ `fieldplot([y,-x+y/3], x=-1..1, y=-1..1);` ◇ See also: `plots[fieldplot3d]`, `plots[gradplot]`.

plots[fieldplot3d]

`fieldplot3d([`$expr_1$`,` $expr_2$`,` $expr_3$`],` *x*`=`*a*`..`*b*`,` *y*`=`*c*`..`*d*`,` *z*`=`*q*`..`*r*`,` *options*`)`
`fieldplot3d([`f_1`,` f_2`,` f_3`],` *a*`..`*b*`,` *c*`..`*d*`,` *q*`..`*r*`,` *options*`)`
Plots a 3d vector field ◇ This function plots the vector field determined by the three expressions or functions. In addition to standard 3d plotting options, this function accepts the options `grid=[`*l*`,`*n*`,`*m*`]` and `arrows=`*style*. The `grid` option sets the mesh of the grid used to make the vector field (default $l = 8$, $n = 8$, $m = 8$), and the arrow style may be `LINE`, `SLIM`, `THICK`, or `THIN` (the default). ◇ `fieldplot3d([y+z,x+z,z+y],`

x=0..1, y=0..1, z=0..1); ◇ See also: plots[fieldplot], plots[gradplot3d].

plots[gradplot]

gradplot(*expr*, *x*=*a*..*b*, *y*=*c*..*d*, *options*)
gradplot(*f*, *a*..*b*, *c*..*d*, *options*)
Plots a 2d gradient vector field ◇ This function plots the gradient field determined by the expression or function. In addition to standard 2d plotting options, this function accepts the options grid=[*n*,*m*] and arrows=*style*. The grid option sets the mesh of the grid used to make the vector field (default $n = 20$, $m = 20$), and the arrow style may be LINE, SLIM, THICK, or THIN (the default). ◇ gradplot(x*y, x=-2..2, y=-2..2); ◇ See also: plots[fieldplot], plots[gradplot3d].

plots[gradplot3d]

gradplot3d(*expr*, *x*=*a*..*b*, *y*=*c*..*d*, *z*=*q*..*r*, *options*)
gradplot3d([*f*, *a*..*b*, *c*..*d*, *q*..*r*, *options*)
Plots a 3d gradient vector field ◇ This function plots the gradient field determined by the expression or function. In addition to standard 3d plotting options, this function accepts the options grid=[*l*,*n*,*m*] and arrows=*style*. The grid option sets the mesh of the grid used to make the vector field (default $l = 8$, $n = 8$, $m = 8$), and the arrow style may be LINE, SLIM, THICK, or THIN (the default). ◇ gradplot3d(x*y*z, x=-2..2, y=-2..2, z=-2..2); ◇ See also: plots[fieldplot3d], plots[gradplot].

plots[implicitplot]

implicitplot(*eqn*, *x*=*a*..*b*, *y*=*c*..*d*, *options*)
implicitplot(*expr*, *x*=*a*..*b*, *y*=*c*..*d*, *options*)
implicitplot(*f*, *a*..*b*, *c*..*d*, *options*)
Two-dimensional implicit plot ◇ The first form plots the curve satisfying the equation in the variables x and y in the given rectangular region. The second form plots the solutions to $expr = 0$, where *expr* is an expression in x and y. The third form plots the set of points satisfying $f = 0$, where f is a function of two variables. In addition to standard 2d plotting options, this function accepts the option grid=[*m*,*n*], specifying that the plot should be constructed on an m by n grid (the default is [25, 25]). ◇ implicitplot(x^3+y^3=1, x=-1..2, y=-1..1.3); ◇ See also: plots[implicitplot3d].

plots[implicitplot3d]

implicitplot3d(*eqn*, *x*=*a*..*b*, *y*=*c*..*d*, *z*=*q*..*r*, *options*)
implicitplot3d(*expr*, *x*=*a*..*b*, *y*=*c*..*d*, *z*=*q*..*r*, *options*)
implicitplot3d(*f*, *a*..*b*, *c*..*d*, *options*, *q*..*r*)
Three-dimensional implicit plotting ◇ The first form plots the surface satisfying the equation in the variables x, y, and z in the given region. The second form plots the solutions to $expr = 0$, where *expr* is an expression in

x, y, and z. The third form plots the set of points satisfying $f = 0$, where f is a function of three variables. In addition to standard 3d plotting options, this function accepts the option `grid=[l,m,n]`, specifying that the plot should be constructed on an l by m by n grid (the default is [10, 10, 10]). ◇ `implicitplot3d(x^2+y^2+z^2-1, x=-1..1, y=-1..1, z=-1..1);` ◇ See also: `plots[implicitplot]`.

plots[loglogplot]

`loglogplot(`*expr*`, `*x*`=`*a*`..`*b*`, `*options*`)`
`loglogplot(`*f*`, `*a*`..`*b*`, `*options*`)`
`loglogplot([`$expr_x$`, `$expr_y$`, `*var*`=`*a*`..`*b*`], `*options*`)`
`loglogplot([`x_1`, `y_1`, `x_2`, `y_2`, ...], style=POINT, `*options*`)`
`loglogplot([[`x_1`, `y_1`], [`x_2`, `y_2`], ...], style=POINT, `*options*`)`
`loglogplot({`$expr_1$`, `$expr_2$`, ...}, `*x*`=`*a*`..`*b*`, `*options*`)`

Creates a two-dimensional log-log plot of functions ◇ This function creates a two-dimensional plot with both the vertical and horizontal axes given in a logarithmic scale. It can plot a curve given by an expression or function, a curve given parametrically, a collection of curves, or a list of points. This function accepts the standard 2d plotting options. ◇ `loglogplot(exp(x),x=1..10);` ◇ See also: `plot`, `plots[logplot]`.

plots[logplot]

`logplot(`*expr*`, `*x*`=`*a*`..`*b*`, `*options*`)`
`logplot(`*f*`, `*a*`..`*b*`, `*options*`)`
`logplot([`$expr_x$`, `$expr_y$`, `*var*`=`*a*`..`*b*`], `*options*`)`
`logplot([`x_1`, `y_1`, `x_2`, `y_2`, ...], style=POINT, `*options*`)`
`logplot([[`x_1`, `y_1`], [`x_2`, `y_2`], ...], style=POINT, `*options*`)`
`logplot({`$expr_1$`, `$expr_2$`, ...}, `*x*`=`*a*`..`*b*`, `*options*`)`

Creates a two-dimensional log plot of functions ◇ This function creates a two-dimensional plot with the vertical axis given in a logarithmic scale. It can plot a curve given by an expression or function, a curve given parametrically, a collection of curves, or a list of points. This function accepts the standard 2d plotting options. ◇ `logplot(exp(x), x=1..10);` ◇ See also: `plot`, `plots[loglogplot]`.

plots[matrixplot]

`matrixplot(`*A*`, `*options*`)`

Three-dimensional plot with z values determined by the matrix entries ◇ This function defines a three-dimensional graph with the x and y coordinates representing row and column indices, respectively. By default, the plot is drawn as a smooth surface. The option `heights=histogram` draws the plot instead as a histogram. When the plot is drawn as a histogram, the option `gap=`*r* leaves a gap of size *r* between the bars. *r* must be between 0 and 0.5; its default value is 0. This function also accepts standard 3d plotting options. ◇ `matrixplot(linalg[toeplitz]([3,2,1]),` `heights=histogram);` ◇ See also: `plot3d`, `plots[sparsematrixplot]`.

plots[odeplot]

`odeplot(`*f*`, `*varlist*`, `*a*`..`*b*`, `*options*`)`
Plots output from `dsolve/numeric` ◇ This function plots the solution curve (either two- or three-dimensional) for a procedure returned by `dsolve` with the `numeric` option. The ordering of the coordinates is given by *varlist*. The function is plotted on the range $[a, b]$. ◇ `f:=dsolve({D(y)(x)=y(x),y(0)=1}, y(x), numeric): odeplot(f, [x,y(x)], 0..2);` ◇ See also: `dsolve`.

plots[pointplot]

`pointplot(`*list*`, `*options*`)`
`pointplot(`*set*`, `*options*`)`
Plots points in three dimensions ◇ The points may be given in a set or a list. This function accepts the standard 3d plotting options. ◇ `pointplot({[1,1,1],[1,2,1]}, axes=FRAME);` ◇ See also: `plot3d`, `plots[polyhedraplot]`.

plots[polarplot]

`polarplot(`*expr*`, `θ`=`*a*`..`*b*`, `*options*`)`
`polarplot(`*f*`, `*a*`..`*b*`, `*options*`)`
`polarplot([`$expr_r$`, `$expr_\theta$`, `*t*`=`*a*`..`*b*`], `*options*`)`
`polarplot({`$expr_1$`, `$expr_2$`, ...}, `θ`=`*a*`..`*b*`, `*options*`)`
Plots a curve in polar coordinates ◇ In the first calling sequence, *expr* is an expression that determines r from the angle θ. If no interval is given, $[-2\pi, 2\pi]$ is used by default. In the second template, f is a function of one variable that gives r from the angle θ. The last two templates also have corresponding functional forms. The third template plots a curve whose polar coordinates are given parametrically. The last form plots multiple polar curves on the same graph. This function accepts the standard 2d plotting options. ◇ See also: `plot`.

plots[polygonplot]

`polygonplot(`*polygon*`, `*options*`)`
`polygonplot([`$polygon_1$`, `$polygon_2$`, ...], `*options*`)`
`polygonplot({`$polygon_1$`, `$polygon_2$`, ...}, `*options*`)`
Plots one or more polygons in the plane ◇ *polygon* is a list of two-dimensional points. These points are the vertices of the polygon. The first form plots a single polygon; the other two graph several polygons at once. The default plotting style is `PATCH`, so polygons are shaded. This function accepts standard 2d plotting options. ◇ `polygonplot([[0,0], [0,1], [1,0]]);` ◇ See also: `plots[polygonplot3d]`, `plots[polyhedraplot]`.

plots[polygonplot3d]

polygonplot3d(*polygon*, *options*)
polygonplot3d([*polygon*$_1$, *polygon*$_2$, ...], *options*)
polygonplot3d({*polygon*$_1$, *polygon*$_2$, ...}, *options*)
Plots one or more polygons in three dimensions ◇ *polygon* is a list of three-dimensional points. These points are the vertices of the polygon. The first form plots a single polygon; the other two graph several polygons at once. This function accepts standard 3d plotting options. ◇ `polygonplot3d([[1,0,0],[0,1,0],[0,0,1]], axes=NORMAL);` ◇ See also: `plots[polygonplot]`, `plots[polyhedraplot]`.

plots[polyhedraplot]

polyhedraplot(*point*, polytype=*shape*, polyscale=*r*, *options*)
polyhedraplot(*pointlist*, polytype=*shape*, polyscale=*r*, *options*)
polyhedraplot(*pointset*, polytype=*shape*, polyscale=*r*, *options*)
Plots one or more polyhedra in three dimensions ◇ This function draws one or more polyhedra of type *shape* and size *r*. These polyhedra are centered at the given points. Each point is a list of three numbers. *shape* may be `tetrahedron` (the default), `hexahedron`, `octahedron`, `dodecahedron`, or `icosahedron`. The default scaling *r* is 1. This function accepts standard 3d plotting options. ◇ `polyhedraplot([seq([t*sin(t/4),t*cos(t/4),t],t=0..40)], style=PATCH);` ◇ See also: `plots[pointplot]`, `plots[polygonplot]`, `plots[polygonplot3d]`.

plots[replot]

replot(*plot*, *options*)
Replots a plot ◇ This function has been superseded by `plots[display]`. ◇ See `plots[display]`.

plots[setoptions]

setoptions(*options*)
Sets default options for two-dimensional plots ◇ These become the default for all subsequent 2d plots in the session. ◇ See also: `plot`, `plots[setoptions3d]`.

plots[setoptions3d]

setoptions3d(*options*)
Sets default options for three-dimensional plots ◇ These become the default for all subsequent 3d plots in the session. ◇ See also: `plot3d`, `plots[setoptions]`.

plots[spacecurve]

`spacecurve(`[*expr*$_x$, *expr*$_y$, *expr*$_z$], *t=a..b*, *options*)
`spacecurve(`{*curve*$_1$, *curve*$_2$, ...}, *t=a..b*, *options*)
Plots curves given parametrically in three dimensions ◇ The first form plots the space curve whose coordinates are given parametrically by the three expressions. The second form plots several space curves on the same graph. If the curves use different variables or are plotted on different ranges, the *t=a..b* argument should be placed at the end of the list defining each curve. This function accepts standard 3d plotting options. ◇ `spacecurve([cos(t),sin(t),t],t=0..2*Pi);` ◇ See also: `plots[tubeplot]`.

plots[sparsematrixplot]

`sparsematrixplot(`*A*, *options*)
Two-dimensional plot of the nonzero values of the matrix ◇ This function defines a two-dimensional graph with the x and y coordinates representing row and column indices, respectively. A point is plotted for each nonzero entry of the matrix. This function accepts standard 2d plotting options. ◇ `sparsematrixplot(array(identity, 1..5, 1..5));` ◇ See also: `plot`, `plots[matrixplot]`.

plots[sphereplot]

`sphereplot(`*expr*, θ*=a..b*, ϕ*=c..d*)
`sphereplot(`*f*, *a..b*, *c..d*)
`sphereplot(`[*expr*$_r$, *expr*$_\theta$, *expr*$_\phi$], *s=a..b*, *t=c..d*)
`sphereplot(`{*expr*$_1$, *expr*$_2$, ...}, θ*=a..b*, ϕ*=c..d*)
Plots a surface in spherical coordinates ◇ In the first calling sequence, *expr* is an expression that determines r from θ and ϕ. In the second template, *f* is a function of two variables that gives r from θ and ϕ. The first range is the range for θ; the second for ϕ. The last two templates also have corresponding functional forms. The third template plots a surface whose spherical coordinates are given parametrically. The last form plots multiple spherical surfaces on the same graph. This function accepts the standard 3d plotting options. ◇ `sphereplot([sin(s)*sin(t),s,t], s=0..Pi, t=0..Pi);` ◇ See also: `plot3d`, `plots[cylinderplot]`.

plots[surfdata]

`surfdata(`[*ptlist*$_1$, *ptlist*$_2$, ...], *options*)
`surfdata(`{*listlist*$_1$, *listlist*$_2$, ...}, *options*)
Creates a surface from a list of data points ◇ The first argument is a list whose elements are lists of points. Each point is itself a list of three real constants. Each *ptlist* determines a row of points on the surface. If the points are initially stored in a matrix, the function `convert/listlist` converts the data to the form required for this function. Multiple surfaces are drawn when the first argument is a set of lists of lists of points. This function accepts standard 3d plotting options. ◇ `surfdata([[[0,0,1],`

[0,1,1]], [[1,0,2],[1,1,2]]]); ◇ See also: convert/listlist, plot3d[options], plot3d[structure].

plots[textplot]

textplot(*textpt*, *options*)
textplot(*list*, *options*)
textplot(*set*, *options*)
Positions text in two-dimensional space ◇ A text point *textpt* is a list of the form [*x*,*y*,*string*]. This function places *string* at the position $[x, y]$. By default, the text is centered at the point. This may be altered with the option align=*pos*, where *pos* is ABOVE, BELOW, LEFT, RIGHT, or a set of these. This function also accepts standard 2d plot options. Multiple strings may be printed by supplying several text points in a list or a set. ◇ textplot([1,1,'Oak']); ◇ See also: plot, plots[textplot3d].

plots[textplot3d]

textplot3d(*textpt*, *options*)
textplot3d(*list*, *options*)
textplot3d(*set*, *options*)
Positions text in three-dimensional space ◇ A text point *textpt* is a list of the form [*x*,*y*,*z*,*string*]. This function places *string* at the position $[x, y, z]$. By default, the text is centered at the point. This may be altered with the option align=*pos*, where *pos* is ABOVE, BELOW, LEFT, RIGHT, or a set of these. This function also accepts standard 3d plot options. Multiple strings may be printed by supplying several text points in a list or a set. ◇ textplot3d([1,0,1,'Elm'], axes=NORMAL); ◇ See also: plot3d, plots[textplot].

plots[tubeplot]

tubeplot([$expr_x$, $expr_y$, $expr_z$], *t*=*a*..*b*, *options*)
tubeplot({$curve_1$, $curve_2$, ...}, *t*=*a*..*b*, *options*)
Draws tubes about space curves ◇ A space curve is defined parametrically by three expressions in a list. This function draws a tube of radius *r* about a space curve, where *r* is given by the option radius=*r*. *r* may be any expression in the variable *t*. Its default value is 1. The option tubepoints=*n* sets the number of points plotted around the tube at each point sampled on the space curve. This function also accepts standard 3d plotting options. The second form plots several tubes on the same graph. If the curves use different options, the variable, radius, and tubepoints arguments should be placed at the end of the list defining each curve. ◇ tubeplot([cos(t),sin(t),t], radius=t/10, t=0..2*Pi); ◇ See also: plots[spacecurve].

plotsetup

plotsetup(*options*)
plotsetup(*devicetype*, *options*)
plotsetup(*devicetype*, *terminaltype*, *options*)
Sets appropriate parameters for plotting ◇ This command sets values for interface variables associated with plotting: plotdevice, plotoptions,

`plotoutput`, `preplot`, and `postplot`. See page 109 for an example, and refer to your on-line documentation for the device types and terminal types supported. ◇ In Release 3, one or more equations may be given as options to set values for these interface variables explicitly, as in `plotoutput=`*filename*. ◇ See also: `interface`, `plot`, `plot3d`.

pointto

`pointto(`*ptr*`)`
Obtains the expression pointed to by an address ◇ This function allows access to the internal representation of an expression. ◇ `map(pointto, [disassemble(addressof(a+2*b))[2..5]]);` $\longrightarrow [a, 1, b, 2]$ ◇ See also: `addressof`, `assemble`, `disassemble`.

poisson

`poisson(`*expr*`, `*vars*`)`
`poisson(`*expr*`, `*vars*`, `*n*`)`
`poisson(`*expr*`, `*vars*`, `*n*`, `*list*`)`
Multivariate Poisson series expansion ◇ *vars* may be a list or set of names or equations. A name is short for the equation *name* = 0. *n* specifies the order of truncation (total degree) for the series. The fourth argument specifies weights (positive integers) for the variables. The result is the same as for the `mtaylor` function, except that sines and cosines in the coefficients are combined into the Fourier canonical form. ◇ You must `readlib(poisson)` before using this command. ◇ See also: `mtaylor`.

polar

`polar(`*z*`)`
`polar(`*r*`, `θ`)`
Converts to polar form ◇ This function converts the complex-valued expression *z* to its representation in polar coordinates. The result is represented as `polar(`*r*`, `θ`)`, where *r* is the modulus and θ is the argument of *z*. ◇ You must `readlib(polar)` before using this command. ◇ `polar(I);` $\longrightarrow \text{polar}\left(1, \frac{1}{2}\pi\right)$ ◇ See also: `argument`, `convert/polar`, `evalc`.

Power

`Power(`*poly*`, `*n*`)`
Inert power function ◇ This function is used in conjunction with `mod` or `modp1` to compute $poly^n$ in a particular domain of coefficients. ◇ `Power(x+3,5) mod 5;` $\longrightarrow x^5 + 3$ ◇ See also: `&^`, `mod`, `modp1`, `Powmod`.

powmod

`powmod(`$poly_1$`, `*n*`, `$poly_2$`, `*var*`)`
Computes $poly_1^n$ mod $poly_2$ efficiently ◇ $poly_1$ and $poly_2$ are polynomials in the variable *var*. ◇ You must `readlib(powmod)` before using this command. ◇ See also: `rem`.

Powmod

`Powmod(`$poly_1$`, `*n*`, `$poly_2$`)`
`Powmod(`$poly_1$`, `*n*`, `$poly_2$`, `*var*`)`
Inert power function with remainder ◇ This function is used in conjunction with `mod` or `modp1` to compute $poly_1^n$ mod $poly_2$ in a particular domain of coefficients. ◇ See also: `&^`, `mod`, `modp1`, `Power`, `Rem`.

powseries Formal power series package ◇ In the descriptions of functions in this package, *ps* represents a formal power series. A formal power series is a procedure that returns the coefficients of the power series it represents: *ps*(*n*) is the coefficient of x^n. All computed coefficients can be seen with `op(4, op(ps))`. The general term of a power series can be displayed by invoking *ps*(`_k`). Each intermediate power series created in a calculation should be named. ◇ See also: `asympt`, `powseries[`*function*`]`, `series`, `with`.

A B C D E F G H I J K L M N O P Q R S T U V W X Y Z

powseries[add]

`add(`ps_1`, `ps_2`, ...)`

Addition of formal power series ◇ This function returns a new power series, the sum of the given power series. ◇ See also: `powseries[multiply]`, `powseries[subtract]`.

powseries[compose]

`compose(`ps_1`, `ps_2`)`

Composition of formal power series ◇ This function returns the formal power series that is $ps_1(ps_2)$. $ps_2(0)$ must be 0, otherwise an error message is given. ◇ See also: `powseries[reversion]`.

powseries[evalpow]

`evalpow(`*expr*`)`

General evaluator for power series expressions ◇ This function evaluates any arithmetic expression involving power series and constants, returning a power series.

powseries[inverse]

`inverse(`*ps*`)`

Multiplicative inverse of a formal power series ◇ This function returns a formal power series. ◇ See also: `powseries[multiply]`, `powseries[quotient]`.

powseries[multconst]

`multconst(`*ps*`, `*c*`)`

Multiplication of a power series by a constant ◇ This function returns a formal power series where each coefficient is *c* times the corresponding coefficient in *ps*. ◇ See also: `powseries[multiply]`.

powseries[multiply]

`multiply(`ps_1`, `ps_2`)`

Multiplication of formal power series ◇ This function returns a new power series, the product of the given power series. ◇ See also: `powseries[add]`, `powseries[multconst]`, `powseries[quotient]`.

powseries[negative]

`negative(ps)`

Negation of a formal power series ◇ This function returns a formal power series where each coefficient is the negative of the corresponding coefficient in *ps*.

powseries[powcreate]

`powcreate(eqn1, eqn2, ...)`

Creates a formal power series from equations defining the coefficients ◇ One equation specifies the nth coefficient, and the others specify initial conditions. The name of the created power series is given by the name used in the left-hand side of the equations. ◇ `powcreate(F(n)=F(n-1)+F(n-2), F(0)=0,F(1)=1);` ◇ See also: `powseries[powpoly]`.

powseries[powdiff]

`powdiff(ps)`

Differentiation of a formal power series ◇ This function returns the formal power series that is the formal derivative of *ps* with respect to its variable. ◇ See also: `powseries[powint]`.

powseries[powexp]

`powexp(ps)`

Exponentiation of a formal power series ◇ This function returns the formal power series exp(*ps*). ◇ See also: `powseries[powlog]`.

powseries[powint]

`powint(ps)`

Integration of a formal power series ◇ This function returns the formal power series that is the formal antiderivative of *ps* with respect to its variable. ◇ See also: `powseries[powdiff]`.

powseries[powlog]

`powlog(p)`

Logarithm of a formal power series ◇ This function returns the formal power series ln(*ps*). *ps* must have a nonzero constant term, otherwise an error message is given. ◇ See also: `powseries[powexp]`.

powseries[powpoly]

`powpoly(poly, var)`

Creates a formal power series from a polynomial ◇ This function returns a formal power series equivalent to *poly*. ◇ `t := powpoly(3*y^7 + 2*y-4, y):` ◇ See also: `powseries[powcreate]`.

powseries[powsolve]

`powsolve(system)`

Solves linear differential equations as power series ◇ *system* may be a set or an expression sequence, with a linear differential equation and possibly some initial conditions. All initial conditions must be at zero. Derivatives in the initial conditions are denoted by the `D` operator. ◇ `a:=powsolve(diff(y(x),x,x)=y(x), y(0)=0, D(y)(0)=1);` ◇ See also: `D`, `dsolve`.

powseries[quotient]

`quotient(`ps_1`, `ps_2`)`
Quotient of two formal power series ◇ This function returns the formal power series equal to ps_1/ps_2. ◇ See also: `powseries[inverse]`, `powseries[multiply]`.

powseries[reversion]

`reversion(`ps_1`, `ps_2`)`
`reversion(`ps_1`)`
Reversion of formal power series ◇ This function returns the formal power series that is the reversion of ps_1 with respect to ps_2. If ps_2 is not specified, it is assumed to be the formal power series with one nonzero coefficient: $ps_2(1) = 1$. Composition of the result into ps_1 gives ps_2. The given power series must satisfy $ps_1(0) = 0$, $ps_1(1) = 1$, and $ps_2(0) = 0$; otherwise an error message is given. ◇ See also: `powseries[compose]`.

powseries[subtract]

`subtract(`ps_1`, `ps_2`)`
Subtraction of two formal power series ◇ This function returns the formal power series in which each coefficient is the corresponding coefficient of ps_1 minus the corresponding coefficient of ps_2. ◇ See also: `powseries[add]`, `powseries[negative]`.

powseries[tpsform]

`tpsform(`*ps*`, `*var*`, `*n*`)`
`tpsform(`*ps*`, `*var*`)`
Truncated power series form ◇ This function creates a truncated power series (a series data structure) in the variable *var* up to order *n* that is equal to the given formal power series up to the order term. If the third argument is missing, the value of the global variable `Order` is used. ◇ `powcreate(t(n)=1/n!): tpsform(t,x,4);` ⟶ $1 + x + \frac{1}{2}x^2 + \frac{1}{6}x^3 + O\left(x^4\right)$ ◇ See also: `type/series`.

prem

`prem(`$poly_1$`, `$poly_2$`, `*var*`)`
`prem(`$poly_1$`, `$poly_2$`, `*var*`, `$name_1$`, `$name_2$`)`
Pseudo-remainder of polynomials ◇ This function returns the pseudo-remainder r of the given multivariate polynomials such that $m\,poly_1 = poly_2 q + r$, where the degree of r in *var* is less than the degree of $poly_2$ in *var*. The multiplier m is the leading coefficient of $poly_2$ raised to the $(\deg(poly_1) - \deg(poly_2) + 1)$ power, where deg(*poly*) is the degree of *poly* in *var*. If $name_1$ is present, it is assigned m. If $name_2$ is present, it is assigned the pseudo-quotient q. ◇ `prem(x^3+y^3,x*y-2,x);` ⟶ $y^6 + 8$ ◇ See also: `Prem`, `rem`, `sprem`.

Prem `Prem(`*poly₁*`,` *poly₂*`,` *var*`)`
`Prem(`*poly₁*`,` *poly₂*`,` *var*`,` *name₁*`,` *name₂*`)`
Inert pseudo-remainder of polynomials ◇ When used with `evala`, `evalgf`, `mod`, or `modp1`, this function computes the pseudo-remainder of $poly_1$ divided by $poly_2$, over the given coefficient domain. ◇ `Prem(x^3+y^3, x*y-2, x) mod 3;` $\longrightarrow y^6 + 2$ ◇ See also: `evala`, `evalgf`, `mod`, `prem`, `RootOf`, `Rem`, `Sprem`.

prevprime `prevprime(`*n*`)`
Computes the largest probable prime smaller than *n* ◇ This function uses `isprime` to compute the previous probable prime. ◇ `prevprime(100);` $\longrightarrow 97$ ◇ See also: `isprime`, `nextprime`.

Primfield `Primfield(`*L*`,` *K*`)`
`Primfield(`*L*`)`
Primitive element of an algebraic extension ◇ When used with `evala`, this function computes an element a of L such that $L = K(a)$. If K is not given, the smallest possible transcendental extension of the rationals is chosen. A list of equations giving a in terms of the generators of L and the generators of L in terms of a is returned. K and L should be sets of `RootOf`s. ◇ See also: `evala`, `RootOf`.

Primitive `Primitive(`*poly*`)`
Primitive polynomial test ◇ When used with `mod`, this function tests whether the polynomial is primitive modulo a prime. It returns *true* or *false*. ◇ `Primitive(x^2+x+1) mod 2;` $\longrightarrow$ *true* ◇ See also: `Irreduc`, `Powmod`.

primpart `primpart(`*poly*`,` *vars*`)`
`primpart(`*poly*`,` *vars*`,` *name*`)`
Primitive part of a multivariate polynomial ◇ The primitive part is *poly*/content(*poly*, *vars*). *vars* may be a single variable name or a list or set of variable names. If *name* is present, it is assigned the content. ◇ `primpart(2*x+4*x*y, y);` $\longrightarrow 1 + 2y$ ◇ See also: `content`, `icontent`.

Primpart `Primpart(`*poly*`,` *vars*`)`
`Primpart(`*poly*`,` *vars*`,` *name*`)`
Inert primitive part function ◇ When used with `evala`, `evalgf`, or `mod`, this function computes the primitive part of the polynomial in the variable or list or set of variables *vars* over the coefficient domain. If *name* is present, it is assigned the content. ◇ See also: `Content`, `evala`, `evalgf`, `mod`.

print `print(`*expr₁*`,` *expr₂*`, ...)`
Prints expressions ◇ The expressions printed are separated by a comma and a blank. The interface variable `prettyprint` determines the format

in which the expressions are printed. This function returns `NULL`. ◇ See also: `interface`, `lprint`, `printf`.

A B C D E F G H I J K L M N O P Q R S T U V W X Y Z

printf `printf(`*formatstring*`,` $expr_1$`,` $expr_2$`, ...)`
Formatted printing of expressions ◇ *formatstring* is composed of literal text together with formatting specifications that describe how each expression is printed. The first specification corresponds to $expr_1$, the second to $expr_2$, and so on. A formatting specification has the form `%`*flags m.n code*. Only the formatting code *code* is required. *m* names the minimal width used in printing the expression; *n* names the number of characters that should appear after the decimal point for floating-point numbers. *code* names the type of expression expected. The following codes may be given:

`a`	any expression
`c`	single character
`d` *or* `i`	signed decimal integer
`e` *or* `E`	floating point number in scientific notation
`f`	fixed point number
`g` *or* `G`	use `d`, `e`, `E`, or `f` format, depending on the value
`o`	unsigned octal integer
`p`	pointer (memory address)
`s`	string
`u`	unsigned decimal integer
`x` *or* `X`	unsigned hexadecimal integer
`%`	literal percent symbol

The *flags* are optional characters that alter the way expressions are formatted:

`+`	put a leading `+` or `-` sign on numeric values
blank	put a leading `-` sign on negative numbers
`#`	affects leading and trailing zeros and decimal points
`-`	left justify the expression within the given width

◇ `printf('%f %4.2f%c %a is %d', 9/7, 9/7, ':', x^2, 9);` ⟶ `1.285714 1.29: x^2 is 9` ◇ See also: `lprint`, `print`, `sscanf`.

printlevel Environment variable controlling the number of levels of results displayed ◇ Statements at a level less than or equal to the setting of this variable have their results displayed to the screen. The top interactive level is level 0. Entering an `if` statement or `do` loop adds 1 to the current level, and entering a procedure adds 5. `printlevel` initially has the value 1. For debugging, set this to a large value. ◇ See also: `infolevel`, `trace`, `mint`.

priqueue `initialize(`*name*`)`
`insert(`*record*`, `*pq*`)`
`extract(`*pq*`)`
Priority queue functions ◇ `initialize` makes a new queue and assigns it to the given name. `insert` adds the record to the priority queue *pq*. The ordering criterion used is *record*`[1]`. `extract` deletes the maximum record in the queue *pq* and returns this record. Also, *pq*`[0]` is the number of entries in the queue *pq*. You must `readlib(priqueue)` before using these commands.

ProbSplit `ProbSplit(`*poly*`, `*n*`, `*var*`) mod `*p*
`ProbSplit(`*poly*`, `*n*`, `*var*`, `*K*`) mod `*p*
Probabilistic splitting of distinct degree factors ◇ When used in conjunction with `mod`, this function factors a monic square-free univariate polynomial that only has factors of degree *n* over a finite field. The optional argument *K* specifies an extension field over which the factorization is performed. The algorithm used is a probabilistic method of Cantor and Zassenhaus. A set of irreducible factors is returned. ◇ This function is new in Release 3. ◇ `ProbSplit(x^4+1, 2, x) mod 3;` $\longrightarrow \{x^2+2x+2, x^2+x+2\}$ ◇ See also: `Berlekamp`, `DistDeg`, `Factors`, `Sqrfree`.

proc `proc(`*args*`) local `*names*`; options `*options*`; `*statements*` end`
Defines a procedure ◇ *args* is the sequence of zero or more formal parameters. The `local` and `options` declarations are both optional. ◇ See also: `args`, `ERROR`, `global`, `local`, `nargs`, `options`, `procname`, `RETURN`.

procbody `procbody(`*procedure*`)`
Creates a "neutralized form" of a procedure ◇ This routine is used in writing routines that manipulate procedures. Do not evaluate the neutral form, since this will almost certainly corrupt it. ◇ You must `readlib(procbody)` before using this command. ◇ `procbody(x->x^2);` $\longrightarrow$ *&proc* $\left([\,x\,], [\,], [operator, arrow], \&args_{[1]}{}^2\right)$ ◇ See also: `procmake`.

procmake `procmake(`*neutralform*`)`
Creates an executable procedure from its neutral form ◇ You must `readlib(procmake)` before using this command. ◇ `procmake(procbody(x->x^2));` $\longrightarrow x \rightarrow x^2$ ◇ See also: `procbody`.

procname The name with which a procedure was invoked ◇ Within a procedure, the value of `procname` is the name with which the procedure was invoked. ◇ See also: `args`, `nargs`, `proc`.

product `product(`*expr*`, `*var*`)`
`product(`*expr*`, `*var*`=`*m*`..`*n*`);`
`product(`*expr*`, `*var*`=`α`)`
Definite and indefinite product ◇ The first form computes the indefinite product of the expression with respect to the variable *var*. The second form computes the definite product of the expression over the given range. The third form computes the definite product over the roots of the polynomial given by the `RootOf` expression α. ◇ `product(k,k);` $\longrightarrow$ $\Gamma(k)$ ◇ `product(k,k=1..7);` $\longrightarrow$ 5040 ◇ `product(k+1, k=RootOf(x^2-x+1));` $\longrightarrow$ 3 ◇ See also: `convert/'*'`, `Product`, `RootOf`, `sum`.

Product `Product(`*expr*`, `*var*`)`
`Product(`*expr*`, `*var*`=`*m*`..`*n*`)`
`Product(`*expr*`, `*var*`=`α`)`
Inert form of product ◇ See also: `product`, `RootOf`, `Sum`.

profile `profile(`f_1`, `f_2`, ...)`
Sets up profiling of procedures ◇ This function names the procedures to be profiled. After executing `profile`, perform the desired calculations, then use `showprofile`. ◇ You must `readlib(profile)` before using this command. ◇ See also: `showprofile`, `showtime`, `time`, `unprofile`.

projgeom Projective geometry ◇ In Release 2 and earlier versions, the functions in this package must first be loaded using `with` before they can be invoked. ◇ See also: `projgeom[`*function*`]`, `geometry`, `geom3d`, `with`.

projgeom[collinear]
`collinear(`pt_1`, `pt_2`, `pt_3`)`
Tests whether three given points are collinear ◇ This function returns *true*, *false*, or a condition. ◇ See also: `projgeom[concur]`.

projgeom[concur]
`concur(`$line_1$`, `$line_2$`, `$line_3$`)`
Tests whether three given lines are concurrent ◇ This function returns *true*, *false*, or a condition. ◇ See also: `projgeom[collinear]`.

projgeom[conic]
`conic(`*name*`, [`*a*`, `*b*`, `*c*`, `*d*`, `*e*`, `*f*`])`
Defines a conic ◇ *name* is the name given to the conic. The elements of the list are the coefficients of the homogeneous equation defining the conic: $a\,x^2 + b\,y^2 + c\,z^2 + d\,y\,z + e\,x\,z + f\,x\,y$ ◇ See also: `projgeom[line]`, `projgeom[point]`.

projgeom[conjugate]
`conjugate(`pt_1`, `pt_2`, `*conic*`)`
Checks whether two given points are conjugate with respect to the given conic ◇ This function returns *true*, *false*, or a condition.

projgeom[ctangent]

`ctangent(`*pt*`,` *conic*`,` *name*$_1$`,` *name*$_2$`)`
Finds the tangent lines to the conic from any point in the complex plane ◇ One or two tangent lines are computed and named *name*$_1$ and *name*$_2$. ◇ See also: `projgeom[ptangent]`, `projgeom[rtangent]`, `projgeom[tangentte]`.

projgeom[fpconic]

`fpconic(`*name*`,` *pt*$_1$`,` *pt*$_2$`,` *pt*$_3$`,` *pt*$_4$`,` *pt*$_5$`)`
Finds the conic that goes through five given points ◇ *name* is the name given to the conic. Use *name*`[coords]` for the coefficients of the equation defining the conic.

projgeom[harmonic]

`harmonic(`*pt*$_1$`,` *pt*$_2$`,` *pt*$_3$`,` *name*`)`
Finds a point that is a harmonic conjugate to another point with respect to two other points ◇ *name* is the name given to the harmonic conjugate of *pt*$_1$ with respect to *pt*$_2$ and *pt*$_3$. `NULL` is returned if such a point does not exist. ◇ See also: `projgeom[tharmonic]`, `geometry[harmonic]`.

projgeom[inter]

`inter(`*line*$_1$, *line*$_2$, *name*`)`
Finds the intersection of the two given lines ◇ *name* is the name given to the point of intersection. ◇ See also: `projgeom[lccutc]`, `projgeom[lccutr]`, `projgeom[lccutr2p]`, `geom3d[inter]`, `geometry[inter]`, `student[intersection]`.

projgeom[join]

`join(`*pt*$_1$`,` *pt*$_2$`,` *name*`)`
Finds the line joining the two given points ◇ *name* is the name given to the line joining the points. ◇ See also: `projgeom[lccutr2p]`.

projgeom[lccutc]

`lccutc(`*conic*`,` *line*`,` *name*$_1$`,` *name*$_2$`)`
Finds the intersection of a conic and a line in a complex plane ◇ *name*$_1$ and *name*$_2$ are the names given to the points of intersection. ◇ See also: `projgeom[lccutc2p]`, `projgeom[lccutr]`.

projgeom[lccutr]

`lccutr(`*conic*`,` *line*`,` *name*$_1$`,` *name*$_2$`)`
Finds the intersection of the given conic and the given line in a real extended plane ◇ *name*$_1$ and *name*$_2$ are the names given to the points of intersection. ◇ See also: `projgeom[lccutr2p]`, `projgeom[lccutc]`.

projgeom[lccutr2p]

`lccutr2p(`*conic*`,` *pt*$_1$`,` *pt*$_2$`,` *name*$_1$`,` *name*$_2$`)`
Finds the intersection of the given conic and the given line joining the two given points in a real extended plane ◇ *name*$_1$ and *name*$_2$ are the names

given to the points of intersection. ◇ See also: `projgeom[join]`, `projgeom[lccutr]`, `projgeom[lccutc]`.

projgeom[line]

`line(`*name*`, [`*a*`, `*b*`, `*c*`])`

Defines a line ◇ *name* is the name given to the line. The elements of the list are the coefficients of the homogeneous equation defining the line: $a\,x + b\,y + c\,z$ ◇ See also: `projgeom[point]`, `projgeom[conic]`.

projgeom[linemeet]

`linemeet(`$line_1$`, `$line_2$`, `*r*`, `*name*`)`

Finds a line concurrent to the two given lines ◇ *name* is the name given to the new line, whose coordinates are $line_1$`[coords]`+r $line_2$`[coords]`. ◇ See also: `projgeom[onsegment]`.

projgeom[midpoint]

`midpoint(`pt_1`, `pt_2`, `*name*`)`

Finds the midpoint of the line joining the two given points ◇ The midpoint is the point that is harmonic to the intersection of the line at infinity and the line joining pt_1 and pt_2. *name* is the name given to the midpoint. ◇ See also: `geom3d[midpoint]`, `geometry[midpoint]`, `student[midpoint]`.

projgeom[onsegment]

`onsegment(`pt_1`, `pt_2`, `*r*`, `*name*`)`

Finds a point that is collinear to the two given points and divides these points by a given ratio *r* ◇ *name* is the name given to the computed point, which divides pt_1 and pt_2 by the ratio *r*. ◇ See also: `projgeom[linemeet]`, `geom3d[onsegment]`, `geometry[onsegment]`.

projgeom[point]

`point(`*name*`, [`*a*`, `*b*`, `*c*`])`

Defines a point in the projective plane ◇ *name* is the name given to the point with homogeneous coordinates *a*, *b*, and *c*. ◇ See also: `projgeom[line]`, `projgeom[conic]`.

projgeom[polarp]

`polarp(`*pt*`, `*conic*`, `*name*`)`

Determines the polar line of the given point with respect to the given conic ◇ *name* is the name given to the polar line. ◇ See also: `projgeom[poleline]`.

projgeom[poleline]

`poleline(`*line*`, `*conic*`, `*name*`)`

Determines the pole of the given line with respect to the given conic ◇ *name* is the name given to the pole point. ◇ See also: `projgeom[polarp]`.

projgeom[ptangent]

`ptangent(`*pt*`, `*conic*`, `*name*`)`
Finds the tangent line to the given conic at the given point on the conic ◇ *name* is the name given to the tangent line. The point must be on the conic. ◇ See also: `projgeom[ctangent]`, `projgeom[rtangent]`, `projgeom[tangentte]`.

projgeom[rtangent]

`rtangent(`*pt*`, `*conic*`, `$name_1$`, `$name_2$`)`
Finds the tangent lines to the given conic at the given point in the real extended plane ◇ $name_1$ and $name_2$ are the names given to the tangent lines. There may be zero, one or two tangent lines. ◇ See also: `projgeom[ctangent]`, `projgeom[ptangent]`, `projgeom[tangentte]`.

projgeom[tangentte]

`tangentte(`*line*`, `*conic*`)`
Checks whether the given line is tangent to the given conic ◇ This function returns *true*, *false*, or a condition. ◇ See also: `projgeom[ctangent]`, `projgeom[ptangent]`, `projgeom[rtangent]`.

projgeom[tharmonic]

`tharmonic(`pt_1`, `pt_2`, `pt_3`, `pt_4`)`
Checks whether the first pair of points is harmonic conjugate to the second pair of points ◇ This function returns *true*, *false*, or a condition. The given points must first be collinear. ◇ See also: `projgeom[harmonic]`.

proot

`proot(`*poly*`, `*n*`)`
Computes the nth root of a polynomial ◇ If the polynomial is not a perfect nth power, the name *_NOROOT* is returned. ◇ You must `readlib(proot)` before using this command. ◇
`proot(x^2+2*x*y+y^2,2);` $\longrightarrow x+y$ ◇ See also: `psqrt`.

protect

`protect(`$name_1$`, `$name_2$`, ...)`
Protects names from being modified ◇ Most system names are protected by default. ◇ This function is new in Release 3. ◇ See also: `type/protected`, `unprotect`.

Psi

`Psi(`*z*`)`
`Psi(`*n*`, `*z*`)`
The digamma and polygamma functions ◇ The first form calculates the digamma function at z. This function is the logarithmic derivative of the gamma function. The second form calculates the nth polygamma function, which is the nth derivative of the digamma function. ◇
`Psi(1);` $\longrightarrow -\gamma$ ◇ See also: `GAMMA`, `lnGAMMA`.

psqrt `psqrt(`*poly*`)`
Square root of a polynomial ◇ If the polynomial is not a perfect square, the name *_NOSQRT* is returned. ◇ You must `readlib(psqrt)` before using this command. ◇ `psqrt(x^2+2*x*y+y^2);` $\longrightarrow x+y$ ◇ See also: `proot`.

quit Quits a Maple session ◇ See also: `done`, `stop`.

quo `quo(`*poly1*`,` *poly2*`,` *var*`)`
`quo(`*poly1*`,` *poly2*`,` *var*`,` *name*`)`
Computes the quotient of $poly_1/poly_2$ ◇ The first two arguments are polynomials in the variable *var*. This function returns the quotient of $poly_1/poly_2$. If *name* is present, it is assigned the remainder. ◇ `quo(x^4+x+1, x^2+1, x);` $\longrightarrow x^2-1$ ◇ See also: `divide`, `iquo`, `Quo`, `rem`.

Quo `Quo(`*poly1*`,` *poly2*`,` *var*`)`
`Quo(`*poly1*`,` *poly2*`,` *var*`,` *name*`)`
Inert polynomial quotient ◇ When used with `evala`, `evalgf`, `mod`, or `modp1`, this function returns the quotient of $poly_1/poly_2$ over the given coefficient domain. If *name* is present, it is assigned the remainder. ◇ `Quo(x^4+x+1, x^2+1, x) mod 2;` $\longrightarrow x^2+1$ ◇ See also: `Divide`, `quo`, `Rem`.

radnormal `radnormal(`*expr*`)`
Simplifies nested radicals ◇ Presently, only algebraic numbers (not algebraic functions) are supported. ◇ You must `readlib(radnormal)` before using this command. ◇ `radnormal(sqrt(4*I*sqrt(5)-1));` $\longrightarrow 2+\sqrt{-5}$

radsimp `radsimp(`*expr*`)`
`radsimp(`*expr*`,` *name*`)`
Simplifies radicals in the expression ◇ This function provides a normal form for expressions involving unnested radicals of rational functions. For nested radicals, the algorithm is applied recursively. If a name is given, it is assigned the simplified expression with its denominator rationalized. ◇ `radsimp((a^3+3*a^2+3*a+1)^(1/3));` $\longrightarrow a+1$ ◇ See also: `simplify`.

rand `rand()`
`rand(`*m*`..`*n*`)`
Random number generator ◇ With no arguments, this function returns a 12 digit nonnegative random integer. Given an integer range, it returns a procedure that generates random integers between m and n. The same sequence of "random" integers is returned in each session unless the global variable `_seed` is reset. ◇ `flip := rand(0..1): flip();` $\longrightarrow$ 1 ◇ See also: `linalg[randmatrix]`, `randpoly`, `_seed`.

randpoly `randpoly(`*vars*`, `*options*`)`
Generates a random polynomial ◇ This function generates a random polynomial in *vars*, a single variable, or a list or set of variables. The following options are recognized. Default values are in parentheses.

`coeffs`	procedure for generating the coefficients (rand(−99..99))
`degree`	total degree for a dense polynomial (5)
`dense`	generate a dense polynomial
`expons`	procedure for generating the exponents (rand(6))
`terms`	number of terms, for a sparse polynomial (6)

◇ `randpoly(x,degree=23,terms=3);` $\longrightarrow 97\,x^{14}+50\,x^{12}+79\,x^{9}$ ◇ See also: `rand`, `Randpoly`.

Randpoly `Randpoly(`*n*`, `*var*`)`
`Randpoly(`*n*`, `*var*`, `*K*`)`
`Randpoly(`*n*`)`
Inert random polynomial ◇ When used in conjunction with `mod`, this function returns a polynomial of degree *n* in the variable *var* whose coefficients are selected at random from the integers modulo a prime *p*. If a single `RootOf` K is supplied, the coefficients are selected from the corresponding extension of the prime field. The third form is used in conjunction with `modp1`. ◇ `Randpoly(4,x) mod 2;` $\longrightarrow x^4+x^3+x^2$ ◇ `modp1(Randpoly(3),7);` $\longrightarrow$ 1000500030002 ◇ See also: `Irreduc`, `randpoly`, `Randprime`, `RootOf`.

Randprime

`Randprime(`*n*`, `*var*`)`
`Randprime(`*n*`, `*var*`, `*K*`)`
`Randprime(`*n*`)`
Inert random monic prime polynomial ◇ When used in conjunction with `mod` or `modp1`, this function returns a monic irreducible polynomial of degree *n* in the variable *var* over the integers modulo a prime *p*. If a single `RootOf` K is supplied, the coefficients are selected from the corresponding algebraic extension of the prime field. The third form is used in conjunction with `modp1`. ◇ `Randprime(4,x) mod 2;` $\longrightarrow x^4+x+1$ ◇ `modp1(Randprime(3),7);` $\longrightarrow$ 1000200020003 ◇ See also: `Randpoly`, `RootOf`.

rationalize

`rationalize(`*expr*`)`
`rationalize(`*list*`)`
`rationalize(`*set*`)`
Rationalizes expressions ◇ A rationalized expression has no radicals in the denominator. The first argument may a single expression or a list or set of expressions. ◇ This function is new in Release 3. ◇ You must `readlib(rationalize)` before using this command. ◇ `rationalize(1/(x+2^(1/3)));` $\longrightarrow \frac{x^2-x\,2^{1/3}+2^{2/3}}{x^3+2}$ ◇ See also: `normal`.

ratrecon `ratrecon(`$poly_1$`, `$poly_2$`, `*var*`, `*N*`, `*D*`, `$name_N$`, `$name_D$`)`
Reconstructs a rational function from its image $poly_1$ mod $poly_2$ ◇ This function finds polynomials p and q in *var* with $\deg p \leq N$ and $\deg q \leq D$ such that $p/q \equiv poly_1$ mod $poly_2$. The solution is unique up to multiplication by a constant if $N + D \geq \deg poly_2$. This function returns *true* if a solution is found and assigns the numerator to $name_N$ and the denominator to $name_D$; otherwise it returns *false*. ◇ You must `readlib(ratrecon)` before using this command. ◇ `ratrecon(x^2-2,x^3+5,x,1,1,'num','den'),num/den;` ⟶ *true*, $-\frac{5+2x}{x}$ ◇ See also: `convert/ratpoly`, `iratrecon`, `Ratrecon`.

Ratrecon `Ratrecon(`$poly_1$`, `$poly_2$`, `*var*`, `*N*`, `*D*`, `$name_N$`, `$name_D$`)`
Inert rational function reconstruction ◇ When used in conjunction with `mod`, this function reconstructs a rational function from its image $poly_1$ mod $poly_2$ over a given domain of coefficients. It returns *true* if a solution is found and assigns the numerator to $name_N$ and the denominator to $name_D$; otherwise it returns *false*. ◇ `Ratrecon(x^2-2,x^3+5,x,1,1,'num','den') mod 2,num/den;` ⟶ *true*, $\frac{1}{x}$ ◇ See also: `ratrecon`.

Re `Re(`*z*`)`
Extracts the real part of the expression ◇ Use `assume` or `evalc` if unknown variables should represent real values. If *expr* includes a function *f*, `Re` tries to execute the procedure `'Re/`*f*`'`. ◇ `Re(exp(2*I));` ⟶ cos(2) ◇ See also: `assume`, `evalc`, `Im`.

read `read` *filename*
Reads the contents of the given file into the session ◇ A file name ending in `.m` is assumed to be in Maple internal format. Otherwise, it should be an ordinary text file. The file name should be enclosed in back quotes. ◇ See also: `convert/hostfile`, `readdata`, `readlib`, `readstat`, `save`.

readdata `readdata(`*filename*`)`
`readdata(`*filename*`, `*type*`)`
`readdata(`*filename*`, `*type*`, `*n*`)`
Reads columns of data from the file ◇ This function reads *n* columns of numeric data from the file, or one column if *n* is absent. The file name should be enclosed in back quotes. *type* identifies the type of data read in: it may be `integer` or `float` (the default). A list of lists containing the data is returned. ◇ You must `readlib(readdata)` before using this command. ◇ See also: `readline`, `sscanf`, `stats[importdata]`.

readlib `readlib(`*f*`)`
`readlib(`*f*`, `$file_1$`, `$file_2$`, ...)`
Reads a library file to define a specified name ◇ The libraries specified by the global variable `libname` are searched in sequence for a file named *f*`.m` giving the definition of the procedure *f*. ◇ See also: `libname`, `read`, `savelib`, `unload`, `with`.

readline

`readline(`*filename*`)`
`readline(terminal)`
`readline()`
Reads a string from the terminal or a file ◇ This function reads the next line from the specified file or a terminal. A file is opened automatically when the first line is read from it. With no arguments, this function reads the next line from the input stream. A string is returned. ◇ See also: `parse`, `readdata`, `readstat`, `sscanf`.

readstat

`readstat(`*string*`)`
Reads a statement from the input stream and returns the result of that statement ◇ *string* is printed as a prompt before reading the statement. The statement must end with a semicolon or colon. ◇ See also: `read`, `readline`.

realroot

`realroot(`*poly*`)`
`realroot(`*poly*`, `*widthgoal*`)`
Finds isolating intervals for real roots of a polynomial ◇ This function returns a list of isolating intervals with rational endpoints for all real roots of the univariate polynomial *poly*. If a second argument is given, the width of each interval is at most *widthgoal*. ◇ You must `readlib(realroot)` before using this command. ◇ `realroot(x^3-2*x,.1);` ⟶ $[[0,0],\left[\frac{11}{8},\frac{23}{16}\right],\left[\frac{-23}{16},\frac{-11}{8}\right]]$ ◇ See also: `fsolve`, `roots`, `Roots`, `solve`, `sturm`, `sturmseq`.

recipoly

`recipoly(`*poly*`, `*var*`)`
`recipoly(`*poly*`, `*var*`, `*name*`)`
Determines whether a polynomial is self-reciprocal ◇ *poly* is a polynomial in the variable *var*. This function returns *true* or *false*. If the degree d of *poly* is even and *true* is returned, then if *name* is present it is assigned the polynomial p such that $var^{d/2}p(var + 1/var) = poly$. If d is odd and *true* is returned, *name* is assigned $poly/(var + 1)$. ◇ You must `readlib(recipoly)` before using this command. ◇ `recipoly(x^4+1,x,'p'),p;` ⟶ *true*, $x^2 - 2$

related

`related(`*topic*`)`
Shows topics related to a topic ◇ The information is extracted from the on-line help text. ◇ This function is new in Release 3. ◇ `related(isprime);` ⟶ `SEE ALSO: nextprime, prevprime, ithprime` ◇ See also: `example`, `help`, `info`, `usage`.

rem

`rem(`*poly₁*`, `*poly₂*`, `*var*`)`
`rem(`*poly₁*`, `*poly₂*`, `*var*`, `*name*`)`
Computes the remainder of $poly_1/poly_2$ ◇ The first two arguments are polynomials in the variable *var*. If *name* is present, it is assigned the quotient. ◇ `rem(x^4+x+1, x^2+1, x);` ⟶ $x + 2$ ◇ See also: `divide`, `irem`, `prem`, `quo`, `Rem`, `sprem`.

Rem `Rem(`*poly₁*`,` *poly₂*`,` *var*`)`
`Rem(`*poly₁*`,` *poly₂*`,` *var*`,` *name*`)`
Inert polynomial remainder ◇ When used with `evala`, `evalgf`, `mod`, or `modp1`, this function returns the remainder of $poly_1/poly_2$ over the given coefficient domain. If *name* is present, it is assigned the quotient. ◇ `Rem(x^4+x+1, x^2+1, x) mod 2;` $\longrightarrow x$ ◇ See also: `Divide`, `Powmod`, `Prem`, `Quo`, `rem`, `Sprem`.

remember `option remember`
`options remember`
Remember option in procedures ◇ If this option is given for a procedure, a "remember table" is created for the procedure. This records the result for each invocation of the procedure, so if it is called more than once with the same arguments, the result is retrieved from this table. The remember table is accessible as operand 4 of the procedure structure. ◇ See also: `forget`, `op`, `options`, `proc`, `table`.

residue `residue(`*expr*`,` *var*`=`*a*`)`
Computes the algebraic residue of the expression in the variable *var* around the point *a* ◇ The residue is the coefficient of $(var - a)^{-1}$ in the Laurent series expansion of *expr*. ◇ You must `readlib(residue)` before using this command. ◇ `residue(csc(x),x=0);` $\longrightarrow 1$ ◇ See also: `numapprox[laurent]`, `series`, `singular`.

restart Clears the internal state of a session so that it appears that Maple has just been started ◇ The settings of all identifiers are forgotten, `libname` is reset, loaded packages are unloaded, and the Maple initialization files are reread. ◇ See also: `unassign`.

resultant `resultant(`*poly₁*`,` *poly₂*`,` *var*`)`
Computes the resultant of two polynomials in the variable *var* ◇ $poly_1$ and $poly_2$ are polynomials in the variable *var*. ◇ `resultant(x^2+1, x^3+1, x);` $\longrightarrow 2$ ◇ See also: `discrim`, `gcd`, `Resultant`, `linalg[bezout]`, `linalg[sylvester]`.

Resultant `Resultant(`*poly₁*`,` *poly₂*`,` *var*`)`
Inert resultant of polynomials ◇ This function is used in conjunction with `evala`, `evalgf`, `mod`, or `modp1` to determine the resultant of the polynomials in the variable *var* over a given coefficient domain. ◇ `Resultant(x^2+1, x^3+1, x) mod 2;` $\longrightarrow 0$ ◇ See also: `Discrim`, `evala`, `evalgf`, `mod`, `modp1`, `resultant`.

RETURN `RETURN(`*expr₁*`,` *expr₂*`, ...)`
Explicit return from a procedure ◇ This function causes an immediate return from a procedure, returning the given expression or expression sequence (possibly null). If there is no `RETURN` statement at the end of a procedure, the value returned is the value of the last statement executed. ◇ See also: `ERROR`, `proc`.

A B C D E F G H I J K L M N O P Q R S T U V W X Y Z

rhs

`rhs(`*expr*`)`
Returns the right-hand side of an expression ◇ The expression may be an equation, an inequality, or a range. ◇ `rhs(3..4);` ⟶ 4 ◇ See also: `lhs`, `op`.

rootbound

`rootbound(`*poly*`,` *var*`)`
Computes a crude bound on the maximum size of any root (real or complex) of the polynomial ◇ You must `readlib(rootbound)` before using this command. ◇ `rootbound(x^20-2,x);` ⟶ 2

RootOf

`RootOf(`*expr*`)`
`RootOf(`*expr*`,` *var*`)`
`RootOf(`*expr*`,` *z*`)`
`RootOf(`*expr*`,` *var*`,` *z*`)`
A representation for roots of equations ◇ This function is a placeholder for representing the roots of an equation in one variable. It is the standard representation for algebraic numbers, algebraic functions, and finite fields. If a variable *var* is not specified, *expr* must be either a univariate polynomial or an expression in `_Z`. ◇ In Release 3, a complex number *z* given as the last argument means that the `RootOf` expression stands for the particular root near *z*. Without this argument, the `RootOf` expression denotes any of the roots (this is always the meaning in Release 2 and earlier versions). ◇ `evalf(RootOf(x^4+1, .7-.7*I), 4);` ⟶ $.7071 - .7071\,I$ ◇ See also: `alias`, `allvalues`, `convert/RootOf`, `evala`, `mod`, `Roots`, `solve`, `type/RootOf`.

roots

`roots(`*poly*`)`
`roots(`*poly*`,` *K*`)`
Roots of a univariate polynomial ◇ This function computes the exact roots of a univariate polynomial over the field determined by the coefficients of *poly* or, in the second case, over the given algebraic number field. *K* may be a single `RootOf` or radical, or a list or set of `RootOf`s or radicals. A list of root and multiplicity pairs is returned. ◇ `roots(x^2+9,I);` ⟶ $[[3\,I, 1], [-3\,I, 1]]$ ◇ See also: `factor`, `realroot`, `Roots`, `RootOf`, `solve`, `sturm`, `sturmseq`.

Roots

`Roots(`*poly*`)`
`Roots(`*poly*`,` *K*`)`
Inert roots of a univariate polynomial ◇ When used with `mod` or `modp1`, the first form computes the roots of the polynomial over a given coefficient domain. The second form, when used with `mod`, computes the roots of the polynomial over the extension given by the `RootOf` *K*. ◇ `Roots(x^4+2, RootOf(t^2+1,t)) mod 3;` ⟶ $[[\mathrm{RootOf}(\_Z^2+1), 1], [2\,\mathrm{RootOf}(\_Z^2+1), 1], [1, 1], [2, 1]]$ ◇ See also: `Factors`, `mod`, `modp1`, `msolve`, `RootOf`, `roots`.

round `round(`*z*`)`
`round(1,` *z*`)`
Returns the nearest integer to *z* if *z* is real ◇ For complex *z*, the function is applied to the real and imaginary parts. The second form returns the first derivative of the round function wherever it is defined. ◇ `round(1.5);` ⟶ 2 ◇ See also: `ceil`, `floor`, `frac`, `trunc`.

rsolve `rsolve(`*eqns*`,` *fcns*`)`
`rsolve(`*eqns*`,` *fcns*`, genfunc(`*var*`))`
`rsolve(`*eqns*`,` *fcns*`, makeproc)`
Solves the recurrence equation or equations ◇ *eqns* is a single equation or a set of equations specifying recurrence relations and boundary conditions. *fcns* is a single function name or a set of function names. This function solves the recurrence relation(s) for the functions. It returns an expression for the general term of the function, or a set of such expressions. With the option `genfunc(`*var*`)`, the function returns the generating functions in the variable *var* of the sequences defined by `eqns`. With the option `makeproc`, the function returns a procedure for evaluating the function defined by *eqns*. ◇ `rsolve({f(n)=f(n-1)+2*f(n-2), f(1..2)=1}, f);` ⟶ $-\frac{1}{3}(-1)^n + \frac{1}{3}2^n$ ◇ `rsolve({g(n)=-2*g(n-1), g(0)=1}, g(n), genfunc(z));` ⟶ $\frac{1}{1+2\,z}$ ◇ See also: `asympt`, `genfunc`, `solve`, `ztrans`.

save `save` *filename*
`save` $name_1$, $name_2$, ..., $name_k$, *filename*
The save statement ◇ The first form saves all assigned names in the current session to the specified file name. The second form saves the specified variable names to the file as a sequence of assignment statements. The file name should be enclosed in back quotes. If *filename* ends with `.m`, the assignment statements are written in Maple internal format. Otherwise, linear printing is used. ◇ See also: `anames`, `convert/hostfile`, `lprint`, `read`, `writeto`.

savelib `savelib(`$name_1$, $name_2$, ..., $name_k$, *filename*`)`
Saves assignments of variables in a library ◇ This function scans the sequence of directory names given in the `libname` global variable for a directory where the user is allowed to save files. An assignment statement for each variable is saved in the given file in the first such directory found. The variable names should be enclosed in right single quotes, and the file name should be enclosed in back quotes. If *filename* ends with `.m`, the assignment statements are written in Maple internal format. Otherwise, linear printing is used. ◇ See also: `libname`, `readlib`, `save`.

search `search(`*string*`,` *targetstring*`)`
`search(`*string*`,` *targetstring*`,` *name*`)`
Substring search ◇ This function returns *true* if the string *targetstring* occurs in the string *string* and *false* otherwise. If the search is successful

and a name is given, it is assigned the index of the first occurrence of the target string. ◇ In Release 3 the new functions `searchtext` and `SearchText` are recommended over `search`. ◇ You must `readlib(search)` before using this command. ◇ `search('mathematics', 'at', r), r;` ⟶ *true*, 2 ◇ See also: `searchtext, SearchText`.

searchtext

`searchtext(`*targetstring*`,` *string*`)`
`searchtext(`*targetstring*`,` *string*`,` *m*`..`*n*`)`
Substring search (case-insensitive) ◇ If the string *targetstring* is found in the string *string*, this function returns the position of the first occurrence of the target string. Otherwise, it returns 0. An optional range restricts the search to the substring beginning at position *m* and ending at position *n*. A positive integer indicates a position counted from the left; a negative integer, from the right. The default range is the entire string (`1..-1`). This function disregards differences between uppercase and lowercase letters. ◇ This function is new in Release 3. ◇ `searchtext('bC', 'abcdABCD');` ⟶ 2 ◇ See also: `SearchText, substring`.

SearchText

`SearchText(`*targetstring*`,` *string*`)`
`SearchText(`*targetstring*`,` *string*`,` *m*`..`*n*`)`
Substring search (case-sensitive) ◇ If the string *targetstring* is found in the string *string*, this function returns the position of the first occurrence of the target string. Otherwise, it returns 0. An optional range restricts the search to the substring beginning at position *m* and ending at position *n*. A positive integer indicates a position counted from the left; a negative integer, from the right. The default range is the entire string (`1..-1`). This function counts differences between uppercase and lowercase letters. ◇ This function is new in Release 3. ◇ `SearchText('BC', 'abcdABCD');` ⟶ 6 ◇ See also: `searchtext, substring`.

sec

`sec(`*z*`)`
The secant function ◇ See also: `arcsec`.

sech

`sech(`*z*`)`
The hyperbolic secant function ◇ See also: `arcsech`.

_seed

Random number seed ◇ Setting this global variable alters the sequence of random numbers generated by routines such as `rand`. The same sequence of "random" numbers is generated in each session unless the seed is reset. ◇ See also: `linalg[randmatrix], rand, randpoly`.

select

`select(`*f*`,` *expr*`)`
`select(`*f*`,` *expr*`,` *args*`)`
Selects the operands of a list, set, sum, or product satisfying *f* ◇ *f* is a Boolean-valued function. An expression of the same type as *expr* is

returned. Additional arguments *args* are passed to f. ◇ `select(type, [-1,4,3.,7,Pi], posint);` ⟶ $[4, 7]$

seq `seq(`*expr*`, `*var*` = `*a*`..`*b*`)`
`seq(`*expr*`, `*var*` = `*x*`)`
Creates a sequence ◇ The first form returns the sequence created by substituting the numbers a, $a + 1$, . . ., b (or up to the last number not exceeding b) for the variable *var* in the given expression. The second form has *var* assume successive operands of the expression x. ◇ `seq(k^2, k=1..4);` ⟶ $1, 4, 9, 16$ ◇ `seq(e+1, e=[1,3,5]);` ⟶ $2, 4, 6$ ◇ See also: `$`, `do`, `op`.

series `series(`*expr*`, `*var*`)`
`series(`*expr*`, `*var*`=`*a*`)`
`series(`*expr*`, `*var*`=`*a*`, `*n*`)`
`series(leadterm(`*expr*`), `*var*`=`*a*`, `*n*`)`
Generalized series expansion ◇ This function computes a truncated series expansion of *expr* with respect to the variable *var* about the point a up to order n. If a is omitted, it is understood to be 0. If a is `infinity`, an asymptotic expansion is given. If n is omitted, the value of the global variable `Order` is used. If the first argument is a function call to `leadterm`, only the lowest degree term of the series of *expr* is computed. ◇ `series(tan(x),x,4);` ⟶ $x + \frac{1}{3}x^3 + O\left(x^4\right)$ ◇ `series(leadterm(tan(x)),x);` ⟶ x ◇ See also: `coeftayl`, `convert/polynom`, `convert/series`, `numapprox`, `Order`, `powseries`, `taylor`, `type/laurent`, `type/series`, `type/taylor`.

shake `shake(`*expr*`)`
`shake(`*expr*`, `*n*`)`
Computes a bounding interval ◇ This function replaces the constants in the expression by an interval with width of order 10^{-n+1}, then uses `evalr` to propagate these intervals up. If n is not given, the value of `Digits` is used. ◇ You must `readlib(shake)` before using this command. ◇ `shake(1,2);` ⟶ $[\,.90\,..\,1.10\,]$ ◇ See also: `Digits`, `evalr`.

Shi `Shi(`*z*`)`
The hyperbolic sine integral ◇ $\mathrm{Shi}(z) = \int_0^z \sinh t/t\, dt$ ◇ You must `readlib(Shi)` before using this command. ◇ See also: `Chi`, `Si`.

showprofile
`showprofile()`
View the results of the profile ◇ This function prints the amount of time and memory that profiled procedures have used. ◇ You must `readlib(profile)` before using this command. ◇ See also: `profile`, `showtime`, `time`, `unprofile`.

showtime `showtime()`
Maintains a history of all values computed, and displays the time and memory used by each command ◇ Enter `showtime()` or `on` to start this

feature. Output of successive statements is saved in the variables `O1`, `O2`, etc. The command `off` turns this feature off. When this feature is on, `"` and `"""` are not available. ◇ You must `readlib(showtime)` before using this command. ◇ See also: `history`, `time`.

Si `Si(`*z*`)`
The sine integral ◇ $\mathrm{Si}(z) = \int_0^z \sin t/t\, dt$ ◇ See also: `Ci`, `Shi`, `Ssi`.

sign `sign(`*poly*`)`
`sign(`*poly*`, ` *varlist*`)`
`sign(`*poly*`, ` *varlist*`, ` *name*`)`
Computes the sign of the leading coefficient of a multivariate polynomial with respect to the given list of variables ◇ If the list of variables is omitted, all of the indeterminates are used. If a name is supplied as a third argument, it is assigned the leading term. ◇ `sign(3*x-2*y, [y,x]);` $\longrightarrow -1$ ◇ See also: `indets`, `lcoeff`, `signum`.

signum `signum(`*z*`)`
`signum(1, ` *z*`)`
`signum(0, ` *z*`, ` z_0`)`
Sign function for real and complex expressions ◇ This function is 1 at any positive real number and −1 at any negative real number. For a nonzero complex number z, it is $z/|z|$. The second form returns the derivative of the signum function at z. ◇ In Release 2 and earlier versions, signum of 0 is 1. In Release 3, signum of 0 is 0 by default. This value may be changed by setting the environment variable `_Envsignum0` or by calling signum with three arguments. In the latter case, the first argument must be 0 and the third argument names the value of signum at 0. ◇ `signum(4*I);` $\longrightarrow I$ ◇ See also: `csgn`, `evalc`, `sign`.

Signum `Signum(`*expr*`)`
Inert sign function ◇ When used with `evalr`, this function determines the sign of the expression using range arithmetic. ◇
`evalr(Signum(Pi^exp(1) - exp(1)^Pi));` $\longrightarrow -1$ ◇ See also: `evalr`.

simplex The simplex linear optimization package ◇ In the descriptions of functions in this package, *C* denotes a list or set of linear constraints (equations or inequalities). This implementation of the simplex algorithm is based on *Linear Programming* by Chvatal, 1983, W. H. Freeman and Company, New York. ◇ See also: `simplex[`*function*`]`, `with`.

simplex[basis]
`basis(`*C*`)`
Computes a list of variables, one per equation, that corresponds to the basis used by the simplex algorithm ◇ *C* is a set of linear equations, in the form produced by `simplex[setup]`. ◇ See also: `simplex[setup]`.

simplex[convert/equality]

`convert(`*list*`, equality)`
`convert(`*set*`, equality)`
Converts inequalities to equations ◇ The first argument may be a list or set of inequalities or equalities. This function can only be used after loading the `simplex` package. ◇ See also: `convert/equality`.

simplex[convert/std]

`convert(`C`, std)`
Converts to standard form for simplex manipulation ◇ This function returns a list or set of constraints obtained by moving all constants to the right-hand side of each equation or inequality in C. It can only be used after loading the `simplex` package. ◇ `convert({y<=3*x+2}, std);`
$\longrightarrow \{y - 3\,x \leq 2\}$

simplex[convert/stdle]

`convert(`C`, stdle)`
Converts constraints to type `<=` ◇ This function returns a set or list of inequalities of the form `<=`, equivalent to the original set of constraints, with all constants on the right-hand side. It can only be used after loading the `simplex` package. ◇ `convert([y=3*x+2], stdle);` $\longrightarrow$
$[\,y - 3\,x \leq 2, -y + 3\,x \leq -2\,]$ ◇ See also: `simplex[standardize]`.

simplex[convexhull]

`convexhull(`*pointset*`)`
Finds the convex hull enclosing the given points ◇ *pointset* is a set of $[x, y]$ pairs. This function returns a list of points in counterclockwise order. ◇ `convexhull({[0,0], [2,0], [1,2], [1,1]});` $\longrightarrow$
$[\,[0, 0], [2, 0], [1, 2]\,]$

simplex[cterm]

`cterm(`C`)`
Computes the list or set of constants from the system of constraints ◇
`cterm([x+y=1, x-2<=0]);` $\longrightarrow [1, 2]$

simplex[define_zero]

`define_zero()`
`define_zero(`ϵ`)`
Defines the zero tolerance for floats ◇ The first form returns the current absolute minimum nonzero quantity. The second form sets the current absolute minimum nonzero quantity to ϵ. ◇ This function is new in Release 3. ◇ See also: `testfloat`.

simplex[display]

`display(`C`)`
`display(`C`, `*varlist*`)`
Displays a linear program in matrix form ◇ C should be in the form produced by `simplex[setup]`. ◇ This function is new in Release 3.

simplex[dual]

`dual(`*expr*`, ` *C*`, ` *name*`)`

Computes the dual of a linear program ◇ The linear program must be in standard inequality form. *expr* is the linear objective function to be maximized, subject to the linear inequalities *C*. *name* is used as a base for constructing names for the dual variables. The result is a sequence consisting of the objective followed by the constraints. ◇ `dual(x+y, {x<=1,y<=1}, t);` $\longrightarrow t1 + t2, \{\, 1 \le t1, 1 \le t2 \,\}$ ◇ See also: `simplex[convert/stdle]`.

simplex[feasible]

`feasible(`*C*`)`

`feasible(`*C*`, ` *vartype*`)`

`feasible(`*C*`, ` *vartype*`, ` $name_s$`, ` $name_t$`)`

Determines whether the system is feasible ◇ This function returns *true* if a feasible solution to the linear system *C* exists, and *false* otherwise. If *vartype* is `NONNEGATIVE`, nonnegativity constraints are imposed on all the variables. No restriction on signs is indicated with `UNRESTRICTED`, the default. If a name is supplied as a third argument, it is assigned the final system found by `feasible`. A fourth argument is assigned the set of variable transformations that occurred.

simplex[maximize]

`maximize(`*expr*`, ` *C*`)`

`maximize(`*expr*`, ` *C*`, ` *vartype*`)`

`maximize(`*expr*`, ` *C*`, ` *vartype*`, ` $name_s$`, ` $name_t$`)`

Maximizes a linear program ◇ *expr* is the linear objective function to be maximized subject to the linear constraints *C*. This function returns a set of equations describing the optimal solution to the specified linear program if it exists; the empty set if no feasible solution exists; or `NULL` if the solution is unbounded. *vartype* may be `NONNEGATIVE` to impose nonnegativity constraints on all variables, or `UNRESTRICTED`, the default. If a name is supplied as a fourth argument, it is assigned the optimal description. A fifth argument is assigned the set of variable transformations that occurred. ◇ `maximize(x+y, {2*x+y<=2, y<=1});` $\longrightarrow \left\{y = 1, x = \frac{1}{2}\right\}$ ◇ See also: `simplex[minimize]`.

simplex[minimize]

`minimize(`*expr*`, ` *C*`)`

`minimize(`*expr*`, ` *C*`, ` *vartype*`)`

`minimize(`*expr*`, ` *C*`, ` *vartype*`, ` $name_s$`, ` $name_t$`)`

Minimizes a linear program ◇ *expr* is the linear objective function to be minimized subject to the linear constraints *C*. This function returns a set of equations describing the optimal solution to the specified linear program if it exists; the empty set if no feasible solution exists; or `NULL` if the solution is unbounded. Other arguments behave as in `simplex[maximize]`. ◇ See also: `simplex[maximize]`.

A B C D E F G H I J K L M N O P Q R S T U V W X Y Z

simplex[pivot]

`pivot(`*C*`,` *var*`,` *eqn*`)`

Constructs a new set of equations given a pivot ◇ This function constructs a new set of equations in a form compatible with the forms used in `simplex[setup]` by solving the specified equation *eqn* for the variable *var*, then substituting the result into *C*. If *eqn* is list, the first element is used. ◇ See also: `simplex[pivoteqn]`, `simplex[pivotvar]`.

simplex[pivoteqn]

`pivoteqn(`*C*`,` *var*`)`

Returns a sublist of equations from *C* given a pivot ◇ *var* is the variable chosen to carry out the pivot operation. ◇ See also: `simplex[pivot]`, `simplex[pivotvar]`.

simplex[pivotvar]

`pivotvar(`*expr*`,` *varlist*`)`

`pivotvar(`*expr*`)`

Returns a variable with positive coefficient ◇ This function returns a variable that has a positive coefficient, or returns *FAIL*. *expr* is an objective function, and the optional argument *varlist* lists the order in which the variables should be examined. ◇ `pivotvar(x-3*y);` $\longrightarrow x$ ◇ See also: `simplex[pivot]`, `simplex[pivoteqn]`.

simplex[ratio]

`ratio(`*C*`,` *var*`)`

Returns a list of ratios ◇ These ratios can be used to determine which equation in *C* is most restrictive with respect to the next simplex pivot.

simplex[setup]

`setup(`*C*`)`

`setup(`*C*`, NONNEGATIVE)`

`setup(`*C*`, NONNEGATIVE,` *name*`)`

Constructs a list or set of equations from *C* with isolated variables on the left-hand side ◇ The isolated variables form a basis for the corresponding linear system. Slack variables of the form *_SLi* are introduced for inequalities. If `NONNEGATIVE` is present, all variables are assumed to be nonnegative. If a name is given as a third argument, it is assigned the transformations used for any unrestricted variable. ◇ `setup({x+y<=2,2*x+y=2});` $\longrightarrow \{y = -2x + 2, \_SL1 = x\}$

simplex[standardize]

`standardize(`*C*`)`

Converts constraints to type `<=` and places constants on the right-hand side ◇ The resulting constraints are equivalent to the original constraints. ◇ `standardize({y=3*x+2});` $\longrightarrow \{y - 3x \le 2, -y + 3x \le -2\}$ ◇ See also: `simplex[convert/std]`, `simplex[convert/stdle]`.

simplex[type/NONNEGATIVE]

`type(`*expr*`, NONNEGATIVE)`
Tests whether the expression is a constraint of the form `a>=0` ◇ This function returns *true* or *false*. It can only be used after loading the `simplex` package. ◇ `type(-t<=0, NONNEGATIVE);` ⟶ *true*

simplify

`simplify(`*expr*`)`
`simplify(`*expr*`,` *name*$_1$`,` *name*$_2$`, ...)`
`simplify(`*expr*`,` *eqns*`)`
`simplify(`*expr*`,` *eqns*`,` *vars*`)`
`simplify(`*expr*`, assume=`*property*`)`
Simplifies an expression ◇ If only one argument is present, this function applies the appropriate simplification rules to the expression. In the second form, the additional arguments name the specific simplifications that should be applied. Simplifications for certain mathematical functions may be requested by supplying one or more of the following names:

arctrig	BesselK	Ei	Heaviside	min	signum
atatsign	BesselY	erf	hypergeom	polar	sqrt
atsign	csgn	erfc	infinity	power	trig
BesselI	dilog	exp	ln	radical	W
BesselJ	Dirac	GAMMA	max	RootOf	

The third and fourth forms simplify *expr* with respect to the side relations given by the list or set of equations *eqns* in the variables *vars* (an expression in *eqns* is understood to be equal to zero). This uses Gröbner basis techniques to compute an equivalent expression that is in normal form with respect to the side relations. If *vars* is given as a set, the total degree ordering is used; if *vars* is a list, the pure lexicographic ordering with respect to the order of the variables in the list is used. ◇ The `assume` option is available in Release 3. With this option, all indeterminates are assumed to possess the given property when simplifying the expression. ◇ `simplify(ln((a*b)^r));` ⟶ $r\ln(a)+r\ln(b)$ ◇ `simplify(cosh(x)^2 - sinh(x)^2);` ⟶ 1 ◇ `simplify(1-sin(x)^2 + ln(x^2), trig);` ⟶ $\cos(x)^2+\ln\left(x^2\right)$ ◇ `simplify(x^6*y-3*x^4*y+3*x^2*y+3, {x^2-y^2=1}, [x,y]);` ⟶ $3+y^7+y$ ◇ `simplify(sqrt(t^2)), assume=negative);` ⟶ $-t$ ◇ See also: `assume`, `collect`, `combine`, `expand`, `factor`, `grobner`, `normal`, `radsimp`.

sin

`sin(`*z*`)`
The sine function ◇ See also: `arcsin`.

singular

`singular(`*expr*`)`
`singular(`*expr*`,` *vars*`)`
Finds singularities of an expression ◇ *vars* may be a variable name or a set of variable names. This function returns an expression sequence of the singularities, with each singularity given as a set of equations. The global variables `_N` or `_NN` may be used in the result to denote any integer or any

natural number, respectively. ◇ You must `readlib(singular)` before using this command. ◇ `singular(csc(t));` ⟶ $\{t = \_N\pi\}, \{t = \infty\}$ ◇ See also: `discont`, `residue`, `roots`, `solve`.

sinh `sinh(`*z*`)`
The hyperbolic sine function ◇ See also: `arcsinh`.

sinterp `sinterp(`*f*`, `*varlist*`, `*m*`, `*p*`)`
Sparse multivariate modular polynomial interpolation ◇ *f* is a procedure that accepts n integers and a prime p and computes a value in the integers modulo p. *varlist* is a list of n variable names. This function returns a polynomial in the listed variables whose values modulo the prime p equal those of f. m is a bound on the total degree of the polynomial returned. *FAIL* is returned if the interpolation does not succeed. ◇ You must `readlib(sinterp)` before using this command. ◇ See also: `interp`.

Smith `Smith(`*A*`, `*var*`, `*name*$_1$`, `*name*$_2$`)`
Inert Smith Normal Form ◇ When used with `mod`, this function computes the Smith Normal Form of a rectangular matrix of univariate polynomials in *var*. In the case of four arguments, the two names are assigned the transformation matrices U and V such that the result is UAV. ◇ See also: `Hermite`, `linalg[smith]`.

solve `solve(`*eqns*`)`
`solve(`*eqns*`, `*vars*`)`
`solve(`*ineq*`, `*var*`)`
`solve(`*eqn*`, `*fcnname*`)`
`solve(identity(`*eqn*`, `*x*`), `*vars*`)`
Solves equations analytically ◇ The first two forms solve an equation or a system of equations for the given variable or variables. An expression is understood to be equal to zero. Multiple equations and multiple unknowns are given as sets. If *vars* is not specified, all indeterminates are used. A sequence of solutions is returned. If there is more than one variable, each solution is given as a set. The maximum number of solutions returned may be controlled by assigning to the environment variable `_MaxSols` (default 100). This function can also solve an inequality and can revert a series. In the fourth form, *fcnname* should only be used as a function in the equation. In this case, `solve` returns a sequence of one or more procedures that solve the equation. The `identity` form attempts to find a solution in *vars* that satisfies *eqn* for any value of the special variable *x*. *x* should not be one of the *vars*. ◇ `solve(x^2=3, x);` ⟶ $\sqrt{3}, -\sqrt{3}$ ◇ `solve({x+y=2,x=y});` ⟶ $\{y = 1, x = 1\}$ ◇ `solve(arctan(x)<0, x);` ⟶ $\{x < 0\}$ ◇ `solve(y=series(x*cos(x),x=0,5), x);` ⟶ $y + \frac{1}{2}y^3 + \mathrm{O}\left(y^5\right)$ ◇ `solve(identity(sin(x+t) = a*sin(x) + b*cos(x), x), {a,b});` ⟶ $\{b = \sin(t), a = \cos(t)\}$ ◇ See also: `assign`, `dsolve`, `fsolve`, `grobner[gsolve]`, `isolate`, `isolve`, `linalg[linsolve]`, `match`, `msolve`, `root`, `rsolve`.

sort

`sort(`*list*`)`
`sort(`*list*`, ` *f*`)`
`sort(`*expr*`)`
`sort(`*expr*`, ` *vars*`)`
`sort(`*expr*`, ` *vars*`, ` *f*`)`
Sorts the elements of a list into ascending order or the terms of polynomials in an algebraic expression into descending order ◇ *f* defines an ordering for sorting. It may be `'<'` or `numeric` for numerical order, `lexorder` or `string` for lexicographic ordering of strings, `address` for sorting by internal address, or a Boolean-valued procedure of two arguments. When sorting a polynomial, *vars* specifies the variable ordering. It may be a variable name or a list or set of variable names. ◇ See also: `lexorder, sortcx, stats[transform, statsort]`.

sortcx

`sortcx(`*list*`)`
`sortcx(`*list*`, ` *method*`)`
Sorts a list of complex numbers ◇ The methods available are listed in the first column of the following table. The second column lists the primary sorting criterion for each method; the third column, the secondary criterion.

`abs`	magnitude	`lexImRe`
`lexImRe`	imaginary part	real part
`lexReIm`	real part	imaginary part
`polar`	argument	magnitude

The default method is `lexReIm`. ◇ You must `readlib(sortcx)` before using this command. ◇ See also: `sort`.

sparse

The sparse indexing function ◇ This indexing function specifies that unassigned entries of an array or a table have the value 0. ◇ `array(sparse,1..2);` $\longrightarrow [0\,0]$ ◇ See also: `array, table`.

spline

`spline(`*xlist*`, ` *ylist*`, ` *var*`)`
`spline(`*xlist*`, ` *ylist*`, ` *var*`, ` *n*`)`
`spline(`*xlist*`, ` *ylist*`, ` *var*`, ` *name*`)`
Computes a natural spline ◇ This function computes a piecewise polynomial approximation of degree n (default 3) in the variable *var* to the given x- and y-values. The degree may also be given as one of the names `linear`, `quadratic`, `cubic`, or `quartic`. The x-values must must be distinct and in ascending order. The result is returned as an `If()` expression whose arguments are conditions and expressions. ◇ You must `readlib(spline)` before using this command. ◇ `spline([0,1,2],[1,0,1], t, 2);` $\longrightarrow$ $\mathrm{If}\left(t < 1, 1 - t^2, 5 - 8\,t + 3\,t^2\right)$ ◇ See also: `bspline, interp, 'spline/makeproc'`.

'spline/makeproc'

`'spline/makeproc'(`*spline*`,` *var*`)`
Converts an `If` expression given by `spline` into a procedure ◇ *spline* is the output from `spline`, and *var* is the main variable of the spline. ◇ You must `readlib(spline)` before using this command. ◇
`'spline/makeproc'(spline([0,1,2],[1,0,1],t,2),t);` ⟶
`proc(t) if t < 1 then 1-t^2 else 5-8*t+3*t^2 fi end` ◇
See also: `spline`.

split

`split(`*poly*`,` *var*`)`
`split(`*poly*`,` *var*`,` *name*`)`
Complete factorization of a univariate polynomial with rational, algebraic number, or polynomial coefficients ◇ The result is given in terms of `RootOf`s. If a name is supplied as a third argument, it is assigned a set of extension of the base field needed to factor the polynomial completely. ◇
You must `readlib(split)` before using this command. ◇
`split(x^2+3,x);` ⟶ $(x - \mathrm{RootOf}(\_Z^2+3))\,(x + \mathrm{RootOf}(\_Z^2+3))$
◇ See also: `factor`, `RootOf`.

sprem

`sprem(`$poly_1$`,` $poly_2$`,` *var*`)`
`sprem(`$poly_1$`,` $poly_2$`,` *var*`,` $name_1$`,` $name_2$`)`
Sparse pseudo-remainder of polynomials ◇ This function is the same as `prem`, except the multiplier m is the smallest power of the leading coefficient of $poly_2$ in *var* such that the division process does not introduce fractions into the pseudo-quotient and remainder. If $name_1$ is present, it is assigned the multiplier. If $name_2$ is present, it is assigned the pseudo-quotient. ◇ See also: `prem`, `rem`, `Sprem`.

Sprem

`Sprem(`$poly_1$`,` $poly_2$`,` *var*`)`
`Sprem(`$poly_1$`,` $poly_2$`,` *var*`,` $name_1$`,` $name_2$`)`
Inert sparse pseudo-remainder of polynomials ◇ When used with `evala`, `evalgf`, `mod`, or `modp1`, this function computes the pseudo-remainder of $poly_1$ divided by $poly_2$, over the given coefficient domain, using as small a multiplier as possible, as in `sprem`. ◇ See also: `Prem`, `Rem`, `sprem`.

sqrfree

`sqrfree(`*poly*`)`
`sqrfree(`*poly*`,` *vars*`)`
Square-free factorization of a multivariate polynomial ◇ *vars* is the name of the main variable or a list or set of the main variables. If it is omitted, all the names in the polynomial are used. The result is a list whose first element is the content and whose second element is a list of factor-exponent pairs giving a square-free factorization. ◇
`sqrfree(x^4+4*x^3+3*x^2-4*x-4);` ⟶
$[1, [[x^2-1, 1], [x+2, 2]]]$ ◇ See also: `convert/sqrfree`, `factors`, `isqrfree`, `Sqrfree`.

A B C D E F G H I J K L M N O P Q R S T U V W X Y Z

Sqrfree

`Sqrfree(`*poly*`)`
Inert square-free factorization ◇ This function is used in conjunction with `evala`, `mod`, or `modp1` and returns a list whose first element is the content and whose second element is a list of factor-exponent pairs giving a square-free factorization. ◇ `Sqrfree(x^4+x^3-x-1) mod 3;` ⟶ $[1,[[x+1,1],[x+2,3]]]$ ◇ See also: `Berlekamp`, `DistDeg`, `Factors`, `ProbSplit`, `sqrfree`.

sqrt

`sqrt(`*expr*`)`
`sqrt(`*expr*`, symbolic)`
Forms the square root of the expression ◇ Square roots of perfect squares of integers and polynomials are extracted. ◇ In Release 3, $\sqrt{x^2}$ does not simplify to x unless x is known to be positive. The `symbolic` option, new in Release 3, requests that this simplification be performed just as in Release 2. ◇ `sqrt(8*x^2, symbolic);` ⟶ $2\sqrt{2}\,x$ ◇ See also: `assume`, `signum`, `type/sqrt`.

sscanf

`sscanf(`*string*`,` *formatstring*`)`
Scans and parses *string* according to the format string ◇ The format string contains conversion specifications that describe how the string is parsed. A list of the parsed objects is returned. A specification has the form `%`*code* or `%`*m code*, where *m* names the maximal number of characters to scan for the object and *code* names the type of object scanned. The following codes may be given:

`a`	any expression
`c`	single character
`d` *or* `D`	signed or unsigned decimal integer
`e`, `f`, *or*, `g`	number in scientific notation
`i`	integer (decimal, octal, or hexadecimal)
`n`	number of characters scanned up to this code
`o` *or* `O`	unsigned octal integer
`p`	pointer (memory address)
`s`	string
`u` *or* `U`	unsigned decimal integer
`x` *or* `X`	unsigned hexadecimal integer
`[`*str*`]`	string consisting of characters in *str*
`[^`*str*`]`	string consisting of characters not in *str*

If an asterisk appears just after the percent sign in a conversion specification, the associated object is not included in the returned list. ◇ `sscanf('3 2.e1 km x+y 19', '%d %f %s %a %*d');` ⟶ $[3, 20., km, x+y]$ ◇ See also: `printf`, `parse`, `readline`.

Ssi

`Ssi(`*z*`)`
The shifted sine integral ◇ $\mathrm{Ssi}(z) = \mathrm{Si}(z) - \pi/2$ ◇ You must `readlib(Ssi)` before using this command. ◇ See also: `Si`.

ssystem `ssystem(`*command*`)`
`ssystem(`*command*`, ` *n*`)`
Invokes a command in the host operating system and returns the result ◇ *command* is a Maple string. If a second argument is given, at most *n* seconds of CPU time may be used executing *command*. A list of two elements is returned. The first element is the return code from executing the command; the second is a string representing the result of the command. This function is not supported on some platforms. ◇ This function is new in Release 3. ◇ See also: `!`, `system`.

stats Maple V Release 3 statistics package ◇ This package has been entirely redesigned in Release 3. Most functions reside in one of six subpackages: `describe`, `fit`, `random`, `statevalf`, `statplots`, and `transform`. The `with` command loads subpackages just as it loads packages. For example, here are three ways to invoke the `mean` function in the `describe` subpackage:

`stats[describe, mean](`*data*`);`
`with(stats): describe[mean](`*data*`);`
`with(stats): with(describe): mean(`*data*`);`

In the descriptions of functions in this package, *data* denotes a list of statistical data. Each data item may be a simple item, a class, a missing observation, or a weighted data item. A simple item is a number or a symbolic expression. A class is denoted by a range *a..b*. It represents a single datum with value greater than or equal to *a* but strictly less than *b*. The class mark of a class is the midpoint of the range. The name `missing` represents a missing observation. An item of the form `Weight(`*item*`, ` *n*`)` represents a data item with weight *n*. ◇ See also: `stats[describe]`, `stats[fit]`, `stats[importdata]`, `stats[random]`, `stats[statevalf]`, `stats[statplots]`, `stats[transform]`, `with`.

stats Maple V Release 2 (and earlier) statistics package ◇ Many functions in this package require a "statistical matrix." This is a matrix of data whose first row has the names of all the columns. This is the "key" of the matrix. In the descriptions of functions in this package, *D* denotes a statistical matrix. ◇ See also: `stats[`*function*`]`, `with`.

stats[addrecord]

`addrecord(`*D*`, ` *list*`)`
`addrecord(`*D*`, ` *A*`)`
Adds records to a statistical matrix ◇ This function creates a new statistical matrix consisting of the given one *D* with the list or matrix of new observations appended. It returns the new statistical matrix. ◇ Release 2 and earlier only. ◇ See also: `stats[getkey]`, `stats[putkey]`, `stats[removekey]`.

A B C D E F G H I J K L M N O P Q R S T U V W X Y Z

stats[average]

`average(`D`, `key_1`, `key_2`, ...)`
`average(`A`)`
`average(`*listlist*`)`
`average(`*list*`)`
`average(`*array*`)`
`average(`x_1`, `x_2`, ...)`
Calculates averages ◇ The average is the arithmetic mean: the sum divided by the number of elements. Given a statistical matrix, this function reports the average of the data columns corresponding to key_1, key_2, ... Given a matrix, each column is averaged individually and the result is returned as a list. A list of lists of the same length is treated as a matrix. A simple list, an array of dimension one, or a sequence may also be given. ◇ Release 2 and earlier only. ◇ `average([[2,3],[4,5]]);` ⟶ [3, 4] ◇ `average(1,2,3);` ⟶ 2 ◇ See also: `stats[mean]`, `stats[median]`, `stats[mode]`, `stats[variance]`.

stats[ChiSquare]

`ChiSquare(`F`, `n`)`
Computes the inverse of the cumulative distribution function for a chi-squared distribution with n degrees of freedom at F ◇ Release 2 and earlier only. ◇ `ChiSquare(0.5, 3);` ⟶ 2.365973884 ◇ See also: `stats[RandChiSquare]`.

stats[correlation]

`correlation(`D`, `key_1`, `key_2`)`
`correlation(`$list_1$`, `$list_2$`)`
Evaluates the correlation coefficient ◇ This function calculates the correlation between two lists of data. The data may also be given by naming two keys in a statistical matrix. ◇ Release 2 and earlier only. ◇ `correlation([2.1,3,5], [5,7,8]);` ⟶ .9188196337 ◇ See also: `stats[covariance]`, `stats[Rsquared]`.

stats[covariance]

`covariance(`D`, `key_1`, `key_2`)`
`covariance(`$list_1$`, `$list_2$`)`
Calculates the covariance ◇ This function calculates the covariance between two lists of data. The data may also be given by naming two keys in a statistical matrix. ◇ Release 2 and earlier only. ◇ `covariance([1,2,3],[2,3,4]);` ⟶ 1 ◇ See also: `stats[correlation]`, `stats[variance]`.

stats[describe]

Data analysis subpackage ◇ The short forms of these functions may be used by entering `with(describe)` after entering `with(stats)`. ◇ This subpackage is new in Release 3. ◇ See also: `stats[describe,` *function*`]`, `with`.

stats[describe, coefficientofvariation]

`coefficientofvariation(`*data*`)`
`coefficientofvariation[`*n*`](`*data*`)`
Computes the coefficient of variation ◇ The coefficient of variation is the standard deviation divided by the mean. Classes are represented by their class mark; `missing` data are ignored. If n is 0 (the default), the standard deviation is computed for the whole population; if n is 1, it is computed for a sample. ◇ `coefficientofvariation([3, Weight(5,2), 7]);` $\longrightarrow \frac{1}{5}\sqrt{2}$

stats[describe, count]

`count(`*data*`)`
Counts data items ◇ Weights are counted appropriately; `missing` data are not counted. ◇ `count([7, Weight(5,9), missing]);` $\longrightarrow 10$ ◇ See also: `stats[describe, countmissing]`.

stats[describe, countmissing]

`countmissing(`*data*`)`
Counts `missing` data items ◇ `count([7, Weight(5,9), missing]);` $\longrightarrow 1$ ◇ See also: `stats[describe, count]`.

stats[describe, covariance]

`covariance(`$data_1$`,` $data_2$`)`
Computes the covariance between two statistical lists ◇ The two lists must have the same number of items, with identical weights on corresponding items. Classes are represented by the class mark. ◇ `covariance([1,2,3], [3,6,9]);` $\longrightarrow 2$ ◇ See also: `stats[describe, linearcorrelation]`.

stats[describe, decile]

`decile[`*n*`](`*data*`)`
`decile[`*n*`](`*data*`,` *gap*`)`
Computes deciles of a statistical list ◇ n should be between 1 and 9. `missing` data are ignored. *gap* may be `false` or a number naming the size of gaps between classes. If `false` (the default), classes are assumed to extend until they touch. Set *gap* to 0 if the boundaries of classes should not be changed. ◇ `decile[4]([$11..20]);` $\longrightarrow 14$ ◇ See also: `stats[describe, quantile]`, `stats[describe, median]`, `stats[describe, quartile]`.

stats[describe, geometricmean]

`geometricmean(`*data*`)`
Computes the geometric mean of the data ◇ The geometric mean is the nth root of the product of the elements of the statistical list, where there are n elements in the list. Classes are represented by their class mark; `missing` data are ignored. ◇ `geometricmean([3, Weight(5,2), 7]);` $\longrightarrow 525^{1/4}$ ◇ See also: `stats[describe, harmonicmean]`, `stats[describe, mean]`, `stats[describe, quadraticmean]`.

stats[describe, harmonicmean]

`harmonicmean(`*data*`)`

Computes the harmonic mean of the data ◇ The harmonic mean is the reciprocal of the arithmetic mean of the reciprocals of the elements in the list. Classes are represented by their class mark; `missing` data are ignored. ◇ `harmonicmean([3, Weight(5,2), 7]);` $\longrightarrow \frac{105}{23}$ ◇ See also: `stats[describe, geometricmean]`, `stats[describe, mean]`, `stats[describe, quadraticmean]`.

stats[describe, kurtosis]

`kurtosis(`*data*`)`
`kurtosis[`*n*`](`*data*`)`

Computes the moment coefficient of kurtosis of the data ◇ This is defined as the fourth moment about the mean divided by the fourth power of the standard deviation. If n is 0 (the default), the standard deviation is computed for the whole population; if n is 1, it is computed for a sample. Classes are represented by their class mark; `missing` data are ignored. ◇ `kurtosis([Weight(1,3),3,5]);` $\longrightarrow \frac{133}{64}$ ◇ See also: `stats[describe, moment]`, `stats[describe, skewness]`.

stats[describe, linearcorrelation]

`linearcorrelation(`$data_1$`, `$data_2$`)`

Computes the coefficient of linear correlation between two lists of data ◇ The two lists must have the same number of items, with identical weights on corresponding items. Classes are represented by their class mark. ◇ `linearcorrelation([1,2,3], [3,6,9]);` $\longrightarrow 1$ ◇ See also: `stats[describe, covariance]`.

stats[describe, mean]

`mean(`*data*`)`

Computes the arithmetic mean of the data ◇ The arithmetic mean is the sum of the terms divided by the number of terms. Classes are represented by their class mark; `missing` data are ignored. ◇ `mean([3, Weight(5,2), 7]);` $\longrightarrow 5$ ◇ See also: `stats[describe, geometricmean]`, `stats[describe, harmonicmean]`, `stats[describe, median]`, `stats[describe, mode]`, `stats[describe, quadraticmean]`, `stats[describe, standarddeviation]`.

stats[describe, meandeviation]

`meandeviation(`*data*`)`

Computes the mean deviation of the data ◇ This is the average absolute deviation from the mean. Classes are represented by their class mark; `missing` data are ignored. ◇ `meandeviation([3, Weight(5,2), 7]);` $\longrightarrow 1$

stats[describe, median]

`median(`*data*`)`
`median(`*data*`, `*gap*`)`
Computes the median of the data ◇ Either the middle value or the arithmetic mean of the two middle values is returned. `missing` data are ignored. *gap* may be `false` or a number naming the size of gaps between classes. If `false` (the default), classes are assumed to extend until they touch. Set *gap* to 0 if the boundaries of classes should not be changed. ◇ `median([1,3]);` $\longrightarrow 2$ ◇ See also: `stats[describe, decile]`, `stats[describe, mean]`, `stats[describe, mode]`, `stats[describe, quantile]`, `stats[describe, quartile]`.

stats[describe, mode]

`mode(`*data*`)`
Computes the mode of the data ◇ The mode is the most frequently appearing element. If there is more than one element with the modal frequency, all these elements are returned in an expression sequence. `missing` data are ignored. ◇ `mode([2,2,6,8,6]);` $\longrightarrow 2, 6$ ◇ See also: `stats[describe, mean]`, `stats[describe, median]`.

stats[describe, moment]

`moment[`*m*`](`*data*`)`
`moment[`*m*`, `*a*`](`*data*`)`
`moment[`*m*`, `*a*`, `*n*`](`*data*`)`
Computes the *m*th moment about *a* ◇ *a* may be a number or a descriptive statistic such as `mean`. Its default value is 0. If *n* is 0 (the default), the moment is computed for the whole population; if *n* is 1, it is computed for a sample. Classes are represented by their class mark; `missing` data are ignored. ◇ `moment[3,mean]([Weight(1,3),3,5]);` $\longrightarrow \frac{432}{125}$ ◇ See also: `stats[describe, kurtosis]`, `stats[describe, skewness]`.

stats[describe, percentile]

`percentile[`*n*`](`*data*`)`
`percentile[`*n*`](`*data*`, `*gap*`)`
Finds the data item corresponding to the given percentile ◇ If the percentile falls between data items, the result is computed by interpolation. `missing` data are ignored. *n* should be an integer between 1 and 99. *gap* may be `false` or a number naming the size of gaps between classes. If `false` (the default), classes are assumed to extend until they touch. Set *gap* to 0 if the boundaries of classes should not be changed. ◇ `percentile[75]([$11..20]);` $\longrightarrow \frac{35}{2}$ ◇ See also: `stats[describe, decile]`, `stats[describe, quantile]`, `stats[describe, quartile]`.

stats[describe, quadraticmean]

`quadraticmean(`*data*`)`

Computes the quadratic mean of the data ◇ The quadratic mean is the square root of the arithmetic mean of the squares of the data elements. Classes are represented by their class mark; `missing` data are ignored. ◇ `quadraticmean([3, Weight(5,2), 7]);` $\longrightarrow 3\sqrt{3}$ ◇ See also: `stats[describe, geometricmean]`, `stats[describe, harmonicmean]`, `stats[describe, mean]`.

stats[describe, quantile]

`quantile[`*q*`](`*data*`)`
`quantile[`*q*`](`*data*`, `*gap*`)`

Finds the specified quantile of the data ◇ If the quantile falls between data items, the result is computed by interpolation. `missing` data are ignored. *q* should be a fraction between 0 and 1. *gap* may be `false` or a number naming the size of gaps between classes. If `false` (the default), classes are assumed to extend until they touch. Set *gap* to 0 if the boundaries of classes should not be changed. ◇ `quantile[3/4]([$11..20]);` $\longrightarrow \frac{35}{2}$ ◇ See also: `stats[describe, decile]`, `stats[describe, percentile]`, `stats[statplots, quantile]`, `stats[describe, quartile]`.

stats[describe, quartile]

`quartile[`*n*`](`*data*`)`
`quartile[`*n*`](`*data*`, `*gap*`)`

Computes the *n*th quartile of the data ◇ *n* should be 1, 2, or 3. *gap* may be `false` or a number naming the size of gaps between classes. If `false` (the default), classes are assumed to extend until they touch. Set *gap* to 0 if the boundaries of classes should not be changed. ◇ `quartile[3]([$11..20]);` $\longrightarrow \frac{35}{2}$ ◇ See also: `stats[describe, decile]`, `stats[describe, median]`, `stats[describe, percentile]`, `stats[describe, quantile]`.

stats[describe, range]

`range(`*data*`)`

Returns the smallest range that includes all the data ◇ `missing` data are ignored. ◇ `range([1..3, 4, 6]);` $\longrightarrow 1..6$

stats[describe, skewness]

`skewness(`*data*`)`
`skewness[`*n*`](`*data*`)`

Computes the moment coefficient of skewness ◇ This is defined as the third moment about the mean divided by the third power of the standard deviation. If *n* is 0 (the default), the standard deviation is computed for the whole population; if *n* is 1, it is computed for a sample. Classes are represented by their class mark; `missing` data are ignored. ◇ `skewness([Weight(1,3),3,5]);` $\longrightarrow \frac{27}{32}$ ◇ See also: `stats[describe, kurtosis]`, `stats[describe, moment]`.

A B C D E F G H I J K L M N O P Q R S T U V W X Y Z

stats[describe, standarddeviation]

`standarddeviation(`*data*`)`
`standarddeviation[`*n*`](`*data*`)`
Computes the standard deviation of the data ◇ If n is 0 (the default), the standard deviation is computed for the whole population (division by n); if n is 1, it is computed for a sample (division by $n - 1$). Classes are represented by their class mark; `missing` data are ignored. ◇ `standarddeviation([3, Weight(5,2), 7]);` $\longrightarrow \sqrt{2}$ ◇ See also: `stats[describe, mean]`.

stats[describe, variance]

`variance(`*data*`)`
`variance[`*n*`](`*data*`)`
Computes the variance of the data ◇ The variance is the square of the standard deviation. If n is 0 (the default), the standard deviation is computed for the whole population; if n is 1, it is computed for a sample. Classes are represented by their class mark; `missing` data are ignored. ◇ `variance([3, Weight(5,2), 7]);` $\longrightarrow 2$ ◇ See also: `stats[describe, standarddeviation]`.

stats[evalstat]

`evalstat(`D`, ` eqn_1`, ` eqn_2`, ...)`
Adds data to a statistical matrix ◇ Each equation gives a new column in terms of the other columns. This function returns a new statistical matrix formed by adding a column to D for each equation or equations given. ◇ Release 2 and earlier only.

stats[Exponential]

`Exponential(`λ`)`
`Exponential(`λ`, ` *bound*`)`
The exponential distribution ◇ Given a single argument λ, this function returns a procedure that computes the corresponding exponential distribution. The second form computes the value of the exponential distribution with parameter λ at the given upper bound. ◇ Release 2 and earlier only. ◇ See also: `stats[RandExponential]`.

stats[Fdist]

`Fdist(`F`, ` r`, ` s`)`
Computes the inverse of the cumulative distribution function for a variance ratio distribution with r numerator degrees of freedom and s denominator degrees of freedom at F ◇ Release 2 and earlier only. ◇ `Fdist(.8,2,6);` $\longrightarrow 2.129927840$ ◇ See also: `stats[RandFdist]`, `stats[Ftest]`.

stats[fit] Regression subpackage ◇ This subpackage is new in Release 3. The `leastsquare` function in this subpackage replaces three functions in Release 2: `stats[linregress]`, `stats[multregress]`, and `stats[regression]`. ◇ Enter `with(fit)` after entering `with(stats)` in order to use the `leastsquare` function without any prefix. ◇ See also: `stats[fit,` *function*`]`, `with`.

stats[fit, leastsquare]

`leastsquare[`*varlist*`](`*listdata*`)`

`leastsquare[`*varlist*`,` *eqn*`,` *coefset*`](`*listdata*`)`

Fits a curve to data using the method of least squares ◇ The first form finds a linear equation in the given variables that best fits the given data in the least-squares sense. The second form allows you to specify an equation and a set of coefficients. ◇ `leastsquare[[x,y]]([[1,2,3], [3,5,7]]);` ⟶ $y = 1 + 2x$ ◇ `leastsquare[[x,y], y=m*x+b, {m,b}]([[1,2,3], [3,5,7]]);` ⟶ $y = 1 + 2x$

stats[Ftest]

`Ftest(`*x*`,` *r*`,` *s*`)`

Computes the value of the cumulative distribution function for a variance ratio distribution with *r* numerator degrees of freedom and *s* denominator degrees of freedom at *x* ◇ Release 2 and earlier only. ◇ `Ftest(2.13,2,6);` ⟶ .1999915605 ◇ See also: `stats[Fdist]`.

stats[getkey]

`getkey(`*D*`)`

Returns the key from a statistical matrix ◇ The key is returned as a list. ◇ Release 2 and earlier only. ◇ See also: `stats[putkey]`, `stats[removekey]`.

stats[importdata]

`importdata(`*filename*`)`

`importdata(`*filename*`,` *n*`)`

Reads statistical data from the file ◇ Numbers in the file should be separated by spaces or line breaks. They are read as floating-point numbers. Missing data may be denoted with a "*" in the file. These are represented in the result with the word `missing`. If *n* is 1 (the default), the data are returned in an expression sequence. If *n* is greater than 1, the data are returned in *n* lists. ◇ This function is new in Release 3. ◇ See also: `readdata`.

stats[linregress]

`linregress(`*D*`,` key_1`=`key_2`)`

`linregress(`*ylist*`,` *xlist*`)`

Linear regression ◇ This function fits the data given in two columns of a statistical matrix or in two lists to a line. It returns a list with the intercept and the slope. ◇ Release 2 and earlier only. ◇ `linregress([3, 5, 7, 9], [1, 2, 3, 4]);` ⟶ [1.,2.] ◇ See also: `stats[multregress]`, `stats[regression]`.

stats[mean]

mean(*dat*)
mean(*dat*, geometric)
mean(*dat*, harmonic)
mean(*dat*, quadratic)
mean(*list₁*, *list₂*, discrete)
mean([[*expr*, *a*..*b*], ...], *var*, continuous)

Calculates means ◇ *dat* may be given as a list or a vector. The first form computes the arithmetic mean of the data. The second argument may request calculation of the geometric, harmonic, or quadratic mean. The fifth form, with the `discrete` option, computes the sum of the products of the corresponding elements of $list_1$ and $list_2$. The last form requires a list of lists as its first argument. Each sublist has two elements: an expression in the variable *var*, and a range. The weighted integral $\int_a^b x f(x)\,dx$ is computed for each pair [$f(x)$, $a..b$], and these values are then summed. ◇ Release 2 and earlier only. ◇ `mean([2,3,4], geometric);` $\longrightarrow 24^{1/3}$ ◇ `mean([1,2,3], [4,5,6], discrete);` $\longrightarrow 32$ ◇ `mean([[cos(x),0..Pi/2]],x,continuous);` $\longrightarrow$ .5707963268 ◇ See also: `stats[average]`, `stats[median]`, `stats[mode]`.

stats[median]

median(*D*, key_1, key_2, ...)
median(*A*)
median(*listlist*)
median(*list*)
median(*array*)
median(x_1, x_2, ...)

Calculates the median ◇ The median is the round($n/2$) entry in a sorted list of n items. Given a statistical matrix, this function reports the median of the data columns corresponding to key_1, key_2, ... Given a matrix, the median of each column is computed and the result is returned as a list. A list of lists of the same length is treated as a matrix. A simple list, an array of dimension one, or a sequence may also be given. ◇ Release 2 and earlier only. ◇ `median([7,2,1,9]);` $\longrightarrow 2$ ◇ `dat := array([[x,y], [1,2], [7,8], [3,4]]): median(dat, x,y);` $\longrightarrow [3,4]$ ◇ See also: `stats[average]`, `stats[mean]`, `stats[mode]`.

stats[mode]

mode(*A*, *lowbnd*, *incr*)
mode(*listlist*, *lowbnd*, *incr*)
mode(*list*, *lowbnd*, *incr*)
mode(*array*, *lowbnd*, *incr*)
mode(x_1, x_2, ...)

Calculates the mode ◇ The mode is the most frequently occurring element. Given a matrix, the mode of each column is computed and the result is returned as a list. A list of lists of the same length is treated as a matrix. An array of dimension one, a simple list, or a sequence may also be given. If there is more than one mode, all of them are returned in a list. If a lower bound and a range increment are specified, the range of numbers with the maximum frequency of observations is the mode. ⋄ Release 2 and earlier only. ⋄ `mode([1,2,2,4,4]);` ⟶ [2,4] ⋄ `mode([1,2,2,4,4],1,2);` ⟶ [1..3] ⋄ See also: `stats[average]`, `stats[mean]`, `stats[median]`.

stats[multregress]

`multregress(`*D*`,` *eqn*`)`
`multregress(`*D*`,` *eqn*`, const)`
Multiple regression ⋄ *eqn* is an equation of the form $y = [x_1, x_2, \ldots]$, where y and the x_i are key names in *D*. The first form fits the data to an equation of the form $y = \sum a_i x_i$. The second form allows a constant term as well. It returns a list of the a_i, with the constant term first if it was requested. ⋄ Release 2 and earlier only. ⋄ See also: `stats[linregress]`, `stats[projection]`, `stats[regression]`.

stats[N]

`N(`*x*`)`
`N(`*x*`,` *μ*`,` *v*`)`
Computes the value of the cumulative distribution function for a normal distribution with mean μ and variance v at x ⋄ The default values of μ and v are 0 and 1, respectively. ⋄ Release 2 and earlier only. ⋄ `N(2.5,4.1,7.3);` ⟶ .2768628314 ⋄ See also: `stats[Q]`, `stats[RandNormal]`.

stats[projection]

`projection(`*D*`,` *eqn*`)`
`projection(`*D*`,` *eqn*`, const)`
Creates a projection matrix ⋄ *eqn* is an equation of the form $y = [x_1, x_2, \ldots]$, where the x_i are key names in *D*. This function constructs a projection matrix according to $X(X^t X)^{-1} X^t$ The projection matrix projects the actual y-values onto the predicted y-values. If `const` is given as a third argument, the projection matrix is constructed as if a constant existed in the equation. ⋄ Release 2 and earlier only. ⋄ See also: `stats[multregress]`.

stats[putkey]

`putkey(`*A*`,` *key*`)`
Adds a key to a matrix ⋄ *key* is a list of names forming the key for the data. This function adds the key to A as the first row, returning a new matrix: a statistical matrix. ⋄ Release 2 and earlier only. ⋄ See also: `stats[getkey]`, `stats[removekey]`.

stats[Q] `Q(x)`

Tail of a standard normal distribution ◇ This function finds the probability that $X > x$ for a standard normal distribution. ◇ Release 2 and earlier only. ◇ `Q(3);` ⟶ .001349898032 ◇ See also: `stats[N]`.

stats[RandBeta]

`RandBeta(a, b)`
`RandBeta(a, b, n)`

Returns a random number generator for the Beta distribution of order *a* and *b* ◇ *n* gives the number of digits of precision. A procedure is returned. ◇ Release 2 and earlier only. ◇ See also: `stats[RandFdist]`, `stats[RandGamma]`.

stats[RandChiSquare]

`RandChiSquare(v)`
`RandChiSquare(v, n)`

Returns a random number generator for the chi-squared distribution with *v* degrees of freedom ◇ *n* gives the number of digits of precision. A procedure is returned. ◇ Release 2 and earlier only. ◇ See also: `stats[ChiSquare]`.

stats[RandExponential]

`RandExponential(u)`
`RandExponential(u, n)`

Returns a random number generator for the exponential distribution with mean time between arrivals *u* ◇ *n* gives the number of digits of precision. A procedure is returned. ◇ Release 2 and earlier only. ◇ See also: `stats[RandPoisson]`.

stats[RandFdist]

`RandFdist(v1, v2)`
`RandFdist(v1, v2, n)`

Returns a random number generator for the *F* distribution with v_1 and v_2 degrees of freedom ◇ *n* gives the number of digits of precision. A procedure is returned. ◇ Release 2 and earlier only. ◇ See also: `stats[Fdist]`, `stats[RandBeta]`, `stats[RandGamma]`.

stats[RandGamma]

`RandGamma(a)`
`RandGamma(a, n)`

Returns random number generator for the Gamma distribution of order *a* ◇ *n* gives the number of digits of precision. A procedure is returned. ◇ Release 2 and earlier only. ◇ See also: `stats[RandBeta]`, `stats[RandFdist]`.

stats[RandNormal]

`RandNormal(`μ`, `σ`)`
`RandNormal(`μ`, `σ`, `n`)`
Returns a random number generator for the normal distribution with mean μ and standard deviation σ ◇ n gives the number of digits of precision. A procedure is returned. ◇ Release 2 and earlier only. ◇ See also: `stats[N]`.

stats[random]

`random[`*distribution*`]()`
`random[`*distribution*`](`n`)`
`random[`*distribution*`](`n`, `f`)`
`random[`*distribution*`](`n`, `f`, `*methodname*`)`
`random[`*distribution*`](generator)`
`random[`*distribution*`](generator[`d`])`
Subpackage for random number generators ◇ This subpackage generates random numbers according to specified distributions. *distribution* may be any of the distributions listed under `stats[statevalf]`. n is the number of random numbers desired (the default is 1). f is a procedure that generates random numbers between 0 and 1 uniformly; `default` may be given for the default method. *methodname* names the method for transforming the uniform stream. The default is `inverse` (use the inverse cumulative distribution function). Other methods are `builtin` (use a built-in method, only available for certain distributions) or `auto` (use a built-in method if it exists). The last two templates return a procedure that generates random numbers according to the specified distribution. d names the number of digits desired (the default is the value of `Digits`). The short forms of these functions may be used by entering `with(random)` after entering `with(stats)`. ◇ This subpackage is new in Release 3. ◇ `random[normald[0,1]](2);` ⟶ 1.175839568, −.5633641309 ◇ See also: `stats[statevalf]`, `with`.

stats[RandPoisson]

`RandPoisson(`λ`)`
`RandPoisson(`λ`, `n`)`
Returns a random number generator for the Poisson distribution where λ is the mean number of occurrences of an event per unit time ◇ n gives the number of digits of precision. A procedure is returned. ◇ Release 2 and earlier only. ◇ See also: `stats[RandExponential]`.

stats[RandStudentsT]

`RandStudentsT(`ν`)`
`RandStudentsT(`ν`, `n`)`
Returns a random number generator for the Student's T distribution with ν degrees of freedom ◇ n gives the number of digits of precision. A procedure is returned. ◇ Release 2 and earlier only. ◇ See also: `stats[StudentsT]`.

stats[RandUniform]

`RandUniform(`a`..`b`)`
`RandUniform(`a`..`b`, `n`)`
Returns a uniform random number generator for the range $[a, b)$ ◇ n gives the number of digits of precision. A procedure is returned. ◇ Release 2 and earlier only. ◇ See also: `stats[Uniform]`.

stats[regression]

`regression(`D`, `*eqn*`)`
Generalized regression routine ◇ *eqn* is an equation in the key names of the statistical matrix and some unknown coefficients. The equation should be linear in the unknown coefficients. This function determines values for the coefficients and returns a set of equations giving these values. ◇ Release 2 and earlier only. ◇ `dat := array([[y,x], [10,1], [50,2], [250,4]]): regression(dat, y=a*exp(x)+b);` ⟶ $\{a = 4.467757980, b = 6.970513279\}$ ◇ See also: `stats[linregress]`, `stats[multregress]`, `stats[statplot]`.

stats[removekey]

`removekey(`D`)`
Removes the key from a statistical matrix ◇ This function removes the first row of a statistical matrix and returns a new matrix full of the data. ◇ Release 2 and earlier only. ◇ See also: `stats[getkey]`, `stats[putkey]`.

stats[Rsquared]

`Rsquared(`$list_1$`, `$list_2$`)`
`Rsquared(`D`, `key_1`, `key_2`)`
Computes the square of the correlation coefficient ◇ This data may be given as two lists or by naming two keys in a statistical matrix. ◇ Release 2 and earlier only. ◇ `Rsquared([1,2,4],[5,6,7]);` ⟶ $\frac{27}{28}$ ◇ See also: `stats[correlation]`, `stats[variance]`.

stats[sdev]

`sdev(`D`, `key_1`, `key_2`, ...)`
`sdev(`A`)`
`sdev(`*listlist*`)`
`sdev(`*array*`)`
`sdev(`x_1`, `x_2`, ...)`
Computes the standard deviation ◇ Given a statistical matrix, this function reports the standard deviation of the data columns corresponding to $key_1, key_2, \ldots$. Given a matrix, the standard deviation of each column is computed and the result is returned as a list. A list of lists of the same length is treated as a matrix. A simple list, an array of dimension one, or a sequence may also be given. ◇ Release 2 and earlier only. ◇ `sdev(1,3,5);` ⟶ 2 ◇ See also: `stats[variance]`, `stats[serr]`.

stats[serr] `serr(`D`, `key_1`, `key_2`, ...)`
`serr(`A`)`
`serr(`*listlist*`)`
`serr(`v`)`
`serr(`x_1`, `x_2`, ...)`
Computes the standard error ◇ Given a statistical matrix, this function reports the standard error of the data columns corresponding to key_1, key_2, ... Given a matrix, the standard error of each column is computed and the result is returned as a list. A list of lists of the same length is treated as a matrix. A simple list, a vector, or a sequence may also be given. ◇ Release 2 and earlier only. ◇ `serr([1,7,4]);` $\longrightarrow \sqrt{3}$ ◇ See also: `stats[sdev]`, `stats[variance]`.

stats[statevalf]

Subpackage for numerical evaluation of distributions ◇ Many distributions are supported in this subpackage as well as the `random` subpackage. Continuous distributions:

beta[n_1, n_2]	fratio[n_1, n_2]	logistic[a, b]	studentst[n]
cauchy[a, b]	gamma[a, b]	lognormal[μ, σ]	uniform[a, b]
chisquare[n]	laplaced[a, b]	normald[μ, σ]	weibull[a, b]
exponential[α, a]			

Discrete distributions:

binomiald[n, p]	hypergeometric[n_1, n_2, n]
discreteuniform[a, b]	negativebinomial[n, p]
empirical[*problist*]	poisson[μ]

The short forms of the functions in this subpackage may be used by entering `with(statevalf)` after entering `with(stats)`. ◇ This subpackage is new in Release 3. ◇ See also: `stats[random]`, `stats[statevalf,` *function*`]`, `with`.

stats[statevalf, cdf]

`cdf[`*distribution*`](`*expr*`)`
Computes the value of the cumulative density function of the given continuous distribution at *expr* ◇ `cdf[uniform[0,4]](3);` $\longrightarrow$ .7500000000

stats[statevalf, dcdf]

`dcdf[`*distribution*`](`*expr*`)`
Computes the value of the cumulative probability function of the given discrete distribution at *expr* ◇ `dcdf[poisson[2]](3);` $\longrightarrow$ .8571234607

stats[statevalf, icdf]

`icdf[`*distribution*`](`*expr*`)`
Computes the inverse of the cumulative density function of the given continuous distribution at *expr* ◇ `icdf[uniform[0,4]](3/4);` $\longrightarrow$ 3.000000000

stats[statevalf, idcdf]

`idcdf[`*distribution*`](`*expr*`)`
Computes the inverse of the cumulative probability function of the given discrete distribution at *expr* ◇ `idcdf[poisson[2]](0.857);` ⟶ 2.

stats[statevalf, pdf]

`pdf[`*distribution*`](`*expr*`)`
Computes the value of the probability density function of the given continuous distribution at *expr* ◇ `pdf[uniform[0,4]](3);` ⟶ .2500000000

stats[statevalf, pf]

`pf[`*distribution*`](`*expr*`)`
Computes the value of the probability function of the given discrete distribution at *expr* ◇ `pf[poisson[2]](3);` ⟶ .1804470442

stats[statplot]

`statplot(`*D*`, `*product*`)`
`statplot(`*D*`, `*eqn*`)`
Statistical plotting ◇ The second argument may be a product or an equation. A product like *x***y* gives the orientation *x* versus *y* to the plot of the data points in *D*. An equation of the form $y = f(x)$ plots $f(x)$ and the data points on the same graph. ◇ Release 2 and earlier only. ◇ See also: `stats[regression]`.

stats[statplots]

Statistical plotting subpackage ◇ The short forms of these functions may be used by entering `with(statplots)` after entering `with(stats)`. ◇ This subpackage is new in Release 3. ◇ See also: `stats[statplots, `*function*`]`, `with`.

stats[statplots, boxplot]

`boxplot(`*data*`)`
`boxplot[`*xvalue*`, `*width*`](`*data*`)`
Draws a box plot summarizing the data ◇ `missing` data are ignored. The box plot displays the median, the first and third quartiles, and outliers. *xvalue* is the x-coordinate of the center of the box (default 0) and *width* is the width of the box (default 1). ◇ See also: `stats[statplots, notchedbox]`.

stats[statplots, changecolour]

`changecolour[`*color*`](`*plot*`)`
Changes the color of a plot ◇ This function operates on any plot. *color* may be any predefined color name, or a user-defined color. See page 27 for a list of colors.

stats[statplots, histogram]

`histogram(`*data*`)`

Draws a histogram summarizing the data ◇ Data points are plotted as vertical lines; data classes are plotted as boxes. Heights correspond to weights. `missing` data are ignored. ◇ See also: `stats[transform, tally]`, `stats[transform, tallyinto]`.

stats[statplots, notchedbox]

`notchedbox(`*data*`)`

`notchedbox[`*xvalue*`,` *width*`](`*data*`)`

Draws a notched box plot summarizing the data ◇ `missing` data are ignored. The notched box plot displays the median, the first and third quartiles, and outliers. *xvalue* is the x-coordinate of the center of the box (default 0) and *width* is the width of the box (default 1). ◇ See also: `stats[statplots, boxplot]`.

stats[statplots, quantile]

`quantile(`*data*`)`

Plots each data item in the list against its quantile value. ◇ Classes are plotted as rectangles; `missing` data are ignored. ◇ See also: `stats[describe, quantile]`.

stats[statplots, quantile2]

`quantile2(`*data*$_1$`,` *data*$_2$`)`

Quantile-quantile plot ◇ After transforming the lists to have the same total weight, the datum corresponding to quantile t in $list_2$ is plotted against the datum corresponding to quantile t in $list_1$. Classes versus classes are plotted as rectangles; `missing` data are ignored. ◇ See also: `stats[describe, quantile]`.

stats[statplots, scatter1d]

`scatter1d(`*data*`)`

`scatter1d[`*stylename*`](`*data*`)`

Draws a one-dimensional scatter plot of the data ◇ Classes are plotted as lines; `missing` data are ignored. *stylename* affects the presentation of the data: `projected` (the default) plots all data at the same y-value; `stacked` stacks repeated data as in a histogram; `jittered` plots the data at scattered y-values. ◇ See also: `stats[statplots, scatter2d]`.

stats[statplots, scatter2d]

`scatter1d(`*data*$_x$`,` *data*$_y$`)`

Draws a two-dimensional scatter plot of the data ◇ Corresponding elements in the lists are drawn as points, lines, or rectangles. Points versus points are plotted as points; points versus classes are plotted as lines; classes versus classes are plotted as rectangles. ◇ See also: `stats[statplots, scatter1d]`.

stats[statplots, symmetry]

`symmetry(`*data*`)`

Draws a symmetry plot of the data ◇ For a simple list of data, the line $y = x$ is drawn together with points of the form $(m - a_i, a_{n-i+1} - m)$, where a_i is the ith element of the sorted data and m is the median. `missing` data are ignored; classes are not handled.

stats[statplots, xscale]

`xscale[`*r*`](`*plot*`)`

Scales the x-values of a two-dimensional plot by *r* ◇ *r* must be positive. This function operates on any two-dimensional plot. ◇ See also: `stats[statplots, xshift]`, `stats[statplots, xyexchange]`.

stats[statplots, xshift]

`xshift[`*r*`](`*plot*`)`

Shifts the x-values of a two-dimensional plot by *r* ◇ This function operates on any two-dimensional plot. ◇ See also: `stats[statplots, xscale]`, `stats[statplots, xyexchange]`.

stats[statplots, xyexchange]

`xyexchange(`*plot*`)`

Exchanges the x- and y-coordinates in a two-dimensional plot ◇ This function operates on any two-dimensional plot. ◇ See also: `stats[statplots, xscale]`, `stats[statplots, xshift]`.

stats[StudentsT]

`StudentsT(`*F*`, `*n*`)`

`StudentsT(`*x*`, `*n*`, area)`

The Student's *T* distribution ◇ The first form computes the inverse of the cumulative distribution function for a Student's *T* distribution with *n* degrees of freedom at *F*. The second form computes the value of the cumulative distribution function at *x*. ◇ Release 2 and earlier only. ◇ `StudentsT(.7,2);` ⟶ .6172133998 ◇ `StudentsT(.6172133998, 2, area);` ⟶ .6999999998 ◇ See also: `stats[RandStudentsT]`.

stats[transform]

Data manipulation subpackage ◇ The short forms of these functions may be used by entering `with(transform)` after entering `with(stats)`. ◇ This subpackage is new in Release 3. ◇ See also: `stats[transform,` *function*`]`, `with`.

stats[transform, apply]

`apply[`*f*`](`*data*`)`

Applies the function *f* to each element of *data* ◇ *f* should be a function of one argument. ◇ `apply[t->t/2]([3,2,6..8]);` ⟶ $\left[\frac{3}{2}, 1, 3..4\right]$ ◇ See also: `stats[transform, multiapply]`.

stats[transform, classmark]

`classmark(`*data*`)`

Replaces classes by their class mark ◇ The class mark of a class is its midpoint. Other data are left unchanged. ◇ `classmark([1..5, 7, missing]);` ⟶ [3, 7, *missing*]

stats[transform, cumulativefrequency]

`cumulativefrequency(`*data*`)`

Computes partial sums of the frequencies in the given data ◇ A list is returned. ◇ `cumulativefrequency([3, Weight(4,2), missing]);` ⟶ [1, 3, 4] ◇ See also: `stats[transform, frequency]`.

stats[transform, deletemissing]

`deletemissing(`*data*`)`

Removes missing data ◇ Missing data is identified by the string `missing`. ◇ `deletemissing([3, Weight(4,2), missing, 1..2]);` ⟶ [3, Weight(4, 2), 1..2]

stats[transform, divideby]

`divideby[`*num*`](`*data*`)`
`divideby[`*f*`](`*data*`)`

Divides each data element by the given number or descriptive statistic ◇ If the indexed argument is not numeric, it is assumed to be a procedure name in the subpackage `stats[describe]`. In this case, the divisor is the result of `stats[describe,` *f*`](`*data*`)`. `missing` data remain unchanged. ◇ `divideby[2]([3,2,6..8]);` ⟶ $\left[\frac{3}{2}, 1, 3..4\right]$ ◇ See also: `stats[describe]`, `stats[transform, apply]`, `stats[transform, remove]`.

stats[transform, frequency]

`frequency(`*data*`)`

Replaces each data point by its frequency ◇ A list is returned. ◇ `frequency([3, Weight(4,2), missing]);` ⟶ [1, 2, 1] ◇ See also: `stats[transform, cumulativefrequency]`.

stats[transform, moving]

`moving[`*n*`](`*data*`)`
`moving[`*n*`,` *f*`](`*data*`)`
`moving[`*n*`,` *f*`,` *weightlist*`](`*data*`)`

Replaces each data element by a function of its neighboring elements ◇ The descriptive statistic *f* is applied to each data element and its n neighbors. Only complete neighborhoods are included, so the returned list has $n-1$ fewer elements. `missing` data is included. If *f* is omitted, `mean` is assumed. A list of numbers may be given to specify weights for elements of neighborhoods. ◇ `moving[2]([1,3,7]);` ⟶ [2, 5] ◇ See also: `stats[describe]`.

stats[transform, multiapply]

`multiapply`[*f*](*listdata*)
Applies the function *f* across lists of statistical lists ◇ Each sublist must contain the same number of elements. If there are n sublists, *f* should be a function of n arguments. Data from the first list are used as the first argument for *f*, data from the second list as the second argument, and so on. ◇ `multiapply[(s,t)->2*s*t]([[2,5], [3,4]]);` ⟶ [12, 40] ◇ See also: `stats[transform, apply]`.

stats[transform, remove]

`remove`[*num*](*data*)
`remove`[*f*](*data*)
Subtracts the given number or descriptive statistic from each data element ◇ If the indexed argument is not numeric, it is assumed to be a procedure name in the subpackage `stats[describe]`. In this case, the number subtracted is the result of `stats[describe,` *f*`]`(*data*). `missing` data remain unchanged. ◇ `remove[2]([6, 1..2, Weight(3,6)]);` ⟶ [4, −1 .. 0, Weight(1, 6)] ◇ See also: `stats[describe]`, `stats[transform, apply]`, `stats[transform, divideby]`.

stats[transform, scaleweight]

`scaleweight`[*expr*](*data*)
Multiplies the weights of the data by the given expression ◇ `scaleweight[2]([6, 1..2, Weight(3,6)]);` ⟶ [Weight(6, 2), Weight(1 .. 2, 2), Weight(3, 12)]

stats[transform, split]

`split`[*n*](*data*)
Splits the data list into *n* lists of the same weight ◇ A list of lists is returned. ◇ `split[2]([1,2,3]);` ⟶ $\left[\left[1, \text{Weight}(2, \tfrac{1}{2})\right], \left[\text{Weight}(2, \tfrac{1}{2}), 3\right]\right]$

stats[transform, standardscore]

`standardscore`(*data*)
`standardscore`[*n*](*data*)
Replaces each data item by its standard score ◇ The standard score of item x is $(x - \mu)/\sigma$, where μ is the mean and σ the standard deviation of the data. If *n* is 0 (the default), the standard deviation is computed for the whole population; if *n* is 1, it is computed for a sample. `missing` data remain unchanged. ◇ `standardscore([3, Weight(5,2), 7]);` ⟶ $[-\sqrt{2}, \text{Weight}(0, 2), \sqrt{2}]$

stats[transform, statsort]

`statsort`(*data*)
Sorts the data into ascending order ◇ `missing` data is placed at the end; classes must not overlap. ◇ `statsort([1, missing, 2..4, 9, 1]);` ⟶ [1, 1, 2 .. 4, 9, *missing*] ◇ See also: `sort`.

A B C D E F G H I J K L M N O P Q R S T U V W X Y Z

stats[transform, statvalue]

`statvalue(`*data*`)`

Resets the weight of each data item to 1 ◇ `missing` data remain unchanged. ◇ `statvalue([4, Weight(5..6, 3), Weight(7,2)]);` ⟶ [4, 5 .. 6, 7]

stats[transform, tally]

`tally(`*data*`)`

Groups data of the same value ◇ `tally([3, 4..5, 3, 3, 7, 4..5, 3]);` ⟶ [Weight(3, 4), 7, Weight(4 .. 5, 2)]

stats[transform, tallyinto]

`tallyinto(`*data*`,` *partition*`)`

`tallyinto[`*name*`](`*data*`,` *partition*`)`

Groups data into a given pattern ◇ *partition* is a list of disjoint classes and numbers. `missing` data remain unchanged. If a name is given, data items not fitting any partition item are gathered in a list and assigned to this variable. ◇ `tallyinto([1,2,3,4], [1..3,3..5]);` ⟶ [Weight(1 .. 3, 2), Weight(3 .. 5, 2)]

stats[Uniform]

`Uniform(`*a*`,` *b*`)`

The uniform distribution ◇ This function returns a procedure that computes the uniform distribution on $[a, b]$. ◇ Release 2 and earlier only. ◇ `Uniform(1,5)(4.2);` ⟶ $\frac{1}{4}$ ◇ See also: `stats[RandUniform]`.

stats[variance]

`variance(`D`,` key_1`,` key_2`, ...)`

`variance(`A`)`

`variance(`*listlist*`)`

`variance(`*array*`)`

`variance(`x_1`,` x_2`, ...)`

Computes the variance ◇ Given a statistical matrix, this function reports the variance of the data columns corresponding to key_1, key_2, ... Given a matrix, the variance of each column is computed and the result is returned as a list. A list of lists of the same length is treated as a matrix. A simple list, an array of dimension one, or a sequence may also be given. ◇ Release 2 and earlier only. ◇ `variance([1,3,5]);` ⟶ 4 ◇ See also: `stats[covariance]`, `stats[sdev]`.

status — Global variable giving memory and CPU usage statistics for the session ◇ Its value is a sequence of eight numbers:

`status[1]`	total number of words requested to the storage allocator
`status[2]`	total number of words allocated
`status[3]`	number of CPU seconds used
`status[4]`	words increment for "bytes used" messages
`status[5]`	words increment for automatic garbage collection
`status[6]`	number of words returned in last garbage collection
`status[7]`	number of words available after last garbage collection
`status[8]`	number of times that garbage collection was executed

The status variable is updated whenever a "bytes used" message is produced and whenever a garbage collection occurs. ◇ See also: `gc`, `time`, `words`.

stop Quits a Maple session ◇ See also: `done`, `quit`.

student The student package ◇ This package is a collection of routines designed to carry out step-by-step solutions to problems. ◇ See also: `student[`*function*`]`, `with`.

student[changevar]

`changevar(`*eqn*`, `*expr*`)`
`changevar(`*eqn*`, `*expr*`, `*var*`)`
`changevar(`*eqnset*`, `*expr*`, `*varlist*`)`

Performs a change of variables ◇ This function performs a change of variables for integrals, sums, or limits. The first argument is an equation defining the new variable in terms of the old variable, or a set of such equations in the case of a double or triple integral. If no integral, sum, or limit appears in the expression, this function acts like `student[powsubs]`. ◇ See also: `Int`, `Limit`, `student[powsubs]`, `student[Doubleint]`, `student[TripleInt]`, `subs`, `Sum`.

student[combine]

`combine(`*expr*`)`
`combine(`*expr*`, `*name*`)`

Combines terms into a single term ◇ See `combine`.

student[completesquare]

`completesquare(`*poly*`)`
`completesquare(`*poly*`, `*vars*`)`

Completes the square of a polynomial of degree 2 ◇ For a univariate polynomial, the variable name need not appear. For a polynomial in several variables, a variable name should be specified, or a list or set of variable names may be given. In this case, the function is applied for each variable in succession. ◇ `completesquare(z^2+2*z);` $\longrightarrow (z+1)^2 - 1$

student[convert]

`convert(`*expr*`, '@');`
`convert(`*expr*`, nested);`

Convert functional composition notation ◇ The first function converts a nested composition of functions to an equivalent expression using the

A B C D E F G H I J K L M N O P Q R S T U V W X Y Z

composition operator, `@`. The second function converts an `@`-expression to nested form. These forms can only be used after loading the `student` package. ◇ `convert(f@g, nested);` ⟶ f(g) ◇ See also: `@`, `convert`.

student[D]

`D(`*f*`)`
`D[`*i*`](`*f*`)`
`D[`*i*`,` *j*`, ...](`*f*`)`
Differential operator ◇ See `D`.

student[distance]

`distance(`*pt*$_1$`,` *pt*$_2$`)`
`distance(`*expr*$_1$`,` *expr*$_2$`)`
Computes the distance between points in one or higher dimensions ◇ A point is a list of values. In one dimension, the list brackets may be omitted. ◇ `distance([0,0], [3,4]);` ⟶ 5 ◇ See also: `geom3d[distance]`, `geometry[distance]`, `student[midpoint]`, `student[type/Point]`.

student[Doubleint]

`Doubleint(`*expr*`,` *x*`,` *y*`)`
`Doubleint(`*expr*`,` *x*`,` *y*`,` *name*`)`
`Doubleint(`*expr*`,` *x*`=`*a*`..`*b*`,` *y*`=`*c*`..`*d*`)`
Inert form of double integration ◇ *name* identifies a region of integration. Use `value` to force this function to evaluate when both ranges are given. ◇ `Doubleint(f(x,y),y,x,S);` ⟶ $\int\int_D \mathrm{f}(x, y)\, dy\, dx$ ◇ See also: `Int`, `student[Tripleint]`, `value`.

student[equate]

`equate(`*left*`,` *right*`)`
Creates a set of equations from lists, arrays, and tables ◇ This function constructs a set of equations by equating corresponding components of *left* and *right*. These arguments may be single expressions, or lists, vectors, matrices, or tables of expressions. If *right* is a single expression, it is used as the right-hand side of all equations. If *right* is missing, 0 is used. ◇ `equate([a,b],1);` ⟶ $\{ a = 1, b = 1 \}$ ◇ `equate([a,b], [1,2]);` ⟶ $\{ a = 1, b = 2 \}$ ◇ See also: `subs`.

student[extrema]

`extrema(`*expr*`,` *constraints*`)`
`extrema(`*expr*`,` *constraints*`,` *vars*`)`
`extrema(`*expr*`,` *constraints*`,` *vars*`,` *name*`)`
Finds relative extrema of the multivariate expression subject to constraints using the method of Lagrange multipliers ◇ This function is identical to `extrema`. ◇ See `extrema`.

A B C D E F G H I J K L M N O P Q R S T U V W X Y Z

student[Int]

`Int(`*expr*`,` *var*`)`
`Int(`*expr*`,` *var*`=`*a*`..`*b*`)`
Inert integration function ◇ See `Int`.

student[integrand]

`integrand(`*expr*`)`
Returns the integrand of an unevaluated integral ◇ If several or no integrands are found in the expression, the result is a set. ◇ `integrand(Int(f(x),x));` ⟶ $f(x)$ ◇ See also: `Int`, `student[Doubleint]`, `student[Tripleint]`.

student[intercept]

`intercept(`*eqn*$_1$`,` *eqn*$_2$`, {`*var*$_1$`,` *var*$_2$`})`
`intercept(`*eqn*$_1$`)`
Computes the points of intersection of the two given curves ◇ If only one equation is given, the y-intercept is returned. In this case, the dependent variable must appear alone on the left side of the equation. A third argument indicates the set of coordinate variables. ◇ `intercept(x+y=1,3*x-y=1);` ⟶ $\left\{x = \frac{1}{2}, y = \frac{1}{2}\right\}$ ◇ See also: `geom3d[inter]`, `geometry[inter]`, `projgeom[inter]`, `student[slope]`.

student[intparts]

`intparts(`*integral*`,` *u*`)`
Performs integration by parts on an unevaluated integral ◇ *integral* has the form `Int(`*u*`*`*dv*`,` *var*`)`. It returns *u*`*`*v* `-` `Int(`*du*`*`*v*`,` *var*`)`. ◇ `intparts(Int(x*exp(x), x), x);` ⟶ $x\,e^x - \int e^x\,dx$ ◇ See also: `student[changevar]`, `student[powsubs]`.

student[isolate]

`isolate(`*eqn*`,` *expr*`)`
`isolate(`*eqn*`,` *expr*`,` *n*`)`
Isolates a subexpression to the left side of an equation ◇ This function solves *eqn* for *expr* using at most *n* transformation steps. It is identical to `isolate`. ◇ See `isolate`.

student[leftbox]

`leftbox(`*expr*`,` *x*`=`*a*`..`*b*`,` *n*`,` *options*`)`
Graphs an approximation to an integral using left endpoints as heights ◇ This function plots the curve and a sequence of rectangles that approximate the definite integral of the expression on the interval. The height of each rectangle is determined by the value of the function at the left endpoint of each interval. *n* gives the number of rectangles (the default is 4). Standard 2d plot options may be used. ◇ `leftbox(sqrt(2*x), x=0..5, 10);` ◇ See also: `plot`, `student[leftsum]`, `student[middlebox]`, `student[rightbox]`.

student[leftsum]

leftsum(*expr*, *x*=*a*..*b*)
leftsum(*expr*, *x*=*a*..*b*, *n*)
Computes a numerical approximation to an integral using left endpoints of intervals ◇ The result is returned as an inert Sum. *n* gives the number of intervals (the default is 4). ◇ **leftsum(sqrt(2*x), x=0..5, 10);** $\longrightarrow \left(\frac{1}{2}\sum_{i=0}^{9}\sqrt{i}\right)$ ◇ See also: student[leftbox], student[middlesum], student[rightsum], student[simpson], student[trapezoid], Sum.

student[Limit]

Limit(*expr*, *var*=*a*)
Limit(*expr*, *var*=*a*, *dir*)
Limit(*expr*, *set*)
Limit(*expr*, *set*, *dir*)
Inert limit function ◇ See also: limit, Limit, value.

student[Lineint]

Lineint(*expr*, *y*, *x*)
Lineint(*expr*, *y*, *x*=*a*..*b*)
Lineint(*expr*, *y*=*y*(*x*), *x*)
Lineint(*expr*, *x*, *y*, *t*)
Lineint(*expr*, *x*=*x*(*t*), *y*=*y*(*t*))
Inert form of line integral ◇ The independent variable is given last. A range may be given for the last variable, as in the second template. Use value to force this function to evaluate. ◇ This function is new in Release 3. ◇ **Lineint(f(x,y),x,y,t);** $\longrightarrow$ $\int \mathrm{f}(x(t), y(t))\sqrt{\left(\frac{\partial}{\partial t}x(t)\right)^2 + \left(\frac{\partial}{\partial t}y(t)\right)^2}\,dt$ ◇ See also: Int, value.

student[makeproc]

makeproc(*expr*, *x*)
makeproc([x_1, y_1], [x_2, y_2])
makeproc([*x*, *y*], slope=*m*)
Converts an expression into a Maple procedure ◇ The first form returns a function that, when evaluated at *x*, returns *expr*. The second form returns a function that calculates points on the line connecting the two given points. The third form returns a function that calculates points on the line with slope m passing through the given point. ◇ **makeproc([0,3], slope=2);** $\longrightarrow x \rightarrow 2\,x + 3$ ◇ See also: unapply.

student[maximize]

maximize(*expr*)
maximize(*expr*, *vars*)
Computes the maximum ◇ This function is identical to **maximize**. ◇ See maximize.

student[middlebox]

`middlebox(`*expr*`, `*x*`=`*a*`..`*b*`, `*n*`, `*options*`)`

Graphs an approximation to an integral using midpoints as heights ◇ This function plots the curve and a sequence of rectangles that approximate the definite integral of the expression on the interval. The height of each rectangle is determined by the value of the function at the midpoint of each interval. *n* gives the number of rectangles (the default is 4). Standard 2d plot options may be used. ◇ `middlebox(sqrt(2*x), x=0..5, 10);` ◇ See also: `plot`, `student[leftbox]`, `student[middlesum]`, `student[rightbox]`.

student[middlesum]

`middlesum(`*expr*`, `*x*`=`*a*`..`*b*`)`
`middlesum(`*expr*`, `*x*`=`*a*`..`*b*`, `*n*`)`

Computes a numerical approximation to an integral using midpoints of intervals ◇ The result is returned as an inert `Sum`. *n* gives the number of intervals (the default is 4). ◇ `evalf(middlesum(sqrt(2*x), x=0..5, 10));` ⟶ 10.56807645 ◇ See also: `student[leftsum]`, `student[middlebox]`, `student[rightsum]`, `student[simpson]`, `student[trapezoid]`, `Sum`.

student[midpoint]

`midpoint(`pt_1`, `pt_2`)`
`midpoint(`$expr_1$`, `$expr_2$`)`

Computes the midpoint of the line segment defined by the two points ◇ The arguments should be either lists of length 2 or expressions. ◇ `midpoint([1,3],[5,7]);` ⟶ $[3,5]$ ◇ See also: `geom3d[midpoint]`, `geometry[midpoint]`, `projgeom[midpoint]`, `student[distance]`, `student[type/Point]`.

student[minimize]

`minimize(`*expr*`)`
`minimize(`*expr*`, `*vars*`)`

Computes the minimum ◇ This function is identical to `minimize`. ◇ See `minimize`.

student[powsubs]

`powsubs(`*eqn*`, `*expr*`)`
`powsubs(`eqn_1`, `eqn_2`, ..., `*expr*`)`

Substitutes for factors of an expression ◇ This function substitutes the right-hand side of the equation for each occurrence of the left-hand side as a subexpression of *expr*. If more than one substitution is specified, they are processed sequentially, not simultaneously. This function differs from `subs` in that it is defined in terms of algebraic factors rather than underlying data structures. ◇ `powsubs(x+1=z, ln=log10, ln((x+1)^2+3));` ⟶ $\log 10\left(z^2+3\right)$ ◇ See also: `asubs`, `subs`.

student[rightbox]

`rightbox(`*expr*`, `*x*`=`*a*`..`*b*`, `*n*`, `*options*`)`

Graphs an approximation to an integral using right endpoints as heights ◇ This function plots the curve and a sequence of rectangles that approximate the definite integral of the expression on the interval. The height of each rectangle is determined by the value of the function at the right endpoint of each interval. n gives the number of rectangles (the default is 4). Standard 2d plot options may be used. ◇ `rightbox(sqrt(2*x), x=0..5, 10);` ◇ See also: `plot`, `student[leftbox]`, `student[middlebox]`, `student[rightsum]`.

student[rightsum]

`rightsum(`*expr*`, `*x*`=`*a*`..`*b*`)`

`rightsum(`*expr*`, `*x*`=`*a*`..`*b*`, `*n*`)`

Computes a numerical approximation to an integral using right endpoints of intervals ◇ The result is returned as an inert `Sum`. n gives the number of intervals (the default is 4). ◇ `leftsum(sqrt(2*x), x=0..5, 10);` $\longrightarrow \left(\frac{1}{2}\sum_{i=1}^{10}\sqrt{i}\right)$ ◇ See also: `student[leftsum]`, `student[middlesum]`, `student[rightbox]`, `student[simpson]`, `student[trapezoid]`, `Sum`.

student[showtangent]

`showtangent(`*expr*`, `*var*` = `*a*`)`

Plots a function and its tangent line ◇ This function generates a graph of the function defined by the expression *expr* in the variable *var* and its tangent line at the point $var = a$. ◇ `showtangent(x^2+8, x=2);` ◇ See also: `plot`.

student[simpson]

`simpson(`*expr*`, `*x*`=`*a*`..`*b*`)`

`simpson(`*expr*`, `*x*`=`*a*`..`*b*`, `*n*`)`

Computes a numerical approximation to an integral using Simpson's rule ◇ The result is returned as an inert `Sum`. n gives the number of intervals (the default is 4). ◇ `evalf(simpson(sqrt(2*x), x=0..5, 10));` $\longrightarrow$ 10.50033649 ◇ See also: `student[leftsum]`, `student[middlesum]`, `student[rightsum]`, `student[trapezoid]`, `Sum`.

student[slope]

`slope(`*eqn*`)`

`slope(`*eqn*`, `*y*`, `*x*`)`

`slope(`*eqn*`, `*y*`(`*x*`))`

`slope(`pt_1`, `pt_2`)`

Computes the slope of the line given by the equation or the pair of two-dimensional points ◇ y is the dependent variable and x the independent variable. In the first form, the equation must be in the form $y = f(x)$, and y is taken to be the dependent variable. ◇

`slope(9*x-3*y=1,y(x));` ⟶ 3 ◇ See also: `student[intersect]`, `student[type/Point]`.

student[Sum]

`Sum(`*expr*`, `*var*`)`
`Sum(`*expr*`, `*var*`=`*m*`..`*n*`)`
`Sum(`*expr*`, `*var*`=`α`)`
Inert summation function ◇ See also: `sum`, `Sum`, `value`.

student[trapezoid]

`trapezoid(`*expr*`, `*x*`=`*a*`..`*b*`)`
`trapezoid(`*expr*`, `*x*`=`*a*`..`*b*`, `*n*`)`
Computes a numerical approximation to an integral using the trapezoidal rule ◇ The result is returned as an inert `Sum`. *n* gives the number of intervals (the default is 4). ◇ `trapezoid(sqrt(2*x), x=0..5, 10);` ⟶ $\left(\frac{1}{2}\sum_{i=1}^{9}\sqrt{i}\right)+\frac{1}{4}\sqrt{5}\sqrt{2}$ ◇ See also: `student[leftsum]`, `student[middlesum]`, `student[rightsum]`, `student[simpson]`, `Sum`.

student[Tripleint]

`Tripleint(`*expr*`, `*x*`, `*y*`, `*z*`)`
`Tripleint(`*expr*`, `*x*`, `*y*`, `*z*`, `*name*`)`
`Tripleint(`*expr*`, `*x*` = `*a*`..`*b*`, `*y*` = `*c*`..`*d*`, `*z*` = `*q*`..`*r*`)`
Inert form of triple integration ◇ *name* identifies a region of integration. Use `value` to force this function to evaluate when all ranges are given. ◇ `Tripleint(x*y*z, x=0..1, y=0..1, z=0..1);` ⟶ $\int_0^1\int_0^1\int_0^1 x\,y\,z\,dx\,dy\,dz$ ◇ See also: `Int`, `student[Doubleint]`, `value`.

student[type/Point]

`type(`*pt*`, Point)`
`type(`*pt*`, 'student/Point')`
Checks for type Point ◇ A Point is defined as a list of coordinate values (such as `[a,b,c]`). The length of the list determines the dimension of the space. The short form of the type check is available for interactive use only after the `student` package has been loaded. Procedures must use the full type name `'student/Point'`. ◇ `type([1,2], Point);` ⟶ *true* ◇ See also: `student[distance]`, `student[midpoint]`, `student[slope]`, `type/point`.

student[value]

`value(`*expr*`)`
Evaluates inert functions ◇ This function is identical to `value`. ◇ See `value`.

sturm `sturm(`*sturmseq*`, `*var*`, `*a*`, `*b*`)`
Calculates the number of real roots of a polynomial in the interval $(a,b]$ ◇ *sturmseq* is a sturm sequence for the polynomial in the variable *var*. ◇

You must `readlib(sturm)` before using this command. ◇ See also: `realroot`, `roots`, `solve`, `sturmseq`.

sturmseq `sturmseq(`*poly*`,` *var*`)`
Computes a Sturm sequence for the polynomial *poly* in the variable *var* ◇ This function returns a list of polynomials. ◇ You must `readlib(sturm)` before using this command. ◇ See also: `sturm`.

Subres `Subres(`$poly_1$`,` $poly_2$`,` *var*`)`
Subresultant polynomial remainder sequence ◇ When used in conjunction with `evala`, this function computes the subresultant remainder sequence of $poly_1$ and $poly_2$. It returns a table of polynomials indexed by degree. ◇ See also: `Resultant`.

subs `subs(`eqn_1`,` eqn_2`, ...,` *expr*`)`
`subs(`*eqns*`,` *expr*`)`
Substitutes subexpressions into an expression ◇ Each equation specifies a substitution. Every occurrence of the left-hand side of an equation appearing in *expr* is replaced by the right-hand side of the equation. A sequence of substitutions is performed sequentially; a list or set of substitutions is done simultaneously. ◇ `subs(x+y=z, (x+y)^4);` $\longrightarrow z^4$ ◇ See also: `asubs`, `op`, `student[powsubs]`, `subsop`, `trigsubs`.

subsop `subsop(`eqn_1`,` eqn_2`, ...,` *expr*`)`
Substitutes for specified operands in an expression ◇ Each equation has the form *n=ex*, specifying a substitution for the *n*th operand. ◇ `subsop(1=s+1, 2=t^2, x+y);` $\longrightarrow s + 1 + t^2$ ◇ See also: `subs`, `op`.

substring `substring(`*string*`,` *m*`..`*n*`)`
Extracts a substring from a string ◇ The substring begins with the mth character and ends with the nth character. ◇ `substring(abcde, 3..6);` $\longrightarrow$ *cde* ◇ See also: `cat`, `length`, `searchtext`, `SearchText`.

sum `sum(`*expr*`,` *var*`)`
`sum(`*expr*`,` *var*`=`*m*`..`*n*`)`
`sum(`*expr*`,` *var*`=`α`)`
Definite and indefinite summation ◇ The first form computes the indefinite sum of *expr* with respect to the variable *var*. The second form computes the definite sum over the range of integers from m to n. The third form computes the sum over all the roots of a polynomial, where α is a `RootOf` expression. ◇ `sum(6*k^2, k);` $\longrightarrow 2k^3 - 3k^2 + k$ ◇ `sum(2^n, n=0..4);` $\longrightarrow 31$ ◇ See also: `Sum`.

Sum `Sum(`*expr*`,` *var*`)`
`Sum(`*expr*`,` *var*`=`*m*`..`*n*`)`
`Sum(`*expr*`,` *var*`=`α`)`
Inert summation function ◇ This function simply returns unevaluated. ◇ See also: `sum`, `value`.

surd `surd(`*expr*`, ` *n*`)`
Computes the *n*th real root of the real expression *expr* ◇ This function is new in Release 3. ◇ You must `readlib(surd)` before using this command. ◇ `surd(5,2);` ⟶ $\sqrt{5}$

Svd `Svd(`*A*`)`
`Svd(`*A*`, ` $name_L$`, left)`
`Svd(`*A*`, ` $name_R$`, right)`
`Svd(`*A*`, ` $name_L$`, ` $name_R$`)`
Computes the singular values and vectors of a numeric matrix ◇ The first form returns a $1 \times \min(n, m)$ array of the singular values of *A*, a $n \times m$ numeric matrix. If $name_L$ is present, it is assigned the left singular vectors. $name_R$ is assigned the right singular vectors. This is an inert function. Use `evalf` to compute the singular values. ◇ See also: `linalg[singularvals]`.

symmdiff `symmdiff(`set_1`, ` set_2`)`
`symmdiff(`set_1`, ` set_2`, ` set_3`, ...)`
The set symmetric difference function ◇ The first form returns the set of elements that appear in precisely one of the two sets: $(set_1 \cup set_2) - (set_1 \cap set_2)$ More generally, an element is included in the resulting set if and only if it appears in an odd number of the given sets. ◇ You must `readlib(symmdiff)` before using this command. ◇ `symmdiff({a,b},{b,c});` ⟶ $\{a, c\}$ ◇ See also: `intersect`, `minus`, `union`.

symmetric The symmetric indexing function for arrays and tables ◇ This indexing function arranges the components of an index into a system-determined canonical ordering. A common application is for symmetric matrices, where the (i, j)th element equals the (j, i)th. ◇ See also: `antisymmetric`, `array`, `geometry[symmetric]`, `table`.

system `system(`*command*`)`
Invokes a command in the host operating system ◇ *command* is a Maple string. The result is the return code or return status of the command performed. This function is not supported on some platforms. ◇ See also: `!`, `ssystem`.

table `table(`*indexfcn*`, ` *list*`)`
Creates a table ◇ This function creates a table explicitly. *indexfcn* names an indexing function for the table. Built-in indexing functions are `antisymmetric`, `diagonal`, `identity`, `sparse`, and `symmetric`. The elements of *list* specify initial values. They may be equations of the form *index=entry*, or simply expressions, in which case the corresponding indices are taken to be 1, 2, 3, and so on. Both arguments are optional. A table may also be created implicitly by assigning to an indexed name. ◇ See also: `antisymmetric`, `array`, `copy`, `diagonal`, `entries`, `identity`, `indices`, `sparse`, `symmetric`.

tan

`tan(`*z*`)`
The tangent function ◇ See also: `arctan`.

tanh

`tanh(`*z*`)`
The hyperbolic tangent function ◇ See also: `arctanh`.

taylor

`taylor(`*expr*`, `*var*`=`*a*`, `*n*`)`
`taylor(`*expr*`, `*var*`, `*n*`)`
Taylor series expansion ◇ This function computes the Taylor series expansion of *expr*, with respect to the variable *var*, about the point *a*, up to order *n*. If *a* is missing, *var* = 0 is assumed. ◇ `taylor(exp(x), x=0,4);` $\longrightarrow 1 + x + \frac{1}{2}x^2 + \frac{1}{6}x^3 + O\left(x^4\right)$ ◇ See also: `coeftayl`, `mtaylor`, `numapprox[laurent]`, `series`, `type/taylor`.

tcoeff

`tcoeff(`*poly*`)`
`tcoeff(`*poly*`, `*vars*`)`
`tcoeff(`*poly*`, `*vars*`, `*name*`)`
Trailing coefficient of a multivariate polynomial ◇ *vars* may be a single indeterminate or a list or set of indeterminates. If *vars* is not specified, all the indeterminates of *poly* are used. The polynomial must be `collected` with respect to the appropriate variables. If a name is given as a third argument, it is assigned the trailing term of *poly*. ◇ `tcoeff(2*x^3-4);` $\longrightarrow -4$ ◇ See also: `coeff`, `coeffs`, `collect`, `indets`, `lcoeff`, `ldegree`.

Tcoeff

`Tcoeff(`*poly*`)`
Inert trailing coefficient ◇ When used with `modp1`, this computes the trailing coefficient of a univariate polynomial over a given domain. ◇ See also: `Coeff`, `Lcoeff`, `tcoeff`.

tensor

A collection of procedures for computing curvature tensors in a coordinate basis ◇ The global variable `Ndim` controls the number of dimensions. Its initial value is 4. The following procedures are available. All return `NULL`.

invmetric()	contravariant metric tensor	`h`*ij*
d1metric()	first partials of covariant metric tensor	*gijk*
d2metric()	second partials of covariant metric tensor	*gijkl*
Christoffel1()	Christoffel symbols of the first kind	`c`*ijk*
Christoffel2()	Christoffel symbols of the second kind	`C`*ijk*
Riemann()	covariant components of the Riemann tensor	`R`*ijkl*
Ricci()	covariant components of the Ricci tensor	`R`*ij*
Ricciscalar()	Ricci scalar	`R`
Einstein()	covariant components of the Einstein tensor	`G`*ij*
Weyl()	covariant components of the Weyl tensor	`C`*ijkl*
tensor()	calls all of the above procedures	
display(*name*)	prints nonzero components of the named tensor	
display()	prints nonzero components of all tensors	

The third column names the global variables that are assigned the results of each computation. You must `readlib(tensor)` before using these functions. ◇ See also: `cartan`, `debever`.

testeq

`testeq(`*expr*$_1$ `=` *expr*$_2$`)`
`testeq(`*expr*$_1$`,` *expr*$_2$`)`
`testeq(`*expr*$_1$`)`
Tests equivalence of expressions probabilistically ◇ The result *false* is always correct; the result *true* may be incorrect with very low probability. If the second expression is missing, it is understood to be 0. *FAIL* is returned if the expression does not lie in the class of expressions handled or if an appropriate modulus is not found. ◇ `testeq(sin(2*t), 2*sin(t)*cos(t));` ⟶ *true*

testfloat

`testfloat(`*expr*$_1$`,` *expr*$_2$`,` *bound*`,` *options*`)`
Compares expressions involving floating-point numbers ◇ This function verifies that the floating-point values appearing in the computed expression *expr*$_1$ are within the given bound of the corresponding values in the reference expression *expr*$_2$. *bound* is a nonnegative real number given in ulps. One ulp relative to a number s is one unit in the nth significant digit of s, where n is the value of `Digits`. A list of test results is returned. Options available are:

`digits=`*n*	use this value for digits
`model=1`	treat complex numbers as single numbers (the default)
`model=2`	treat complex numbers as pairs of reals
`test=1`	use a relative error test (the default)
`test=2`	use an absolute error test

◇ You must `readlib(testfloat)` before using this command. ◇ `testfloat([2.1,3.2], [2,3], 1, digits=2);` ⟶ [*true*, [*false*, 2., *ulps*]] ◇ See also: `comparray`, `fnormal`.

Testzero

Environment variable ◇ Its value is a procedure for testing whether an expression is zero. It is initially set to:
`proc() evalb(normal(args[1]) = 0) end`

TEXT

`TEXT(`*string*$_1$`,` *string*$_2$`, ...)`
Prints a sequence of strings ◇ Each string is prettyprinted on a new line. When called under the worksheet interface, a separate window displays the `TEXT` object. ◇ See also: `type/TEXT`.

thaw

`thaw(`*var*`)`
Thaws an expression frozen as a variable ◇ The original expression is returned. ◇ You must `readlib(freeze)` before using this command. ◇ See also: `freeze`.

thiele

`thiele(`*xlist*`,` *ylist*`,` *expr*`)`
Thiele's continued fraction interpolation formula ◇ *xlist* is a list of distinct x-values and *ylist* a list of corresponding y-values. This function computes a rational function in continued fraction form that interpolates

the points (x[1],y[1]), . . ., (x[n],y[n]). It returns the value of this function at the expression *expr*, which is often simply a variable name. ◇ You must `readlib(thiele)` before using this command. ◇ `thiele([1,2,3],[3,4,1],x);` $\longrightarrow 3 + \frac{x-1}{5-2x}$ ◇ See also: `interp`, `sinterp`.

time `time()`
Total CPU time used since the start of the session ◇ This function returns a floating-point number giving the CPU usage in seconds. This number is stored in `status[3]`. ◇ See also: `profile`, `showtime`, `status`, `words`.

totorder The total ordering of names package ◇ In Release 2, the functions in this package must first be loaded using `with` before they can be invoked. ◇ See also: `totorder[`*function*`]`, `with`.

totorder[forget]

`forget(`*rel₁*`,` *rel₂*`, ...)`
Deletes relationships within the ordering ◇ See also: `forget`, `totorder[tassume]`.

totorder[init]

`init()`
Initializes the total ordering, forgetting any previously defined ordering ◇ See also: `totorder[forget]`.

totorder[ordering]

`ordering()`
Prints the current total ordering ◇ See also: `totorder[tassume]`.

totorder[tassume]

`tassume(`rel_1`,` rel_2`, ...)`
Defines an ordering on a sequence of names ◇ `tassume(a<b, b<c);` $\longrightarrow$ *assumed*, $a < b, b < c$ ◇ See also: `assume`, `totorder[forget]`.

totorder[tis]

`tis(`*rel*`)`
Tests relationships in the ordering ◇ `tassume(a<b, b<c): tis(a<c);` $\longrightarrow$ *true* ◇ See also: `is`, `totorder[tassume]`.

trace `trace(`f_1`,` f_2`, ...)`
Turns on tracing of the named procedures for debugging ◇ When the procedures are subsequently invoked, the entry points, intermediate results calculated, and exit points are printed. ◇ See also: `debug`, `infolevel`, `linalg[trace]`, `printlevel`, `profile`, `untrace`, `userinfo`.

Trace `Trace(`a`,` L`,` K`)`
Trace of an algebraic number (or function) ◇ This function is a placeholder for representing the trace of an algebraic number (or function),

which is the sum of its conjugates. ◇ See also: `evala`, `RootOf`, `mod`, `trace`, `linalg[trace]`, `sum`.

translate `translate(`*poly*`, `*var*`, `*a*`)`
Linear translation of a polynomial ◇ This function translates the polynomial by replacing *var* by *var* + a. ◇ You must `readlib(translate)` before using this command. ◇ `translate(x^2-2*x+1, x,1);` ⟶ x^2 ◇ See also: `expand`, `subs`.

traperror `traperror(`*expr*`)`
`traperror(`*expr*$_1$`, `*expr*$_2$`, ...)`
Traps an error condition ◇ If an error occurs during either evaluation or simplification of the expression or expressions, this function returns a string corresponding to the error message of the first error occurring. If no error occurs, it returns the result of the expressions. One way to trap an error in a procedure is to compare the result of `testerror` with the global variable `lasterror`. ◇ See also: `ERROR`, `lasterror`.

trigsubs `trigsubs(`*expr*`)`
`trigsubs(`*eqn*`)`
`trigsubs(`*eqn*`, `*expr*`)`
`trigsubs(0)`
Manages a table of valid trigonometric identities ◇ The first form returns a list of trigonometric expressions equal to *expr*. The second form tests whether the identity *eqn* is in the table. It returns the string *'found'* or *'not found'*. The third form applies *eqn* to *expr* if *eqn* is in the table. The last form returns the set of functions known to `trigsubs`. ◇ You must `readlib(trigsubs)` before using this command. ◇ `trigsubs(sin(t+Pi/2) = cos(t));` ⟶ *'found'* ◇ See also: `subs`.

true Boolean constant.

trunc `trunc(`*z*`)`
`trunc(1, `*z*`)`
Truncates z to the next integer toward 0 if z is real ◇ For real z, this is the integer part of z. For complex z, the function is applied to the real and imaginary parts. The second form returns the first derivative of the truncation function wherever it is defined. ◇ `trunc(5.1+5.9*I);` ⟶ $5 + 5I$ ◇ See also: `ceil`, `floor`, `frac`, `round`.

tutorial `tutorial()`
`tutorial(`*n*`)`
On-line tutorial introduction to Maple ◇ The optional argument *n* names the starting chapter number.

type `type(`*expr*`, `*type*`)`
`type(`*expr*`, `*typeset*`)`
Type-checking function ◇ This function returns *true* if *expr* is of type *type* and *false* otherwise. If the second argument is a set of type names,

true is returned if *expr* is any of the types in the set. ⋄ See also: `convert`, `hastype`, `type/`*name*, `whattype`.

type/'!' `type(`*expr*`, '!')`
Checks for a factorial expression ⋄ The type of an arithmetic expression is the last operation Maple would perform when evaluating the fully expanded operation.

type/'*' `type(`*expr*`, '*')`
Checks for a multiplication (or division) expression ⋄ The type of an arithmetic expression is the last operation Maple would perform when evaluating the fully expanded operation. ⋄ `type(a*b+1, '*');` ⟶ *false*

type/''** `type(`*expr*`, '**')`
Checks for an exponentiation expression ⋄ This is equivalent to `type/'^'`. ⋄ `type(a^b, '**');` ⟶ *true*

type/'+' `type(`*expr*`, '+')`
Checks for an addition (or subtraction) expression ⋄ The type of an arithmetic expression is the last operation Maple would perform when evaluating the fully expanded operation. ⋄ `type(a*b-1, '+');` ⟶ *true*

type/'.' `type(`*expr*`, '.')`
Checks for a concatenation expression ⋄ See also: `cat`.

type/'..' `type(`*expr*`, '..')`
Checks for a range expression ⋄ This is equivalent to `type/range`. ⋄ `type(a..b, '..');` ⟶ *true*

type/'<' `type(`*expr*`, '<')`
Checks for a strong inequality ⋄ `type(a>b, '<');` ⟶ *true* ⋄ See also: `type/relation`.

type/'<=' `type(`*expr*`, '<=')`
Checks for a weak inequality ⋄ `type(a>=b, '<=');` ⟶ *true* ⋄ See also: `type/relation`.

type/'<>' `type(`*expr*`, '<>')`
Checks for a "does-not-equal" expression ⋄ See also: `type/relation`.

type/'=' `type(`*expr*`, '=')`
Checks for an equation ⋄ See also: `type/relation`.

type/'^' `type(`*expr*`, '^')`
Checks for an expression of type '^' ⋄ The type of an arithmetic expression is the last operation Maple would perform when evaluating the fully expanded operation. This is equivalent to `type/'**'`. ⋄ `type(a**(b+1), '^');` ⟶ *true*

type/algebraic

`type(`*expr*`, algebraic)`
Checks for an algebraic expression ◇ An algebraic expression is a number, name, function call, series, or an arithmetic, dot, indexed, or unevaluated expression. ◇ `type(sin(t), algebraic);` ⟶ *true*

type/algext

`type(`*expr*`, algext))`
`type(`*expr*`, algext(`*type*`))`
Checks for a `RootOf` expression ◇ The first form is equivalent to `type/RootOf`. The second form checks for a `RootOf` expression over a given coefficient domain such as `integer` or `rational`. ◇ See also: `RootOf`, `type/algnum`, `type/algnumext`, `type/radext`.

type/algfun

`type(`*expr*`, algfun)`
`type(`*expr*`, algfun(`*type*`))`
`type(`*expr*`, algfun(`*type*`, `*vars*`))`
Checks for an algebraic function ◇ An algebraic function is an expression in the variables *vars* over the domain *type* extended by `RootOf`s. *vars* may be a single name or a list or set of names. ◇ See also: `RootOf`, `type/algext`, `type/radfun`.

type/algnum

`type(`*expr*`, algnum)`
Checks for an algebraic number in terms of `RootOf`s ◇ This function tests for a rational number, a `RootOf` a univariate polynomial with algebraic number coefficients, or a sum, product, or quotient of these. ◇ See also: `RootOf`, `type/algfun`, `type/radnum`.

type/algnumext

`type(`*expr*`, algnumext)`
Checks for an algebraic number extension ◇ An algebraic number extension is a root of a univariate polynomial with algebraic number coefficients, specified by a `RootOf`. This is equivalent to testing both `type/algext` and `type/algnum`. ◇ See also: `RootOf`, `type/algext`, `type/algnum`, `type/radnumext`.

type/'and'

`type(`*expr*`, 'and')`
Checks for an `and` expression ◇ See also: `type/logical`, `type/'not'`, `type/'or'`.

type/anyfunc

`type(`*expr*`, anyfunc(`$type_1$`, `$type_2$`, ...))`
Checks for a function call with arguments of the given types in the given order ◇ `type(g(2,1.3), anyfunc(integer, float));` ⟶ *true* ◇ See also: `type/function`, `type/specfunc`.

type/anything

`type(`*expr*`, anything)`
This function always returns *true* ◇ See also: `type/nothing`.

type/arctrig

`type(`*expr*`, arctrig)`
`type(`*expr*`, arctrig(`*var*`))`
Checks for an inverse trigonometric or hyperbolic inverse trigonometric function call ◇ The inverse trig function must be the last operation performed when evaluating the expression. The second form tests that the function call contains the variable *var*. ◇ `type(2*arcsinh(x), arctrig);` ⟶ *false* ◇ See also: `type/trig`, `type/function`, `type/mathfunc`, `type/trig`.

type/array

`type(`*expr*`, array)`
`type(`*expr*`, 'array'(`*type*`))`
Checks for an array ◇ The second form checks for an array with entries of the given type. ◇ See also: `array`, `type/matrix`, `type/scalar`, `type/table`, `type/vector`.

type/boolean

`type(`*expr*`, boolean)`
Checks for type Boolean ◇ A Boolean expression is an expression of type `relation`, type `logical`, or one of the Boolean constants `true`, *false*, or *FAIL*. ◇ See also: `type/logical`, `type/relation`.

type/complex

`type(`*expr*`, complex)`
`type(`*expr*`, complex(`*type*`))`
Checks for a complex constant ◇ The second form checks that the real and imaginary parts are of the specified type. ◇ See also: `evalc`, `type/numeric`, `type/realcons`.

type/constant

`type(`*expr*`, constant)`
Checks for a complex constant or one of the constants given by the global variable `constants` ◇ See also: `constants`, `type/numeric`, `type/realcons`.

type/cubic

`type(`*expr*`, cubic)`
`type(`*expr*`, cubic(`*vars*`))`
Checks for a cubic function in the specified variables ◇ *vars* may be a single variable or a list or set of variables. If *vars* is not given, the expression must be of total degree 3 in its indeterminates. ◇ See also: `type/linear`, `type/polynom`, `type/quadratic`, `type/quartic`.

A B C D E F G H I J K L M N O P Q R S T U V W X Y Z

type/equation

`type(`*expr*`, equation)`

Checks for an equation ◇ This function checks whether *expr* is of type `equation`, that is, whether it is an equality. ◇ See also: `type`, `equation`.

type/even

`type(`*expr*`, even)`

Checks for an even integer ◇ See also: `type/odd`.

type/evenfunc

`type(`*expr*`, evenfunc(`*var*`))`

Checks for an even expression in the variable *var* ◇ See also: `type/oddfunc`.

type/expanded

`type(`*expr*`, expanded)`

Checks for an expanded polynomial ◇ returns *true* if *expr* is an expanded polynomial, *false* otherwise. ◇ See also: `expand`, `type/polynom`.

type/facint

`type(`*expr*`, facint)`

Tests for a factored integer ◇ A factored integer is an expression of the form returned by `ifactor`. ◇ See also: `ifactor`, `type/integer`.

type/float

`type(`*expr*`, float)`

Checks for a floating-point number ◇ See also: `evalf`, `Float`, `type/integer`, `type/numeric`, `type/realcons`.

type/fraction

`type(`*expr*`, fraction)`

Checks for a fraction ◇ A fraction has an integer numerator and a positive integer denominator. ◇ See also: `type/integer`, `type/numeric`, `type/rational`.

type/function

`type(`*expr*`, function)`

Checks for a function call ◇ A function call has the form *f*`(`*args*`)`. ◇ `type(g(x,y), function);` ⟶ *true* ◇ See also: `type/anyfunc`, `type/evenfunc`, `type/mathfunc`, `type/oddfunc`, `type/specfunc`.

type/identical

`type(`$expr_1$`, identical(`$expr_2$`))`

Checks whether $expr_1$ is identical to $expr_2$

type/indexed

`type(`*expr*`, indexed)`

Checks for an indexed name ◇ An indexed name has the form *name*`[`*exprseq*`]`. ◇ See also: `type/name`.

type/infinity

`type(`*expr*`, infinity)`
Checks whether *expr* evaluates to plus or minus infinity ◇ This option is new in Release 3.

type/integer

`type(`*expr*`, integer)`
Checks for an integer ◇ See also: `type/facint`, `type/negint`, `type/nonnegint`, `type/posint`, `type/primeint`.

type/'intersect'

`type(`*expr*`, 'intersect')`
Checks for an `intersect` expression ◇ See also: `type/'minus'`, `type/'union'`.

type/laurent

`type(`*expr*`, laurent)`
Checks for a Laurent series (with finite principal part) ◇ A Laurent series is a series that contains only integer power terms in the variable in which it is expanded. ◇ See also: `numapprox[laurent]`, `type/series`, `type/taylor`.

type/linear

`type(`*expr*`, linear)`
`type(`*expr*`, linear(`*vars*`))`
Checks for a linear function in the specified variables ◇ *vars* may be a single variable or a list or set of variables. If *vars* is not given, the expression must of total degree 1 in its indeterminates. ◇ See also: `type/cubic`, `type/polynom`, `type/quadratic`, `type/quartic`.

type/list

`type(`*expr*`, list)`
`type(`*expr*`, list(`*type*`))`
Checks for a list ◇ The second form tests for a list with elements of the given type. ◇ `type([1,2,3.], list(integer));` ⟶ *false* ◇ See also: `type/listlist`, `type/set`.

type/listlist

`type(`*expr*`, listlist)`
Checks for a list of lists, with each sublist having the same number of elements ◇ `type([[1,2],[3]], listlist);` ⟶ *false* ◇ See also: `type/list`.

type/logical

Checks for an expression of type `'and'`, `'or'`, or `'not'` ◇ See also: `type/'and'`, `type/boolean`, `type/'not'`, `type/'or'`.

type/mathfunc

`type(`*expr*`, mathfunc)`
Checks for known mathematical functions ◇ The names of functions are recognized (not function calls). ◇ `type(Zeta, mathfunc);` ⟶ *true* ◇ See also: `type/function`.

type/matrix

`type(`*expr*`, matrix)`
`type(`*expr*`, 'matrix'(`*type*`))`
`type(`*expr*`, 'matrix'(`*type*`, square))`
Checks for a matrix ◇ A matrix is a two-dimensional array with indices starting at 1. The second form checks for a matrix with entries of the given type. The third form also checks for a square matrix. ◇ See also: `linalg[matrix]`, `type/array`, `type/scalar`, `type/vector`.

type/\`minus\`

`type(`*expr*`, `minus`)`
Checks for a `minus` expression ◇ See also: `type/`intersect``, `type/`union``.

type/monomial

`type(`*expr*`, monomial)`
`type(`*expr*`, monomial(`*type*`))`
`type(`*expr*`, monomial(`*type*`, `*vars*`))`
Checks for a monomial ◇ *type* names the type of the coefficient domain, and *vars* names the indeterminates. *vars* may be a single variable or a list or set of variables. ◇ See also: `type/polynom`.

type/name

`type(`*expr*`, name)`
Checks for a name ◇ A name may be a string or an indexed name. ◇ `type(a[1], name);` ⟶ *true* ◇ See also: `type/indexed`, `type/string`.

type/negative

`type(`*expr*`, negative)`
Checks for a negative number ◇ See also: `type/negint`, `type/nonneg`, `type/numeric`, `type/positive`.

type/negint

`type(`*expr*`, negint)`
Checks for a negative integer ◇ See also: `type/integer`, `type/negative`, `type/nonnegint`, `type/posint`.

type/nonneg

`type(`*expr*`, nonneg)`
Checks for a nonnegative number ◇ See also: `type/negative`, `type/nonnegint`, `type/numeric`, `type/positive`.

type/nonnegint

`type(`*expr*`, nonnegint)`
Checks for a nonnegative integer ◇ See also: `type/integer`, `type/negint`, `type/nonneg`, `type/posint`.

type/'not'

`type(`*expr*`, 'not')`
Checks for a `not` expression ◇ See also: `type/'and'`, `type/logical`, `type/'or'`.

type/nothing

`type(`*expr*`, nothing)`
This function always returns *false* ◇ See also: `type/anything`.

type/numeric

`type(`*expr*`, numeric)`
Checks for an integer, fraction, or float ◇ See also: `type/complex`, `type/float`, `type/fraction`, `type/integer`, `type/rational`, `type/realcons`.

type/odd

`type(`*expr*`, even)`
Checks for an odd integer ◇ See also: `type/even`.

type/oddfunc

`type(`*expr*`, oddfunc(`*var*`))`
Checks for an odd expression in the variable *var* ◇ See also: `type/evenfunc`.

type/operator

`type(`*f*`, operator)`
Checks for a functional operator ◇ `type(x->x^2, operator);` ⟶ *true* ◇ See also: `type/procedure`.

type/'or'

`type(`*expr*`, 'or')`
Checks for an `or` expression ◇ See also: `type/'and'`, `type/logical`, `type/'not'`.

type/PLOT

`type(`*expr*`, PLOT)`
Checks for a PLOT data structure ◇ A `PLOT` data structure is returned by a two-dimensional plotting command such as `plot`. ◇ See also: `plot`, `type/PLOT3D`.

type/PLOT3D

`type(`*expr*`, PLOT3D)`
Checks for a PLOT3D data structure ◇ A `PLOT3D` data structure is returned by a three-dimensional plotting command such as `plot3d`. ◇ See also: `plot3d`, `type/PLOT`.

type/point

`type(`*expr*`, point)`
Checks for a point ◇ This function returns *true* if *expr* is an equation with a name on the left-hand side and an algebraic expression on the right, or if *expr* is a set of such equations. ◇ `type({a=x+y,b=1}, point);` ⟶ *true* ◇ See also: `student[type/Point]`.

type/polynom

`type(`*expr*`, polynom)`
`type(`*expr*`, polynom(`*type*`))`
`type(`*expr*`, polynom(`*type*`, `*vars*`))`
Checks for a polynomial ◇ *type* names the type of the coefficient domain, and *vars* names the indeterminates. *vars* may be a single variable or a list or set of variables. ◇ `type(x^5+2*ln(y), polynom(anything,x));` ⟶ *true* ◇ See also: `type/cubic`, `type/linear`, `type/monomial`, `type/quadratic`, `type/quartic`, `type/ratpoly`.

type/posint

`type(`*expr*`, posint)`
Checks for a positive integer ◇ See also: `type/integer`, `type/positive`, `type/negint`, `type/nonnegint`.

type/positive

`type(`*expr*`, positive)`
Checks for a positive number ◇ See also: `type/negative`, `type/nonneg`, `type/numeric`, `type/posint`.

type/primeint

`type(`*expr*`, primeint)`
Probable prime test ◇ This function returns *true* if *expr* is an integer and `isprime(`*expr*`)` is true; otherwise it returns *false*. ◇ `type(211,primeint);` ⟶ *true* ◇ See also: `isprime`, `type/integer`.

type/procedure

`type(`*expr*`, procedure)`
Checks for a procedure ◇ `type(type, procedure);` ⟶ *true* ◇ See also: `type/operator`.

type/protected

`type(`*expr*`, protected)`
Checks whether *expr* is a protected name ◇ Most system names are protected by default. ◇ This option is new in Release 3. ◇ See also: `protect`, `unprotect`.

type/quadratic

`type(`*expr*`, quadratic)`
`type(`*expr*`, quadratic(`*vars*`))`
Checks for a quadratic function in the specified variables ◇ *vars* may be a single variable or a list or set of variables. If *vars* is not given, the expression must be of total degree 2 in its indeterminates. ◇ See also: `type/cubic`, `type/linear`, `type/polynom`, `type/quartic`.

type/quartic

`type(`*expr*`, quartic)`
`type(`*expr*`, quartic(`*vars*`))`
Checks for a quartic function in the specified variables ◇ *vars* may be a single variable or a list or set of variables. If *vars* is not given, the expression must be of total degree 4 in its indeterminates. ◇ See also: `type/cubic`, `type/linear`, `type/polynom`, `type/quadratic`.

type/radext

`type(`*expr*`, radext)`
`type(`*expr*`, radext(`*type*`))`
Checks for a radical expression ◇ This first form is equivalent to `type/radical`. The second form checks for a radical expression where the subexpression under the root sign has the given type. ◇ See also: `type/algext`, `type/radnum`, `type/radnumext`.

type/radfun

`type(`*expr*`, radfun)`
`type(`*expr*`, radfun(`*type*`))`
`type(`*expr*`, radfun(`*type*`, `*vars*`))`
Checks for a radical function ◇ An radical function is an expression in the variables *vars* over the domain *type* extended by radicals. *vars* may be a single name or a list or set of names. ◇ See also: `type/algfun`, `type/radext`.

type/radfunext

`type(`*expr*`, radfunext)`
Checks for a radical function extension ◇ A radical function extension is a root of a combination of rational functions and roots of rational functions, with roots specified in terms of radicals. This is equivalent to testing both `type/radext` and `type/radfun`. ◇ `type(x^(1/2), radfunext);` ⟶ *true* ◇ See also: `type/radext`, `type/radfun`, `type/radical`, `type/radnumext`.

type/radical

`type(`*expr*`, radical)`
Checks for fractional powers ◇ See also: `` type/`^` ``, `type/fraction`, `type/radnum`, `type/radext`, `type/sqrt`.

A B C D E F G H I J K L M N O P Q R S T U V W X Y Z

type/radnum

`type(expr, radnum)`
Checks for an algebraic number in terms of radicals ◇ This function tests for a rational number, a root of a rational number specified in terms of radicals, or a sum, product, or quotient of these. ◇ See also: `type/algnum`, `type/radfun`, `type/radical`.

type/radnumext

`type(expr, radnumext)`
Checks for a radical number extension ◇ A radical number extension is a root of a combination of rational numbers and roots of rational numbers specified in terms of radicals. This is equivalent to testing both `type/radext` and `type/radnum`. ◇ See also: `type/algnumext`, `type/radext`, `type/radfunext`, `type/radnum`.

type/range

`type(expr, range)`
Checks for a range ◇ This is equivalent to `type/'..'`. An expression of type `range` (also called type `'..'`) has two operands: the expression on the left-hand side and the expression on the right-hand side.

type/rational

`type(expr, rational)`
Checks for a fraction or an integer ◇ See also: `type/integer`, `type/fraction`, `type/numeric`.

type/ratpoly

`type(expr, ratpoly)`
`type(expr, ratpoly(type))`
`type(expr, ratpoly(type, vars))`
Checks for a rational function ◇ *type* names the type of the coefficient domain, and *vars* names the indeterminates. *vars* may be a single variable or a list or set of variables. ◇ See also: `type/polynom`.

type/realcons

`type(expr, realcons)`
Checks for a real constant ◇ A real constant is an expression for which `evalf` returns a floating-point number, or `infinity` or `-infinity`. ◇ See also: `type/complex`, `type/constant`.

type/relation

Checks for an expression of type `'<'`, `'<='`, `'<>'` or `'='` ◇ See also: `type/'='`, `type/'<>'`, `type/'<'`, `type/'<='`, `type/boolean`.

type/RootOf

`type(expr, RootOf)`
Checks for a `RootOf` expression ◇ This is equivalent to `type/algext`. ◇ See also: `RootOf`, `type/algext`, `type/algnum`.

type/scalar

`type(`*expr*`, scalar)`
Checks for a scalar ◇ A scalar is anything that is not an array. ◇ See also: `type/array`, `type/matrix`, `type/vector`.

type/series

`type(`*expr*`, series)`
Checks for a series ◇ See also: `series`, `type/laurent`, `type/taylor`.

type/set

`type(`*expr*`, set)`
`type(`*expr*`, set(`*type*`))`
Checks for a set ◇ The second form tests for a set with elements of the given type. ◇ `type({1,2,3.}, set(integer));` ⟶ *false* ◇ See also: `type/list`.

type/specfunc

`type(`*expr*`, specfunc(`*type*`, `*name*`))`
Checks for a specific function call with arguments of the given type ◇ `type(g(2,1), specfunc(integer, g));` ⟶ *true* ◇ See also: `type/anyfunc`, `type/function`.

type/sqrt

`type(`*expr*`, sqrt)`
`type(`*expr*`, 'sqrt'(`*type*`))`
Checks for a square root ◇ A square root is a radical with the denominator of the exponent equal to 2. The second form checks that the subexpression under the radical is of the given type. ◇ `type((x+1)^(5/2), sqrt);` ⟶ *true* ◇ See also: `sqrt`, `type/radical`.

type/square

`type(`*expr*`, square)`
Checks for a perfect square ◇ `type(121, square);` ⟶ *true* ◇ `type(ln(x)^4, square);` ⟶ *true*

type/string

`type(`*expr*`, string)`
Checks for a string ◇ See also: `type/name`.

type/table

`type(`*expr*`, table)`
Checks for a table ◇ See also: `table`, `type/array`.

type/taylor

`type(`*expr*`, taylor)`
Checks for a Taylor series ◇ A Taylor series is a series that contains only nonnegative integer power terms in the variable in which it is expanded. ◇ See also: `series`, `taylor`, `type/laurent`, `type/series`.

type/TEXT

`type(`*expr*`, TEXT)`
Checks for a TEXT object ◇ See also: `TEXT`, `type/string`.

type/trig `type(`*expr*`, trig(`*var*`))`
`type(`*expr*`, trig)`
Checks for a trigonometric or hyperbolic trigonometric function ◇ The trig function must be the last operation performed when evaluating the expression. The second form tests that the function call contains the variable *var*. ◇ See also: `type/arctrig`, `type/mathfunc`.

type/type `type(`*expr*`, type)`
Checks for a type expression ◇ A type expression is an expression that might be used as the second argument of a `type` command. ◇ See also: `type`.

type/uneval
`type(`*expr*`, uneval)`
Checks for an unevaluated expression ◇ An unevaluated expression is enclosed in single right quotes, ' '

type/'union'
`type(`*expr*`, 'union')`
Checks for a `union` expression ◇ See also: `type/'intersect'`, `type/'minus'`.

type/vector
`type(`*expr*`, vector)`
`type(`*expr*`, 'vector'(`*type*`))`
Checks for a vector ◇ A vector is a one-dimensional array indexed from 1. The second form checks for a vector with entries of the given type. ◇ See also: `linalg[vector]`, `type/array`, `type/matrix`, `type/scalar`.

unames `unames()`
Unassigned names ◇ This function returns an expression sequence of all the active names in the current Maple session that are "unassigned names." An unassigned name is a name that has been used, but has no value other than its own name. ◇ See also: `anames`, `assigned`.

unapply `unapply(`*expr*`, `*var*`)`
`unapply(`*expr*`, `var_1`, `var_2`, ...)`
Creates a function from an expression ◇ This function returns an operator that computes the expression. The number and order of arguments required by the operator is given by the sequence of variables. ◇ `unapply(a*tan(x), x);` $\longrightarrow x \to a\ \tan(x)$ ◇ `unapply(a*tan(x), a,x);` $\longrightarrow (a, x) \to a\ \tan(x)$

unassign `unassign(`$name_1$`, `$name_2$`, ...)`
Unassigns names ◇ The arguments are fully evaluated, so names are usually enclosed in single right quotes. This function returns `NULL`. ◇ You must `readlib(unassign)` before using this command. ◇ See also: `assign`.

undebug `undebug(`f_1`, `f_2`, ...)`
Turns off tracing of the named procedures ◇ Synonym for `untrace`. ◇ See also: `debug`, `untrace`.

union set_1 `union` set_2
`` `union`( ``set_1`, `set_2`, `set_3`, ...)`
The set union operator ◇ This function returns the set of all the elements that appear in at least one of the given sets. ◇ `{a,b} union {b,c};` $\longrightarrow \{a, b, c\}$ ◇ See also: `intersect`, `minus`, `symmdiff`.

unload `unload(`*f*`)`
Unloads the function *f* from the current session ◇ This function returns the name *f*. ◇ You must `readlib(unload)` before using this command. ◇ See also: `forget`, `readlib`, `unassign`, `with`.

unprofile `unprofile(`f_1`, `f_2`, ...)`
Turns off profiling of the named procedures ◇ You must `readlib(profile)` before using this command. ◇ See also: `profile`, `showprofile`.

unprotect `unprotect(`$name_1$`, `$name_2$`, ...)`
Releases protection on names ◇ Once unprotected, a name may be assigned a new meaning. It is not recommended to modify Maple system names. Names may also be unprotected by unassigning them. ◇ This function is new in Release 3. ◇ See also: `protect`, `type/protected`, `unassign`.

untrace `untrace(`f_1`, `f_2`, ...)`
Turns off tracing of the named procedures ◇ See also: `trace`, `undebug`.

updtsrc `updtsrc` *filename*
`updtsrc` *options filename*
Updates Maple V Release 2 source files for Release 3 ◇ This external command adds `global` declarations to Release 2 procedures so that they behave the same under Release 3. Three options may be given:

- `-h` Suppress printing of the file name header
- `-p` Allow printing of copyright procedures
- `-w` Suppress warning messages of deleted procedures

◇ This command is new in Release 3. ◇ See also: `global`, `local`, `m2src`.

usage `usage(`*name*`)`
Shows the calling sequence and descriptions of parameters for a function ◇ The information is extracted from the on-line help text. The recommended way to invoke this function is with the double question mark. ◇ This function is new in Release 3. ◇ See also: `??`, `example`, `help`, `info`, `related`.

userinfo `userinfo(`n, f, $expr_1$, $expr_2$ `...)`
`userinfo(`n, *fset*, $expr_1$, $expr_2$ `...)`
Evaluates and prints the expressions when `infolevel[`f`]` or `infolevel[all]` is n or larger ◇ The second argument may also be a set of procedure names for which the information should be printed. n is an integer traditionally between 1 and 5. Larger n corresponds to more detailed information. ◇ See also: `infolevel`.

value `value(`*expr*`)`
Evaluates inert functions ◇ This function forces inert functions such as `Int`, `Limit`, `Diff`, `Sum`, and `Product` to be evaluated. The same effect could be accomplished by substituting the upper case names by their lower case equivalents and then invoking `eval`. ◇ `value(Int(x^2, x=0..1));` ⟶ $\frac{1}{3}$ ◇ See also: `diff`, `eval`, `limit`, `product`, `int`, `sum`.

verify `verify('`*f*(*args*)`', ` *expr*`)`
`verify('`*f*(*args*)`')`
`verify(`*expr*`)`
`verify(`*eqn*`)`
`verify(`*list*`)`
`verify(`*set*`)`
Verifies certain Maple computations ◇ The first form verifies that the function call evaluates to *expr*. The second form verifies the computation of the function call by evaluating the function, then calling the inverse function on the result. In both these forms, single right quotes should be used around the function call to prevent premature evaluation. The third form checks that *expr* is equivalent to zero. The fourth form checks that the given equation holds. The last two forms verify each element of a list or a set. The result of this function is *true*, *false*, or *FAIL*. ◇ You must `readlib(verify)` before using this command. ◇ `verify('int(2*t,t)');` ⟶ *true*

W `W(`z`)`
`W(`n, z`)`
Lambert's W (or omega) function ◇ `W(`z`)` satisfies $W(z)\exp(W(z)) = z$. The branch chosen is the unique branch that is analytic at 0. The second form computes other branches of the solution to this equation. n may be any integer.

whattype `whattype(`*expr*`)`
Queries the basic data type of the expression ◇ `whattype(a+2*b);` ⟶ + ◇ `whattype(a,b,c);` ⟶ *exprseq* ◇ See also: `hastype`, `type`.

while `while` *cond* `do` *statements* `od`
Executes the statements as long as the condition is true ◇ The condition is tested at the beginning of each pass through the loop. ◇ See `do`.

A B C D E F G H I J K L M N O P Q R S T U V W X Y Z

with `with(`*package*`)`
`with(`*package*`,` f_1`,` f_2`, ...)`
`with(share)`
Defines functions from a library package ◇ This function allows the use of short names of the form *f* rather than full names of the form *package*`[`*f*`]`. The first form defines all the functions in the package in the current session. The second form defines only the named functions from the package. The libraries specified by the global variable `libname` are searched in sequence for the package. The special form `with(share)` updates `libname` so that share library functions may be accessed. It does not define any functions. If the share library is not in the expected location, assign its directory to the variable `sharename` before invoking `with(share)`. ◇ `with(plots):` ◇ See also: `libname, readlib`.

words `words()`
`words(`*n*`)`
Queries memory usage ◇ This function returns the total number of words used in the session. This number is stored in `status[1]`. The second form requests that a memory usage statement be produced after every *n* words of storage are used. The default value of *n* varies by system. Its current value is stored in `status[4]`. ◇ See also: `gc, status, time`.

write `write(`$expr_1$`,` $expr_2$`, ...)`
Writes a sequence of expressions to a file opened by `open` ◇ Expressions are line printed to the file. ◇ You must `readlib(write)` before using this command. ◇ See also: `appendto, close, lprint, open, save, writeln, writeto`.

writeln `writeln(`$expr_1$`,` $expr_2$`)`
Writes a sequence of expressions followed by a newline to a file opened by `open` ◇ Expressions are line printed to the file. ◇ You must `readlib(write)` before using this command. ◇ See also: `appendto, close, lprint, open, save, write, writeto`.

writeto `writeto(`*filename*`)`
`writeto(terminal)`
Writes all successive output to the file ◇ Under the command-line interface, prompts and input statements are also echoed to the file, and prompts are not displayed on the screen while writing to a file. The second form restores printing to the screen. If the file already exists, the old contents are lost. ◇ See also: `appendto, save, write`.

Zeta

`Zeta(`*z*`)`
`Zeta(`*n*`, `*z*`)`
`Zeta(`*n*`, `*z*`, `*q*`)`
The Riemann Zeta function ◇ The first form evaluates the `Zeta` function at z:

$$\zeta(z) = \sum_{i=1}^{\infty} \frac{1}{i^z}$$

The second form gives the nth derivative of the ζ function. The third parameter changes the sum:

$$\zeta(0, z, q) = \sum_{i=0}^{\infty} \frac{1}{(i+q)^z}$$

and $\zeta(n, z, q)$ gives the nth derivative of this function. ◇ In Release 3, z may be complex; in earlier versions, z must be real. ◇ `Zeta(2);` $\longrightarrow \frac{1}{6}\pi^2$

zip

`zip(`*f*`, `*list₁*`, `*list₂*`)`
`zip(`*f*`, `*list₁*`, `*list₂*`, `*default*`)`
Zips together two lists or vectors ◇ This function applies the binary function f to corresponding components of two lists or two vectors and returns a new list or vector. If a fourth argument is given, it is used as a default value when one of the lists or vectors is shorter than the other. ◇ `zip(f, [a,b], [c,d,e]);` $\longrightarrow$ $[\,\mathrm{f}(a,c), \mathrm{f}(b,d)\,]$ ◇ See also: `map`.

ztrans

`ztrans(`*expr*`, `*n*`, `*z*`)`
Z transform ◇ This function finds the Z transformation of an expression *expr* in the variable n, yielding an expression in z. It recognizes the delta function `Delta(...)` and the step function `Step(...)`. This transform is defined as: $\sum_{n=0}^{\infty}(expr)/z^n$ ◇ `ztrans(a^n,n,z);` $\longrightarrow \frac{z}{z-a}$ ◇
See also: `invztrans`, `rsolve`.

Chapter 5
Electronic Resources

A large amount of information on Maple is available on-line through electronic mail, Internet file transfer, and electronic bulletin boards. You can access this information with an account on Internet or Bitnet or with a modem hooked up to a computer.

1 THE SHARE LIBRARY

The Maple share library is a repository of contributed Maple code and documentation. It is distributed electronically and is freely available. The share library contains dozens of applications of Maple in many areas of mathematics, science, and engineering. These applications are donated by Maple users around the world. The share library also includes tutorials on Maple, summaries of new features in recent versions, and documents on specific applications. The developers of Maple also post bug fixes and enhancements to Maple library routines here.

The share library is available over the Internet via anonymous FTP from the five sites listed in the following table. The last section of this chapter (page 418) describes how to use anonymous FTP. The share library is updated approximately three times per year.

Internet Name	**Internet Address**	**Directory**	**Location**
`daisy.uwaterloo.ca`	129.97.140.58	maple	Canada
`ftp.maplesoft.on.ca`	192.139.233.5	pub/maple	Canada
`ftp.inria.fr`	128.93.2.54	lang/maple	France
`canb.can.nl`	192.16.187.2	pub/maple-ftplib	Holland
`neptune.inf.ethz.ch`	129.132.101.33	maple	Switzerland

The share library is partitioned by version number: Maple V Release 2 contributions are in the "5.2" subdirectory of the master directory listed in the third column of the table. For example, on `daisy.uwaterloo.ca`, they are in maple/5.2. Inside 5.2 are four main subdirectories:

share	Contributed Maple code and associated documentation.
lib	Fixes and enhancements to Maple library routines and help files.
doc	General articles on Maple: many in TEX or LATEX, others in PostScript or plain text.
test	Test cases for many contributed routines.

If you do not have access to FTP and the Internet, you can still obtain files from the share library through electronic mail. Send mail to `maple-netlib@can.nl` and include the line "send info." A program receives your message and sends you further instructions.

A portion of the share library is shipped with Maple. Enter the help command `?share[contents]` to view a list of the share library routines that came with your system. It is very easy to load and use these share library functions under Maple V Release 2 and

Release 3.

The share library is usually installed in a directory named `share` adjacent to the main Maple library directory. If the share library is installed elsewhere, you must first assign its location to the variable `sharename`. Then set up your session to use the share library with the command:

```
> with(share);
```

This changes the `libname` variable so that Maple searches the share library after searching the main library whenever you use `readlib` or `with`. For example, you can now load the share library function `fit` for computing a least-squares fit to some data with the command `readlib(fit)`.

Most share library functions come with Maple-style help pages. These are accessed like other help topics after you load a share library file. For example, look at the help page for `fit` by entering `?fit`.

1.1 Selected Share Library Contributions

This section describes selected functions, worksheets, and documents in the share library. Some of these may already be installed on your system, since a portion of the share library is packaged with the academic edition of Maple V Release 2 and Release 3. All are available by anonymous FTP or electronic mail, as described above. At the end of each short description is the name of the author(s) and the name and location of the file containing the program or document. Worksheets have file names ending with ".ms" and may be used on any Maple platform supporting the worksheet interface.

The contributions are listed in nine categories: Algebra, Analysis, Combinatorics, Geometry, Linear Algebra, Number Theory, Science and Engineering, Utilities, and Miscellaneous Documents.

Algebra

algcurve — A package of routines for computing with algebraic curves in two dimensions. Includes facilities for computing singular points, tangents, inflection points, intersection points, and a routine for plotting curves. S. Schwendimann. (`algcurve`)

charsets — A package implementing Ritt-Wu's characteristic sets method. Includes characteristic sets of (multivariate) polynomial sets, decomposing polynomial sets into ascending sets and irreducible ascending sets, decomposing algebraic varieties into irreducible components, factoring polynomials over algebraic number fields, and solving systems of polynomial equations. D. Wang. (`algebra/charsets`)

compoly — A faster algorithm for polynomial decomposition (Gutierrez et al.) that does not require factorization. B. Salvy and F. Morain. (`compoly`)

coxeter
weyl
: Packages for studying root systems and finite coxeter groups. The weyl package requires the coxeter package and contains procedures for manipulating weight vectors and computing multiplicities for irreducible representations of semisimple Lie algebras. J. Stembridge. (`coxeter/coxeter`, `coxeter/weyl`)

fields
: A package of routines for determining the transcendence degree of a field extension, or the degree of an algebraic extension, of two fields lying over the rationals or an algebraic number field. G. Kemper. (`algebra/fields`, `algebra/fields.ms`)

fullparfrac
: Full partial fraction decomposition of a rational function in $\mathbf{Q}(x)$ over the algebraic closure of $\mathbf{Q}$ without factorization. B. Salvy. (`fparfrac`)

galois
: Computes the Galois group of a polynomial of degree at most 7. Extends the standard library routine `galois` from $\mathbf{Q}[x]$ to $\mathbf{Q}[t_1, t_2, \ldots, t_m][x]$, that is, to polynomials with polynomial coefficients. R. Sommeling, T. Mattman, and J. McKay. (`algebra/galois`)

GB
: Computes a reduced, minimal Gröbner basis of a set of polynomials over a field of prime order using pure lexicographic or total degree term ordering. Includes the routine NormalForm for reducing a polynomial over such a field modulo an ideal. D. Gruntz. (`mod/GB`)

group.ms
: A worksheet answering questions about the small Rubik's cube. (`algebra/group.ms`)

Hurwitz
: Tests whether a polynomial is of Hurwitz type; that is, whether the real part of each root is negative. R. Corless. (`Hurwitz`)

IF
: Package of routines for working with real algebraic numbers and solving systems of equations and inequalities. An application of these procedures computes a cylindrical algebraic decomposition of a plane algebraic curve. F. Cucker. (`IF`)

IntBasis
: Routines for quickly computing an integral basis for an algebraic function field (integral_basis), computing Puiseux series expansions (puiseux), and computing the genus of an algebraic curve (genus). M. van Hoeij. (`IntBasis`)

invar
: A package of routines for computing the invariant ring of permutation groups or finite linear groups over $\mathbf{Q}$ or an algebraic number field. G. Kemper. (`algebra/invar`)

posets
: A package of routines for manipulating partially ordered sets, including routines to enumerate all nonisomorphic posets and lattices on n vertices. J. Stembridge. (`posets`)

SF — A package of routines for manipulating symmetric functions. The package can express the symmetric functions in the bases: monomial symmetric functions, elementary symmetric functions, complete homogeneous symmetric functions, power sum symmetric functions, and Schur functions. J. Stembridge. (`SF`)

sglplot — Displays a subgroup lattice in three dimensions. M. Monagan. (`plots/sglplot`)

SqrFree — Computes the square-free factorization of a multivariate polynomial over a field of prime order. (Maple's `Sqrfree` only works for univariate polynomials.) S. Swanson. (`mod/SqrFree`)

subres — Computes the subresultant polynomial remainder sequence. G. Labahn and M. Monagan. (`subres`)

Subres — Computes the subresultant polynomial remainder sequence over the integers modulo m. G. Labahn and M. Monagan. (`mod/Subres`)

symmpoly — Generates elementary symmetric polynomials. M. Monagan. (`combinat/symmpoly`)

traubjen — Traub-Jenkins algorithm for computing the complex roots of a polynomial. B. Salvy. (`traubjen`)

trinom.ms — Computes primitive trinomials over finite fields, for use in error correcting codes and random number generators. This shows how to use Maple to compute with polynomials over finite fields. M. Monagan. (`algebra/trinom.ms`)

Analysis

approx.ms — A Maple worksheet showing various numerical approximation techniques for developing efficient polynomial and rational function approximations, including Taylor and Chebyshev series and Pade, Chebyshev-Pade, and minimax approximations. K. Geddes. (`numerics/approx.ms`)

autodiff — A collection of routines for automatic (algorithmic) differentiation. W. Neuenschwander and M. Monagan. (`autodiff/DIFF`, `GRADIENT`, `HESSIAN`, `JACOBIAN`, `TAYLOR`)

balloon.ms — A long flexible rope is coiled neatly on the ground and a helium balloon is attached at one end. The motion of the balloon rising is modeled by a differential equation, solved numerically and plotted. This worksheet requires the ODE plots package from the share library. D. Schwalbe. (`plots/balloon.ms`)

BesselH — Implementation of Hankel functions in terms of BesselJ and BesselY. D. Meade. (`hankel`, `hankel.ms`)

fft — Calculates the Fast Fourier Transform. S. Earl. (`numerics/fft`)

fht — Calculates the Fast Hartley Transform. S. Earl. (`numerics/fht`)

fit — A (linear) least-squares fit of a curve to a set of data. D. Gruntz. (`fit`)

fjeforms — A revised version of the `difforms` package for computing with differential forms. FJE Enterprises. (`fjeforms/fjeforms`)

gdev
glimit — General series expansions and symbolic limits. Uses a more powerful model for asymptotic series expansions than Maple's `asympt` and `series` commands. B. Salvy. (`gdev`)

gfun — A package for computing with generating functions. Converts implicit equations into differential equations, differential equations into recurrences, and vice-versa, ordinary into exponential recurrences. Also recovers linear recurrences or differential equations from finite lists of coefficients. B. Salvy and P. Zimmermann. (`gfun`)

GRADIENT.ms — A worksheet showing examples from automatic (algorithmic) differentiation using Maple's `D` operator and the `GRADIENT` function from the share library packaged with Release 2 to compute partial derivatives, gradients, and Jacobians of functions given as Maple procedures. M. Monagan. (`autodiff/GRADIENT.ms`)

guesss — A routine for guessing the next values in a sequence. This routine searches for more general differential equations than the gfun package does. It uses a generalized continued fraction method to find the differential equation. H. Derksen. (`guesss`)

ilp — An integer linear programming algorithm for maximizing a linear function given a set of linear constraints. A. Pathria. (`ilp`)

intpak — Experimental interval arithmetic package. A. Connell and R. Corless. (`numerics/intpak`)

LambertW.ps — A PostScript file documenting Lambert's W function (the `W` function in Maple). Includes properties, integration formulae and techniques, asymptotics, and applications. R. Corless, G. Gonnet, D. Hare, and D. Jeffrey. (`LambertW.ps`)

logistic.ms — A worksheet solving the logistics law of population growth: $P' = aP - bP^2$. D. Schwalbe. (`plots/logistic.ms`)

logmap.ms — A worksheet exploring the period-doubling bifurcation sequence of the discrete logistic map. R. Corless. (`calculus/logmap.ms`)

MatPade — Functions for computing matrix-type Pade approximants, including Hermite-Pade, Simultaneous Pade, and right and left matrix Pade approximants. G. Labahn. (`algebra/MatPade`, `algebra/MatPade.ms`)

ODE — A collection of routines for numerical solution and plotting of differential equations, from Chapter 5 of the *Maple V Flight Manual*. This code is also included in the share library of the student edition of Maple V Release 2. D. Schwalbe. (`plots/ODE`)

pade2 — Generalized Pade approximations. H. Derksen. (`numerics/pade2`)

parfrac.ms — A worksheet showing partial fraction decomposition for rational function integration. M. Monagan. (`calculus/parfrac.ms`)

polycon — A package for analysis of polynomial and rational control systems. K. Forsman. (`engineer/polycon`)

PS — A formal power series package with many improvements over the library package `powseries` for manipulating power series. D. Gruntz. (`PS`)

ratinterp — Computes a rational function that interpolates the given data points. C. von Achenbach. (`numerics/rfinterp`)

ratlode — Finds the rational solutions to a linear n^{th} order ODE with rational coefficients. M. Bronstein. (`calculus/ratlode`)

trans — A package of routines for rational function approximations to rational points, Taylor and asymptotic series, and functions. J. Grotendorst. (`numerics/trans`)

Combinatorics

coxpoly — Computes the coxeter polynomial of a rooted cycletree. A. Boldt. (`combinat/coxpoly`)

perm — A package of routines for computing with permutation groups, including routines for computing orbits, subgroups, and testing for conjugacy. There are also procedures used to represent and manipulate combinatorial structures. Y. Chiricota. (`combinat/perm`)

relpoly — Computes the all terminal reliability polynomial for a graph with a given probability of edge failure. M. Monagan. (`combinat/relpoly`)

Geometry

billiard.ms — A worksheet showing how to solve the billiard problem for a circular billiard table. The numerical solution exhibits the use of algorithmic differentiation. W. Gander and D. Gruntz. (`geometry/billiard.ms`)

surfaces — A package for calculating basic differential-geometric quantities of parametric surfaces in space, including first and second fundamental forms, metric determinant, Gauss curvature, and Mean curvature. T. Murdoch. (`geometry/surfaces`)

Linear Algebra

autoevalm A utility routine for automatic evaluation of matrix expressions. E. Johnson. (`linalg/autoevalm`)

classical Rational canonical form for a matrix of rational numbers. Includes a utility routine for constructing the hyper-companion matrix of a univariate polynomial. E. Johnson. (`linalg/class`)

eigen.ms A worksheet tutorial on **`eigenvals`**, **`eigenvects`**, and related commands to compute exact symbolic values for eigenvalues and eigenvectors of numeric and symbolic matrices. Includes a tutorial on computing with roots of polynomials using `RootOf` notation. M. Monagan. (`linalg/eigen.ms`)

IntSolve A linear integral equation solver (requires `linalg/Echelon`). H. Ye and R. Corless. (`linalg/IntSolve`)

lapack.ms A worksheet investigating special properties of the 6×6 matrix on the front cover of the LAPACK User's Guide. M. Monagan. (`linalg/lapack.ms`)

magic Creates a magic square. Includes the routine ismagic for testing whether a square matrix of integers is a magic square. D. Gruntz. (`linalg/magic`)

normform A collection of routines for computing matrix normal forms. Includes ismithex, smithex, frobenius, jordan, jordansymbolic, and ratjordan for computing Smith, Frobenius, and Jordan normal forms. Typically faster and more general than similar routines in the Maple library. T. Mulders and A. Levelt. (`linalg/normform`)

Normform A collection of routines for computing normal forms of matrices over the integers mod p. Includes Frobenius, Ratjordan, and Jordansymbolic. T. Mulders and A. Levelt. (`mod/Normform`)

pascal Creates a Pascal matrix. M. Monagan. (`linalg/pascal`)

pffge Primitive fraction-free Gaussian elimination for a rectangular matrix of multivariate polynomials. After a row operation, the gcd of the polynomials in the row is divided out, minimizing the size of the intermediate polynomials. M. Monagan. (`linalg/pffge`)

RowEchelon Returns the reduced row echelon form R of a rectangular matrix A. Also computes multiplier matrices P, L, and U such that $A = PLUR$ where P is a permutation matrix, L is lower triangular, and U is upper triangular. R. Corless and D. Jeffrey. (`linalg/Echelon`)

rrefshow A teaching tool for viewing each step of matrix row reduction to echelon form. E. Johnson. (`linalg/rrefshow`)

sffdet Computes the determinant of a square matrix using a variant of Bareiss's fraction-free algorithm, particularly suited to sparse matrices with integer or polynomial entries. Q. Nguyen. (`linalg/sffdet`) sffge — Performs Bareiss' fraction free Gaussian elimination algorithm on a rectangular matrix of polynomials. This routine has the same functionality as `linalg[ffgausselim]` but is suited to sparse matrices. I. Berchtold and M. Monagan. (`linalg/sffge`)

Number Theory

DistTo1 — The $3n+1$ conjecture, or Collatz's problem or Ulam's conjecture. Computes the number of iterations to 1 in an efficient way. G. Gonnet. (`numtheor/Collatz`)

elliptic — Determines the order of the group of points on a nonsingular elliptic curve over a finite field of prime order. E. Von York. (`numtheor/elliptic`)

irredheu — A heuristic irreducibility test of polynomials in $\mathbf{Q}[x]$ via finding prime evaluations and integer primality testing. This is conjectured to be a polynomial time irreducibility test. M. Monagan. (`numtheor/irredheu`)

isprime — An improved probabilistic primality testing routine, using one modified strong pseudoprime test base 2 and one Lucas test. No composite number that satisfies both tests has ever been found, although such a counterexample may exist. This improved `isprime` is incorporated in Maple V Release 3. G. Gonnet with M. Schoenert. (`isprime`)

Science and Engineering

AFA.ms — A worksheet for algebraic function approximation. A procedure reconstructs the expression of the Dirac energies from the series coefficients using a heuristic polynomial fit. G. Fee and T. Scott. (`science/AFA.ms`)

AnalSPT.in
Anal3PTB.in — Harmonic analysis of single phase and 3-phase waveforms of thyristor circuits. A. Rough and J. Richardson. (`engineer/AnalSPT.in, Anal3PTB.in, harmonal`)

biology.ms — A worksheet applying the ODE package to a problem from biology where two populations compete with each other for resources to survive. The biological system is modeled by two first order differential equations, and phase plots of the solutions are plotted showing examples of stable and unstable populations. D. Schwalbe. (`plots/biology.ms`)

bohratom.ms — A worksheet solving three nonlinear equations arising from Bohr theory applied to the hydrogen atom. D. Redfern, T. Scott, and R. Pavelle. (`science/bohratom.ms`)

chemeqn.ms — A worksheet that balances the coefficients in a chemical reaction. D. Redfern, T. Scott, and R. Pavelle. (`science/chemeqn.ms`)

doubpend.ms A worksheet that approximates the solution to the differential equation associated to a double pendulum and animates its motion. D. Schwalbe. (`plots/doubpend.ms`)

filter.ms A worksheet analyzing the effects of a low-pass filter on a change in signal or power frequency. R. I. Adare and T. Lee. (`engineer/filter.ms`)

flash.ms A worksheet performing flash calculations to determine the phase condition of a mixture at a specified temperature and pressure. R. Taylor. (`science/flash.ms`)

heatcap.ms A worksheet obtaining the mean energy and heat capacity of an Einstein solid. T. Scott. (`science/heatcap.ms`)

kinetics Procedures for determining the system of differential equations for a given reaction scheme, the associated conservation laws, and some of the species that have a zero steady state. M. Holmes. (`science/kinetics`)

lagrange.ms A worksheet applying lagrange multipliers to minimize a function subject to a constraint. A. Rough. (`engineer/lagrange.ms`)

Maxgas.ms A worksheet that computes the most probable speed of the Maxwell-Boltzman distribution. D. Redfern, T. Scott, and R. Pavelle. (`science/Maxgas.ms`)

McConnel.ms A worksheet on using matrix algebra, in particular matrix exponentials, for the symbolic solution of a system of first order differential equations in biochemistry; namely, the McConnell equations. J. Grotendorst, P. Jansen, and S. M. Schoberth. (`science/McConnel.ms`)

nonlin.ms A worksheet exhibiting geometric iteration and the Newton-Raphson iteration for approximating roots of nonlinear equations in one variable. A. Rough. (`engineer/nonlin.ms`)

pendulum.ms A worksheet modeling the motion of a pendulum in a fluid as a second order ODE, solving the ODE analytically, and plotting its motion for different fluid resistances. D. Harper. (`science/pendulum.ms`)

phase.ms A worksheet performing simple thermodynamic calculations, including bubble and dew point determinations, creating and plotting phase diagrams, and flash calculations. R. Taylor. (`science/phase.ms`)

planck.ms A worksheet computing a definite integral, the Stefan-Boltzman Law (of Blackbody radiation). T. Scott. (`science/planck.ms`)

poten.ms Potential fields using Green's functions in two dimensions. The Neumann and Dirichlet problems. A. Rough. (`engineer/poten.ms`)

quantopt.ms A worksheet application for optical transitions between two orbitals on different atoms. This involves three-dimensional integration of the product of Gaussian distributions. D. Redfern, T. Scott, and R. Pavelle. (`science/quantopt.ms`)

robotarm.ms A worksheet solving the path planning problem for a simple two-link planar robot arm, common in assembly line operations. Animates two specific solutions. G. P. Porciello and T. Lee. (`engineer/robotarm.ms`)

shear.ms A worksheet presenting two design problems for a machine element, solving equilibrium equations to obtain general solutions for shear and normal stresses. J. Argent and T. Lee. (`engineer/shear.ms`)

shottraj.ms A worksheet that solves a second order ODE describing the motion of a bullet shot straight into the air and plots the solution. M. Monagan. (`science/shottraj.ms`)

statval.ms A worksheet with exercises for finding maxima, minima, and saddle points on surfaces in two dimensions. A. Rough. (`engineer/statval.ms`)

stepresp.ms A worksheet analyzing the step response of an RC circuit. R. I. Adare and T. Lee. (`engineer/stepresp.ms`)

stream.ms A worksheet that illustrates streamlines in two-dimensional vector fields. A. Rough. (`engineer/stream.ms`)

wheatsto.ms A worksheet applying Kirchoff's laws to the "Wheatstone Bridge" electrical circuit. T. Scott and M. Monagan. (`science/wheatsto.ms`)

Utilities

convert/arctanh Converts all logarithms and inverse hyperbolic trigonometric functions in an expression to a corresponding form in `arctanh`. V. Broman. (`convert/arctanh`)

eps Generates encapsulated postscript for a Maple plot, for placing Maple graphics in other applications. Provides control over a number of parameters, including scaling, rotation, and translation. J. S. Devitt. (`eps`)

fortran.ms A discussion of the pitfalls of naive use of symbolic computation when generating Fortran code. The example is of generating Fortran code for computing the inverse of a 3 by 3 symmetric matrix. M. Monagan. (`numerics/fortran.ms`)

gdegree Computes the degree of an expression, allowing rational and symbolic exponents. This is an example of how to write a Maple procedure that steps through an expression recursively. M. Monagan. (`gdegree`)

maclaurinsubs Replaces functions in an expression with formal Maclaurin series. V. Broman. (`macsubs`) macroC — A Maple package for generating C language code. All C structures can be generated and the resulting code can be optimized. The Maple library routine `C` handles only expressions. P. Capolsini. (`macroC`)

macrofort — A collection of routines for generating FORTRAN code, allowing for the generation of complete FORTRAN programs, including declarations and control structures. C. Gomez. (`macrofor`)

mathematica — Converts a Maple expression to an equivalent Mathematica expression. Includes translation of integrals, matrices, simple operators, polynomials, and formulae. D. Gruntz. (`math`)

readshare — Simplifies loading of code from the share library. M. Monagan. (`readshare`)

reorder — Reorders sums, limits, or integrals in an expression. V. Broman. (`reorder`)

saveglob restglob — Routines to save and restore global variables so that one can make changes when debugging but restore the state. D. Gruntz. (`saveglob`)

sprint — Short print utility for displaying large expressions. M. Monagan. (`sprint`)

tex — Package for generating LaTeX, plain TeX, and AMS-TeX output with line breaking, including line breaking of vectors and output for procedures. Y. Yu. (`tex`)

undistribute — Moves constant factors out of integrands, summands, and limits. V. Broman. (`undist`)

xdvi — Generates LaTeX output for an expression, formats it with `latex`, and previews the resulting .dvi file with `xdvi`. This command runs only on Unix systems that support `latex` and `xdvi`. M. Monagan. (`xdvi`)

Miscellaneous Documents

New Features in Maple V Release 2, Symbolic Computation Group, 32 pages.
An accompanying worksheet (M. Monagan and R. Neumann) gives examples of these changes in an interactive session. (`doc/summary.tex`, `doc/summary.ms`)

Programming in Maple: The Basics, M. Monagan, 53 pages. (`program.tex`)

The Maple Computer Algebra System, M. Monagan, 7 pages. (`doc/Maple.tex`)

The document section of the Release 2 share library also includes several LaTeX style files that many authors have used to write articles that include Maple commands.

2 THE INFO SERVER AND USER GROUPS

The Maple Info Server

The Maple Info server is an FTP site supported by Waterloo Maple Software. This archive contains a copy of the share library, plus a great deal of information on Maple. It includes Maple demonstrations, lists of Maple features, a list of platforms on which Maple is available, information on particular platforms, a current book list, press releases related to Maple and Waterloo Maple Software, and answers to some common technical questions and problems with Maple.

You can reach the Info Server via anonymous FTP at `ftp.maplesoft.on.ca`.

The Maple User Group

The Maple User Group is an electronic forum for discussing Maple that links hundreds of Maple users worldwide. Messages are distributed by electronic mail. Many questions to the group ask whether Maple can do a particular calculation; others request assistance with a problem. Announcements are circulated about additions to the share library, new platforms supporting Maple, and new releases of the software. Waterloo Maple Software sometimes solicits comments on Maple functions and documentation. Though the Maple User Group is sponsored by Waterloo Maple Software, it operates independently of the company. Maple users respond to other users' questions.

To join the Maple User Group, send an electronic mail message to

```
maple_group@daisy.waterloo.edu
```

and ask to subscribe. There is no fee.

News Groups

The USENET electronic bulletin board `sci.math.symbolic` is another great source of information. Subscribers from all over the world discuss computer algebra systems like Maple. Many universities and corporations receive the USENET bulletin boards. You can obtain your own subscription through companies such as America On-line, Compuserve, Delphi, and UUNET Technologies, to name just a few.

3 USING ANONYMOUS FTP

The Internet file transfer program `ftp` copies files between computers linked on the Internet. Some sites have archives that anyone can access by using a special mode called "anonymous FTP."

Start an FTP session from a computer on the Internet with the command `ftp` *hostname*. For example, to access the share library at `daisy.uwaterloo.ca`, type:

```
ftp daisy.uwaterloo.ca
```

If your computer responds with an error message, try the host's numerical address in place of the host name:

```
ftp 129.97.140.58
```

When you connect, the remote host prompts you for a name and password. Log in as `anonymous` and type your electronic address as the password.

Some commands for browsing through the archive and transferring selected files are listed in the following table.

FTP Command	**Description**
`ls` *or* `dir`	List files in the current directory
`cd` *dir*	Change directory
`cd ..`	Change to the parent directory
`get` *file*	Transfer *file* to your computer
`mget` *files*	Transfer multiple files
`quit` *or* `bye`	Terminate an `ftp` session
`ascii`	Set ASCII (text) transfer
`binary`	Set binary transfer
`help`	List all commands

Chapter 6
Frequently Asked Questions

This chapter answers questions that users frequently ask about Maple. The first section answers some general questions on Maple, and the second section answers some common technical questions.

1 GENERAL QUESTIONS

What can Maple do?

Maple can do advanced symbolic manipulations, precise numerical calculations, and graphs of functions and data in two and three dimensions. Maple can manipulate expressions in one or several variables over the rational numbers, algebraic number fields, or finite fields. It can do many operations in calculus: limits, differentiation, numeric and symbolic integration, taylor series, differential equations, and asymptotic analysis. Maple solves equations and recurrence relations. It can calculate exact numerical results or compute answers as approximations to very high precision. Maple supports many kinds of two- and three-dimensional graphs, including implicit plots, log plots, parametric curves and surfaces, contour plots, conformal maps, vector fields, polygons and polyhedra, data plots, and animation. Maple includes packages of mathematical functions encoding hundreds of operations and algorithms in linear algebra, combinatorics and graph theory, geometry, number theory, statistics, mathematical physics, group theory, Gröbner bases, formal logic, linear programming, and more.

Maple has a powerful programming language similar to Pascal for writing your own mathematical applications. You can call any Maple function from a Maple program. Most of Maple is written in this programming language, and you can readily inspect and modify this code.

Who uses Maple?

Maple is standard software at many universities and businesses throughout the world. Mathematicians, scientists, engineers, educators, and students are the primary users of Maple.

Is Maple used in high schools?

Yes, Maple is being used in high schools in North America and in Europe. In fact, there are more than 700 high schools in the province of Ontario using Maple. The price of Maple and its low hardware requirements make the software attractive to high schools.

On which computers does Maple run?

Maple runs on a wide range of computers including 386- and 486-based computers (DOS, Windows), 88000-based systems (Unix), Amiga, Apollo, Apple Macintosh, Atari, Convex C2 and C3, Cray XMP and YMP, DEC Alpha OSF/1, DECstation, Personal DECstation, VAX, VAXstation and MicroVAX (Ultrix and VMS), Fujitsu VP 2000, HP 9000, IBM RS/6000, S/370 and personal computers, Intergraph 2000 and 6000, MIPS, NeXT, Sequent Symmetry, Silicon Graphics Indy, Indigo, Iris and Crimson and Sun 3, 4, ELC, IPC, and SPARCstation (SunOS or Solaris). Maple runs on DOS personal computers with or without Microsoft Windows. Maple commands and programs are portable: you can run the same Maple program on a personal computer, workstation, mainframe, and supercomputer.

How much memory do you need to run Maple? How much disk space is required?

It depends on your platform and whether you have the student edition or the academic edition.

The academic edition on the Macintosh requires 2.5 megabytes of memory and 12 to 17 megabytes of free disk space, depending on whether you want to install some optional components like the share library and an interactive tutorial. The academic edition on a personal computer requires 10 to 16 megabytes of free disk space. Four megabytes of memory are needed for the Microsoft Windows interface; as little as two suffice for the DOS interface.

The student edition alters these numbers in three ways. First, only a small portion of the share library is distributed with the student edition, making the maximum disk space requirement about 3.5 megabytes smaller. Second, the Maple library is stored in a compressed format, reducing the disk space required by a couple of megabytes. Third, more memory is needed for expanding the library routines as you run Maple. One or two additional megabytes of memory are recommended for the student version.

How does the student edition of Maple differ from the academic edition?

The section entitled "Student and Academic Editions" on page 80 in Chapter 1 compares the two and lists the limits of both.

Where can I purchase Maple?

To order Maple, visit your local software dealer, or contact one of the companies listed in the last section of Chapter 1, "How to Order Maple" (page 81). A list of addresses and phone numbers appears there.

Do you have any advice for learning Maple?

The best way to learn Maple is by using it. We encourage you to work through the examples and exercises in Chapter 1. There are several excellent tutorial books on Maple, too. They are listed in Chapter 8 beginning on page 465.

Where can I get help with Maple if I have problems with it?

When you have problems or questions, we suggest reading books and articles on Maple, experimenting with the system, and talking with other users. You can reach other Maple users electronically if you have access to a modem or to a network like Internet or Bitnet. See Chapter 5 for more information on electronic correspondence and the Maple User Group (page 418). Also, see Chapter 8 for information on Maple's technical support (page 469).

Is there an Internet news group for Maple users?

The USENET news group `sci.math.symbolic` contains discussions on computer algebra and computer algebra systems like Maple.

2 TECHNICAL QUESTIONS

This section answers some common technical questions about Maple. It gives examples of some often overlooked functions and options, describes how to work around some common errors, gives tips on timing and optimizing calculations, and answers some more advanced questions on programming.

2.1 Mathematical Functions

How can I obtain different sequences of random numbers?

Every time you start a Maple session, `rand` gives the same sequence of "random" numbers.

```
% maple
> rand();
```

$$427419669081$$

```
> rand();
```

$$321110693270$$

```
> quit
```

In a new session, we get the same numbers:

```
% maple
> rand();
```

$$427419669081$$

```
> rand();
```

$$321110693270$$

To alter the sequence of random numbers, set the global variable `_seed` to an arbitrary nonzero integer.

```
% maple
> _seed := 2;
```

$$\_seed := 2$$

```
> rand();
```

$$854839338162$$

```
> rand();
```

$$642221386540$$

How do I fit some data to a curve?

In Maple V Release 3, use the `leastsquare` function in the `fit` subpackage of the `stats` package. For example, we fit a cubic curve to data points sampled from the exponential function between 0 and 1.

```
> with(stats): with(fit):
> xdata := [seq(i/10., i=0..10)]:
> ydata := map(exp, xdata):
> leastsquare[[x,y], y=c0+c1*x+c2*x^2+c3*x^3]([xdata, ydata]);
```

$$y = .9995154052 + 1.016158410\,x + .4229242923\,x^2 + .2791658137\,x^3$$

In Maple V Release 2 and earlier, the `regression` function in the `stats` package fits data to a curve, but you must first format your data as a statistical matrix. We solve the same example with this command.

```
> with(stats):
> data := array([seq([i/10., exp(i/10.)], i=0..10)]):
> statmatrix := putkey(data, [x,y]):
> regression(statmatrix, y = c0 + c1*x + c2*x^2 + c3*x^3);
```

$$\{c2 = .4229245377, c1 = 1.016158301, c3 = .2791656638, c0 = .9995154153\}$$

The `fit` function in the share library provides another method of fitting data to a curve. It too uses the least-squares algorithm.

```
> with(share):
> readlib(fit):
> xdata := [seq(i/10., i=0..10)]:
> ydata := map(exp, xdata):
> fit(xdata, ydata, [1, t, t^2, t^3], t);
```

$$.999515409 + 1.016158418\,t + .4229242375\,t^2 + .279165859\,t^3$$

See Chapter 5 (page 407) for more information on the share library and how to use it.

2.2 Manipulating Maple Results

How can I prevent Maple from substituting `%1`, `%2`, and so on for subexpressions in long answers?

Set the `interface` variable `labeling` to `false`.

```
> interface(labeling=false);
```

See page 19 for an example of labeling in the output of a `solve` calculation.

How can I determine the number of elements in an expression sequence or pick individual expressions out of a sequence?

Every function in Maple accepts its arguments in the form of an expression sequence. So calling `nops` on an expression sequence gives an error, since `nops` expects a single argument. A similar problem occurs with `op`.

```
> myseq := v, w, x, y, z:
> nops(myseq);
  Error, wrong number (or type) of parameters in function nops;
> op(2..3, myseq);
  Error, wrong number (or type) of parameters in function op;
```

To resolve this problem, make the sequence into a list by enclosing it in square brackets.

```
> nops([myseq]);
```

$$5$$

```
> op(2..3, [myseq]);
```

$$w, x$$

The selection operator picks elements out of expression sequences directly.

```
> myseq[2..3];
```

$$w, x$$

Is there a way to map a function onto the coefficients of a polynomial? I tried mapping the absolute value function onto a polynomial to make all the coefficients positive, but that did not work.

The fourth argument to the `collect` command specifies a function to apply to all the coefficients of a polynomial. It can be any function that requires a single argument, such as `abs`, `floor`, `round`, or `c -> c^2`.

```
> p := x^3 - x^2 + x - 3:
> collect(p, x, distributed, abs);
```

$$x^3 + x^2 + x + 3$$

The same technique works for polynomials in several variables.

```
> q := 4.7*x^3*y^2 + 3.01*x*y + 2.*x + 6.9:
> collect(q, [x,y], distributed, round);
```

$$7 + 2x + 5x^3y^2 + 3xy$$

Another use of the `collect` function turns a polynomial in `x` and `y` into a polynomial in `x` with factored polynomials in `y` as coefficients. We list `x` as the only indeterminate so that the polynomials in `y` are treated as the coefficients.

```
> r := x^2*y^2 + 2*x^2*y + x^2 + x*y^3 + x + 1:
> factor(r);
```

$$xy^3 + x + x^2 + 2x^2y + x^2y^2 + 1$$

```
> collect(r, x, distributed, factor);
```

$$1 + (y + 1)(y^2 - y + 1)x + (y + 1)^2x^2$$

How can I evaluate a series at a point?

Direct substitution does not work. You must convert the series to an ordinary expression first, without the big O.

```
> s := taylor(tanh(x), x=0, 9);
```

$$s := x - \frac{1}{3}x^3 + \frac{2}{15}x^5 - \frac{17}{315}x^7 + \mathrm{O}\left(x^9\right)$$

```
> subs(x=0.5, s);
  Error, invalid substitution in series
```

Convert a power series to an ordinary polynomial with the `convert` command.

```
> convert(s, polynom);
```

$$x - \frac{1}{3}x^3 + \frac{2}{15}x^5 - \frac{17}{315}x^7$$

```
> subs(x=0.5, ");
```

$$.4620783730$$

This method also works for Laurent series, even though the result may not be a polynomial (a Laurent series may start with a few terms having x in the denominator). However, it does not work with many asymptotic series. We compute an asymptotic expansion to the natural log of $n!$ as n goes to infinity and attempt to drop the O term with the `convert` command:

```
> asympt(ln(n!), n);
```

$$(\ln(n) - 1)\,n + \frac{1}{2}\ln(n) + \ln\left(\sqrt{2}\sqrt{\pi}\right) + \frac{1}{12}\frac{1}{n} - \frac{1}{360}\frac{1}{n^3} + \mathrm{O}\left(\frac{1}{n^5}\right)$$

```
> convert(", polynom);
```

$$(\ln(n) - 1)\,n + \frac{1}{2}\ln(n) + \ln\left(\sqrt{2}\sqrt{\pi}\right) + \frac{1}{12}\frac{1}{n} - \frac{1}{360}\frac{1}{n^3} + \mathrm{O}\left(\frac{1}{n^5}\right)$$

Substitute a zero for the capital O to get rid of the last term. This makes a zero function operating on $1/n^5$, and the zero function always returns 0, regardless of its argument. The `simplify` function with the `atsign` option detects this and brings the expression into its final form.

```
> subs(O=0, ");
```

$$(\ln(n) - 1)\,n + \frac{1}{2}\ln(n) + \ln\left(\sqrt{2}\sqrt{\pi}\right) + \frac{1}{12}\frac{1}{n} - \frac{1}{360}\frac{1}{n^3} + 0\left(\frac{1}{n^5}\right)$$

```
> simplify(", atsign);
```

$$(\ln(n) - 1)\,n + \frac{1}{2}\ln(n) + \ln\left(\sqrt{2}\sqrt{\pi}\right) + \frac{1}{12}\frac{1}{n} - \frac{1}{360}\frac{1}{n^3}$$

```
> evalf(subs(n=200, "));
```

$$863.2319872$$

```
> evalf(ln(200!));
```

$$863.2319872$$

2.3 Reading Maple Files

I get an error message about an incorrect or outdated .m file when I try to read a file into my Maple session. How can I fix this?

There are two possible sources of an error message like this:

```
Error, filename.m is an incorrect or outdated .m file
```

Maple expects a file ending with ".m" to have a special binary format. Maple processes a file in this format very quickly. If your file is a plain text file (i.e., you can read it), just rename it and try again.

If your file really is in the binary format, it may have been created under another version of Maple. If you are using Maple V Release 3, the `m2src` command can convert a Release 2 ".m" file to a plain text file. Another utility, `updtsrc`, can then update the Release 2 text file for Release 3.

The formats in Maple V and Maple V Release 2 are also incompatible. In this case, obtain the original text version and read it into Maple. Then save a new ".m" file for future use.

2.4 Timing and Optimization

How do I time a Maple calculation?

We show three ways to do this. Below, we factor a large integer using three different algorithms and time the calculation in three different ways. The algorithms used are Morrison and Brillhart's continued fraction algorithm (the default), Lenstra's elliptic curve method, and Pollard's rho method. (These timing results should not be used to judge the overall strength of these methods, since one algorithm may work better than the others in certain situations.)

First, the `time` function returns the total amount of CPU time in seconds used since the start of your Maple session. Sample its value before and after a calculation to calculate the time used by a sequence of commands.

```
> start := time():
> ifactor(10^31 - 1);
```

$$(3)^2\,(6943319)\,(57336415063790604359)\,(2791)$$

```
> time() - start;
```

$$133.083$$

Second, the `timing` command is available to time calculations in the `history` environment.

```
> readlib(history):
> history();
O1 := timing(ifactor(10^31 - 1, lenstra));
```

$$(3)^2 (6943319) (57336415063790604359) (2791)$$

```
time = 41.68
O2 := off;
```

Third, the `showtime` environment displays the time and memory used by each command you enter.

```
> readlib(showtime):
> showtime();
O1 := ifactor(10^31 - 1, pollard);
```

$$(3)^2 (6943319) (57336415063790604359) (2791)$$

```
time   8.66   words   324926
O2 := off;
```

Maple seems to take an inordinately long time to compute an expression of the form a^b mod m, when a and b are medium-size integers. In fact, sometimes Maple refuses to do the computation at all. What can I do about this?

Use the neutral exponentiation operator, `&^`. With the usual exponentiation operator, Maple computes a^b first, then reduces the result modulo m. This greatly increases the cost of the computation, and may imply an intermediate result so large that Maple refuses the calculation. The intermediate result in the first example below has over thirty thousand digits; the second would have over five million!

```
> start := time():
> 8765 ^ 7832 mod 5367;
```

$$5308$$

```
> time() - start;
```

$$27.767$$

```
> 6578493 ^ 735493 mod 73638;
  Error, object too large
```

The neutral exponentiation operator in conjunction with `mod` avoids this nearsightedness and calculates the answer in a much more efficient manner.

```
> start := time():
> 8765 &^ 7832 mod 5367;
```

$$5308$$

```
> time() - start;
```

$$.017$$

```
> 6578493 &^ 735493 mod 73638;
```

$$19935$$

```
> time() - .017 - start;
```

$$.016$$

I've written some numerical routines in Maple that require only a modest amount of precision, but they seem significantly slower than some similar compiled programs. Is there any way to speed them up?

Numerical routines in Maple are slower than compiled programs for two reasons. First, Maple programs are interpreted at run time, not compiled ahead of time. Second, floating-point arithmetic is done in the software, even if the required precision is small enough for direct evaluation in your computer's hardware. There is little you can do about the cost of interpreting, but you can speed up your application considerably by requesting evaluation in the hardware.

The `evalhf` function evaluates expressions using the hardware's floating-point arithmetic. Most computers calculate floating-point expressions in hardware to about fifteen decimal digits of precision. The `Digits` option to `evalhf` returns the number of digits that your hardware handles.

```
> evalhf(Digits);
```

$$15.$$

If your application requires this amount of precision or less, then adjust your code to use `evalhf` to improve its performance.

The `evalhf` function requires each operand in an expression to be a number, a variable assigned a number, or a reference to a numeric array. It does not accept a symbolic expression or a reference to another data structure, like a set or a list. Because of this restriction, you may have to save an intermediate result before calling `evalhf`.

```
> nums := [10, 20, 30, 40]:
> evalhf(nums[3]/100);
  Error, unable to evaluate expression to hardware floats
> temp := nums[3]:
> evalhf(temp/100);
```

$$.3000000000000000$$

Even with `evalhf`, a Maple program is slower than an analogous compiled program. Still, many people prefer Maple for experimenting with calculations and developing prototypes for numerical work, then switch to a compiled language for intense computations. Maple's code translation routines (see page 76) facilitate a transition to C or FORTRAN.

2.5 Programming

Can I call Maple functions from external programs?

Not with the usual version of Maple. However, there is another version available on many platforms for calling Maple routines directly from C language programs. Contact your Maple software distributor for information on this "C-callable" version of Maple.

Can I call external programs from within Maple?

The `system` command and escape-to-host character `!` are available on many platforms for calling an external program from within Maple. If you want to run a program using values generated by Maple as input, you have to write these values first to a file using a function like `writeto` or `write`. Be sure to close the file after writing all the data by using (respectively) `writeto(terminal)` or `close()`. Then use `system` to call the external program with this file as its input. If you need to load the results of an external program into your Maple session, you must again use an intermediate file. The Maple V Release 2 commands `readline` and `readdata` facilitate reading data from files.

Is there a way to evaluate a function repeatedly without writing a cumbersome expression like f(f(f(...f(x)...)))? Can I compose two different functions without writing f(g(x))?

The repeated composition operator, `@@`, denotes a nested evaluation of a single function. This binary operator has the form f@@n, where f is the function you want to iterate, and n is the desired number of iterations. Thus, `ln@@3` evaluated at `x` is `ln(ln(ln(x)))`. You must enclose an iterated function in parentheses when you invoke it:

```
> (cos@@50)(1.);
```

$$.7390851340$$

The composition operator, `@`, composes any two functions. The function f@g evaluated at x computes $f(g(x))$. For example, we define a function `logsec` that computes the natural logarithm of the absolute value of the secant of a number:

```
> logsec := ln@abs@sec;
```

$$logsec := \ln@\text{abs}@\text{sec}$$

```
> logsec(Pi);
```

$$0$$

As with repeated composition, you must enclose a composition of functions in parentheses if you invoke it directly.

```
> (ln@ln)(1000.);
```

$$1.932644734$$

Can I look at Maple's source code?

All of the functions in the Maple library are written in Maple's own programming language. You can inspect the source code of any function in the library. The library contains the vast majority of Maple's mathematical routines, but some basic functions are defined in the kernel. The kernel is a compiled C program, and you cannot look at its source code.

To see the source code of a library function, set the `interface` variable `verboseproc` to 2. Then print the body of the function with `eval` or `print`. In this verbose mode, you also see the body of any library function you read in with `readlib`.

In the following example, we first set `verboseproc`, then read in the definition for `copy`, a routine for creating a copy of an array or table. Maple displays its source code. We then try to display the source code for `evalf` but fail, since `evalf` is defined in the Maple kernel (the `options builtin` identifies it as such). Several additional examples appear in Chapter 1, beginning on page 62.

```
> interface(verboseproc = 2);
> readlib(copy);
  proc(A)
  options `Copyright 1993 by Waterloo Maple Software`;
      if type(A,{array,table}) then
          if type(A,name) then map(proc() args end,eval(A))
          else map(proc() args end,A)
          fi
      else A
      fi
  end

> print(evalf);
  proc() options builtin,remember; 76 end

> interface(verboseproc = 1);
```

The master version of the Maple library contains explanatory comments written by Maple's programmers. These comments have been removed in the standard commercial version. You can purchase a master copy of the source of the Maple library, complete with comments, from your Maple distributor.

What are Maple's scoping rules? Can I declare one function inside of another function, and use all the variables from the outer function in the inner one, as I can in Pascal?

Each variable in a Maple procedure is either a formal parameter, a local variable, or a global variable. Prior to Release 3, the difference between local and global variables is simple: a variable is local if and only if it is declared as a local variable. In Release 3, a variable is local if it is declared local or if it is undeclared and one of the following conditions is satisfied.

- It is the indexing variable of a `for` or `seq` statement.
- It appears on the left-hand side of an assignment statement.

Unlike Pascal, if one procedure is declared inside of another, the inner procedure cannot access the local variables or formal parameters of the outer procedure. We illustrate this with a couple of examples.

In this program, `g` is defined inside the body of `f`, but the `a` inside the body of `g` refers to the global variable `a`, not to the local variable of `f`. Calling `f` returns the global value of `a`.

```
> f := proc()
>      local a, g;
>      a := 1;
>      g := proc() a end;
>      g()
> end:
> a := 2:
> f();
```

$$2$$

Next, we attempt to define a function `addn` of a single variable `n` that returns another function. This returned function should add `n` to its argument. Again, we are stymied when the inner function cannot see the formal parameter of the outer function.

```
> addn := n -> (x -> x + n);
```

$$addn := n \rightarrow x \rightarrow x + n$$

```
> addn(5);
```

$$x \rightarrow x + n$$

Other programming languages, like Pascal, use a more sophisticated scoping rule, nested lexical scoping, which resolves this problem. Maple will use nested lexical scoping in a future release.

With the current version of Maple, we resolve these problems by simulating nested lexical scoping with a creative substitution.

```
> f := proc()
>     local a, g;
>     a := 1;
>     g := subs(aa=a, proc() aa end);
>     g()
> end:
> a := 2:
> f();
```

$$1$$

```
> addn := n -> subs(nn=n, (x -> x + nn)):
> addn(5);
```

$$x \to x + 5$$

```
> "(10);
```

$$15$$

Another possibility is using `unapply` in a procedure body to create a function whenever the procedure is run.

```
> addn := n -> unapply(x+n, x):
> addn(17);
```

$$x \to x + 17$$

How do I create my own package of Maple functions? How can I make my own library?

We demonstrate both of these tasks in this example. A package in Maple is just a table with procedures for its entries. This explains why calling a Maple function in a package resembles the syntax for table lookups (e.g., `combinat[fibonacci](15)`). To make your own package, type all its functions into a file, and name each one as an entry in a table. The name of the table is the name of your package. Below, we have coded several operations on polynomials that are not available in the Maple library. We would like to put them all in a new package called `polyops`. Here is the code for our package, which we type into a file called "polyops".

```
polyops[len] := proc(p:polynom(constant), x:name)
    convert(map(abs, [coeffs(expand(p), x)]), `+`)
end:
```

```
polyops[height] := proc(p:polynom(constant), x:name)
    max(coeffs(collect(p, x, distributed, abs), x))
end:

polyops[qnorm] := proc(p:polynom(constant), x:name)
    sqrt(convert(map(z->z*conjugate(z), [coeffs(expand(p), x)]),`+`))
end:


polyops[boundRoots] := proc(p:polynom(constant), x:name)
    local q, d, s, n, t, b, b1, b2, b3;
    q := collect(p, x, distributed, evalf);
    d := degree(q, x); s := lcoeff(q, x); n := nops(q);
    if d=ldegree(q, x) then RETURN(0) fi;
    if s <> 1. then q := expand(q/s) fi;
    b1 := 1 + polyops[height](q - x^d, x);
    b2 := max(1., polyops[len](q - x^d, x));
    b := min(b1, b2); b3 := 0;
    for t in q-x^d while b3<b do
        b3 := max(b3, (n*abs(lcoeff(t,x)))^(1./(d-degree(t,x))))
    od;
    userinfo(3, polyops, [b1, b2, b3]);
    min(b, b3)
end:

polyops[boundRootsDense] := proc(p:polynom(constant), x:name)
    local q, d, s, k, b, b4, b5;
    q := collect(p, x, distributed, evalf);
    if ldegree(q)>0 then q := expand(q/x^ldegree(q)) fi;
    if type(q, constant) then RETURN(0) fi;
    d := degree(q, x); s := lcoeff(q, x);
    if s <> 1. then q := expand(q/s) fi;
    b := polyops[boundRoots](q, x);
    b4 := convert([abs(coeff(q,x,0)),
                   seq(abs(coeff(q,x,k-1)-coeff(q,x,k)), k=1..d)], `+`);
    if nops(q)<d+1 then b5 := infinity else
        if type(q, polynom(positive, x)) then
            b5 := max(seq(coeff(q,x,k-1)/coeff(q,x,k), k=1..d))
        else
            b5 := max(abs(coeff(q,x,0)/coeff(q,x,1)),
                      seq(2*abs(coeff(q,x,k-1)/coeff(q,x,k)), k=2..d))
        fi
    fi;
    userinfo(3, polyops, [b, b4, b5]);
```

```
    min(b, b4, b5)
end:
```

This package provides a nice example of programming in Maple, so we digress briefly to describe it. If you only need to know about the mechanics of making packages, skip the next few paragraphs to the "end of digression" remark.

Start of digression

Each of these routines requires two arguments: a polynomial in one variable, and the name of the variable. The polynomial may have real or complex coefficients, but all the coefficients must be constants.

The first three functions compute some measures of the complexity of a polynomial. Given the polynomial

$$p(x) = a_0 + a_1 x + \cdots + a_n x^n$$

these functions calculate:

$$\begin{aligned} \text{len}(p, x) &= |a_0| + |a_1| + \cdots + |a_n| \\ \text{height}(p, x) &= \max\{|a_0|, |a_1|, \ldots, |a_n|\} \\ \text{qnorm}(p, x) &= \left(|a_0|^2 + |a_1|^2 + \cdots + |a_n|^2\right)^{1/2} \end{aligned}$$

The `len` function is the *length* of the polynomial. We have named it `len` to avoid conflicting with Maple's `length` command. The `qnorm` function computes the *quadratic norm* of $p(x)$.

The last two functions compute an upper bound for the size of the biggest root of the polynomial. So, for example, if `polyops[boundRoots]` returns 5.7 for a polynomial $p(x)$, then all of the roots of $p(x)$ (complex as well as real) lie inside the circle of radius 5.7 in the complex plane. The function `polyops[boundRoots]` uses three methods to compute a bound, returning the best of the three intermediate results `b1`, `b2`, and `b3`. The first two bounds are based on the length and height of the polynomial. None of the methods attempts to locate any of the roots. Indeed, this function never plugs a single value into the polynomial. This is a great advantage when trying to bound the roots of a polynomial with very large degree. This function works best on *sparse* polynomials – polynomials with only a few nonzero coefficients.

The last function, `polyops[boundRootsDense]`, calls `polyops[boundRoots]`, then calculates two more bounds, `b4` and `b5`. These bounds are only worth computing if the polynomial is *dense*; that is, all or almost all of its coefficients are nonzero. It returns the best of these five bounds.

If you want to see all the bounds these functions compute, enter the command

```
> infolevel[polyops] := 3:
```

before calling them. This activates the `userinfo` statements in these two routines.

Methods for bounding roots, along with many other interesting facts about polynomials, are described in *Mathematics for Computer Algebra* by Maurice Mignotte (see the book list on page 467).

For completeness, we note that Maple has a miscellaneous library function `rootbound` that bounds roots of polynomials, but it is presently undocumented. Inspecting the source code reveals that `rootbound` uses an entirely different algorithm. It simply plugs larger and larger numbers into the polynomial until the leading term is so large that the lower-order terms could not possibly cancel it to make zero.

End of digression

After reading this file into a Maple session, we can call any of these functions, but we must use the long, table-lookup form of the name.

```
> read polyops;
> p := x^10 + x^9 - x^7 - x^6 - x^5 - x^4 - x^3 + x + 1:
> polyops[qnorm](p, x);
```

$$3$$

```
> polyops[boundRootsDense](p, x);
```

$$2.$$

Functions in standard library packages can be invoked with short names (without the table lookup) once they've been loaded into your session with the `with` command. We would like to do the same thing with our `polyops` package. This requires two steps: we must create our own library and put `polyops` in it, and we must modify our Maple environment so that the `with` command searches for packages in this library, too.

First, pick a location for your Maple library. In this example, we use the Unix directory `/usr/myname/mymaplelib`. After making this directory, start Maple in the directory containing your Maple code and enter these commands:

```
> read polyops;
> save `/usr/myname/mymaplelib/polyops.m`;
> quit
```

Naturally, you would substitute the name of your package and the location of your library, but it is essential that the saved file name end with ".m". This saves the file in Maple's internal format. All files in a Maple library must be in this format.

Next, add the following line to the end of your Maple initialization file (see page 66), substituting, of course, for your library directory. (This syntax is new in Maple V Release 2. Under Maple V, use the variable name `_liblist` instead of `libname`, and put the paths in a list.)

```
>  libname := libname, `/usr/myname/mymaplelib`:
```

Now whenever you use `readlib` or `with` in Maple to load a function or a package into your session, Maple will search your library for the function or package automatically if it does not find it in the standard library.

The `libname` global variable can list as many library directories as you like. The libraries are always searched in the order that they appear in this sequence. For example, `libname` may have the form

bug-fix-library, standard-library, group-project-library, personal-library

In a new Maple session, we load our package using `with` and call its functions using the short names.

```
> libname;
```

/usr/local/mapleV.2/lib, /usr/myname/mymaplelib

```
> with(polyops);
```

[*boundRoots, boundRootsDense, height, len, qnorm*]

```
> p := x^10 + x^9 - x^7 - x^6 - x^5 - x^4 - x^3 + x + 1:
> qnorm(p, x);
```

3

```
> boundRootsDense(p, x);
```

2.

There is one disadvantage to this method: all of the functions in a package are loaded at once into your session. This is not a problem with the short example presented here, but loading a function only when it is required is important for larger packages. We illustrate this technique with the `polyops` package.

First, make a directory called `polyops` in your personal library directory. Next, change the names of its functions to `` `polyops/``*fcnname*`` ` ``. Save each function in its own ".m" file in the `polyops` directory. The simplest way to do this is to add `save` statements to your original code. For example, our modified `polyops` package now begins:

```
`polyops/len` := proc(p:polynom(constant), x:name)
    convert(map(abs, [coeffs(expand(p), x)]), `+`)
end:
save `polyops/len`, `/usr/myname/mymaplelib/polyops/len.m`;
```

Now read this file into a Maple session to make all the ".m" files. Finally, make the master file for `polyops` that reads in the members of the package as they are needed. It has the following form:

```
polyops[len] := 'readlib(`polyops/len`)':
polyops[height] := 'readlib(`polyops/height`)':
polyops[qnorm] := 'readlib(`polyops/qnorm`)':
polyops[boundRoots] := 'readlib(`polyops/boundRoots`)':
polyops[boundRootsDense] := 'readlib(`polyops/boundRootsDense`)':
```

(Under Maple V, `readlib` searches only the standard library, so each of these statements requires a second argument, the full path of the file where the function is defined.) Use Maple again to save this as "polyops.m" and add it to your library (at the top level, not in the `polyops` directory). It replaces the old `polyops.m` file. Now when we issue the command `with(polyops)`, none of the functions is actually loaded, but each of their names is assigned a `readlib` command that loads the function automatically the first time it is called. This does not slow down later invocations of the same function, since `readlib` has no effect when the function has already been loaded (`readlib` has `option remember`).

We include one final example using this new scheme. The changes we've made do not affect the way you load the package or call its functions. To make things interesting, we use a much larger polynomial.

```
> with(polyops);
```

$$[boundRoots, boundRootsDense, height, len, qnorm]$$

```
> p := x^10000 + randpoly(x, degree=9999, terms=30,
>        coeffs=rand(-10^6..10^6)) + rand(-10^6..10^6)():
> height(p, x);
```

$$974460$$

```
> boundRoots(p, x);
```

$$1.205553735$$

How do I make help text for my own Maple functions and packages?

To make a help page for a function or a package, first write a plain text version of the help page in an editor, mimicking the style of the Maple help pages. The simplest way to convert this text file into a Maple help page is with the `makehelp` command. This miscellaneous library function installs your help text in the "help" subdirectory of a Maple library. Your help page is then available even if you haven't loaded the function or package.

However, many people prefer to store the help text for a function or a package in the same place as the code. This makes it easier to transport the code and help text to other computers, or to share it with other users. We describe here a method to save your help text in this manner. First, edit your help file to make a `TEXT` object out of the help text, enclosing each line in back quotes and placing a comma between lines. If the help page is an introduction to a package, or help for a function that is not part of a package, assign the `TEXT` object to the variable `` `help/text/name` ``, where *name* is the name of the function or package. For example, we create a help page for the `polyops` package developed in the previous section:

```
`help/text/polyops` := TEXT(
`HELP FOR: Introduction to the polyops package`,
``,
```

```
`CALLING SEQUENCE:`,
`  <function>(args)`,
`  polyops[<function>](args)`,
``,
`SYNOPSIS:`,
` - The polyops package contains functions for doing special`,
`   operations on univariate polynomials with real or complex`,
`   coefficients.`,
``,
` - The functions available are:`,
``,
`         boundRoots  boundRootsDense  height  len  qnorm`,
``,
` - For help with a particular function use ?polyops[<function>]`,
``,
` - For example, to compute the height of a polynomial p in the`,
`   variable x, use`,
`         with(polyops); height(p, x);`,
``,
`SEE ALSO: with, polyops[<function>]`):
```

To make a help page for a function *name* defined in a package *pkg*, assign the TEXT object to the variable `` `help/pkg/text/name` ``. For example, here is a help page for the height function in the polyops package:

```
`help/polyops/text/height` := TEXT(
`HELP FOR: polyops[height] - compute the height of a polynomial`,
``,
`CALLING SEQUENCE:`,
`  height(p, x)`,
``,
`PARAMETERS:`,
`  p - a polynomial in one variable with constant coefficients`,
`  x - the indeterminate of the polynomial`,
``,
`SYNOPSIS:`,
` - The function height computes the height of a polynomial in one`,
`   variable.`,
``,
` - The height of a polynomial of one variable is the largest coef-`,
`   ficient in absolute value of the polynomial.`,
``,
` - This function is part of the polyops package and must be defined`,
`   using with(polyops) before using.`,
``,
`EXAMPLES:`,
```

```
`> with(polyops):`,
`> height(x^3 + 2*x + 1, x);`,
`                                    2`,
``,
`> height(5.3*y^14 + (4+3*I)*y^7 + 16/3, y);`,
`                                  16/3`,
``,
`SEE ALSO:  polyops`):
```

Add the help text definitions to the file containing your source code. In our own example from the last question, these definitions would go in the polyops file, and would get translated into `polyops.m`. The help text is available only after you load the function or package.

```
> with(polyops):
> ?polyops
HELP FOR: Introduction to the polyops package

CALLING SEQUENCE:
  <function>(args)
  polyops[<function>](args)

SYNOPSIS:
 - The polyops package contains functions for doing special
   operations on univariate polynomials with real or complex
   coefficients.

 - The functions available are:

          boundRoots  boundRootsDense  height  len  qnorm

 - For help with a particular function use ?polyops[<function>]

 - For example, to compute the height of a polynomial p in the
   variable x, use
          with(polyops); height(p, x);

SEE ALSO: with, polyops[<function>]
```

Chapter 7
Mathematica to Maple

Maple and Mathematica are the two most popular computer algebra systems on the market today. This chapter compares and contrasts the two systems. The first section discusses some of the main similarities of the two systems as well as some important differences. The second section lists common Mathematica functions, options, and symbols together with their Maple analogues. It is written to help people who have experience with Mathematica to learn and use Maple.

1 COMPARISON

Maple and Mathematica have similar capabilities. They both do symbolic manipulations, numeric computations to arbitrarily high precision, and two- and three-dimensional graphics. Both are programmable. Both run on a wide variety of computers, including many personal computers and workstations, as well as some mainframes and supercomputers. They differ, however, in the way they are designed.

Size

Maple has a small kernel and a modular design that loads functions into memory only as they are needed. It requires two megabytes of memory to run effectively. Mathematica, on the other hand, has a large kernel and requires eight megabytes of memory to run reasonably. Because of its size, it can take up to a minute to load the Mathematica kernel on some systems, such as the Macintosh.

Source Code and On-line Help

Most of Maple is written in Maple's own programming language. You can inspect and even modify this code. On the other hand, most of Mathematica is written in C. You cannot examine the code, nor can you modify it.

Both Maple and Mathematica provide on-line help for finding and using the functions you need. On-line help in Maple is more detailed than the help in Mathematica. Maple help usually includes examples and cross-references, unlike Mathematica. Both Maple and Mathematica include an interactive help topic browser that assists you in finding information about a command even if you don't know its name.

Programmability

Maple and Mathematica are both programmable. Programming in Maple is similar to programming in Pascal. Mathematica supports several styles of programming: procedural, functional, rule-based, and object-oriented.

Naming Conventions

Maple is less consistent in its naming conventions than Mathematica. Every built-in function name in Mathematica begins with a capital letter. In Maple, some names begin with a capital letter, some begin with a lowercase letter, and some names are composed entirely of uppercase letters.

Maple tends to use abbreviations. Mathematica rarely uses abbreviations.

Features

Below are several important features of Maple and Mathematica.

Maple V Release 2 and Release 3

- Worksheet interface (as described in Chapter 2)
- Output in mathematical notation (using symbols such as $\sqrt{\ }$, $\int$, and $\sum$)
- Small kernel and modular design (small memory requirements, starts up rapidly on all systems)
- Open system (source code for most functions in Maple can be viewed and modified)
- On-line documentation with examples

Mathematica 2.2

- Notebook front end
- Pattern matching
- Rule-based programming
- Sound generation

The Notebook front end is similar to the Maple Worksheet interface. This front end, which is found on Macintosh, NeXT, MS-Windows, and X-Windows versions of Mathematica, lets you write Notebooks or files that contain text, Mathematica input, output, and graphics.

The pattern matcher enables you to assign names to parts of expressions and then specify what to do with the names. For example, if you had a list of pairs of numbers where the first number was the year and the second number was the U.S. national debt at that year, then by using pattern matching you can assign the first value to have the name `year` and the second values to have the name `debt`. The rule `{year_, debt_} :> {year, Log[debt]}` instructs Mathematica to return the logarithm of the second value in the pair.

```
In[1]:= {{1950, 256097}, {1970, 370094}, {1990, 3233313}} /.
{year_, debt_} :> {year, Log[debt]}

Out[1]=  {{1950, Log[256097]}, {1970, Log[370094]}, {1990, Log[3233313]}}
```

Version 2.0 and later releases of Mathematica support sound. Just as you can plot functions, you can "play" functions.

Both Maple and Mathematica have been significantly enhanced since they were first introduced. Some of the improvements are in response to users' requests. Other improvements were made because both Waterloo Maple Software and Wolfram Research, Inc., are interested in making their software the best on the market. We expect both companies to continue to strive to produce software that is better than their competitors, so the list of features is likely to grow.

2 CORRESPONDING FUNCTIONS

The following pages list hundreds of Mathematica functions alphabetically with their equivalent or closely related Maple functions. Mathematica functions defined in packages are included in the list. The package name of a function appears just below the name of the function; for example, the `AddEdge` routine in Mathematica is part of the `DiscreteMath'Combinatorica'` package. The notation *package*[*function*] in the Maple column specifies that *function* is part of the Maple package named *package*. If there is more than one Maple function corresponding to a function in Mathematica, they are listed on successive lines; for example, the Mathematica function `Eigenvalues` corresponds to the Maple functions `Eigenvals` and `linalg[eigenvals]`.

Mathematica	Maple
A	
Abort[ ]	ERROR(*message*)
Abs[z]	abs(z)
AddEdge[g, $\{v_1, v_2\}$, Directed] DiscreteMath'Combinatorica'	networks[addedge]([v_1, v_2], g)
AddEdge[g, $\{v_1, v_2\}$] DiscreteMath'Combinatorica'	networks[addedge]($\{v_1, v_2\}$, g)
AddVertex[g] DiscreteMath'Combinatorica'	networks[addvertex](v, g)
AiryAi[x]	Ai(x)
AiryBi[x]	Bi(x)
AllPairsShortestPath[g] DiscreteMath'Combinatorica'	networks[allpairs](g)
AmbientLight->*color*	ambientlight=[r, g, b]
And[$expr_1$, $expr_2$]	$expr_1$ and $expr_2$
Apart[*expr*, x]	convert(*expr*, parfrac, x)
Append[*list*, *elem*]	[op(*list*), *elem*]
AppendColumns[$A_1, A_2, \ldots$] LinearAlgebra'MatrixManipulation'	linalg[stack]($A_1, A_2, \ldots$)
AppendRows[$A_1, A_2, \ldots$] LinearAlgebra'MatrixManipulation'	linalg[augment]($A_1, A_2, \ldots$) linalg[concat]($A_1, A_2, \ldots$)
Apply[Plus, *expr*]	convert(*expr*, '+')
Apply[Times, *expr*]	convert(*expr*, '*')
ArcCos[z]	arccos(z)
ArcCosh[z]	arccosh(z)
ArcCot[z]	arccot(z)
ArcCoth[z]	arccoth(z)

Mathematica	Maple
ArcCsc[*z*]	arccsc(*z*)
ArcCsch[*z*]	arccsch(*z*)
ArcSec[*z*]	arcsec(*z*)
ArcSech[*z*]	arcsech(*z*)
ArcSin[*z*]	arcsin(*z*)
ArcSinh[*z*]	arcsinh(*z*)
ArcTan[*x*, *y*]	arctan(*y*, *x*)
ArcTan[*z*]	arctan(*z*)
ArcTanh[*z*]	arctanh(*z*)
Array[*f*, {*m*, *n*}]	linalg[matrix](*m*, *n*, *f*)
Array[*f*, *dims*]	array(*indexfcn*, *bounds*, *list*)
Array[*f*, *n*]	[*f*(*i*) $ *i*=1..*n*] linalg[vector](*n*, *f*) [seq(*f*(*i*), *i*=1..*n*)]
AspectRatio->Automatic	scaling=CONSTRAINED
Axes->False	axes=NONE
Axes->True	axes=NORMAL
AxesLabel->{$label_x$, $label_y$, $label_z$}	labels=[$label_x$, $label_y$, $label_z$]
AxesLabel->{$label_x$, $label_y$}	labels=[$label_x$, $label_y$]

B

Mathematica	Maple
BaseForm[*number*, 16]	convert(*number*, hex)
BaseForm[*number*, 2]	convert(*number*, binary)
BaseForm[*number*, 8]	convert(*number*, octal)
BaseForm[*number*, *n*]	convert(*number*, base, *n*)
BernoulliB[*n*, *x*]	bernoulli(*n*, *x*) numtheory[B](*n*, *x*)
BernoulliB[*n*]	bernoulli(*n*) numtheory[B](*n*)
BesselI[*n*, *z*]	BesselI(*n*, *z*)
BesselJ[*n*, *z*]	BesselJ(*n*, *z*)
BesselK[*n*, *z*]	BesselK(*n*, *z*)
BesselY[*n*, *z*]	BesselY(*n*, *z*)
Beta[z_1, z_2]	Beta(z_1, z_2)
BetaDistribution[α, β] Statistics'ContinuousDistributions'	beta[α, β] stats[random], stats[statevalf]
Binomial[*n*, *m*]	binomial(*n*, *m*)
BinomialDistribution[*n*, *p*] Statistics'DiscreteDistributions'	binomiald[*n*, *p*] stats[random], stats[statevalf]
Boxed->True	axes=BOXED
Break[ ]	break

C

Mathematica	Maple
Cancel[*expr*]	normal(*expr*)
CartesianMap[*e*, {*a*, *b*}, {*c*, *d*}] Graphics'ComplexMap'	plots[conformal](*e*, *z*=*a*+*c**I..*b*+*d**I)

CartesianProduct[$list_1$, $list_2$] DiscreteMath‘Combinatorica‘	combinat[cartprod]([$list_1$, $list_2$])
Catalan	Catalan
CauchyDistribution[*a*, *b*] Statistics‘ContinuousDistributions‘	cauchy[*a*, *b*] stats[random], stats[statevalf]
CDF[*distribution*, *x*] Statistics‘ContinuousDistributions‘	stats[statevalf, cdf[*distribution*]](*x*)
CDF[*distribution*, *x*] Statistics‘DiscreteDistributions‘	stats[statevalf, dcdf[*distribution*]](*x*)
Ceiling[*x*]	ceil(*x*)
CentralMoment[*list*, *n*] Statistics‘DescriptiveStatistics‘	stats[describe, moment[*n*, mean]](*list*)
CForm[*expr*]	C(*expr*)
ChebyshevT[*n*, *x*]	orthopoly[T](*n*, *x*)
ChebyshevU[*n*, *x*]	orthopoly[U](*n*, *x*)
Check[*expr*, *failexpr*]	traperror(*expr*)
CheckAbort[*expr*, *failexpr*]	traperror(*expr*)
ChineseRemainderTheorem[$list_1$, $list_2$] NumberTheory‘NumberTheoryFunctions‘	chrem($list_1$, $list_2$) numtheory[mcombine](*a*, *ra*, *b*, *rb*)
ChiSquareDistribution[*n*] Statistics‘ContinuousDistributions‘	chisquare[*n*] stats[random], stats[statevalf]
Chop[*expr*]	fnormal(*expr*)
ChromaticPolynomial[*g*, *z*] DiscreteMath‘Combinatorica‘	networks[chrompoly](*g*, *z*)
Clear[*name*]	unassign('*name*') *name* := '*name*'
Close[*stream*]	close( )
Coefficient[*poly*, *expr*]	coeff(*poly*, *expr*)
Coefficient[*poly*, *var*, *n*]	coeff(*poly*, *x*, *n*)
Coefficient[*expr*, *form*]	degree(*a*, *x*) lcoeff(*p*, *x*)
CoefficientList[*poly*, $\{x_1, x_2, \ldots\}$]	coeffs(*poly*, $\{x_1, x_2, \ldots\}$)
CoefficientList[*poly*, *x*]	coeffs(*poly*, *x*)
Collect[*expr*, *x*]	collect(*expr*, *x*)
Complement[$list_1$, $list_2$]	set_1 minus set_2
Complex	complex
ComplexExpand[*expr*]	evalc(*expr*)
ComplexToTrig[*expr*] Algebra‘Trigonometry‘	convert(*expr*, trig)
ComposeSeries[*a*, *b*]	powseries[compose](*a*, *b*)
Composition[*f*, *g*]	*f* @ *g*
Compositions[*n*, *k*] DiscreteMath‘Combinatorica‘	combinat[composition](*n*, *k*)
Conjugate[*z*]	conjugate(*z*)
ConnectedComponents[*g*] DiscreteMath‘Combinatorica‘	networks[components](*g*)

Mathematica	Maple
ConstrainedMax[*expr*, *ineqs*, *vars*]	simplex[maximize](*expr* , *ineqs*, NONNEGATIVE)
ConstrainedMin[*expr*, *ineqs*, *vars*]	simplex[minimize](*expr* , *ineqs*, NONNEGATIVE)
ContextToFilename["*context*`"]	convert(*filename*, hostfile)
ContinuedFraction[*x*, *n*]	convert(*x*, confrac, *n*)
NumberTheory ' ContinuedFractions '	numtheory[cfrac](*x*, *n*, quotients)
ContourPlot[. . . , Contours->*list*]	plots[contourplot](. . . , contours=*list*)
ContourPlot[. . . , Contours->*n*]	plots[contourplot](. . . , contours=*n*)
ContourPlot[. . . , ContourStyle->Dashing[*list*]]	plots[contourplot](. . . , linestyle=*n*)
ContourPlot[. . . , ContourStyle->Thickness[*r*]]	plots[contourplot](. . . , thickness=*n*)
ContourPlot[*expr*, {*x*, *a*, *b*}, {*y*, *c*, *d*}]	plot3d(*expr*, *x*=*a*..*b*, *y*=*c*..*d*, style=CONTOUR, orientation=[0, 0])
	plots[contourplot](*expr*, *x*=*a*..*b*, *y*=*c*..*d*)
Contours->*list*	contours=*list*
Contours->*n*	contours=*n*
ContourStyle->Dashing[*list*]	linestyle=*n*
ContourStyle->Thickness[*r*]	thickness=*n*
Contract[*g*, {*x*, *y*}] DiscreteMath ' Combinatorica '	networks[contract]({*x*, *y*}, *g*)
Convert[*old*, *new*] Miscellaneous ' Units '	convert(*old*, degrees)
	convert(*old*, metric)
	convert(*old*, radians)
ConvexHull[{{x_1, y_1}, {x_2, y_2}, . . .}] DiscreteMath ' ComputationalGeometry '	geometry[convexhull](*points*)
Cos[*z*]	cos(*z*)
Cosh[*z*]	cosh(*z*)
CosIntegral[*z*]	Ci(*z*)
Cot[*z*]	cot(*z*)
Coth[*z*]	coth(*z*)
Count[*expr*, *subexpr*, Infinity]	numboccur(*expr*, *subexpr*)
CrossProduct[*u*, *v*] Calculus ' VectorAnalysis '	linalg[crossprod](*u*, *v*)
Csc[*z*]	csc(*z*)
Csch[*z*]	csch(*z*)
Curl[*f*, Cylindrical[*vars*]] Calculus ' VectorAnalysis '	linalg[curl](*f*, *vars*, coords=cylindrical)
Curl[*f*, Spherical[*vars*]] Calculus ' VectorAnalysis '	linalg[curl](*f*, *vars*, coords=spherical)
Curl[*f*] Calculus ' VectorAnalysis '	linalg[curl](*f*, *vars*)
Cycle[*n*] DiscreteMath ' Combinatorica '	networks[cycle](*n*)
Cyclotomic[*n*, *x*]	numtheory[cyclotomic](*n*, *x*)
CylindricalPlot3D[*expr*, rng_r, rng_ϕ] Graphics ' ParametricPlot3D '	plot3d(*expr*, rng_θ, rng_z, coords=cylindrical)
	plots[cylinderplot](*expr*, rng_θ, rng_z)

D

D[*expr*, x_1, x_2, ...]	diff(*expr*, x_1, x_2, ...)
D[*expr*, *x*]	diff(*expr*, *x*)
DeleteVertex[*g*, *v*] DiscreteMath'Combinatorica'	networks[delete](*v*, *g*)
Delta[*t*] Calculus'Common'Support'	Dirac(*t*)
Denominator[*expr*]	denom(*expr*)
DensityPlot[*expr*, r_x, r_y]	plots[densityplot](*expr*, r_x, r_y)
Det[*A*, Modulus->*p*]	Det(*A*) mod *p*
Det[*A*]	linalg[det](*A*)
DiagonalMatrix[{b_1, b_2, ...}]	linalg[BlockDiagonal](b_1, b_2, ...) linalg[diag](b_1, b_2, ...)
Dijkstra[*g*, *v*] DiscreteMath'Combinatorica'	networks[shortpathtree](*g*, *v*)
Dimensions[*A*]	linalg[coldim](*A*) linalg[rowdim](*A*)
DiscreteUniformDistribution[*a*, *b*] Statistics'DiscreteDistributions'	discreteuniform[*a*, *b*] stats[random], stats[statevalf]
Display[*channel*, *graphics*]	interface(plotdevice=postscript)
Div[*f*, Cylindrical[*vars*]] Calculus'VectorAnalysis'	linalg[diverge](*f*, *vars*, coords=cylindrical)
Div[*f*, Spherical[*vars*]] Calculus'VectorAnalysis'	linalg[diverge](*f*, *vars*, coords=spherical)
Div[*f*] Calculus'VectorAnalysis'	linalg[diverge](*f*, *vars*)
Divisors[*n*, GaussianIntegers->True]	GaussInt[GIdivisor](*n*)
Divisors[*n*]	numtheory[divisors](*n*)
DivisorSigma[*k*, *n*]	numtheory[sigma][*k*](*n*)
DivisorSigma[*n*]	numtheory[sigma](*n*)
Do[*expr*, {*i*, i_0, i_1, *d*}]	for *i* from i_0 to i_1 by *d* do *body* od
Do[*expr*, {*i*, i_0, i_1}]	for *i* from i_0 to i_1 do *body* od
Do[*expr*, {*i*, i_1}]	for *i* to i_1 do *body* od
Dot[*A*, *B*]	linalg[multiply](*A*, *B*)
Dot[*u*, *v*]	linalg[dotprod](*u*, *v*)
Drop[*list*, *n*]	[op(*n*+1..nops(*list*), *list*)] [*list*[*n*+1..nops(*list*)]]
Drop[*list*, −*n*]	[op(1..nops(*list*)−*n*, *list*)] [*list*[1..nops(*list*)−*n*]]
DSolve[*eqn*, *y*[*x*], *x*]	dsolve(*eqn*, *y*(*x*))
Dt[*f*]	D(*f*)

E

E	E
Edges[*g*]	networks[adjacency](*g*)

Mathematica	Maple
DiscreteMath'Combinatorica'	
Edit[*expr*]	edit(*expr*)
Eigenvalues[*A*]	Eigenvals(*A*) linalg[eigenvals](*A*)
Eigenvectors[*A*]	Eigenvals(*A*, *vecs*) linalg[eigenvects](*A*)
EllipticE[ϕ, *m*]	LegendreE(sin(ϕ), sqrt(*m*))
EllipticE[*m*]	LegendreEc(sqrt(*m*))
EllipticF[ϕ, *m*]	LegendreF(sin(ϕ), sqrt(*m*))
EllipticK[*m*]	LegendreKc(sqrt(*m*))
EllipticPi[*n*, ϕ, *m*]	LegendrePi(sin(ϕ), *n*, sqrt(*m*))
EllipticPi[*n*, *m*]	LegendrePic(*n*, sqrt(*m*))
Erf[*x*]	erf(*x*)
Erfc[*x*]	erfc(*x*)
EulerE[*n*, *x*]	euler(*n*, *x*)
EulerE[*n*]	euler(*n*)
EulerGamma	gamma
EulerPhi[*n*]	numtheory[phi](*n*)
Evaluate[*expr*]	eval(*expr*)
EvenQ[*n*]	type(*n*, even)
Exit[ ]	done quit stop
Exp[*z*]	exp(*z*)
Expand[*expr*, Trig–>True]	combine(*expr*, trig)
Expand[*expr*]	expand(*expr*)
ExpandDenominator[*expr*]	normal(*expr*, expanded)
ExpandNumerator[*expr*]	normal(*expr*, expanded)
ExpIntegralEi[*x*]	Ei(*x*)
ExponentialDistribution[λ]	exponential[λ, 0]
Statistics'ContinuousDistributions'	stats[random], stats[statevalf]
ExtendedGCD[*n*, *m*]	igcdex(*n*, *m*, '*s*', '*t*')

F

Mathematica	Maple
Factor[*poly*, GaussianIntegers–>True]	GaussInt[GIfacpoly](*poly*)
Factor[*poly*, Modulus–>*p*]	Factor(*poly*) mod *p*
Factor[*poly*]	factor(*poly*)
Factor[*expr*, Trig–>True]	expand(*expr*)
Factorial[*n*]	factorial(*n*)
FactorInteger[*n*, GaussianIntegers–>True]	GaussInt[GIfactor](*n*) GaussInt[GIfactors](*n*)
FactorInteger[*n*]	ifactor(*n*) ifactors(*n*)
FactorIntegerECM[*n*]	ifactor(*n*, lenstra)
NumberTheory'FactorIntegerECM'	
FactorList[*poly*]	factors(*poly*)

FactorSquareFree[*poly*]	convert(*poly*, sqrfree, *x*)
FactorSquareFreeList[*poly*, Modulus->*p*]	Sqrfree(*poly*) mod *p*
FactorSquareFreeList[*poly*]	sqrfree(*poly*)
False	false
Fibonacci[*n*]	combinat[fibonacci](*n*)
DiscreteMath'CombinatorialFunctions'	
FindMinimum[*expr*, {*x*, x_0}]	extrema(*expr*, { }, *x*)
	minimize(*expr*, {*x*})
First[*list*]	op(1, *list*)
	list[1]
Fit[*data*, *fcns*, *vars*]	linalg[leastsqrs](*A*, *b*)
	stats[fit, leastsquare[*vars*, *eqn*, *cfs*]](*data*)
Floor[*x*]	floor(*x*)
FontForm[*str*, {*font*, *n*}]	axesfont=[*family*, *style*, *n*]
	font=[*family*, *style*, *n*]
	labelfont=[*family*, *style*, *n*]
	titlefont=[*family*, *style*, *n*]
FortranForm[*expr*]	fortran(*expr*)
Fourier[{z_1, z_2, ...}]	FFT(*m*, *x*, *y*)
FourierTransform[*expr*, *t*, *w*]	fourier(*expr*, *t*, *w*)
Calculus'FourierTransform'	
Frame->True	axes=FRAME
FRatioDistribution[n_1, n_2]	fratio[n_1, n_2]
Statistics'ContinuousDistributions'	stats[random], stats[statevalf]
Frequencies[*list*]	stats[transform, tally](*list*)
Statistics'DescriptiveStatistics'	
FresnelC[*x*]	FresnelC(*x*)
FresnelS[*x*]	FresnelS(*x*)
Function[*body*]	proc(*args*) *body* end

G

Gamma[*a*, *z*]	GAMMA(*a*, *z*)
Gamma[*z*]	GAMMA(*z*)
GammaDistribution[α, β]	gamma[α, β]
Statistics'ContinuousDistributions'	stats[random], stats[statevalf]
GCD[n_1, n_2, ...]	GaussInt[GIgcd](n_1, n_2, ...)
	igcd(n_1, n_2, ...)
	numtheory[GIgcd](n_1, n_2, ...)
GegenbauerC[*n*, *m*, *x*]	orthopoly[G](*n*, *m*, *x*)
GeometricMean[*list*]	stats[describe, geometricmean](*list*)
Statistics'DescriptiveStatistics'	
Get["*filename*"]	read '*filename*'
Grad[*expr*, Cylindrical[*vars*]]	linalg[grad](*expr*, *vars*, coords=cylindrical)
Calculus'VectorAnalysis'	
Grad[*expr*, Spherical[*vars*]]	linalg[grad](*expr*, *vars*, coords=spherical)
Calculus'VectorAnalysis'	

Mathematica	Maple
Grad[*expr*]	linalg[grad](*expr*, *vars*)
Calculus'VectorAnalysis'	
GramSchmidt[{v_1, v_2, ...}]	linalg[GramSchmidt]([v_1, v_2, ...])
LinearAlgebra'Orthogonalization'	
Graph[*adjmat*, *vlist*]	networks[graph](*vset*, *eset*)
DiscreteMath'Combinatorica'	
GraphUnion[*g*, *h*]	networks[gunion](*g*, *h*)
DiscreteMath'Combinatorica'	
GrayCode[*list*]	combinat[graycode](*n*)
DiscreteMath'Combinatorica'	
GroebnerBasis[{$poly_1$, ...}, {x_1, ...}]	grobner[gbasis]([$poly_1$, ...], [x_1, ...])

H

Mathematica	Maple
HarmonicMean[*list*]	stats[describe, harmonicmean](*list*)
Statistics'DescriptiveStatistics'	
Head[*expr*]	op(0, *expr*)
	whattype(*expr*)
HermiteH[*n*, *x*]	orthopoly[H](*n*, *x*)
HiddenSurface–>False	style=WIREFRAME
HiddenSurface–>True	style=HIDDEN
HilbertMatrix[*n*, *n*]	linalg[hilbert](*n*)
LinearAlgebra'MatrixManipulation'	
Hold[*expr*]	freeze(*expr*)
	'*expr*'
Hypergeometric0F1[*a*, *z*]	hypergeom([ ], [*a*], *z*)
Hypergeometric1F1[*a*, *b*, *z*]	hypergeom([*a*], [*b*], *z*)
Hypergeometric2F1[*a*, *b*, *c*, *z*]	hypergeom([*a*, *b*], [*c*], *z*)
HypergeometricDistribution[*n*, n_s, n_t]	hypergeometric[n_s, n_f, *n*]
Statistics'DiscreteDistributions'	stats[random], stats[statevalf]
HypergeometricF[{n_1, ...}, {d_1, ...}, *z*]	hypergeom([n_1, ...], [d_1, ...], *z*)
DiscreteMath'RSolve'	
HypergeometricPFQ[{n_1, ...}, {d_1, ...}, *z*]	hypergeom([n_1, ...], [d_1, ...], *z*)

I

Mathematica	Maple
I	I
IdentityMatrix[*n*]	array(identity, 1..*n*, 1..*n*)
If[*condition*, $expr_t$, $expr_f$]	if *condition* then $expr_t$ else $expr_f$ fi
Im[*z*]	Im(*z*)
ImplicitPlot[*eqn*, {*x*, *a*, *b*}]	plots[implicitplot](*eqn*, *x*=*a*..*b*)
Graphics'ImplicitPlot'	
IncidenceMatrix[*g*]	networks[incidence](*g*)
DiscreteMath'Combinatorica'	
Infinity	infinity
Input["*prompt*"]	readstat('*prompt*')
InputForm[*expr*]	lprint(*expr*)
Insert[*l*, *e*, *n*]	[op(1..*n*−1, *l*), *e*, op(*n*..nops(*l*), *l*)]

Integer	integer
IntegerQ[*n*]	type(*n*, integer)
Integrate[*expr*, {*x*, *a*, *b*}]	int(*expr*, *x*=*a*..*b*)
Integrate[*expr*, *x*]	int(*expr*, *x*)
InterpolatingPolynomial[{{x_1, y_1}, . . .}, *x*]	interp([x_1, . . .], [y_1, . . .], *x*)
Intersection[$list_1$, $list_2$]	set_1 intersect set_2
InverseFourier[{z_1, z_2, . . .}]	iFFT(*m*, *x*, *y*)
InverseFourierTransform[*expr*, *t*, *w*] Calculus'FourierTransform'	invfourier(*expr*, *t*, *w*)
InverseFunction[*f*]	invfunc[*f*]
InverseLaplaceTransform[*expr*, *s*, *t*] Calculus'LaplaceTransform'	invlaplace(*expr*, *s*, *t*)

J

JacobiSymbol[*n*, *m*]	numtheory[J](*n*, *m*) numtheory[jacobi](*n*, *m*)
Join[$list_1$, $list_2$, . . .]	[op($list_1$), op($list_2$), . . .]
JordanDecomposition[*A*]	linalg[jordan](*A*)

K

KSubsets[*l*, *k*] DiscreteMath'Combinatorica'	combinat[choose](*l*, *k*)
Kurtosis[*list*] Statistics'DescriptiveStatistics'	stats[describe, kurtosis](*list*)

L

LaguerreL[*n*, *a*, *x*]	orthopoly[L](*n*, *a*, *x*)
LaguerreL[*n*, *x*]	orthopoly[L](*n*, *x*)
LaplaceDistribution[μ, β] Statistics'ContinuousDistributions'	laplaced[μ, β] stats[random], stats[statevalf]
LaplaceTransform[*expr*, *t*, *s*] Calculus'LaplaceTransform'	laplace(*expr*, *t*, *s*)
Laplacian[*expr*] Calculus'VectorAnalysis'	linalg[laplacian](*expr*, *vars*)
Laplacian[*ex*, Cylindrical[*vrs*]] Calculus'VectorAnalysis'	linalg[laplacian](*ex*, *vrs*, coords=cylindrical)
Laplacian[*ex*, Spherical[*vrs*]] Calculus'VectorAnalysis'	linalg[laplacian](*ex*, *vrs*, coords=spherical)
Last[*list*]	op(nops(*list*), *list*) *list*[nops(*list*)]
LatticeReduce[{v_1, v_2, . . .}]	lattice([v_1, v_2, . . .])
LCM[n_1, n_2, . . .]	ilcm(n_1, n_2, . . .)
LegendreP[*n*, *x*]	orthopoly[P](*n*, *x*)
Length[*expr*]	linalg[vectdim](*v*) nops(*expr*)
LightSources->{{pos_1, c_1}, . . .}	light=[ϕ, θ, *r*, *g*, *b*]

Limit[*expr*, *x* –> x_0, Direction–>1]	limit(*expr*, *x*=x_0, left)
Limit[*expr*, *x* –> x_0, Direction–>-1]	limit(*expr*, *x*=x_0, right)
Limit[*expr*, *x* –> x_0]	limit(*expr*, *x*=x_0)
LinearProgramming[*c*, *m*, *b*]	simplex[minimize](*expr* , *ineqs*, NONNEGATIVE)
LinearSolve[*A*, *b*]	linalg[linsolve](*A*, *b*)
ListPlot3D[*A*]	plots[matrixplot](*A*)
	plots[surfdata](*list*)
ListPlot[{{x_1, y_1}, . . .}, PlotJoined–>True]	plot([x_1, y_1, . . .])
ListPlot[{{x_1, y_1}, {x_2, y_2}, . . .}]	plot([x_1, y_1, x_2, y_2, . . .], style=POINT)
ListQ[*expr*]	type(*expr*, list)
Log[10, *z*]	log10(*z*)
Log[*b*, *z*]	log[b](*z*)
Log[*z*]	ln(*z*)
	log(*z*)
LogGamma[*z*]	lnGAMMA(*z*)
LogicalExpand[*expr*]	evalb(*x*)
LogIntegral[*x*]	Li(*x*)
LogisticDistribution[μ, β] Statistics'ContinuousDistributions'	logistic[μ, β] stats[random], stats[statevalf]
LogLogPlot[*expr*, {*x*, *a*, *b*}] Graphics'Graphics'	plots[loglogplot](*expr*, *x*=*a*..*b*)
LogNormalDistribution[μ, σ] Statistics'ContinuousDistributions'	lognormal[μ, σ] stats[random], stats[statevalf]
LogPlot[*expr*, {*x*, *a*, *b*}] Graphics'Graphics'	plots[logplot](*expr*, *x*=*a*..*b*)

M

MakeSimple[*g*] DiscreteMath'Combinatorica'	networks[gsimp](*g*)
Map[*f*, *expr*]	map(*f*, *expr*)
MatchQ[*expr*, *pattern*]	match(*expr*=*pattern*, *v*, '*s*')
MatrixExp[*A*]	linalg[exponential](*A*)
MatrixQ[*expr*]	type(*expr*, listlist)
	type(*expr*, matrix)
Max[x_1, x_2, . . .]	max(x_1, x_2, . . .)
Mean[*list*] Statistics'DescriptiveStatistics'	stats[describe, mean](*list*)
MeanDeviation[*list*] Statistics'DescriptiveStatistics'	stats[describe, meandeviation](*list*)
Median[*list*] Statistics'DescriptiveStatistics'	stats[describe, median](*list*)
MemberQ[*expr*, *item*]	has(*expr*, *item*)
	member(*item*, *expr*)
MemoryInUse[ ]	words( )
MeshStyle–>Dashing[*list*]	linestyle=*n*

MeshStyle->Thickness[*r*]	thickness=*n*
Message[*symbol*::*tag*, e_1, e_2, . . .]	userinfo(*level*, *name*, e_1, e_2 . . .)
Min[x_1, x_2, . . .]	min(x_1, x_2, . . .)
MinimumSpanningTree[*g*] DiscreteMath‘Combinatorica‘	networks[spantree](*g*)
MKS[*expr*] Miscellaneous‘Units‘	convert(*expr*, metric)
Mod[*m*, *n*]	irem(*m*, *n*) *m* mod *n* modp(*m*, *n*)
Mode[*list*] Statistics‘DescriptiveStatistics‘	stats[describe, mode](*list*)
Module[{v_1, v_2, . . .}, *body*]	proc(*args*) local v_1, v_2, . . . ; *body* end
MoebiusMu[*n*]	numtheory[mobius](*n*)
MovieParametricPlot[{e_1, e_2}, r_x, r_t] Graphics‘Animation‘	plots[animate]([e_1, e_2, r_x], r_t)
MoviePlot3D[*expr*, r_x, r_y, r_t] Graphics‘Animation‘	plots[animate3d](*expr*, r_x, r_y, r_t)
MoviePlot[*expr*, r_x, r_t] Graphics‘Animation‘	plots[animate](*expr*, r_x, r_t)
MovingAverage[*list*, *n*] Statistics‘DescriptiveStatistics‘	stats[transform, moving[$n + 1$]](*list*)
Multinomial[n_1, n_2, . . .]	combinat[multinomial](n, n_1, n_2, . . .)

N

N[*expr*, *n*]	evalf(*expr*, *n*)
N[*expr*]	convert(*expr*, float) evalf(*expr*)
Names[ ]	anames( )
NDSolve[*eqns*, *vars*, {*x*, *xmin*, *xmax*}]	dsolve(*eqns*, *vars*, numeric)
Needs["*context*‘"]	with(*package*)
Negative[*x*]	type(*x*, negative)
NegativeBinomialDistribution[*n*, *p*] Statistics‘DiscreteDistributions‘	negativebinomial[*n*, *p*] stats[random], stats[statevalf]
Nest[*f*, *x*, *n*]	(f@@n)(x)
NetworkFlow[*g*, *source*, *sink*] DiscreteMath‘Combinatorica‘	networks[flow](*g*, *source*, *sink*)
NextComposition[*l*] DiscreteMath‘Combinatorica‘	combinat[nextpart](*l*)
NextPrime[*n*] NumberTheory‘NumberTheoryFunctions‘	nextprime(*n*)
NIntegrate[*expr*, {*x*, *a*, *b*}]	evalf(Int(*expr*, *x*=*a*..*b*)) ‘evalf/int‘(*expr*, *x*=*a*..*b*) student[simpson](*expr*, *x*=*a*..*b*) student[trapezoid](*expr*, *x*=*a*..*b*)
NonNegative[*x*]	type(*x*, nonneg)

Mathematica	Maple
Normal[*series*]	convert(*series*, polynom)
NormalDistribution[μ, σ]	normal[μ, σ]
Statistics'ContinuousDistributions'	stats[random], stats[statevalf]
Normalize[v]	linalg[normalize](v)
LinearAlgebra'Orthogonalization'	
Not[*expr*]	not *expr*
NSolve[*eqns*, *vars*]	fsolve(*eqns*, *vars*)
NullSpace[A]	linalg[kernel](A)
	linalg[nullspace](A)
NumberOfCompositions[n, k]	combinat[numbcomp](n, k)
DiscreteMath'Combinatorica'	
NumberOfPartitions[n]	combinat[numbpart](n)
DiscreteMath'Combinatorica'	
NumberQ[x]	type(x, numeric)
Numerator[*expr*]	numer(*expr*)

O

Mathematica	Maple
O[x]^n	O(x^n)
OddQ[n]	type(n, odd)
Off[$f_1, f_2, \ldots$]	undebug($f_1, f_2, \ldots$)
	untrace($f_1, f_2, \ldots$)
On[$f_1, f_2, \ldots$]	debug($f_1, f_2, \ldots$)
	trace($f_1, f_2, \ldots$)
OpenAppend["*filename*"]	appendto('*filename*')
OpenRead["*filename*"]	readline('*filename*')
OpenWrite["*filename*"]	open('*filename*')
	writeto('*filename*')
Or[$expr_1$, $expr_2$]	$expr_1$ or $expr_2$
OrderedQ[{$string_1$, $string_2$}]	lexorder($string_1$, $string_2$)
OutputForm[*expr*]	print(*expr*)

P

Mathematica	Maple
Pade[*expr*, {x, a, m, n}]	numapprox[pade](*expr*, x=a, [m, n])
Calculus'Pade'	
ParametricPlot3D[{e_x, e_y, e_z}, r_t, r_u]	plot3d([e_x, e_y, e_z], r_t, r_u)
ParametricPlot3D[{e_x, e_y, e_z}, r_t]	plots[spacecurve]([e_x, e_y, e_z], r_t)
ParametricPlot[{e_x, e_y}, {t, t_0, t_1}]	plot([e_x, e_y, t=t_0..t_1])
Part[*expr*, i]	op(i, *expr*)
Partitions[n]	combinat[partition](n)
DiscreteMath'Combinatorica'	
PartitionsP[n]	combinat[numbpart](n)
PDF[*distribution*, x]	stats[statevalf, pdf[*distribution*]](x)
Statistics'ContinuousDistributions'	
PDF[*distribution*, x]	stats[statevalf, pf[*distribution*]](x)
Statistics'DiscreteDistributions'	
Permutations[*list*]	combinat[permute](*list*)

Pi	Pi
PlanarQ[g] DiscreteMath'Combinatorica'	networks[isplanar](g)
Plot3D[. . . , AmbientLight->*color*]	plot3d(. . . , ambientlight=[r, g, b])
Plot3D[. . . , Boxed->True]	plot3d(. . . , axes=BOXED)
Plot3D[. . . , HiddenSurface->False]	plot3d(. . . , style=WIREFRAME)
Plot3D[. . . , HiddenSurface->True]	plot3d(. . . , style=HIDDEN)
Plot3D[. . . , LightSources->$\{\{pos_1, c_1\}, \ldots\}$]	plot3d(. . . , light=[ϕ, θ, r, g, b])
Plot3D[. . . , MeshStyle->Dashing[*list*]]	plot3d(. . . , linestyle=n)
Plot3D[. . . , MeshStyle->Thickness[r]]	plot3d(. . . , thickness=n)
Plot3D[. . . , PlotPoints->$\{n, m\}$]	plot3d(. . . , grid=[n, m])
Plot3D[. . . , PlotRange->$\{\{a, b\}, \{c, d\}, \{r, s\}\}$]	plot3d(. . . , view=[$a..b$, $c..d$, $r..s$])
Plot3D[. . . , PlotRange->$\{r, s\}$]	plot3d(. . . , view=$r..s$)
Plot3D[. . . , Shading->True]	plot3d(. . . , style=PATCH)
Plot3D[. . . , Ticks->$\{ticks_x, ticks_y, ticks_z\}$]	plot3d(. . . , tickmarks=[n_1, n_2, n_3])
Plot3D[. . . , ViewPoint->$\{x, y, z\}$]	plot3d(. . . , orientation=[θ, ϕ])
Plot3D[$\{expr_1, \ldots\}$, $\{x, a, b\}$, $\{y, c, d\}$]	plot3d($\{expr_1, \ldots\}$, x=$a..b$, y=$c..d$)
Plot3D[$\{expr, s\}$, $\{x, a, b\}$, $\{y, c, d\}$]	plot3d(*expr*, *rx*, *ry*, color=s)
Plot3D[*expr*, $\{x, a, b\}$, $\{y, c, d\}$]	plot3d(*expr*, x=$a..b$, y=$c..d$)
Plot[. . . , AspectRatio->Automatic]	plot(. . . , scaling=CONSTRAINED)
Plot[. . . , Axes->False]	plot(. . . , axes=NONE)
Plot[. . . , Axes->True]	plot(. . . , axes=NORMAL)
Plot[. . . , AxesLabel->$\{label_x, label_y\}$]	plot(. . . , labels=[$label_x$, $label_y$])
Plot[. . . , Frame->True]	plot(. . . , axes=FRAME)
Plot[. . . , PlotDivision->d]	plot(. . . , resolution=n)
Plot[. . . , PlotLabel->*label*]	plot(. . . , title=*label*)
Plot[. . . , PlotPoints->n]	plot(. . . , numpoints=n)
Plot[. . . , PlotRange->$\{\{a, b\}, \{c, d\}\}$]	plot(. . . , view=[$a..b$, $c..d$])
Plot[. . . , PlotRange->$\{c, d\}$]	plot(. . . , $c..d$)
Plot[. . . , PlotStyle->Dashing[*list*]]	plot(. . . , linestyle=n)
Plot[. . . , PlotStyle->Hue[h]]	plot(. . . , color=COLOR(HUE, h))
Plot[. . . , PlotStyle->RGBColor[r, g, b]]	plot(. . . , color=COLOR(RGB, r, g, b))
Plot[. . . , PlotStyle->Thickness[r]]	plot(. . . , thickness=n)
Plot[. . . , Ticks->$\{ticks_x, ticks_y\}$]	plot(. . . , tickmarks=[n_1, n_2])
	plot(. . . , xtickmarks=n_1, ytickmarks=n_2)
Plot[$\{expr_1, expr_2, \ldots\}$, $\{x, a, b\}$]	plot($\{expr_1, expr_2, \ldots\}$, x=$a..b$)
Plot[*expr*, $\{x, a, b\}$]	plot(*expr*, x=$a..b$)
PlotDivision->d	resolution=n
PlotGradientField3D[*expr*, r_x, r_y, r_z] Graphics'PlotField3D'	plots[gradplot3d](*expr*, r_x, r_y, r_z)
PlotGradientField[*expr*, r_x, r_y] Graphics'PlotField'	plots[gradplot](*expr*, r_x, r_y)
PlotLabel->*label*	title=*label*
PlotPoints->$\{n, m\}$	grid=[n, m]
PlotPoints->n	numpoints=n
PlotRange->$\{\{a, b\}, \{c, d\}, \{r, s\}\}$	view=[$a..b$, $c..d$, $r..s$]

PlotRange->$\{\{a, b\}, \{c, d\}\}$	view=[*a*..*b*, *c*..*d*]
PlotRange->$\{c, d\}$	*c*..*d*
PlotRange->$\{r, s\}$	view=*r*..*s*
PlotStyle->Dashing[*list*]	linestyle=*n*
PlotStyle->Hue[*h*]	color=COLOR(HUE, *h*)
PlotStyle->RGBColor[*r*, *g*, *b*]	color=COLOR(RGB, *r*, *g*, *b*)
PlotStyle->Thickness[*r*]	thickness=*n*
PlotVectorField3D[$\{e_x, e_y, e_z\}, r_x, r_y, r_z$]	plots[fieldplot3d]([e_x, e_y, e_z], r_x, r_y, r_z)
Graphics'PlotField3D'	
PlotVectorField[$\{e_x, e_y\}, r_x, r_y$]	plots[fieldplot]([e_x, e_y], r_x, r_y)
Graphics'PlotField'	
PoissonDistribution[μ]	poisson[μ]
Statistics'DiscreteDistributions'	stats[random], stats[statevalf]
PolarPlot[*expr*, r_t]	plot([$expr_r$, $expr_\theta$, r_t], coords=polar)
Graphics'Graphics'	plots[polarplot](*expr*, r_t)
PolyGamma[*n*, *z*]	Psi(*n*, *z*)
PolyGamma[*z*]	Psi(*z*)
Polygon[$\{pt_1, pt_2, \ldots\}$]	plots[polygonplot3d]([$pt_1, pt_2, \ldots$])
	plots[polygonplot]([$pt_1, pt_2, \ldots$])
Polyhedron[*name*]	plots[polyhedraplot](*p*, polytype=*name*)
Graphics'Polyhedra'	
PolynomialDivision[*p*, *q*, *x*]	quo(*p*, *q*, *x*, '*r*')
PolynomialGCD[$poly_1$, $poly_2$, Modulus->*n*]	Gcd($poly_1$, $poly_2$) mod *n*
PolynomialGCD[$poly_1$, $poly_2$]	gcd($poly_1$, $poly_2$)
PolynomialLCM[$poly_1$, $poly_2$, Modulus->*n*]	Lcm($poly_1$, $poly_2$) mod *n*
PolynomialLCM[$poly_1$, $poly_2$]	lcm($poly_1$, $poly_2$)
PolynomialMod[$poly_1$, $poly_2$]	rem($poly_1$, $poly_2$, *x*)
PolynomialMod[*poly*, *m*]	*poly* mod *m*
PolynomialQ[*expr*, *x*]	type(*expr*, polynom(anything, *x*))
PolynomialQuotient[$poly_1$, $poly_2$, *x*]	quo($poly_1$, $poly_2$, *x*)
PolynomialRemainder[$poly_1$, $poly_2$, *x*]	rem($poly_1$, $poly_2$, *x*)
Positive[*x*]	type(*x*, positive)
PowerExpand[*expr*]	expand(*expr*)
PowerMod[*a*, *b*, *n*]	Power(*a*, *b*) mod *n*
	a &^ *b* mod *n*
Prepend[*list*, *elem*]	[*elem*, op(*list*)]
Prime[*n*]	ithprime(*n*)
PrimeQ[*n*, GaussianIntegers->True]	GaussInt[GIprime](*n*)
PrimeQ[*n*]	isprime(*n*)
	type(*n*, primeint)
PrimitiveRoot[*n*]	numtheory[primroot](*n*)
NumberTheory'NumberTheoryFunctions'	
Print[*expr*]	print(*expr*)
Product[*expr*, $\{i, a, b\}$]	Product(*expr*, *i*=*a*..*b*)
	product(*expr*, *i*=*a*..*b*)
Protect[*name*]	protect(*name*)

Put[*expr*, "*filename*"]	writeln('*filename*') writeto('*filename*')
PutAppend[*expr*, "*filename*"]	appendto('*filename*')

Q

Quantile[*list*, *q*] Statistics'DescriptiveStatistics'	stats[describe, quantile[*q*]](*list*)
Quartiles[*list*] Statistics'DescriptiveStatistics'	stats[describe, quartile[*n*]](*list*)
Quit[ ]	done quit stop
Quotient[*m*, *n*]	floor(*m*/*n*)

R

Random[Integer, 999999999999]	rand( )
Random[*distribution*] Statistics'ContinuousDistributions'	stats[random, *distribution*]()
Random[*distribution*] Statistics'DiscreteDistributions'	stats[random, *distribution*]()
RandomGraph[*n*, *p*] DiscreteMath'Combinatorica'	networks[random](*n*, prob=*p*)
RandomPartition[*n*] DiscreteMath'Combinatorica'	combinat[randpart](*n*)
RandomPermutation1[*n*] DiscreteMath'Combinatorica'	combinat[randperm](*n*)
RandomPermutation2[*n*] DiscreteMath'Combinatorica'	combinat[randperm](*n*)
Range[*a*, *b*]	[$*a*..*b*] [seq(*i*, *i*=*a*..*b*)]
Rational[*n*, *d*]	type(*x*, fraction) type(*x*, rational)
Rational	fraction
Rationalize[*x*]	convert(*x*, fraction) convert(*x*, rational)
Re[*z*]	Re(*z*)
Read[*stream*, String]	readline('*filename*')
Read[*stream*]	readstat('*prompt*')
ReadList["*filename*", *type*]	readdata('*filename*', *type*, *cols*)
Real	float
RealInterval[{*a*, *b*}]	*a*..*b*
Recognize[*r*, *n*, *t*] NumberTheory'Recognize'	minpoly(*r*, *n*)
Regress[*data*, *fcns*, *vars*] Statistics'LinearRegression'	stats[fit, leastsquare[*vars*]](*data*)
ReleaseHold[*expr*]	thaw(*var*)

Mathematica	Maple
ReplaceAll[*expr*, *rules*]	asubs(*eqn*, *expr*)
	subs(*subs*, *expr*)
ReplacePart[*expr*, *new*, *n*]	subsop(*n*=*new*, *expr*)
Residue[*expr*, {*x*, *a*}]	residue(*expr*, *x*=*a*)
Rest[*list*]	op(2..nops(*list*), *list*)
	list[2..nops(*list*)]
Resultant[$poly_1$, $poly_2$, *x*, Modulus->*p*]	Resultant($poly_1$, $poly_2$, *x*) mod *p*
Resultant[$poly_1$, $poly_2$, *x*]	resultant($poly_1$, $poly_2$, *x*)
Return[*expr*]	RETURN(*expr*)
RootMeanSquare[*list*]	stats[describe, quadraticmean](*list*)
Statistics‘DescriptiveStatistics‘	
RotateLeft[*l*]	[op(2..nops(*l*), *l*), op(1, *l*)]
RotateRight[*l*]	[op(nops(*l*), *l*), op(1..nops(*l*)−1, *l*)]
Round[*x*]	round(*x*)
RowReduce[*A*, Modulus->*p*]	Gaussjord(*A*) mod *p*
RowReduce[*A*]	linalg[gaussjord](*A*)
	linalg[rref](*A*)
RSolve[*eqns*, *names*, *var*]	rsolve(*eqns*, *names*)
DiscreteMath‘RSolve‘	
Run[*expr*]	!*command*
	system(*command*)

S

Mathematica	Maple
SampleRange[*list*]	stats[describe, range](*list*)
Statistics‘DescriptiveStatistics‘	
Save["*filename*", $name_1$, $name_2$, . . .]	save $name_1$, $name_2$, . . ., ‘*filename*‘
ScatterPlot3D[{{x_1, y_1, z_1}, . . .}]	plots[pointplot]({[x_1, y_1, z_1], . . .})
Graphics‘Graphics3D‘	
Sec[*z*]	sec(*z*)
Sech[*z*]	sech(*z*)
SeedRandom[*n*]	_seed := *n*
Select[*list*, *test*]	select(*test*, *list*)
SequenceForm[$expr_1$, $expr_2$, . . .]	‘‘ . $expr_1$. $expr_2$
	cat($expr_1$, $expr_2$, . . .)
Series[*expr*, {*x*, Infinity, *n*}]	asympt(*expr*, *x*, *n*)
Series[*expr*, {*x*, x_0, *n*}]	series(*expr*, *x*=x_0, *n*)
	taylor(*expr*, *x*=x_0, *n*)
Set[*name*, *value*]	assign(*name*, *value*)
	assign(*name*=*value*)
	name := *value*
SetDelayed[*name*, *value*]	*name* := '*value*'
SetOptions[Plot, *options*]	plots[setoptions](*options*)
SetOptions[Plot3D, *options*]	plots[setoptions3d](*options*)
SetOptions[*stream*, PageHeight->*n*]	interface(screenheight=*n*)
SetOptions[*stream*, PageWidth->*n*]	interface(screenwidth=*n*)
Shading->True	style=PATCH

Show[$g_1, g_2, \ldots$]	plots[display3d]([$g_1, g_2, \ldots$]) plots[display]([$g_1, g_2, \ldots$])
Show[*graphics*]	plots[display3d](*graphics*) plots[display](*graphics*)
ShowAnimation[{$g_1, g_2, \ldots$}] Graphics\`Animation\`	plots[display]([$g_1, g_2, \ldots$], insequence=true)
Sign[z]	signum(z)
Simplify[*expr*]	simplify(*expr*)
Sin[z]	sin(z)
SingularValues[A]	Svd(A)
Sinh[z]	sinh(z)
SinhIntegral[z]	Shi(z)
SinIntegral[z]	Si(z)
Skewness[*list*] Statistics\`DescriptiveStatistics\`	stats[describe, skewness](*list*)
Solve[{*eqns*, Modulus==p}, Mode->Modular]	msolve(*eqns*, p)
Solve[*eqns*, *vars*]	solve(*eqns*, *vars*)
Sort[*list*]	sort(*list*)
Sort[*list*, f]	sort(*list*, f)
SphericalPlot3D[*expr*, r_θ, r_ϕ] Graphics\`ParametricPlot3D\`	plot3d(*expr*, r_θ, r_ϕ, coords=spherical) plots[sphereplot](*expr*, r_θ, r_ϕ)
Spline[*points*, Cubic] Graphics\`Spline\`	spline($points_x$, $points_y$, x, cubic)
Sqrt[z]	sqrt(z)
SqrtMod[d, n] NumberTheory\`NumberTheoryFunctions\`	numtheory[msqrt](d, n)
SquareFreeQ[n] NumberTheory\`NumberTheoryFunctions\`	numtheory[issqrfree](n)
StandardDeviation[*list*] Statistics\`DescriptiveStatistics\`	stats[describe, standarddeviation[1]](*list*)
StandardDeviationMLE[*list*] Statistics\`DescriptiveStatistics\`	stats[describe, standarddeviation](*list*)
StirlingFirst[n, m] DiscreteMath\`Combinatorica\`	combinat[stirling1](n, m)
StirlingS1[n, m]	combinat[stirling1](n, m)
StirlingS2[n, m]	combinat[stirling2](n, m)
StirlingSecond[n, m] DiscreteMath\`Combinatorica\`	combinat[stirling2](n, m)
StringDrop["*str*", n]	substring(\`*str*\`, n+1..length(\`*str*\`))
StringJoin["$string_1$", "$string_2$"]	\`$string_1$\` . \`$string_2$\` cat(\`$string_1$\`, \`$string_2$\`)
StringLength["*string*"]	length(\`*string*\`)
StringPosition["*string*", "*sub*"]	search(\`*string*\`, \`*sub*\`)
StringQ[*expr*]	type(*expr*, string)
StringTake["*string*", {m, n}]	substring(\`*string*\`, $m..n$)
StudentTDistribution[n]	studentst[n]

Mathematica	Maple
Statistics'ContinuousDistributions'	stats[random], stats[statevalf]
Sum[*expr*, {*i*, *a*, *b*}]	sum('*expr*', '*i*'=*a*..*b*)
	Sum('*expr*', '*i*'=*a*..*b*)
Symbol	name

T

Mathematica	Maple
Table[*expr*, {*i*, 1, *m*}, {*j*, 1, *n*}]	linalg[matrix](*m*, *n*, *f*)
Table[*expr*, {*i*, 1, *n*}]	linalg[vector](*n*, *f*)
Table[*expr*, {*i*, *a*, *b*}]	[*expr* $ *i*=*a*..*b*]
	[seq(*expr*, *i*=*a*..*b*)]
Take[*A*, {*m*, *n*}]	linalg[row](*A*, *m*..*n*)
Take[*list*, {*m*, *n*}]	[op(*m*..*n*, *list*)]
	[*list*[*m*..*n*]]
Take[*list*, *n*]	[op(1..*n*, *list*)]
	[*list*[1..*n*]]
TakeColumns[*A*, {*m*, *n*}] LinearAlgebra'MatrixManipulation'	linalg[col](*A*, *m*..*n*)
TakeMatrix[*A*, {r_1, r_2}, {c_1, c_2}] LinearAlgebra'MatrixManipulation'	linalg[submatrix](*A*, $r_1..r_2$, $c_1..c_2$)
TakeRows[*A*, {*m*, *n*}] LinearAlgebra'MatrixManipulation'	linalg[row](*A*, *m*..*n*)
Tan[*z*]	tan(*z*)
Tanh[*z*]	tanh(*z*)
TeXForm[*expr*]	latex(*expr*)
Text[*expr*, {*x*, *y*, *z*}]	plots[textplot3d]([*x*, *y*, *z*, *string*])
Text[*expr*, {*x*, *y*}]	plots[textplot]([*x*, *y*, *string*])
Ticks->{$ticks_x$, $ticks_y$, $ticks_z$}	tickmarks=[n_1, n_2, n_3]
Ticks->{$ticks_x$, $ticks_y$}	tickmarks=[n_1, n_2]
	xtickmarks=*n*
	ytickmarks=*n*
TimeUsed[ ]	time( )
Timing[*expr*]	history( ); timing(*expr*)
	showtime( ); *expr*
ToAdjacencyLists[*g*] DiscreteMath'Combinatorica'	networks[adjacency](*g*)
ToExpression["*string*"]	parse('*string*')
Together[*expr*]	normal(*expr*)
ToString[*expr*]	convert(*expr*, name)
	convert(*expr*, string)
Trace[*expr*]	debug(*f*)
	trace(*f*)
Transpose[*A*]	linalg[transpose](*A*)
TrigReduce[*expr*] Algebra'Trigonometry'	expand(*expr*)
TrigToComplex[*expr*] Algebra'Trigonometry'	convert(*expr*, exp)

True	true
TrueQ[*expr*]	evalb(*expr*)
U	
UniformDistribution[*a*, *b*]	uniform[*a*, *b*]
Statistics 'ContinuousDistributions'	stats[random], stats[statevalf]
Union[$list_1$, $list_2$]	set_1 union set_2
UnitStep[*x*]	Heaviside(*x*)
Calculus 'DiracDelta'	
Unprotect[*name*]	unprotect(*name*)
Unset[*name*]	unassign('*name*') *name* := '*name*'
V	
ValueQ[*name*]	assigned(*name*)
Variables[*expr*]	indets(*expr*)
Variance[*list*]	stats[describe, variance[1]](*list*)
Statistics 'DescriptiveStatistics'	
VarianceMLE[*list*]	stats[describe, variance](*list*)
Statistics 'DescriptiveStatistics'	
VectorQ[*expr*]	type(*expr*, vector)
ViewPoint->{*x*, *y*, *z*}	orientation=[θ, ϕ]
W	
WeibullDistribution[α, β]	weibull[α, β]
Statistics 'ContinuousDistributions'	stats[random], stats[statevalf]
While[*test*, *body*]	while *test* do *body* od
Write["*filename*", *expr*]	write(*expr*) writeln(*expr*) writeto('*filename*')
Z	
ZeroMatrix[*m*, *n*]	linalg[matrix](*m*, *n*, 0)
LinearAlgebra 'MatrixManipulation'	
Zeta[*s*, *a*]	Zeta(0, *s*, *a*)
Zeta[*s*]	Zeta(*s*)
$	
$Display	interface(plotoutput='*filename*')
$Failed	FAIL
$MachinePrecision	evalhf(Digits)
$Path	libname
$Version	interface(version)
$a + b$	$a + b$
$a - b$	$a - b$

<table>
<tr><td>a b</td><td>a * b</td></tr>
<tr><td>a * b</td><td>a * b</td></tr>
<tr><td>a ** b</td><td>a &* b</td></tr>
<tr><td>a / b</td><td>a / b</td></tr>
<tr><td>a ^ b</td><td>a ^ b</td></tr>
<tr><td></td><td>a ** b</td></tr>
<tr><td>n!</td><td>n!</td></tr>
<tr><td>+a</td><td>+a</td></tr>
<tr><td>-a</td><td>-a</td></tr>
<tr><td>u . v</td><td>linalg[dotprod](u, v)</td></tr>
<tr><td>a == b</td><td>a = b</td></tr>
<tr><td>a != b</td><td>a <> b</td></tr>
<tr><td>a < b</td><td>a < b</td></tr>
<tr><td>a <= b</td><td>a <= b</td></tr>
<tr><td>a > b</td><td>a > b</td></tr>
<tr><td>a >= b</td><td>a >= b</td></tr>
<tr><td>a && b</td><td>a and b</td></tr>
<tr><td>a || b</td><td>a or b</td></tr>
<tr><td>!a</td><td>not a</td></tr>
<tr><td>"string"</td><td>`string`</td></tr>
<tr><td>"$string_1$" <> "$string_2$"</td><td>`$string_1$` . `$string_2$`</td></tr>
<tr><td>{e_1, e_2, . . .}</td><td>[e_1, e_2, . . .]</td></tr>
<tr><td>expr[[i]]</td><td>op(i, expr)</td></tr>
<tr><td>list[[i]]</td><td>list[i]</td></tr>
<tr><td>name = value</td><td>name := value</td></tr>
<tr><td></td><td>assign(name, value)</td></tr>
<tr><td>name := value</td><td>name := 'value'</td></tr>
<tr><td>name =.</td><td>unassign('name')</td></tr>
<tr><td></td><td>name := 'name'</td></tr>
<tr><td>expr /. old->new</td><td>asubs(old=new, expr)</td></tr>
<tr><td></td><td>subs(old=new, expr)</td></tr>
<tr><td>statement (* show output *)</td><td>statement; # show output</td></tr>
<tr><td>statement; (* suppress output *)</td><td>statement: # suppress output</td></tr>
<tr><td>body&</td><td>proc(args) body end</td></tr>
<tr><td>name_head</td><td>name:type</td></tr>
<tr><td>%</td><td>"</td></tr>
<tr><td>%%</td><td>""</td></tr>
<tr><td>%%%</td><td>"""</td></tr>
<tr><td><< "filename"</td><td>read `filename`</td></tr>
<tr><td>expr >> "filename"</td><td>writeln(`filename`)</td></tr>
<tr><td></td><td>writeto(`filename`)</td></tr>
<tr><td>expr >>> "filename"</td><td>appendto(`filename`)</td></tr>
<tr><td>(* comment *)</td><td># comment</td></tr>
<tr><td>?name</td><td>?topic</td></tr>
<tr><td>!command</td><td>!command</td></tr>
<tr><td></td><td>system(command)</td></tr>
</table>

Chapter 8
Learning More

1 BOOKS AND JOURNALS

The books and journals listed in this section may be of interest to Maple users, as well as those interested in symbolic computation and computer algebra systems in general.

1.1 On Maple

The number of books on Maple is growing steadily. This section lists books and journals on Maple in three main categories: general tutorials and reference books, mathematics tutorials and course supplements, and applications.

General Tutorials and Reference Books

Abell, Martha L., and James P. Braselton, *Maple V by Example* (Academic Press, forthcoming) ISBN 0-12-041545-3

Abell, Martha L., and James P. Braselton, *The Maple V Handbook* (Academic Press, forthcoming) ISBN 0-12-041542-9

Blachman, Nancy, *Using Maple* (Brooks/Cole, forthcoming)

Char, Bruce W., Keith O. Geddes, Gaston H. Gonnet, Benton L. Leong, Michael B. Monagan, and Stephen M. Watt, *First Leaves: A Tutorial Introduction to Maple V* (Springer-Verlag, 1992) ISBN 0-387-97621-3

Char, Bruce W., Keith O. Geddes, Gaston H. Gonnet, Benton L. Leong, Michael B. Monagan, and Stephen M. Watt, *Maple V Language Reference Manual* (Springer-Verlag, 1991) ISBN 0-387-97622-1

Char, Bruce W., Keith O. Geddes, Gaston H. Gonnet, Benton L. Leong, Michael B. Monagan, and Stephen M. Watt, *Maple V Library Reference Manual* (Springer-Verlag, 1991) ISBN 0-387-97592-6

Heck, André, *Introduction to Maple* (Springer-Verlag, 1993) ISBN 0-387-97662-0

Kamerich, Ernic, *A Guide to Maple* (Springer-Verlag, forthcoming) ISBN 0-387-94116-9

Redfern, Darren, *The Maple Handbook* (Springer-Verlag, 1993) ISBN 0-387-94054-5

Mathematics Tutorials and Course Supplements

Abell, Martha L., and James P. Braselton, *Differential Equations with Maple V* (Academic Press, forthcoming) ISBN 0-12-041548-8

Auer, John W., *Maple Solutions Manual for Linear Algebra with Applications* (Prentice-Hall, 1991) ISBN 0-13-535-5508

Bauldry, William C., and Joseph R. Fielder, *Calculus Laboratories with Maple* (Brooks/Cole, 1991) ISBN 0-534-13788-1

Burbella, D. C. M., and C. T. J. Dodson, *Self-Tutor for Computer Calculus Using Maple* (Prentice-Hall Canada, 1993) ISBN 0-13-063926-5

Cheung, C. K., and John Harer, *Multivariable Calculus with Maple V* (John Wiley & Sons, 1994) ISBN 0-471-59835-6

Devitt, John S., *Calculus with Maple V* (Brooks/Cole, 1993) ISBN 0-534-16362-9

Ellis, Wade Jr., Eugene W. Johnson, Ed Lodi, and Daniel Schwalbe, *Maple V Flight Manual: Tutorials for Calculus, Linear Algebra, and Differential Equations* (Brooks/Cole, 1992) ISBN 0-534-17338-1

Ellis, Wade Jr., and Ed Lodi, *Maple for the Calculus Student: A Tutorial* (Brooks/Cole, 1989) ISBN 0-534-11874-7

Fattahi, Abi, *Maple V Calculus Labs* (Brooks/Cole, 1992) ISBN 0-534-19272-6

Geddes, K. O., B. Marshman, I. McGee, P. Ponzo, and B. Char, *Maple Calculus Workbook: Problems and Solutions* (Waterloo Maple Software, 1989)

Harris, Kent, *Discovering Calculus with Maple* (John Wiley & Sons, 1992) ISBN 0-471-55156-2

Holmes, M. H., J. G. Ecker, W. E. Boyce, and W. L. Siegmann, *Exploring Calculus with Maple* (Addison-Wesley, 1993) ISBN 0-201-52616-6

Johnson, Eugene, *Linear Algebra with Maple V* (Brooks/Cole, 1993) ISBN 0-534-13069-0

Kreyszig, Erwin, and E. J. Normington, *Maple Computer Manual for Advanced Engineering Mathematics* (John Wiley & Sons, 1994) ISBN 0-471-31126-X

Mathews, David M., and Keith E. Schwingendorf, *Precalculus Investigations Using Maple* (HarperCollins, 1994) ISBN 0-673-99410-4

McLaughlin, R., *Calculus and Maple V* (Saunders College Publishing, 1993) ISBN 0-03-096778-3

Schwalbe, Daniel, *Differential Equations Using Maple* (Brooks/Cole, forthcoming)

Small, Donald B., and John M. Hosack, *Explorations in Calculus with a Computer Algebra System* (McGraw-Hill, 1990) ISBN 0-07-058267-X

Applications

Baylis, W. E., *Theoretical Methods in the Physical Sciences, An Introduction to Problem Solving Using Maple V* (Birkhäuser, forthcoming) ISBN 0-8176-3715-X

Devitt, J. S., and G. J. Fee, *Tackling Mathematical Problems with Maple* (Oxford, forthcoming)

Gander, Walter, and Jiří Hřebíček, eds., *Solving Problems in Scientific Computing Using Maple and Matlab* (Springer-Verlag, 1993) ISBN 0-387-57329-1

Lee, Thomas, ed., *Mathematical Computation with Maple V – Ideas and Applications: Proceedings of the Maple Summer Workshop and Symposium, University of Michigan, Ann Arbor, 1993* (Birkhäuser, 1993) ISBN 0-8176-3724-9

Maple Technical Newsletter, Birkhäuser, Boston

Scott, Tony, *Computer Algebra and the Magic of Maple* (Birkhäuser, forthcoming)

1.2 On Computer Algebra

The following books, conference proceedings, and journals discuss algorithms and issues in computer algebra and computer algebra systems.

Akritas, Alkiviadis G., *Elements of Computer Algebra with Applications* (John Wiley & Sons, 1989) ISBN 0-471-61163-8

Becker, Thomas, and Volker Weispfenning, in cooperation with Heinz Kredel, *Gröbner Bases, A Computational Approach to Commutative Algebra* (Springer-Verlag, 1993) ISBN 0-387-97971-9

Buchberger, Bruno, George E. Collins, and Rüdiger Loos, in cooperation with Rudolph Albrecht, eds., *Computer Algebra: Symbolic and Algebraic Computation* (Springer-Verlag, 2nd Edition, 1983) ISBN 0-387-81776-X

Davenport, J. H., Y. Siret, and E. Tournier, *Computer Algebra: Systems and Algorithms for Algebraic Computation* (Academic Press, 2nd Edition, 1993) ISBN 0-12-209232-8

Geddes, Keith O., Stephen R. Czapor, and George Labahn, *Algorithms for Computer Algebra* (Kluwer Academic Publishers, 1992) ISBN 0-7923-9259-0

Harper, David, Chris Wooff, and David Hodgkinson, *A Guide to Computer Algebra Systems* (John Wiley & Sons, 1991) ISBN 0-471-92910-7

Journal of Symbolic Computation, Academic Press

Knuth, Donald E., *The Art of Computer Programming*, Vol. 2, *Seminumerical Algorithms* (Addison-Wesley, 2nd Edition, 1981) ISBN 0-201-03822-6

Mignotte, Maurice, *Mathematics for Computer Algebra* (Springer-Verlag, 1992) ISBN 0-387-97675-2

SIGSAM Bulletin, A Quarterly Publication of the Special Interest Group on Symbolic & Algebraic Manipulation, ACM Press

Zippel, Richard, *Effective Polynomial Computation* (Kluwer Academic Publishers, 1993) ISBN 0-7923-9375-9

Conference Proceedings

Computational Algebra, Klaus G. Fischer, Philippe Loustaunau, Jay Shapiro, Edward L. Green, and Daniel Farkas, eds. (Marcel Dekker, 1993) ISBN 0-8247-9070-7

Computer Algebra, David V. Chudnovsky and Richard D. Jenks, eds. (Marcel Dekker, 1989) ISBN 0-8247-8038-8

Computer Algebra and Parallelism, J. Della Dora and J. Fitch, eds. (Academic Press, 1989) ISBN 0-12-209042-X

Computer Algebra and Parallelism, Second International Workshop, Richard E. Zippel, ed. (Springer-Verlag, 1992) ISBN 0-387-55328-2

Computer Algebra in Industry: Problem Solving in Practice, Arjeh M. Cohen, ed. (John Wiley & Sons, 1993) ISBN 0-471-93829-7

Computers and Mathematics, Erich Karltofen and Stephen M. Watt, eds. (Springer-Verlag, 1989) ISBN 0-387-97019-3

Computers in Mathematics, David V. Chudnovsky and Richard D. Jenks, eds. (Marcel Dekker, 1990) ISBN 0-8247-8341-7

Design and Implementation of Symbolic Computation Systems: International Symposium DISCO '90, Capri, Italy, Alfonso Miola, ed. (Springer-Verlag, 1990) ISBN 0-387-52531-9

ISSAC '88, Proceedings of the International Symposium on Symbolic and Algebraic Computation, Rome, Italy, Patrizia Gianni, ed. (Springer-Verlag, 1989) ISBN 0-387-51084-2

ISSAC '89, Proceedings of the ACM-SIGSAM 1989 International Symposium on Symbolic and Algebraic Computation, Portland, Oregon, Gaston H. Gonnet, ed. (ACM Press, 1989)

ISSAC '90, Proceedings of the International Symposium on Symbolic and Algebraic Computation, Tokyo, Japan, Shunro Watanabe and Morio Nagata, eds. (ACM Press, 1990) ISBN 0-89791-401-5

ISSAC '91, Proceedings of the 1991 International Symposium on Symbolic and Algebraic Computation, Bonn, Germany, Stephen M. Watt, ed. (ACM Press, 1991) ISBN 0-89791-437-6

ISSAC '92, Proceedings of the 1992 International Symposium on Symbolic and Algebraic Computation, Berkeley, California, Paul S. Wang, ed. (ACM Press, 1992) ISBN 0-89791-489-9

ISSAC '93, Proceedings of the 1993 International Symposium on Symbolic and Algebraic Computation, Kiev, Ukraine, (ACM Press, 1993) ISBN 0-89791-604-2

Trends in Computer Algebra, Rainer Janssen, ed. (Springer-Verlag, 1988) ISBN 0-387-18928-9

Before 1988, the major symbolic and algebraic computation symposia were named EUROCAL, EUROCAM, EUROSAM, SYMSAC or SYMSAM. These were unified under ISSAC in 1988.

1.3 On Graphics

The following book is an excellent source of curves and surfaces you can make with a system like Maple.

von Seggern, David, *CRC Standard Curves and Surfaces* (CRC Press, 1993) ISBN 0-8493-0196-3

2 TECHNICAL SUPPORT

For technical support, contact the supplier from whom you purchased your copy of Maple. Waterloo Maple Software, Brooks/Cole, MathSoft, and other suppliers each support their own customers. A telephone number and electronic mail address for support should be provided with your installation instructions. Addresses for some suppliers appear on page 81.

If you obtained your copy of Maple through a site license, contact your local Maple technical representative or site administrator for technical support.

The Maple Info Server contains answers to some common technical questions and problems with Maple. See Chapter 5 (page 418) for more information.

3 TRAINING

There are workshops on Maple given at conferences, universities, and in industry. Variable Symbols and Waterloo Maple Software are currently offering workshops. Variable Symbols' address appears on page 470 (under "Nancy Blachman"). Waterloo Maple Software's address appears on page 81.

4 CONFERENCES

The Maple Summer Workshop and Symposium (MSWS) is an excellent source of information on Maple. It is sponsored annually by Waterloo Maple Software. This conference normally includes short courses for beginning and experienced Maple users, discussions on the use of Maple in the classroom, invited speakers, presentations of research papers, and interaction with developers and technical support personnel of Waterloo Maple Software. For more information on MSWS and other conferences on Maple, contact Waterloo Maple Software. (Their address appears on page 81.)

The International Symposium on Symbolic and Algebraic Computation (ISSAC) is a large annual conference where researchers discuss algorithms in computer algebra and the development of computer algebra systems.

5 IS ANYTHING MISSING?

We sincerely hope this reference guide helps you use Maple effectively. We have tried to include important information and to organize it logically. Please let us know if we can improve this guide in any way.

Nancy Blachman
Variable Symbols, Inc.
6537 Chabot Road
Oakland, CA 94618-1618
Email: `nb@cs.stanford.edu`
Fax: 510-652-8461
Telephone: 510-652-8462

Michael Mossinghoff
Department of Mathematics
University of Texas at Austin
Austin, TX 78712
Email: `mossingh@math.utexas.edu`

6 ABOUT THE AUTHORS

Nancy Blachman is an internationally known author and instructor on mathematical software. Her book *Mathematica: A Practical Approach* (Prentice Hall, 1992) is the best selling tutorial book on Mathematica. She also wrote the *Mathematica Quick Reference, Version 2* (Variable Symbols, Inc. and Addison-Wesley, 1992) and helped in the development of the Macintosh HyperCard-based *Mathematica Help Stack* and the MS-Windows *Mathematica Help*. Nancy is currently writing the *Mathematica Graphics Guidebook* (Addison-Wesley, 1994) with Cameron Smith. She is the founder of Variable Symbols, Inc., a company that specializes in training and consulting in mathematical software, and she is an instructor in the computer science department at Stanford University. She holds degrees in mathematics, operations research, and computer science from the University of Birmingham (U.K.), the University of California at Berkeley, and Stanford University.

Michael Mossinghoff is a graduate student at the University of Texas at Austin, where he is pursuing a doctoral degree in mathematics. He has taught mathematics classes at the university and has held workshops on Maple and Mathematica for faculty and students. He has a Master of Science in computer science from Stanford University and a Bachelor of Science in mathematics from Texas A & M University. His research interests include computational number theory, Diophantine approximation, and computer algebra systems.

Chapter 9
Glossary

ALGOL 68 An historically important procedural programming language that influenced the design of many modern programming languages, including Pascal, Ada, and Maple's programming language.

Alias An abbreviation used in both Maple input and output of Maple expressions. For example, the imaginary unit `I` in Maple is actually an alias for `(-1)^(1/2)`. You can define an alias with the `alias` command.

Arguments Expressions given as input when invoking a function. For example, in the function call `evalf(Pi, 25)`, the first argument is `Pi` and the second is `25`. Also called actual arguments or actual parameters.

Array A special kind of table indexed with a range of consecutive integers in each dimension. See page 58.

Associativity The direction of evaluation of operators of the same precedence in an expression with no parentheses. For example, `*` and `/` are left associative operators with the same precedence, so the expression `a*b/c` is evaluated as `(a*b)/c`. Nonassociative operators have no default grouping rule, so the expression `a^b^c` is syntactically incorrect and cannot be evaluated by Maple. The table on page 480 lists the type of associativity for each programming language and neutral operator in Maple. This property, together with precedence, determines how expressions are evaluated.

Binary Operator An operator that requires two arguments. Binary programming language operators in Maple are written in infix notation, that is, the operator is written between its two arguments: for example, `a + b`. Neutral operators may also be written in infix notation: for example, `a &^ b`.

Built-in Function A function defined in the Maple kernel, hence always available. Built-in functions implement centrally important Maple routines, like arithmetic and parsing of expressions. You cannot inspect the source code of a built-in function. See page 62.

C A programming language developed by Dennis Ritchie, historically associated with the Unix operating system, but widely used today because of its power, simplicity, and easy portability. Maple can translate expressions to C syntax. See page 76.

Constant A numeric constant (an integer, fraction, or floating-point number), or a symbolic constant. The symbolic constants in Maple are `E`, `Pi`, `gamma`, `Catalan`, `I`, `true`, `false`, `FAIL`, and `infinity`. See page 15.

Environment Variable A special kind of global variable in Maple that you can change locally without affecting the global value. You can reset an environment variable inside a procedure, but it reverts to its original value upon exiting the procedure. The built-in environment variables in Maple are `Digits`, `'mod'`, `Normalizer`, `printlevel`, and `Testzero`. Any variable that begins with `_Env` is also an environment variable.

Eqn A system for typesetting equations and mathematics, used in conjunction with the `troff` program for formatting documents on Unix systems. Maple can translate expressions to `eqn` syntax. See page 76.

Escape Character A character that instructs Maple to treat the remaining characters on the input line in some special way. There are four escape characters in Maple: `#`, `\`, `?`, and `!`. Their meanings are given in the table on page 481.

Expression Sequence A collection of Maple expressions separated by commas. The empty sequence is denoted by `NULL`. See page 52.

Formal Parameters The names that appear between the parentheses after `proc` in a procedure definition, or the names that appear on the left side of an arrow operator for defining functions, or the names that appear after the first vertical bar in an angle-bracket function. Maple uses *call by evaluated name* semantics to replace formal parameters when a procedure or function is called: the actual arguments are first evaluated, then every occurrence of a formal parameter in the body of the procedure or function is replaced by the value of the corresponding actual argument. As a result, you cannot assign to a formal parameter in the body of a procedure or function, unless you pass in an unassigned symbol as the corresponding actual argument.

FORTRAN A programming language widely used for numerical work in science and engineering. Maple can translate expressions to FORTRAN syntax. See page 76.

FTP The File Transfer Protocol, a program for transferring files from one computer to another on the Internet. Chapter 5 describes how to transfer files using FTP. See page 418.

Global Variable A variable whose value is known throughout your Maple session. Every variable you explicitly assign in your session is a global variable. In Maple V Release 2 and earlier versions, a variable used in a procedure is a global variable if and only if it is not explicitly declared `local`. In Maple V Release 3, a variable used in a procedure may be explicitly declared `global`. An undeclared variable in a procedure is considered global if it is not the indexing variable of a `for` or `seq` statement and if it does not appear on the left-hand side of any assignment statement in the procedure. Maple also recognizes several system global variables, like the random number seed `_seed`. A global variable persists until you explicitly free it. See page 47.

Hue One means of specifying a color in Maple graphics. A hue is a color in the rainbow, specified with a real number between 0 and 1. A hue value of 0 represents red;

larger values run through shades of orange, yellow, green, blue and indigo; and 1 represents pure violet. See pages 26 and 34.

Inert Function A formal function representing a mathematical expression. It does not perform any calculations. For example, `Int(cos(x), x)` returns $\int \cos x\, dx$. All inert functions begin with a capital letter in Maple (though not every function that begins with a capital letter is inert). See page 6.

Kernel One of the three main components of the Maple system. The others are the Maple library and the user interface. The kernel is the core computational engine of Maple and consists of compiled C code. It is loaded every time you start a Maple session. Functions implemented in the kernel (built-in functions) are always available, but you cannot display their source code. See page 62.

Keyword A reserved word used in forming Maple statements. A keyword is not a valid variable name. The keywords in Maple are `and`, `by`, `do`, `done`, `elif`, `else`, `end`, `fi`, `for`, `from`, `if`, `in`, `intersect`, `local`, `minus`, `mod`, `not`, `od`, `option`, `options`, `or`, `proc`, `quit`, `read`, `save`, `stop`, `then`, `to`, `union`, and `while`.

LaTeX A special version of TeX written by Leslie Lamport that adds several commands to make typesetting easier. This guide was typeset with LaTeX. Maple can translate expressions to LaTeX syntax. See page 76.

List Zero or more expressions separated by commas and enclosed in square brackets `[ ]`. A list may contain duplicates, and its elements are never reordered by the system. See page 54.

Local Variable In Maple V Release 2 and earlier versions, a variable is local to a procedure if and only if it is explicitly declared `local`. In Maple V Release 3, a variable declared neither `local` nor `global` is a local variable if it is the indexing variable of a `for` or `seq` statement or if it appears on the left-hand side of an assignment statement in the procedure. The value of a local variable is known only within the procedure or function where it is defined and is independent of the value of any global variable with the same name. For example, the following procedure declares two local variables, `a` and `b` (and one formal parameter, `x`):

```
proc(x) local a,b; statement-sequence end;
```

Maple V Release 2 and earlier versions allow local variable declarations in arrow functions as well:

```
x -> local a,b; expression;
```

All versions support local variables in angle-bracket functions:

```
< expression | x | a,b >;
```

Maple Library One of the three main components of the Maple system. The others are the kernel and the user interface. The Maple library contains the vast majority of Maple's mathematical routines and is written in Maple's own programming language. You can inspect the source code of any function in the Maple library. The Maple library is divided into three parts: standard library functions, miscellaneous library functions, and packages. See page 62.

Maple V A version of Maple first released in September of 1991.

Maple V Release 2 A version of Maple first released in November of 1992. Maple V Release 2 features new and expanded mathematical routines, more effective numerical algorithms, mathematical output, the worksheet interface, a help topic browser, many new graphics routines and options, improved input and output, and several new packages.

Maple V Release 3 A version of Maple first released in March of 1994. Maple V Release 3 adds additional help commands, more plotting options, protection against assigning to system names, enhanced string processing, and improvements to many mathematical functions and packages, including integration, algebraic manipulation, differential equations, laplace transforms, vector calculus, solving equations, statistics, and number theory.

March A system utility distributed with Maple V Release 2 and Release 3 for creating and maintaining archives of Maple routines. The Maple library is a single archive file in Maple V Release 2 and Release 3.

Mathematica A computer algebra system developed by Wolfram Research, Inc. Mathematica and Maple have similar capabilities.

Matrix A special kind of two-dimensional array where indexing starts at 1 in both dimensions. See page 58.

Mint A diagnostic program distributed with Maple that spots syntax errors, points out probable programming errors, and tracks the use of global and local variables in Maple programs. See page 70.

Miscellaneous Library Function A function in the Maple library that must be loaded explicitly with `readlib` before you can use it in a session. You can inspect the source code of a miscellaneous library function. See page 65.

Neutral Operator An operator that begins with an ampersand (`&`). The rest may be any simple string, or any sequence of nonalphanumeric characters from the following list: `+-*/^=<>?!".,@%$~`. For example, `&myop` and `&++%` are valid neutral operators. Maple imposes no semantics on neutral operators. Procedures may be defined to manipulate expressions containing such operators. Maple uses the neutral operator `&*` to denote noncommutative multiplication, including matrix multiplication, and uses `&^` in two contexts: as formal exponentiation in conjunction with `mod`, and as the wedge product in `liesymm` and other packages. Neutral operators may be written in unary prefix form, binary infix form, or the usual function-call form.

Nonassociative Operator An operator with no grouping rule. You must supply parentheses in an expression with two or more adjacent nonassociative operators of the same precedence to clarify the order of operation. For example, `a^b^c` is a syntax error; you must write `(a^b)^c` or `a^(b^c)` instead. The type of associativity for each programming language and neutral operator in Maple is recorded in the table on page 480.

Nullary Operator An operator that requires no arguments. There are three nullary programming language operators in Maple: ", "", and """, which are used to refer the three most recent results. See page 479.

Operator A procedure or function name, arrow function expression, angle bracket function expression, neutral operator, programming language operator, or a combination of these.

Option An optional argument to a function. Options appear after required arguments and often are of the form *name*=*value*. For example, `scaling=CONSTRAINED` is an option you can supply to the `plot` function (see page 23). Options can appear in any order.

Package A table with procedures for its entries. The Maple library includes many packages that implement hundreds of common operations and algorithms in many areas of mathematics. You can inspect the source code of any function defined in a package. There are several ways to load and use functions in Maple library packages. For example, to call the `display` function in the `plots` package, you can use the long function name `plots[display]`. You can also load this function first by using `with(plots, display)` or all of the functions in the package at once by using `with(plots)`. Then you can simply call `display`. See pages 64 and 434.

Pascal A procedural programming language designed by Niklaus Wirth and originally intended for teaching structured programming. It is often taught in introductory programming courses and is widely used today, especially on personal computers. The syntax of Maple's programming language is similar in some ways to Pascal.

Portable Easily transferred from one computer to another. Maple commands and programs are designed to be portable across all supported platforms.

Precedence The binding strength of a programming language or neutral operator. Precedence and associativity determine the order of evaluation in an expression without parentheses to provide explicit grouping. Higher precedence operators are evaluated before lower precedence ones, so `a := b . 3 + 2 * c` is evaluated as `a :=` ((`b . 3`) `+` (`2 * c`)). A precedence table for all programming language and neutral operators in Maple is on page 480.

Procedural Programming A technique of programming in which a function or procedure is written as a sequence of steps. Maple has a procedural programming language with a syntax similar to ALGOL 68 and Pascal.

Programming Language Operator A built-in Maple operator for creating expressions, like `+`, `<`, or `union`. A complete list of all the programming language operators appears in the tables beginning on page 479. Programming language operators are either nullary, unary, or binary.

Punctuation Mark A character that delimits expressions, statements, or data structures in Maple. The punctuation marks in Maple are `; : ' ' ( ) [ ] { } < | >`. Their meanings are given in the table on page 481.

Quoted String Any combination of zero or more characters surrounded by back quotes: for example, `name`. Any character may occur in a quoted string, including the newline character. If you want to include the back quote character in a quoted string, type two successive back quotes wherever you want a single back quote in the string. In Maple V Release 2 and earlier versions, the maximum length of a string is 499 characters. There is no restriction on the length of a string in Maple V Release 3. See page 46.

Range A Maple expression of the form *a*..*b*, which represents some set of numbers between *a* and *b*. See page 53.

Recursive Function A function that calls itself, whether directly or through an intermediate function. Maple allows recursive functions.

RGB The red-green-blue method of specifying a color in Maple graphics. You can make any color by specifying its three component intensities *r*, *g*, and *b* as real numbers between 0 and 1, where 0 means no light of that color and 1 denotes full intensity. See pages 27 and 35.

Session An interactive use of Maple, from starting up to quitting. Results you calculate in a session must be saved explicitly before you quit in order to use them in a future session. See page 3.

Set Zero or more expressions separated by commas and enclosed in curly braces { }. A set never contains duplicates, and its elements may be reordered by the system. See page 56.

Share Library A collection of freely available Maple programs and documents. The share library provides the most up-to-date source of bug fixes to Maple library routines, Maple code for many applications contributed by authors throughout the world, and several documents, including introductions and information on the latest version of Maple. Part of the Maple share library is distributed with Maple and is usually installed in a directory called "share." For information on obtaining the most recent copy of the share library, see Chapter 5 (page 407).

Simple String Any combination of letters, numbers, and underscore characters that begins with a letter or an underscore. In Maple V Release 2 and earlier versions, the maximum length of a string is 499 characters. There is no restriction on the length of a string in Maple V Release 3. See page 46.

Standard Library Function A function in the Maple library that is loaded into your session automatically when you need it. You can inspect the source code of any standard library function. See page 63.

Statement Separator Punctuation that separates successive statements in a Maple session or program. There are two statement separators in Maple: the semicolon (;) and the colon (:). If a statement ends with a colon, its result is not printed. See page 49.

String A simple string or a quoted string. Every string is a valid variable name, except for a string that matches a protected name. In Maple V Release 3, most system names are protected; in earlier versions, only keywords and certain symbolic constants are protected. In Release 3, there is no limit on the length of a string; in earlier versions, the maximum length is 499 characters. See page 46.

Subpackage A package defined within a package. In Maple V Release 3, the `stats` package partitions most of its member functions into subpackages. The `with` command loads subpackages just as it loads packages. For example, to call the `mean` function in the `display` subpackage of `stats`, you can use the long function name `stats[describe, mean]`. You can also enter `with(stats)`, then use `describe[mean]`. Alternatively, you can enter `with(stats)` followed by `with(describe)` and then simply call `mean`. See page 65.

Symbolic Computation Group A research group in the Department of Computer Science at the University of Waterloo, founded by Keith Geddes and Gaston Gonnet in 1980. The Symbolic Computation Group develops Maple and is an important center of research in computer algebra.

Table A general multidimensional data structure in Maple that can be indexed by any string or expression. See page 60.

Template The template for a command indicates the number and types of arguments that the function requires. For example, the template `iquo(`m`, `n`)` indicates that the integer quotient function requires two integers as arguments. Chapter 4, which begins on page 143, includes templates for all Maple commands.

TEX A system written by Donald Knuth for typesetting books and documents, especially those containing mathematics.

Token Syntactical component of a Maple expression or statement. Tokens in Maple consist of integers, strings, programming language operators, keywords, and punctuation marks.

Ulps "Units in the last place," a unit for testing approximate equality of floating-point numbers in Maple, used in commands like `testfloat`. For a given floating-point number s, one ulp is a change of ± 1 at the nth significant digit of s, where n is the current value of `Digits`. For example, if `Digits` is 5, then 1.2345 is one ulp away from 1.2344 and 1.2346.

Unary Operator An operator that requires one argument. Most unary programming language operators in Maple are written in prefix notation: for example, `-3`. Some are written in postfix notation: for example, `4!`. Neutral operators may be written in unary prefix notation.

User Interface One of the three main components of the Maple system. Unlike the other two components, the kernel and the Maple library, the user interface may be different on different computers. The user interface determines the style of interaction in a Maple session. In Maple V Release 2 and Release 3, all computers with sufficient graphics capabilities use a worksheet interface. Systems without graphics, such as simple text terminals, use a text-oriented user interface to Maple. Many Maple V platforms also support a graphical interface (for example, under X or SunView), but only a few have a worksheet-style interface (namely, the Macintosh and the NeXT). Chapter 2, which begins on page 83, describes the main kinds of Maple user interfaces.

Variable Symbols, Inc. A company founded by Nancy Blachman in 1989 that provides training and tools to make technical software, such as Maple, easier to use. Variable Symbols, Inc., 6537 Chabot Road, Oakland, CA 94618, fax 510-652-8461, telephone 510-652-8462.

Vector A special kind of one-dimensional array where indexing starts at 1. See page 58.

Worksheet A Maple document that lets you combine Maple commands and results, graphics, and explanatory text. Worksheets also let you manipulate graphics, change fonts, and specify page breaks. Maple worksheets let you create interactive tutorials and high-quality documents. Worksheets provide the standard user interface to Maple V Release 2 and Release 3 on all computers with sufficient graphics capabilities. See page 83.

X A windowing system developed at the Massachusetts Institute of Technology and widely used on many workstations.

Chapter 10
Tables

1 NULLARY OPERATORS

Maple has three programming language operators that require no arguments at all. They are the "ditto" operators.

Operator	Description
"	Previous result
""	Second to last result
"""	Third to last result

2 OPERATOR PRECEDENCE

Maple has many unary and binary operators. The following table lists all the programming language operators and neutral operators in Maple that require one or two arguments. (The neutral operators are those beginning with an ampersand, `&`.) The operators are listed in order of precedence, from highest to lowest. The operator with the highest precedence is concatenation or decimal point; the lowest, assignment. Operators in the same box have the same precedence. The table also lists the associativity rule for each operator. For example, since binary `+` and `-` are left associative and have the same precedence, the expression `x + y - z` is computed as `(x + y) - z`, and `x - y + z` as `(x - y) + z`. Nonassociative operators have no grouping rule, so `x^y^z` is a syntax error. Most unary operators in Maple are prefix operators: for example, negation (`-3`). The only exceptions, factorial (`4!`) and sometimes decimal point (`5.`), are noted in the table.

Operator	Type	Name	Associativity
`.`	binary	Concatenation (left operand is string)	Left associative
`.`	binary	Decimal point	
`.`	unary	Decimal point (prefix or postfix)	
`%`	unary	Label	Nonassociative
`&`*string*	binary	Neutral operator	Left associative
`&`*string*	unary	Neutral operator	
`!`	unary	Factorial (postfix)	Left associative
`^`		Exponentiation	
`**`	binary	Exponentiation	Nonassociative
`@@`		Repeated composition	
`*`		Multiplication	
`&*`		Noncommutative multiplication	
`/`	binary	Division	Left associative
`@`		Composition	
`intersect`		Set intersection	
`+`	binary	Addition	
`+`	unary	Plus	
`-`	binary	Subtraction	Left associative
`-`	unary	Negation	
`union`	binary	Set union	
`minus`	binary	Set subtraction	
`mod`	binary	Modulo	Nonassociative
`..`	binary	Ellipsis (range)	Nonassociative
`<`		Less than	
`<=`		Less than or equal to	
`>`	binary	Greater than	Nonassociative
`>=`		Greater than or equal to	
`=`		Equals	
`<>`		Does not equal	
`$`	binary	Sequence	Nonassociative
`$`	unary	Sequence	
`not`	unary	Logical negation	Right associative
`and`	binary	Logical conjunction	Left associative
`or`	binary	Logical disjunction	Left associative
`->`	binary	Arrow (function)	Right associative
`,`	binary	Comma (expression sequence)	Left associative
`:=`	binary	Assignment	Nonassociative

3 PUNCTUATION MARKS

Maple has many punctuation marks for delimiting statements, expressions, functions, strings, and data structures.

<table>
<tr><th>Punctuation</th><th>Description</th></tr>
<tr><td>;</td><td>Statement separator</td></tr>
<tr><td>:</td><td>Statement separator, do not print result</td></tr>
<tr><td>`<i>chars</i>`</td><td>Literal string</td></tr>
<tr><td>'expr'</td><td>Withhold evaluation of expr</td></tr>
<tr><td>( )</td><td>Grouping of expressions</td></tr>
<tr><td>f(args)</td><td>Function call</td></tr>
<tr><td>[e_1, e_2, ...]</td><td>List</td></tr>
<tr><td>expr[sub]</td><td>Selection operation</td></tr>
<tr><td>{e_1, e_2, ...}</td><td>Set</td></tr>
<tr><td>(| |)</td><td>Alternative to [ ]</td></tr>
<tr><td>(* *)</td><td>Alternative to { }</td></tr>
<tr><td><<i>expr</i>></td><td>Angle bracket function</td></tr>
<tr><td><<i>expr</i> | <i>vars</i>></td><td>Angle bracket function</td></tr>
<tr><td><<i>expr</i> | <i>vars</i> | <i>localvars</i>></td><td>Angle bracket function</td></tr>
</table>

4 ESCAPE CHARACTERS

An *escape character* signals that the text on the line should be treated in a special way.

Character	Position	Description
#	anywhere	Rest of line is a comment
\	end of line	Line continuation
\	not end of line	No effect (for readability)
?	beginning of line	Help
!	beginning of line	Host command

The escape-to-host character ! is not supported on all Maple platforms.

Index

Page numbers in **bold** represent the definition or the main source of information about the entry.